Creative Secondary School Mathematics

125 Enrichment Units for Grades 7 to 12

Problem Solving in Mathematics and Beyond

Print ISSN: 2591-7234
Online ISSN: 2591-7242

Series Editor: Dr. Alfred S. Posamentier
Distinguished Lecturer
New York City College of Technology - City University of New York

---

There are countless applications that would be considered problem solving in mathematics and beyond. One could even argue that most of mathematics in one way or another involves solving problems. However, this series is intended to be of interest to the general audience with the sole purpose of demonstrating the power and beauty of mathematics through clever problem-solving experiences.

Each of the books will be aimed at the general audience, which implies that the writing level will be such that it will not engulfed in technical language — rather the language will be simple everyday language so that the focus can remain on the content and not be distracted by unnecessarily sophiscated language. Again, the primary purpose of this series is to approach the topic of mathematics problem-solving in a most appealing and attractive way in order to win more of the general public to appreciate his most important subject rather than to fear it. At the same time we expect that professionals in the scientific community will also find these books attractive, as they will provide many entertaining surprises for the unsuspecting reader.

Published

Vol. 26 *Creative Secondary School Mathematics: 125 Enrichment Units for Grades 7 to 12*
by Alfred S Posamentier

Vol. 25 *Geometry in Our Three-Dimensional World*
by Alfred S Posamentier, Bernd Thaller, Christian Dorner, Robert Geretschläger, Guenter Maresch, Christian Spreitzer and David Stuhlpfarrer

Vol. 24 *Innovative Teaching: Best Practices from Business and Beyond for Mathematics Teachers*
by Denise H Sutton and Alfred S Posamentier

Vol. 23 *Learning Trigonometry by Problem Solving*
by Alexander Rozenblyum and Leonid Rozenblyum

Vol. 22 *Mathematical Labyrinths. Pathfinding*
by Boris Pritsker

For the complete list of volumes in this series, please visit www.worldscientific.com/series/psmb

Problem Solving in Mathematics and Beyond
Volume 26

Creative Secondary School Mathematics

125 Enrichment Units for Grades 7 to 12

Alfred S. Posamentier
City University of New York, USA

World Scientific

NEW JERSEY • LONDON • SINGAPORE • BEIJING • SHANGHAI • HONG KONG • TAIPEI • CHENNAI • TOKYO

Published by

World Scientific Publishing Co. Pte. Ltd.
5 Toh Tuck Link, Singapore 596224
USA office: 27 Warren Street, Suite 401-402, Hackensack, NJ 07601
UK office: 57 Shelton Street, Covent Garden, London WC2H 9HE

British Library Cataloguing-in-Publication Data
A catalogue record for this book is available from the British Library.

Problem Solving in Mathematics and Beyond — Vol. 26
CREATIVE SECONDARY SCHOOL MATHEMATICS
125 Enrichment Units for Grades 7 to 12

ISBN 978-981-124-042-3 (hardcover)
ISBN 978-981-124-097-3 (paperback)
ISBN 978-981-124-043-0 (ebook for institutions)
ISBN 978-981-124-044-7 (ebook for individuals)

For any available supplementary material, please visit
https://www.worldscientific.com/worldscibooks/10.1142/12373#t=suppl

Typeset by Stallion Press
Email: enquiries@stallionpress.com

About the Author

Alfred S. Posamentier is currently Distinguished Lecturer at New York City College of Technology of the City University of New York. Prior to that he was Executive Director for Internationalization and Funded Programs at Long Island University, New York. This was preceded by 5 years as Dean of the School of Education and Professor of Mathematics Education at Mercy College, New York. The major part of his professional career was the 40 years at the City College of New York (City University of New York) at which he is Professor Emeritus of Mathematics Education and former Dean of the School of Education. He is the author or co-author of more than 75 mathematics books for teachers, secondary and elementary school students, as well as the general readership. Dr. Posamentier is also a frequent commentator in newspapers and journals on topics related to education.

After completing his B.A. degree in mathematics at Hunter College of the City University of New York, he took a position as a teacher of mathematics at Theodore Roosevelt High School (Bronx, New York), where he focused his attention on improving the students' problem-solving skills and at the same time enriching their instruction far beyond what the traditional textbooks offered. During his six-year tenure there, he also developed the school's first mathematics teams (both at the junior and senior level). He is still involved in working with mathematics teachers and supervisors, nationally and internationally, to help them maximize their effectiveness.

Immediately upon joining the faculty of the City College of New York in 1970 (after having received his master's degree there in 1966), he began to develop in-service courses for secondary school mathematics teachers, including such special areas as recreational mathematics and problem solving in mathematics. As Dean of the City College School of Education for 10 years, his scope of interest in educational issues covered the full gamut. During his tenure as dean he took the School from the bottom of the New York State rankings to the top with a perfect NCATE accreditation assessment in 2009. Posamentier repeated this successful transition at Mercy College, which was then the only college to have received both NCATE and TEAC accreditation simultaneously.

In 1973, Dr. Posamentier received his Ph.D. from Fordham University (New York) in mathematics education and has since extended his reputation in mathematics education to Europe. He has been visiting professor at several European universities in Austria, England, Germany, Czech Republic, Turkey and Poland. In 1990, he was Fulbright Professor at the University of Vienna.

In 1989, he was awarded an *Honorary Fellow* position at the South Bank University (London, England). In recognition of his outstanding teaching, the City College Alumni Association named him *Educator of the Year* in 1994, and in 2009. New York City had the day, May 1, 1994, named in his honor by the President of the New York City Council. In 1994, he was also awarded the *Das Grosse Ehrenzeichen für Verdienste um die Republik Österreich*, (Grand Medal of Honor from the Republic of Austria), and in 1999, upon approval of Parliament, the President of the Republic of Austria awarded him the title of *University Professor of Austria*. In 2003, he was awarded the title of *Ehrenbürgerschaft* (Honorary Fellow) of the Vienna University of Technology, and in 2004 was awarded the *Österreichisches Ehrenkreuz für Wissenschaft & Kunst 1.Klasse* (Austrian Cross of Honor for Arts and Science, First Class) from the President of the Republic of Austria. In 2005 he was inducted into the Hunter College Alumni Hall of Fame, and in 2006 he was awarded the prestigious *Townsend Harris Medal* by the City College Alumni Association. He was inducted into the New York State Mathematics Educator's Hall of Fame in 2009, and in 2010 he was awarded the coveted *Christian-Peter-Beuth Prize* from the Technische Fachhochschule — Berlin. In 2017, Posamentier was awarded *Summa Cum Laude nemmine discrepante*, by the Fundacion Sebastian, A.C., Mexico City, Mexico.

He has taken on numerous important leadership positions in mathematics education locally. He was a member of the New York State Education Commissioner's Blue Ribbon Panel on the Math-A Regents Exams, and the Commissioner's Mathematics Standards Committee, which redefined the Mathematics Standards for New York State, and he also served on the New York City schools' Chancellor's Math Advisory Panel.

Dr. Posamentier is still a leading commentator on educational issues and continues his long time passion of seeking ways to make mathematics interesting to both teachers, students and the general public — as can be seen from some of his more recent books, *Geometry in Our Three — Dimensional World* (World Scientific, 2022), *Math Tricks* (Prometheus Books, 2021), *Teaching Secondary School Mathematics* (World Scientific, 2021), *Innovative Teaching: Best Practices from Business and Beyond for Mathematics Teachers* (World Scientific, 2021), *Mathematics Entertainment for the Million* (World Scientific Publishing, 2020). *The Joy of Geometry* (Prometheus Books, 2020), *Math Makers: The Lives and Works of 50 Famous Mathematicians* (Prometheus Books, 2020), *Understanding Mathematics Through Problem Solving* (World Scientific Publishing, 2020), *The Psychology of Problem Solving: The Background to Successful Mathematics Thinking* (World Scientific Publishing, 2020), *Solving Problems in Our Spatial World* (World Scientific Publishing, 2019), *Tools to Help Your Children Learn Math: Strategies, Curiosities, and Stories to Make Math Fund or Parents and Children* (World Scientific Publishing, 2019), *The Mathematics of Everyday Life* (Prometheus, 2018), *The Joy of Mathematics* (Prometheus Books, 2017), *Strategy Games to Enhance Problem — Solving Ability in Mathematics* (World Scientific Publishing, 2017), *The Circle: A Mathematical Exploration Beyond the Line* (Prometheus Books, 2016), *Effective Techniques to Motivate Mathematics Instruction*, 2nd Ed. (Routledge, 2016), *Problem — Solving Strategies in Mathematics* (World Scientific Publishing, 2015), *Numbers: There Tales, Types, and Treasures* (Prometheus Books, 2015), *Mathematical Curiosities* (Prometheus Books, 2014), *Magnificent Mistakes in Mathematics* (Prometheus Books, 2013), *100 Commonly Asked Questions in Math Class: Answers that Promote Mathematical Understanding, Grades 6-12* (Corwin, 2013), *What Successful Math Teachers do — Grades 6-12* (Corwin, 2013), *The Secrets of Triangles: A Mathematical Journey* (Prometheus Books, 2012), *The Glorious Golden Ratio* (Prometheus Books, 2012), *The Art of Motivating Students for Mathematics Instruction* (McGraw-Hill, 2011), *The Pythagorean Theorem: Its Power and Glory* (Prometheus, 2010), *Teaching Secondary Mathematics: Techniques and Enrichment Units,* 9th Ed. (Pearson, 2015), *Mathematical Amazements and Surprises: Fascinating Figures and Noteworthy Numbers* (Prometheus, 2009), *Problem Solving in Mathematics: Grades 3-6: Powerful Strategies to Deepen Understanding* (Corwin, 2009), *Problem-Solving Strategies for Efficient and Elegant Solutions, Grades 6-12* (Corwin, 2008), *The Fabulous Fibonacci Numbers* (Prometheus Books,

2007), *Progress in Mathematics K-9* textbook series (Sadlier-Oxford, 2006-2009), *What successful Math Teacher Do: Grades K-5* (Corwin 2007), *Exemplary Practices for Secondary Math Teachers* (ASCD, 2007), *101 + Great Ideas to Introduce Key Concepts in Mathematics* (Corwin, 2006), π, *A Biography of the World's Most Mysterious Number* (Prometheus Books, 2004), *Math Wonders: To Inspire Teachers and Students* (ASCD, 2003), and *Math Charmers: Tantalizing Tidbits for the Mind* (Prometheus Books, 2003).

Contents

Preface

One wonders why so many people look back at their secondary school mathematics class experience without much enthusiasm. To take this one notch further, some people are proud to say they were not good mathematics students, but still are living a successful life. Casting further negativity on their mathematics class experiences. One reason for this negative feeling towards mathematics instruction is perhaps the pressure put upon teachers to cover the mandated curriculum and the many standardized tests on which their students must perform well. What is missing from this "picture" is that many teachers do not take the time to enrich their instruction with extracurricular topics or activities. Unfortunately, it has gotten so far along this path that many new teachers are not even aware of the trove of ideas and topics that they could use to enrich their instructional program. With that in mind, this book offers 125 topics, which are all easily accessible and easily understood by students, yet, are not part of the general curriculum.

Each unit in this book is presented in a way that a teacher can see exactly where the topic can be used in the curriculum and how it can be best presented. The units offer performance objectives which provide the teacher some guidance as to where this topic could fit and enhance the curriculum. In order to determine student readiness for the topic, a preassessment is offered for each unit as well. The discussion of the topic provides a teacher with a complete description so that those teachers who may be unfamiliar with the topic can easily become comfortable with the topic so that they can present it in an enthusiastic fashion. This is always a key factor in enriching instruction. At the conclusion of each topic there is a post-assessment and often time further references as a guide to pursue the topics further, if so desired.

For example, the method in which a teacher chooses to introduce the Pythagorean theorem can vary tremendously. We offer a unit specifically designed to enrich such an initial presentation. When dealing with topics in arithmetic, this book offers a plethora of themes that use and simplify arithmetic procedures as well as demonstrate the beauty of number relationships, which can surely generate student interest. Even such mundane topics as learning to solve quadratic equations, can be enhanced with the units designed to enrich the presentation of this skill. There is an extremely wide selection of topics available from the beginning of the secondary school curriculum to the more sophisticated, somewhat higher, mathematics topics.

It must be emphasized that the presentation of these topics typically generates a teacher's personal enthusiasm, which is very often transmitted to the students and factors into the enrichment atmosphere that the topics intend. As a result, students generally begin to get a significantly more positive view towards mathematics. Mathematics instruction throughout a student's school career is geared towards exposing its usefulness in our everyday lives. However, teachers should be motivated to demonstrate that there is also an inherent power and beauty in mathematics that is often lost through an overconcentration of the skills required for ensuing tests transmitted in the normal curriculum. The enrichment units presented in this book span a wide variety of topics, which should not only motivate teachers to enhance their instruction, but also provide an opportunity to broaden a student's perspective. Experience has shown that students tend to react very favorably towards mathematics when presented with the topics and ideas included in this collection of enrichment units. Students begin to realize that these topics provide entertainment as well as a new look at the usefulness of mathematics. Often some of these topics are taken home and shared with the rest of the family.

With the ever-increasing role of technology in our society, mathematics plays a very significant role. Students who excel in mathematics, which is often generated by a love of the subject, typically have a much broader selection of career options. So, let's enrich the mathematics classroom to show off the wonders of mathematics that often pass us by unnoticed.

Introduction

A wide variety of mathematics topics are well suited as enrichment for secondary school mathematics courses. We can draw such topics from all branches of mathematics, and many of these clearly reinforce the NCTM standards. Numerical curiosities, algebraic investigations of number relationships, geometric phenomena not usually available to this audience, as well as many other topics generally not found in the secondary school or college curricula, are among the units found in this part of the book.

A very common topic viewed from a not-so-common viewpoint would certainly provide enrichment for the appropriate audience. The trick to providing students with enrichment activities is to present the material in a highly motivating and intelligible manner. This challenging objective will guide us through the units presented here. It should be remembered from the outset that enrichment activities are *not* limited only to gifted students. As you will see throughout this part of the book, many enrichment activities can be used successfully with students of average mathematical ability as well as in remedial classes, provided proper adjustments are made. Naturally, these modifications in the presentation (both in content and in method of presentation) can be made only by the classroom teacher using the material. With this in mind, let us consider the format in which these enrichment units are presented.

Each enrichment unit treats a separate topic. With few exceptions, the topics can be considered in almost any order. Following a brief introduction, *Performance Objectives* for the respective units are stated. Not only do these objectives succinctly foreshadow the content of the unit, but they provide a good indication of the scope of the material that follows. So you can determine better the suitability of the unit for your class, a *Preassessment* section is offered. In addition to assisting you to ascertain your students' readiness for the unit, this section often also serves as a source for motivation you may wish to use in presenting the topic to your class.

In the next section, *Teaching Strategies*, the enrichment topic is presented in a manner that you may use to introduce it to your class. Here the topic is carefully developed with an eye toward anticipating possible pitfalls and hurdles students may encounter along the way. The style is conversational throughout, making the reading more relaxed. Occasionally suggestions for extensions are offered so that the units do not appear to be terminal. Perhaps an underlying goal throughout these enrichment units is to allow them to serve as springboards for further investigation. Where appropriate, additional references are offered for further study.

One efficient way to ascertain whether the objectives for a particular unit have been met is to question the students on the topic presented. Sample questions are provided in the *Postassessment* section at the end of the unit. You are invited to augment these questions with some of your own where needed.

Since many of these enrichment units can be used in a variety of different mathematics classes and with students of different levels of mathematical ability, a cross-cataloging chart is provided. The chart will enable you to select enrichment units according to subject, grade level, and student ability level. Naturally, you will need to make some modifications in the form to make the unit properly suit the intended audience. Many of these enrichment units very clearly support the NCTM standard of making connections, since they demonstrate how topics and concepts traditionally taught in one context can be used fruitfully in an entirely unexpected context.

On the whole, you ought to make every effort to inject enrichment activities into all your mathematics instruction, regardless of the students' mathematical ability. Such activities can be just as rewarding in a remedial class as in one comprised of gifted students. The benefit, although manifested differently, should be about the same for all classes.

Cross-Catalog of Enrichment Units

To facilitate using the enrichment units found in this section, a cross-catalog is provided. The units are listed in the order in which they are presented (page numbers for each Unit are listed in the Table of Contents). The grade level, the ability level, and the branch of mathematics to which each unit is related are provided. These assessments are simply the opinion of the authors and some secondary school teachers. You may, however, try these units with audiences other than those specified.

You will notice that each ability level — Remedial, Average, or Gifted — has been partitioned into four grade-level divisions: 7–8, 9, 10, 11–12.

For the remedial student, grades 7–8 are usually the middle school low-level mathematics courses: at grades 9 and 10, a general mathematics or introductory algebra is assumed. At grades 11–12, there is usually a continuation of the earlier courses, but with a greater degree of sophistication.

The average student partition refers to the middle school prealgebra program for grades 7–8, the elementary algebra course for grade 9, the high school geometry course for grade 10, and the second-year algebra (with trigonometry) course and beyond for grades 11–12.

Although very often the gifted student begins the study of elementary algebra in the eighth grade (or earlier), for the sake of simplicity, we shall use the same course determination for the gifted students as for the average students (above). A greater ability in mathematics is assumed here, however.

The numbers 1 and 2 indicate the primary and secondary audiences for each unit. This implies that variations of these units can (and should) be used at all levels as you deem appropriate. Naturally, some modifications will have to be made. Some units may have to be "watered down" for weaker mathematics students, and for more gifted youngsters some units may serve as springboards for further investigation.

Ratings

1. Primary use (specifically intended for that audience).
2. Secondary use (may be used for that audience with some modifications).

Another important consideration when selecting an enrichment unit is the branch of mathematics to which it is related. For many units this is difficult to isolate, since these units relate to many branches of mathematics. Using the following code, the *Subject* column indicates the related branches of mathematics. Although very often the order could easily be changed with no loss of accuracy, every attempt has been made to list the branches of mathematics in descending order of relevance to each unit, as judged by a group of mathematics teachers.

Subject code

1. Arithmetic
2. Number theory
3. Probability
4. Logic
5. Algebra
6. Geometry
7. Analytic geometry
8. Topology
9. Statistics
10. Problem solving
11. Applications
12. Mathematical curiosities

Technology applications

The chart shows which technology applications might be used with specific enrichment units. However, the reader should make the final evaluation/ decision.

Enrichment Unit Table

Unit	Unit Name	Subject Code	Scientific Calculator	Graphing Calculator or Graphing Software	Spreadsheet	Geometer's Sketchpad	Remedial Classes 7–8	Remedial Classes 9	Remedial Classes 10	Remedial Classes 11–12	Average Classes 7–8	Average Classes 9	Average Classes 10	Average Classes 11–12	Gifted Classes 7–8	Gifted Classes 9	Gifted Classes 10	Gifted Classes 11–12
1	Constructing Odd-Order Magic Squares	1,4,12					1	1	1	2	1	2						
2	Constructing Even-Order Magic Squares	1,12					1	1	1	2	1	2						
3	Introduction to Alphametics	1,4					2	2	1	1	1	1			1	2		
4	A Checkerboard Calculator	1,12					1	1	1	2	1	2			2	2		
5	The Game of Nim	4,1,12					2	1	1	1	1	2			2	2		
6	The Tower of Hanoi	4,1,12					2	1	1	1	1	2			2	2		
7	What Day of the Week Was It?	4,1,12					2	2	1	1	1	1			1	2		2
8	Palindromic Numbers	2,1,12					1	1	2	2	1	1		2	1	1		2
9	The Fascinating Number Nine	1,2,12					2	1	1	1	1	2			1	2		
10	Unusual Number Properties	1,2,12					2	1	1	1	1	2			1	2		
11	Enrichment with a Handheld Calculator	1,2,12					1	1	1	1	1	2			1	2		
12	Symmetric Multiplication	1,2,12					2	1	1	1	1	2		2	1	2		
13	Variations on a Theme — Multiplication	1,2,12					2	1	1	1	1	2		2	1	2		2
14	Ancient Egyptian Arithmetic	2,1						2	2	1	1	2		2	1	1		2
15	Napier's Rods	1,2,11					1	1	1	1	1	2		2	2			
16	Unit Pricing	1,11					1	1	1	1	1	2						
17	Successive Discounts and Increases	1,11,12					2	2	1	1	1	1		2	1	2		

(Continued)

Enrichment Unit Table (*Continued*)

Unit	Unit Name	Subject Code	Scientific Calculator	Graphing Calculator or Graphing Software	Spreadsheet	Geometer's Sketchpad	Remedial Classes				Average Classes				Gifted Classes			
							7–8	9	10	11–12	7–8	9	10	11–12	7–8	9	10	11–12
18	Prime and Composite Factors of a Whole Number	1,2					1	1	1	2	2							
19	Prime Numeration System	2,1,12						2	1	1	2	1		2	1			
20	Repeating Decimal Expansions	2,1								2	2	1		2	2			
21	Peculiarities of Perfect Repeating Decimals	2.1,12									1	1		2	1	2		
22	Patterns in Mathematics	4,5							2	2	1	1			2	2		
23	Googol and Googolplex	1,12					2	1	1	2	1	2			2			
24	Mathematics of Life Insurance	3,9,5							2	2		2		1	2	1		2
25	Geometric Dissections	6,4,10							2	2	1	1	1		1	1	2	
26	The Klein Bottle	8										2	1	2	1	1	1	2
27	The Four-Color Map Problem	8,4									1	1	1	2	1	2		
28	Mathematics on a Bicycle	6,4,1						2	1	1	1	2	2		2			
29	Mathematics and Music	11,1										2	2	2	1	1		2
30	Mathematics in Nature	11,12,2						2	2	2	1	1	1	2	1	1	2	2
31	The Birthday Problem	3,12,11							2	2	2	1	2	1	1	1	1	1
32	The Structure of the Number System	1,5									2	1	2	1	1	1		
33	Excursions in Number Bases	2,1,5						2	2	2	1	1	2	2	1	2		2
34	Raising Interest	5,11										1		1	1	1		2
35	Reflexive, Symmetric, and Transitive Relations	4,2									1	1	1		1	1	2	

(*Continued*)

Enrichment Unit Table (*Continued*)

Unit	Unit Name	Subject Code	Scientific Calculator	Graphing Calculator or Graphing Software	Spreadsheet	Geometer's Sketchpad	Remedial Classes 7–8	Remedial 9	Remedial 10	Remedial 11–12	Average Classes 7–8	Average 9	Average 10	Average 11–12	Gifted Classes 7–8	Gifted 9	Gifted 10	Gifted 11–12
36	Bypassing an Inaccessible Region	6,10										2	1	1		1	1	
37	The Inaccessible Angle	6,10										2	1	1		1	1	
38	Triangle Constructions	6,10										2	1	1		1	1	1
39	The Criterion of Constructibility	6,5											1	1		1	1	1
40	Constructing Radical Lengths	6,5										2	1	1		1	1	2
41	Constructing a Pentagon	6,5										1	1	1		1	1	1
42	Investigating the Isosceles Triangle Fallacy	6,12											1	1		2	1	1
43	The Equiangular Point	6											1	1		2	1	2
44	The Minimum-Distance Point of a Triangle	6											1	1		2	1	2
45	The Isosceles Triangle Revisited	6,10											1			2	1	2
46	Reflective Properties of the Plane	6,11										2	1	1		1	1	2
47	Finding the Length of a Cevian of a Triangle	6,5										2	1	1			1	
48	A Surprising Challenge	6,10											1	2		2	1	
49	Making Discoveries in Mathematics	6									2	1	1	1	1	1	2	
50	Tessellations	6,5									2	2	1	1	2	1	1	2
51	Introducing the Pythagorean Theorem	6,2,5										1	1	2		1	1	2
52	Trisection Revisited	6,5,12									2	1	1	1	2	1	1	2
53	Proving Lines Concurrent	6,10											2	2		2	1	2

(*Continued*)

Enrichment Unit Table (*Continued*)

Unit	Unit Name	Subject Code	Scientific Calculator	Graphing Calculator or Graphing Software	Spreadsheet	Geometer's Sketchpad	Remedial Classes				Average Classes				Gifted Classes			
							7–8	9	10	11–12	7–8	9	10	11–12	7–8	9	10	11–12
54	Squares	6,10											1	2		1	1	2
55	Proving Points Collinear	6,10											2	2		2	1	1
56	Angle Measurement with a Circle	6											1			2	1	
57	Trisecting a Circle	6,12										1	1	2		1	1	2
58	Ptolemy's Theorem	6,10										2	1	1		1	1	2
59	Constructing π	6,1										2	1	1		1	1	2
60	The Arbelos	6,5										2	1	1		1	1	1
61	The Nine-Point Circle	6									2	2	1	2	2	1	1	2
62	The Euler Line	6										2	1	2	2	1	1	2
63	The Simson Line	6									2	2	1	2	2	1	1	2
64	The Butterfly Problem	6,10									2	2	1	2	2	1	1	1
65	Equicircles	6,5											1	2		2	1	1
66	The Inscribed Circle and the Right Triangle	6,5											2	2		2	1	1
67	The Golden Rectangle	6,5,2										2	1	1		1	1	2
68	The Golden Triangle	6,5,2											1	1		1	1	1
69	Geometric Fallacies	6,5											1	1		1	1	2
70	Regular Polyhedra	6,5								2		2	2	1	2	1	1	1
71	An Introduction to Topology	8,4						2	2	1	1	1	1	2	1	1	1	2
72	Angles on a Clock	1,5					1	1	1	1	1	1	1	2	1	1	2	2
73	Averaging Rates — The Harmonic Mean	5,1							2	2	1	1	2	1	1	1	2	2
74	Howlers	1,5					1	1	1	1	1	1	1	1	1	1	1	1
75	Digit Problems Revisited	5,1					1	1	1	1	1	1	1	1	1	1	1	1
76	Algebraic Identities	5,6										1	1	2		1	2	
77	A Method for Factoring Trinomials of the Form $aX^2 + bX + c$	5										1	2	1		1		2

(*Continued*)

Enrichment Unit Table (*Continued*)

Unit	Unit Name	Subject Code	Scientific Calculator	Graphing Calculator or Graphing Software	Spreadsheet	Geometer's Sketchpad	Remedial Classes				Average Classes				Gifted Classes			
							7–8	9	10	11–12	7–8	9	10	11–12	7–8	9	10	11–12
78	Solving Quadratic Equations	5,10										1	2	1		1		1
79	The Euclidean Algorithm	2,5										1		1		1		1
80	Prime Numbers	2,5										2		2		1		1
81	Algebraic Fallacies	5,12										1	2	1		1	2	1
82	Sum Derivations with Arrays	5,6,12								2	2	1	2	1	1	1		1
83	Pythagorean Triples	2,5,6										2	1	1	2	1	1	1
84	Divisibility	2,1,5					2	1	1	1	1	1	2	1	1	1	2	1
85	Fibonacci Sequence	5,2,12			See unit 19				2	2	2	1	2	1	1	1	2	1
86	Diophantine Equations	5,2										2		1	2	1	2	1
87	Continued Fractions and Diophantine Equations	5,2										2		1	2	1	2	1
88	Simplifying Expressions Involving Infinity	5,2										2		1	2	1	2	1
89	Continued Fraction Expansion of Irrational Numbers	5										2		1		1	2	1
90	The Farey Sequence	1,5						2	2	2	2	1	2	1	1	1	2	2
91	The Parabolic Envelope	6,7							2	2	2	2	1	1	1	1	1	1
92	Application of Congruence to Divisibility	2,5												2	2	1	2	1
93	Problem Solving — A Reverse Strategy	10,6,5										1	1	2	2	1	1	2
94	Decimals and Fractions in Other Bases	1,5										2	2	1	2	1	2	1
95	Polygonal Numbers	2,5								2		1	1	1	2	1	1	1
96	Networks	4,5,11,8							2	1	2	1	1	1	2	1	1	1

(*Continued*)

Enrichment Unit Table (*Continued*)

Unit	Unit Name	Subject Code	Scientific Calculator	Graphing Calculator or Graphing Software	Spreadsheet	Geometer's Sketchpad	Remedial Classes				Average Classes				Gifted Classes			
							7–8	9	10	11–12	7–8	9	10	11–12	7–8	9	10	11–12
97	Angle Trisection — Possible or Impossible?	5,6,2											2	1		2	1	1
98	Comparing Means	5,2,6										1	1	1		1	1	1
99	Pascal's Pyramid	5			See Unit 119									1		2	1	1
100	The Multinomial Theorem	5,1,4										2		1		1	2	1
101	Algebraic Solution of Cubic Equations	5												2		1		1
102	Solving Cubic Equations	5										2		2		1		1
103	Calculating Sums of Finite Series	5										2		1		1		1
104	A General Formula for the Sum of Series of the Form $\sum_{t=1}^{n} t^r$	5												2		2		1
105	A Parabolic Calculator	7,5								2		2	2	1		1	1	1
106	Constructing Ellipses	6,5,12,7										2	2	1		1	1	1
107	Constructing the Parabola	6,7,5							2	2	2	2	1	1	1	1	1	2
108	Using Higher Plane Curves to Trisect an Angle	7,5,6											2	1		1	1	1
109	Constructing Hypocycloid and Epicycloid Circular Envelopes	6,5,7											2	1		1	1	1
110	The Harmonic Sequence	5,6										2	1	1		1	1	1
111	Transformations and Matrices	5,7												2		2	2	1

(*Continued*)

Enrichment Unit Table (*Continued*)

Unit	Unit Name	Subject Code	Scientific Calculator	Graphing Calculator or Graphing Software	Spreadsheet	Geometer's Sketchpad	Remedial Classes				Average Classes				Gifted Classes			
							7–8	9	10	11–12	7–8	9	10	11–12	7–8	9	10	11–12
112	The Method of Differences	1,5												1		2	2	1
113	Probability Applied to Baseball	3,5,4								2		1	1	1	2	1	1	1
114	Introduction to Geometric Transformations	6,5											1	1		2	1	1
115	The Circle and the Cardioid	6,5,11										2	1	1		1	1	1
116	Complex-Number Applications	6,5,11											2	1		2	1	1
117	Hindu Arithmetic	2,1						2	2	1	1	2		2	1	1		2
118	Proving Numbers Irrational	5														11	2	1
119	How to Use a Computer Spreadsheet to Generate Solutions to Certain Mathematics Problems	1,2,10			****							2	1	1	1	1	2	2
120	The Three Worlds of Geometry	6														2	1	1
121	π Mix	5																1
122	Graphical Iteration	5															2	1
123	The Feigenbaum Plot	5															2	1
124	The Sierpinski Triangle	6													1		2	3
125	Fractals	6														1	1	1

Unit 1

Constructing Odd-Order Magic Squares

This unit is intended for enrichment of students who have already mastered the fundamentals of elementary algebra. Carefully chosen parts of this unit may also prove effective in remedial classes, where students would appreciate some "recreational" mathematics.

Performance Objectives

1. *Students will construct magic squares of any odd order required.*
2. *Students will discover properties of given odd-order magic squares.*
3. *Students will determine the sum of the elements of any row (or column, or diagonal) of any magic square, given only its order.*

Preassessment

Challenge students to form a 3×3 matrix with the numbers 1–9 so that the sum of the elements in each row, column, or diagonal is the same. Indicate to them that such a matrix is called a *magic square* (of order 3).

Teaching Strategies

After students have had enough time to be either successful with, or thoroughly frustrated by, the challenge (usually less than 15 minutes), you may begin to attack the problem with them. Have them realize the advantage of knowing beforehand the sum of each row (or column, or diagonal).

To develop a formula for the sum of the elements in any row, column, or diagonal of a magic square*, students must be familiar with the formula for the sum of an arithmetic series, $S = \frac{n}{2}(a_1 + a_n)$. If they are not familiar with this formula, it can be very easily related to them by telling the story of young Carl Friedrich Gauss (1777–1855) who, at the age of 10, successfully responded to his teacher's challenge.

---

*Unless stated otherwise, this unit will be concerned with magic squares of consecutive natural numbers beginning with 1.

His teacher had a habit of providing rather lengthy chores for the students to complete (while he knew of a shortcut formula). One day this teacher told the class to add a series of numbers of the sort: $1+2+3+4+\cdots+97+98+100$. As the teacher finished stating the problem, young Gauss submitted the answer. In amazement, the teacher asked Gauss to explain his rapid solution. Gauss explained that rather than merely adding the 100 numbers in the order presented, he considered the following pairs $1+100=101$; $2+99=101$; $3+98=101$; $4+97=101;\ldots;$ $50+51=101$. Since there were 50 pairs of numbers whose sum was 101, his answer was $50\times 101=5,050$. In effect, he multiplied one-half the number of numbers to be added $\left(\frac{n}{2}\right)$ by the sum of the first and last numbers in the series (a_1+a_n) to obtain the sum of the entire series.

From this formula, the sum of natural numbers from 1 to n^2 (the numbers used in an $n\times n$ magic square) is $S=\frac{n^2}{2}(1+n^2)$. However, if it is required that each row must have the same sum, then the sum is $\frac{S}{n}$. (From here on, the expression "the sum of a row" will actually refer to "the sum of the numbers in a row.") Therefore, the sum of any row is $\frac{n}{2}(n^2+1)$. You might want to have students consider why the sum of a diagonal is also $\frac{n}{2}(n^2+1)$.

Students are now ready to begin to systematically consider the original problem. Have them consider the following matrix of letters representing the numbers 1–9.

d_1	c_1	c_2	c_3	d_2
r_1	a	b	c	
r_2	d	e	f	
r_3	g	h	i	

Using the formula developed earlier, $S=\frac{n}{2}\times(n^2+1)$, we find that the sum of a row of a third order (3×3) magic square is $\frac{3}{2}(3^2+1)=15$. Therefore, $r_2+c_2+d_1+d_2=4\cdot 15=60$. However, $r_2+c_2+d_1+d_2=(d+e+f)+(b+e+h)+(a+e+i)+(c+e+g)=3e+(a+b+c+d+e+f+g+h+i)=3e+45$ (since the sum of $1+2+3+\cdots+9=\frac{9}{2}(1+9)=45$). Therefore, $3e+45=60$ and $e=5$. Thus, it is established that the center position of a third-order magic square must be occupied by 5.

Since the sum of each row, column, and diagonal in this magic square is $15, a+i=g+c=b+h=d+f=15-5=10$. (*Note*: Two numbers of an nth order magic square are said to be complementary if their sum is n^2+1; thus a and i are complementary.) Now lead your students through the following argument.

The number 1 cannot occupy a corner position. Suppose $a = 1$; then $i = 9$. However, 2, 3, and 4 cannot be in the same row (or column) as 1, since there is no natural number less than 10 which would be large enough to occupy the third position of such a row (or column). This would leave only two positions (the nonshaded squares below) to accommodate these three numbers (2, 3, and 4). Since this cannot be the case, the numbers 1 and 9 may occupy only the middle positions of a row (or column).

1		
	5	
		9

The number 3 cannot be in the same row (or column) as 9, for the third number in such a row (or column) would then have to be 3, to obtain the required sum of 15. This is not possible because a number can be used only once in the magic square.

Now have students realize that neither 3 nor 7 may occupy comer positions. They should then use the above criteria to construct a magic square of order 3. Students should get any of the following magic squares.

2	7	6
9	5	1
4	3	8

4	3	8
9	5	1
2	7	6

8	1	6
3	5	7
4	9	2

6	1	8
7	5	3
2	9	4

2	9	4
7	5	3
6	1	8

4	9	2
3	5	7
8	1	6

8	3	4
1	5	9
6	7	2

6	7	2
1	5	9
8	3	4

Students might now want to extend this technique to constructing other odd-order magic squares. However, this scheme becomes somewhat tedious. Following is a rather mechanical method for constructing an odd-order magic square.

Begin by placing a 1 in the first position of the middle column. Continue by placing the next consecutive numbers successively in the cells of the (positive slope) diagonal. This, of course, is impossible since there are no cells "above" the square.

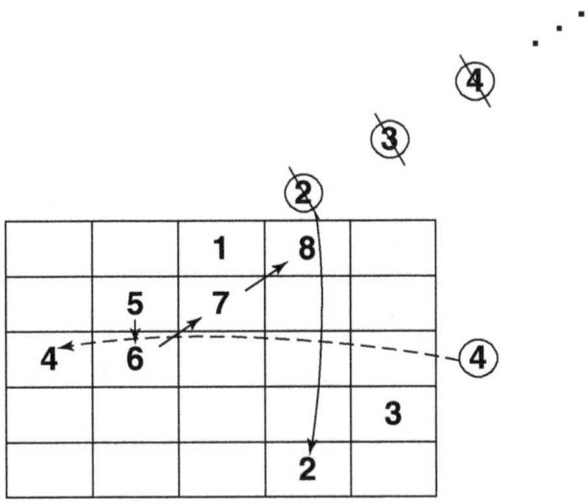

When a number must be placed in a position "above" the square, it should instead be placed in the last cell of the next column to the right. Then the next numbers are placed consecutively in this new (positive slope) diagonal. When (as in the figure above) a number falls outside the square to the right, it should be placed in the first (to the left) cell of the next row above the row whose last (to the right) cell was just filled (as illustrated). The process then continues by filling consecutively on the new cell until an already occupied cell is reached (as is the case with 6, above). Rather than placing a second number in the occupied cell, the number is placed below the previous number. The process continues until the last number is reached.

After enough practice, students will begin to recognize certain patterns (e.g., the last number always occupies the middle position of the bottom row).

This is just one of many ways of constructing odd-order magic squares. More adept students should be urged to justify this rather mechanical technique.

Postassessment

Have students do the following exercises:

1. Find the sum of row of a magic square of order (a) 4; (b) 7; (c) 8.
2. Construct a magic square of order 11.
3. State some properties common to magic squares of odd order less than 13.

References

Posamentier, A. S. and B. Thaller, Numbers: Their Tales, Types, and Treasures, Amherst, New York: Prometheus Books, 2015.

Unit 2

Constructing Even-Order Magic Squares

This topic can be used with a remedial class in high school as well as with a more advanced class at any secondary school grade level. In the former case, only magic squares of doubly-even order should be considered, while in the latter case, singly-even order magic squares may be included. When used with a remedial class, the development of doubly-even order magic squares may serve as motivation for drill of arithmetic fundamentals.

Performance Objectives

1. *Students will construct magic squares of any even order required.*
2. *Students will discover properties of given even-order magic squares.*

Preassessment

Begin your introduction with a historical note. Mention the German artist (and mathematician) Albrecht Dürer (1471–1528), who did considerable work with mathematics related to his artwork. One of the more curious aspects of his work was the appearance of a magic square in an engraving of 1514 entitled "Melancholia" (Figure 1).

At the upper right-hand corner of the engraving is the magic square (Figure 2). It is believed that this was one of the first appearances of magic squares in Western civilization. Of particular interest are the many unusual properties of this magic square. For example, the two center positions of the bottom row indicate the year the engraving was made, 1514. Offer your students some time to find other unusual properties (other than merely a constant sum of rows, columns, and diagonals).

Teaching Strategies

Students will probably enjoy discussing the many properties of this magic square, some of which are as follows:

1. The four corner positions have a sum of 34.
2. The four corner 2×2 squares each has a sum of 34.
3. The center 2×2 square has a sum of 34.

Figure 1.

Figure 2.

4. The sum of the numbers in a diagonal equals the sum of those not in a diagonal.
5. The sum of the squares of the numbers in the diagonals (748) equals the sum of the squares of the numbers not in the diagonals.
6. The sum of the cubes of the numbers in the diagonals (9,248) equals the sum of the cubes of the numbers not in the diagonals.
7. The sum of the squares of the numbers in both diagonals equals the sum of the squares of the numbers in the first and third rows (or columns),

which equals the sum of the squares of the numbers in the second and fourth rows (or columns).

8. Note the following symmetries:

$$2+8+9+15=3+5+12+14=34$$

$$2^2+8^2+9^2+15^2=3+5^2+12^2+14^2=374$$

$$2^3+8^3+9^3+15^3=3^3+5^3+12^3+14^3=4{,}624$$

9. The sum of each adjacent upper and lower pair of numbers vertically or horizontally produces an interesting symmetry.

$$\text{vertically:} \quad \begin{array}{cccc} 21 & 13 & 13 & 21 \\ 13 & 21 & 21 & 13 \end{array}$$

$$\text{horizontally:} \quad \begin{array}{cc} 19 & 25 \\ 15 & 19 \\ 15 & 19 \\ 19 & 15 \end{array}$$

Consider first constructions of magic squares whose *order is a multiple of 4* (sometimes referred to as *doubly-even order*). Have students construct the square below with the diagonals as shown in Figure 3.

1	2	3	4
5	6	7	8
9	10	11	12
13	14	15	16

Figure 3.

Then have them replace each number in a diagonal with its complement (i.e., that number which will give a sum of $n^2+1=16+1=17$). This will yield a 4×4 magic square (Figure 4). (*Note*: Dürer simply interchanged columns 2 and 3 to obtain his magic square.)

16	2	3	13
5	11	10	8
9	7	6	12
4	14	15	1

Figure 4.

A similar process is used to construct larger doubly-even order magic squares. To construct an 8×8 magic square, divide the square into 4×4 magic squares (Figure 5) and then replace the numbers in the diagonals of each of the 4×4 squares with their complements.

1	2	3	4	5	6	7	8
9	10	11	12	13	14	15	16
17	18	19	20	21	22	23	24
25	26	27	28	29	30	31	32
33	34	35	36	37	38	39	40
41	42	43	44	45	46	47	48
49	50	51	52	53	54	55	56
57	58	59	60	61	62	63	64

Figure 5.

The resulting magic square is shown as Figure 6. Now have students construct a magic square of order 12.

64	2	3	61	60	6	7	57
9	55	54	12	13	51	50	16
17	47	46	20	21	43	42	24
40	26	27	37	36	30	31	33
32	34	35	29	28	38	39	25
41	23	22	44	45	19	18	48
49	15	14	52	53	11	10	56
8	58	59	5	4	62	63	1

Figure 6.

A different scheme is used to construct magic squares of singly-even order (i.e., those whose order is even but *not* a multiple of 4). Any singly-even order (say, of order n) magic square may be separated into quadrants (Figure 7). For convenience label them A, B, C, and D.

A	C
D	B

Figure 7.

Students should now be instructed to construct four *odd-order* magic squares in the order A, B, C, and D (refer to the accompanying model

"Constructing Odd-Order Magic Squares"). That is, square A will be an odd-order magic square using the first $\frac{n^2}{4}$ natural numbers; square B will be an odd-order magic square beginning with $\frac{n^2}{4}+1$ and ending with $\frac{n^2}{2}$; square C will be an odd-order magic square beginning with $\frac{n^2}{2}+1$ and ending with $\frac{3n^2}{4}$; square D will be an odd-order magic square beginning with $\frac{3n^2}{4}+1$ and ending with n^2. (Figure 8 illustrates the case where $n = 6$.)

Have students notice the relation of the four magic squares of Figure 8 to the first magic square in the upper left position, A (Figure 9).

8	1	6	26	19	24
3	5	7	21	23	25
4	9	2	22	27	20
35	28	33	17	10	15
30	32	34	12	14	16
31	36	29	13	18	11

Figure 8.

$0+8$	$0+1$	$0+6$	$18+8$	$18+1$	$18+6$
$0+3$	$0+5$	$0+7$	$18+3$	$18+5$	$18+7$
$0+4$	$0+9$	$0+2$	$18+4$	$18+9$	$18+2$
$27+8$	$27+1$	$27+6$	$9+8$	$9+1$	$9+6$
$27+3$	$27+5$	$27+7$	$9+3$	$9+5$	$9+7$
$27+4$	$27+9$	$27+2$	$9+4$	$9+9$	$9+2$

Figure 9.

Now only some minor adjustments need be made to complete the construction of the magic squares. Let $n = 2(2m + 1)$. Take the numbers in the first m positions in each row of A (except the middle row, where you skip the first position and take the next m positions) and interchange them with the numbers in the corresponding positions of square D. Then take the numbers in the last $m - 1$ positions of square C and interchange them with the number in the corresponding positions of square B. Notice that for $n = 6$ (Figure 10) $m - 1 = 0$, hence squares B and C remain unaltered.

Have students apply this technique to the construction of a magic square of order 10 ($n = 10$ and $m = 2$; see Figures 11 and 12).

35	1	6	26	19	24
3	32	7	21	23	25
31	9	2	22	27	20
8	28	33	17	10	15
30	5	34	12	14	16
4	36	29	13	18	11

Figure 10.

17	24	1	8	15	67	74	51	58	65
23	5	7	14	16	73	55	57	64	66
4	6	13	20	22	54	56	63	70	72
10	12	19	21	3	60	62	69	71	53
11	18	25	2	9	61	68	75	52	59
92	99	76	83	90	42	49	26	33	40
98	80	82	89	91	48	30	32	39	41
79	81	88	95	97	29	31	38	45	47
85	87	94	96	78	35	37	44	46	28
86	93	100	77	84	36	43	50	27	34

Figure 11.

92	99	1	8	15	67	74	51	58	40
98	80	7	14	16	73	55	57	64	41
4	81	88	20	22	54	56	63	70	47
85	87	19	21	3	60	62	69	71	28
86	93	25	2	9	61	68	75	52	34
17	24	76	83	90	42	49	26	33	65
23	5	82	89	91	48	30	32	39	66
79	6	13	95	97	29	31	38	45	72
10	12	94	96	78	35	37	44	46	53
11	18	100	77	84	36	43	50	27	59

Figure 12.

Postassessment

As a formal postassessment, have students do the following tasks:

1. Construct a magic square of order (a) 12; (b) 16.
2. Construct a magic square of order (a) 14; (b) 18.
3. Find additional properties of magic squares of order (a) 8; (b) 12.

Reference

Posamentier, A. S. and B. Thaller, Numbers: Their Tales, Types, and Treasures, Amherst, New York: Prometheus Books, 2015.

Unit 3

Introduction to Alphametics

This unit can be used to reinforce the concept of addition.

Performance Objective

Given alphametic problems, students will solve them in a systematic fashion.

Preassessment

Have students solve the following addition problems, either by simple addition in (a) and by filling in the missing digits in (b).

$$
\begin{array}{lr@{\qquad}lr}
\text{a.} & 562 & \text{b.} & 5\;6\;7\;\_ \\
 & 3{,}943 & & \_\;8\;\_\;9 \\
 & \underline{8{,}807} & & \overline{\_\;3\;\_\;3\;3}
\end{array}
$$

Teaching Strategies

The preceding problems should serve as a motivation for this lesson. Alphametics are mathematical puzzles that appear in several disguises. Sometimes the problem is associated with the restoration of digits in a computational problem; at other times, the problem is associated with decoding the complete arithmetical problem where letters of the alphabet represent all the digits. Basically, construction of this type of puzzle is not difficult, but the solution requires a thorough investigation of all elements. Every clue must be tested in all phases of the problem and carefully followed up. For example, suppose we were to eliminate certain digits in problem (a) above and supply the answer with some digits missing. Let us also assume that we do not know what these digits are. We may then be left with the following skeleton problem:

$$
\begin{array}{ccccc}
\textcircled{1} & \textcircled{2} & \textcircled{3} & \textcircled{4} & \textcircled{5} \\
 & & \_ & 6 & 2 \\
 & 3 & 9 & 4 & \_ \\
 & \_ & 8 & \_ & 7 \\
\hline
\_ & 3 & 3 & 1 & 2
\end{array}
$$

Have students analyze the problem. Lead them through the reconstruction as follows. From column five, $2 + \_ + 7 = 12$. Therefore, the missing digit in the fifth column must be 3. In the fourth column, we have $1 + 6 + 4 + \_ = 1$, or $11 + \_ = 1$, therefore the digit must be zero. In the third column, we have $1 + \_ + 9 + 8 = 23$, and the missing digit must be 5. Now, from the second column, we have now $2 + 3 + \_ = 13$. This implies that the digit must be 8, and therefore, the digit to the left of 3 in the first column, bottom row must be 1. Thus, we have reconstructed the problem. Students should now be able to find the missing digits in the second problem of the preassessment (if they haven't already solved it). The completed solution is

$$\begin{array}{r} 5\ 6\ 7\ \textcircled{4} \\ \textcircled{7}\ \textcircled{8}\ 5\ 9 \\ \hline \textcircled{1}\ 3\ \textcircled{5}\ 3\ 3 \end{array}$$

Have students create their own problems and then interchange these with others in their class. So far, we have considered such problems that have exactly one solution. The following example will show a problem that has more than one solution.

$$\begin{array}{r} \_\ 8\ 7 \\ 3\ \_\ 1 \\ +\ 5\ 6\ \_ \\ \hline \_\ 3\ \_\ 0 \end{array}$$

In the units column $7 + 1 + \_ = 10$, the missing digit must be 2.

$$\begin{array}{r} \_\ 8\ 7 \\ 3\ \_\ 1 \\ +\ 5\ 6\ 2 \\ \hline \_\ 3\ \_\ 0 \end{array}$$

In the tens column, $1+8+\_+6 = \_$, or $15+\_ = \_$. An inspection must now be made of the hundreds column so that all possible outcomes are considered. In the hundreds column, we have $\_+3+5 = 13$. Thus, if we assigned any of the digits 5, 6, 7, 8, or 9, for the value of the missing number (second row) in the tens column (second row position), we would have $15 + 5 - 20$, or $15 + 6 = 21$, or $15 + 7 = 22$, or $15 + 8 = 23$, or $15 + 9 = 24$.

$$\begin{array}{r} 3\ 8\ 7 \\ 3\ \_\ 1 \\ +\ 5\ 6\ 2 \\ \hline 1\ 3\ \_\ 0 \end{array}$$

This will then make the digit in the hundreds column equal to 3, since a 2 is being carried. Hence, we have as possible solutions:

$$\begin{array}{r} 387 \\ 351 \\ 562 \\ \hline 1{,}300 \end{array} \quad \text{or} \quad \begin{array}{r} 387 \\ 361 \\ 562 \\ \hline 1{,}310 \end{array} \quad \text{or} \quad \begin{array}{r} 387 \\ 371 \\ 562 \\ \hline 1{,}320 \end{array} \quad \text{or} \quad \begin{array}{r} 387 \\ 381 \\ 562 \\ \hline 1{,}330 \end{array} \quad \text{or} \quad \begin{array}{r} 387 \\ 391 \\ 562 \\ \hline 1{,}340 \end{array}$$

On the other hand, if we were to assign values for the missing digit in the second row of the tens column to be 0, 1, 2, 3, or 4, then the digit in the first row of the hundreds column would have to be a 4, since 1 is now being carried over from the tens column (rather than the 2 as before). These additional solutions would be acceptable:

$$\begin{array}{r} 487 \\ 301 \\ 562 \\ \hline 1{,}350 \end{array} \quad \text{or} \quad \begin{array}{r} 487 \\ 311 \\ 562 \\ \hline 1{,}360 \end{array} \quad \text{or} \quad \begin{array}{r} 487 \\ 321 \\ 562 \\ \hline 1{,}370 \end{array} \quad \text{or} \quad \begin{array}{r} 487 \\ 331 \\ 562 \\ \hline 1{,}380 \end{array} \quad \text{or} \quad \begin{array}{r} 487 \\ 341 \\ 562 \\ \hline 1{,}390 \end{array}$$

Therefore, 10 different solutions result from having two missing digits in the same column.

In the second type of problem, where all digits are represented by letters (hence the name alphametics), the problem is quite different from the preceding ones. Here, the clues from the "puzzle" must be analyzed for all different possible values to be assigned to the letters. No general rule can be given for the solution of alphametic problems. What is required is an understanding of basic arithmetic, logical reasoning, and plenty of patience.

One fine example of this type is the following addition problem:

$$\begin{array}{ccccc} ① & ② & ③ & ④ & ⑤ \\ F & O & R & T & Y \\ & & T & E & N \\ & & T & E & N \\ \hline S & I & X & T & Y \end{array}$$

Since the first line and the fourth line have T Y repeated, this would imply that the sum of both the Es and the Ns in columns four and five must end in zero. If we let N = 0, then E must equal 5, and 1 is carried over to column three. We now have

$$\begin{array}{ccccc} F & O & R & T & Y \\ & & T & 5 & 0 \\ & & T & 5 & 0 \\ \hline S & 1 & X & T & Y \end{array}$$

Since there are two spaces before each T E N, the 0 in F O R T Y must be 9, and with 2 carried over from the hundred's place (column three), the I must be one. And a 1 is carried to column one, making F+1 = S. Ask the students why 2 and not 1 was carried over to the second column. The reason 2 must be carried from column three is that if a 1 were carried, the digits I and N would both be zero. We are now left with the following numbers 2, 3, 4, 6, 7, 8 unassigned.

$$\begin{array}{r} \mathrm{F\,9\,R\,T\,Y} \\ \mathrm{T\,5\,0} \\ \mathrm{T\,5\,0} \\ \hline \mathrm{S\,1\,X\,T\,Y} \end{array}$$

In the hundreds column, we have $2T + R + 1$ (the 1 being carried over from column four), whose sum must be equal to or greater than 22; which implies T and R must be greater than 5. Therefore, F and S will be either 2, 3, or 4. Now X cannot be equal to 3, otherwise F and S would not be consecutive numbers. Then X equals 2 or 4, which is impossible if T is equal to or less than 7. Hence, T must be 8, with R equal to 7 and X equal to 4. Then $F = 2$, $S = 3$, leaving $Y = 6$. Hence, the solution to the problem is

$$\begin{array}{r} 29,786 \\ 850 \\ 850 \\ \hline 31,486 \end{array}$$

Postassessment

Have students solve the following alphametic problems.

1.
$$\begin{array}{r} 4\,\_\,\_\,3 \\ \_\,\_\,1\,4\,\_ \\ \hline \_\,\_\,3\,7\,4\,6 \end{array}$$

Answer
4,603
99,143
103,746

2.
$$\begin{array}{r} 5\,\_\,4\,\_ \\ \_\,4\,5\,\_\,8 \\ 6\,\_\,2\,5\,9 \\ \hline 9\,4\,1\,9\,6 \end{array}$$

Answer
5,349
24,588
64,259
94,196

3.

T R I E D	Answer
D R I V E	17,465
	57,496
R I V E R	74,961

4.

S E N D	Answer
M O R E	9,567
	1,085
M O N E Y	10,652

5.

A L L S	Answer
W E L L	9,332
T H A T	8,433
E N D S	6,596
	4,072
S W E L L	28,433

For more on this topic and other related ideas, see *Math Wonders to Inspire Teachers and Students*, by Alfred S. Posamentier (Association for Supervision and Curriculum Development, 2003).

Unit 4

A Checkerboard Calculator

This enrichment unit will give students an easy, enjoyable method of operation with binary numerals.

Performance Objective

Students will be able to use a checkerboard calculator to do addition, subtraction, multiplication, and division with binary numerals.

Preassessment

Have students find:

a. $1100_2 + 110_2 = \_$ b. $12 + 6 = \_$

c. $111_2 \times 10_2 = \_$ d. $7 \times 2 = \_$

Teaching Strategies

John Napier, the 16th-century mathematician who developed logarithms and Napier's Bones (the calculating rods), also described in his work *Rabdologia* a method for calculating by moving counters across a chessboard. Besides being the world's first binary computer, it is also a valuable teaching aid. Although use of checkered boards was common in the Middle Ages and Renaissance period, by adopting a binary system and basing algorithms on old methods of multiplying by "doubling," Napier's Counting Board became much more efficient than any previous device.

Have students bring to school a standard chessboard or checkerboard. Begin by having students label rows and columns with the doubling series: $1, 2, 4, 8, 16, 32, 64, 128$.

Now show how the board can be used for addition and subtraction. Every number is expressed by placing counters on a row. Each counter has the value of its column. For example, ask the students to add $89 + 41 + 52 + 14$. The fourth row (89) will show $64 + 16 + 8 + 1$ (Figure 1).

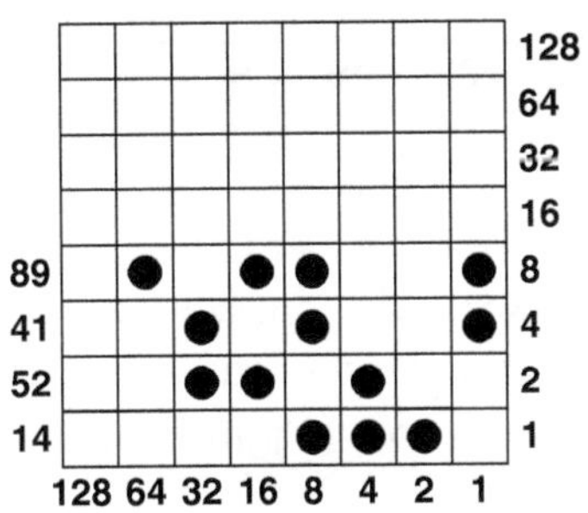

Figure 1.

If the students think of each counter as a 1 and each empty space as a 0, then 89 is represented in binary notation as 1011001_2.

The counters are positioned by starting at the left and putting a counter on the column of the largest number less than or equal to the number the student is representing. Place subsequent counters on the next largest number that when added to the previous number will not exceed the desired total, and so on.

To add, have students move all counters straight down (Figure 2).

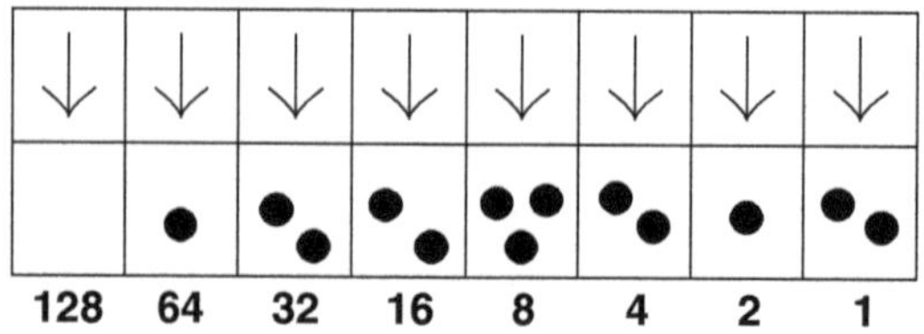

Figure 2.

Adding the values of these counters will give the correct sum, but to use the board for binary notation, we must first "clear" the row of multiple counters on one cell. Have students start at the right, taking each cell in turn. Remove every PAIR of counters on a cell and replace them with a single counter on the next cell to the left. Assure students that this will not affect the sum as every two counters having value n are replaced by one counter having value $2n$. In our example, the final result is the binary number 11000100_2 (Figure 3).

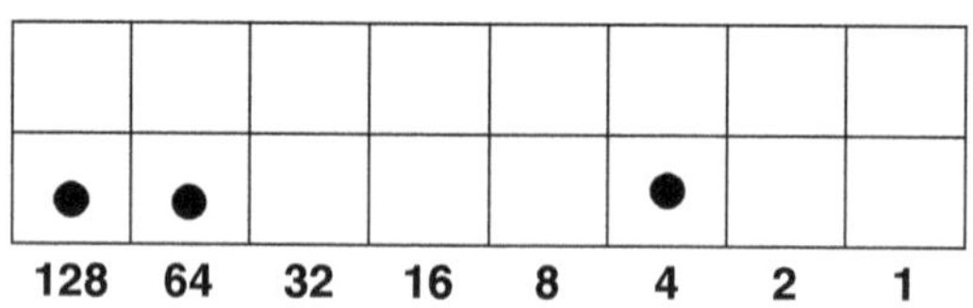

Figure 3.

Subtraction is almost as simple. Suppose students want to take 83 from 108. Have them represent the larger number on the second row and the smaller number on the bottom row (Figure 4).

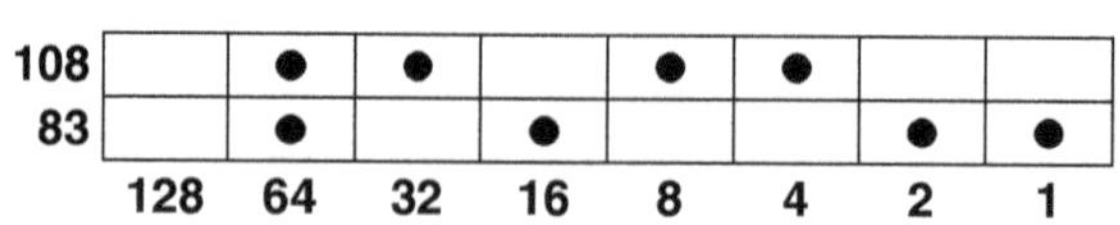

Figure 4.

Students can now do subtraction in the usual manner, starting at the right and borrowing from cell to cell. Or instead, students can alter the entire second row until each counter on the bottom row has one or two counters above it, and no empty cell on the bottom row has more than one counter above it. This can be done by "doubling down" on the second row, removing a counter, and replacing it with two counters on the next cell to the right (Figure 5).

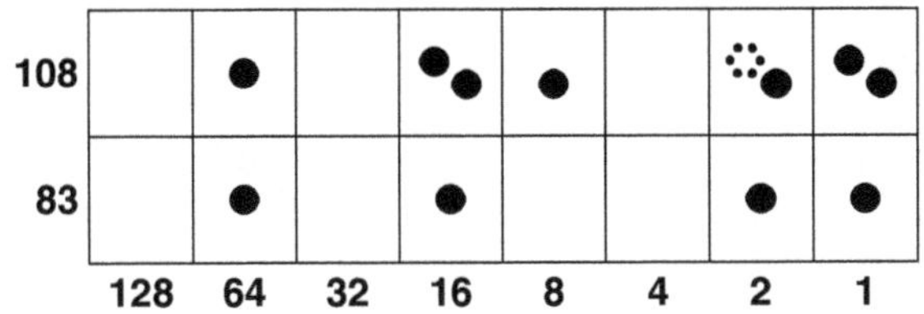

Figure 5.

After this, "king" each counter in the bottom row by moving a counter on top of it from the cell directly above (Figure 6).

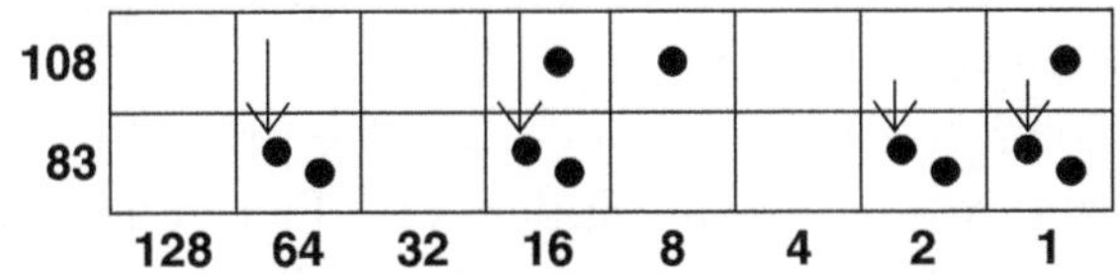

Figure 6.

The top row now shows the difference of the two numbers in binary notation ($11001_2 = 25_{10}$).

Multiplication is also very simple. As an example, use $19 \times 13 = 247$. Have students indicate one number, say 19, by marking below the board under the proper *columns* and the other number, 13, by marking the proper *rows*. Place a counter on every intersection of a marked column and marked row (Figure 7a). Every counter not on the extreme right-hand column is next moved diagonally up and to the right as a bishop in chess (Figure 7b).

Clear the column by having up as in addition, and the desired product is expressed in binary notation as 11110111_2 or 247_{10}, which students can quickly confirm.

Students will want to know how this works. Counters on the first row keep their values when moved to the right; counters on the second row double in value; counters on the third row quadruple in value; and so on. The procedure can be shown to be equivalent to multiplying with powers of the base 2. Nineteen is expressed in our example as $2^4 + 2^1 + 2^0$ and 13 as $2^3 + 2^2 + 2^0$. Multiplying the two trinomials gives us $2^7 + 2^6 + 2 \cdot 2^4 + 2 \cdot 2^3 + 2^2 + 2^1 + 2^0 = 247$.

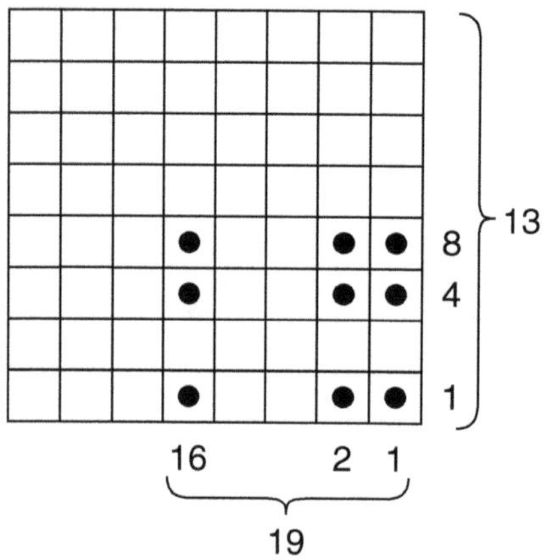

Figure 7a

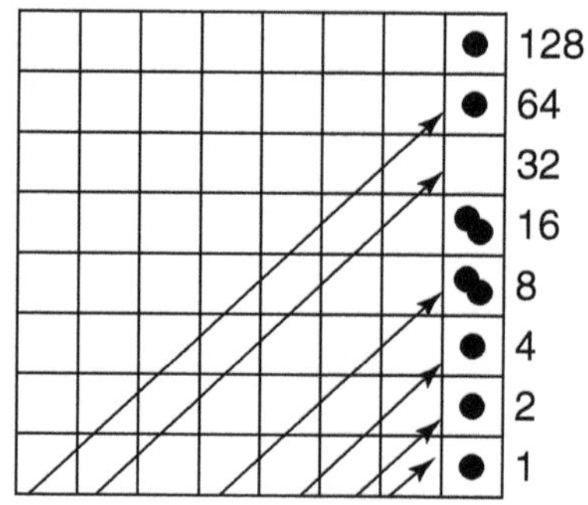

Figure 7b

Moving the counters is equivalent to multiplying. We are, in effect, *multiplying* powers by *adding* exponents.

As an example of division, use $250 \div 13$. The procedure, as students may be expecting, is the reverse of multiplication. The divisor, in this case, 13, is marked at the bottom of the board and the dividend by counters on the column at the extreme right (Figure 8a). The dividend counters now move down and to the left, again like chess bishops, but in the opposite direction to multiplication. This procedure produces a pattern that has counters (one to a cell) only on marked columns, and each marked column must have its counters on the same rows. Only one such pattern can be formed. To do so it is necessary at times to double down on the right column; that is, remove single counters, replacing each with a pair of counters on the next lower cell. Have students start with the top counter and move it diagonally to the leftmost marked column. If the counter cannot proceed, have students return it to the original cell, double down, and try again. Have them continue in this way, gradually filling in the pattern until the unique solution is achieved (Figure 8b).

After the final counter is in place, students should note that three counters are left over. This represents the remainder (3 or 11_2). The value of the right margin is now 10011_2 or 19_{10}, with $\frac{3}{13}$ left over.

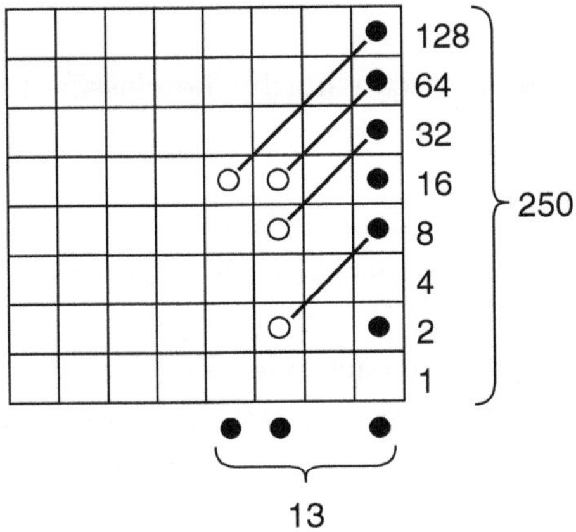

Figure 8a

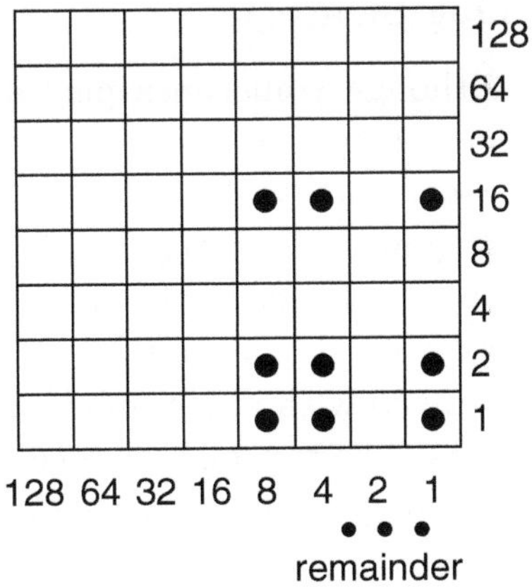

Figure 8b

Postassessment

Have students solve the following problems by using the checkerboard methods:

a. $27 \cdot 64 =$ b. $194 - 63 =$
c. $54 + 43 =$ d. $361 \div 57 =$

Unit 5

The Game of Nim

This unit will present an application of the binary system through the playing of a simple game called Nim.

Performance Objective

Students will play the game of Nim using binary notation system strategy to win.

Preassessment

Have students represent in binary notation:

a. 14 b. 7 c. 13

Teaching Strategies

Nim, although sometimes played for money, can hardly be classified as a gambling game. This is because a player who knows the "secret" of the game can virtually always win.

The game of Nim may be played with sticks, pebbles, coins, or any other small objects. Describe the game to students as played with toothpicks. Have students arrange the toothpicks in three piles (other numbers of piles may be used, also) with any number of toothpicks in each pile. Have two students be players. The two players take turns making their moves. A move consists of taking away toothpicks according to certain rules. The rules are:

1. In each move, a student may take away toothpicks from only one pile.
2. Each player may take any number of toothpicks, but must take at least one, and may take an entire pile at one time.
3. The player who takes away the last toothpick wins.

The "secret" of winning is quite simple, but practice is necessary to accurately perform mentally the arithmetic involved. Therefore, it is probably easier to start with a small number of toothpicks. The winning technique is based on choosing a move so that your opponent must draw from an *even set.*

First, it is necessary to learn how to identify an even set and an odd set. Suppose, for example, that the toothpicks are divided into three piles of (14), (7), and (13) toothpicks. Have students express each of these numbers in binary notation, and add the digits in each column in the same manner as when the decimal base is used. If at least one of the individual sums or digits is an odd number, the distribution is called an odd set. This example is an odd set because one sum is an odd number.

$$\begin{array}{lcr}
\text{Fourteen} & = & 1\,1\,1\,0 \\
\text{Seven} & = & 1\,1\,1 \\
\text{Thirteen} & = & 1\,1\,0\,1 \\
\hline
 & & 2\,3\,2\,2
\end{array} \qquad \text{(odd set)}$$

If the toothpicks are divided into the piles (9), (13), and (4) toothpicks, each individual sum is even and so it is considered an even set.

$$\begin{array}{lcr}
\text{Nine} & = & 1\,0\,0\,1 \\
\text{Thirteen} & = & 1\,1\,0\,1 \\
\text{Four} & = & 1\,0\,0 \\
\hline
 & & 2\,2\,0\,2
\end{array} \qquad \text{(even set)}$$

If a student draws from any even set, he necessarily must leave an odd set, for, considering the representation of the set in the binary scale, any draw will remove a one from at least one column, and the sum of the column will no longer be even.

On the other hand, if a player draws from an odd set, he can leave either an odd set or an even set. There are, however, usually only a few moves that can be made which will change an odd set into an even set. Therefore, a drawing at random from an odd set will very likely result in leaving an odd set.

Explain to the students that the object of the game is to try to force your opponent to draw from an even set, and his drawing will then leave an odd set. There are two winning end distributions that are even sets:

a. Two piles of two toothpicks each, designated symbolically (2), (2).
b. Four piles of one toothpick each, designated (1), (1), (1), (1).

If the student can leave an even set each time he plays, he is eventually able to force his opponent to draw from one of the above even sets, and the game is won. If at the start of the game the student has an even set before him, the best procedure is to draw a single toothpick from the largest pile leaving an odd set. If the opponent does not know the "secret" of the game, he or she will probably draw, leaving an odd set and you will then be able to force a win.

Have students follow moves in a sample game. Put toothpicks in piles of (7), (6), and (3) toothpicks each.

/////// //////		///	
Seven	=	1 1 1	
Six	=	1 1 0	
Three	=	1 1	(odd set)
		2 3 2	

To leave an even set, the first student must draw two toothpicks from any one pile. Drawing from the first pile would give:

///// //////		///	
Five	=	1 0 1	
Six	=	1 1 0	
Three	=	1 1	(even set)
		2 2 2	

No matter how the second student moves, he is forced to leave an odd set. For instance, let him remove three toothpicks from the second pile.

/////	///	///
Five	=	1 0 1
Three	=	1 1
Three	=	1 1
		1 2 3

(odd set)

At this point, the first student should draw all five toothpicks from the first pile.

	///	///
Three	=	1 1
Three	=	1 1
		2 2

(even set)

Now, regardless of how the second student chooses the first will win. Students should now be permitted to play each other. This will provide reinforcement of the binary numeration system. After they have mastered the game as presented above, have them reverse the objective. (That is, let the loser be the player who must pick the last toothpick.)

Postassessment

Have a student, who has been taught the strategy, play Nim against a student who only knows the rules. Use any (or all) of the following choices of piles of toothpicks:

a. (17), (15), (4); b. (18), (15), (4); c. (18), (15), (3)

The student who has been taught the strategy of the game should always win.

References

Posamentier, A. S. and S. Krulik, *Strategy Games to Enhance Problem Solving Ability in Mathematics*, Hackensack, New Jersey: World Scientific Publishing, 2017.

Unit 6

The Tower of Hanoi

This unit provides students with an opportunity to construct and solve an ancient puzzle using the binary system of numeration. The puzzle, known as the Tower of Hanoi, was invented in 1883 by the French mathematician Edward Lucas.

Performance Objective

Each student will build and solve his own Tower of Hanoi puzzle using the binary system, and making use of knowledge required in this lesson.

Preassessment

Prior to this lesson students should be able to convert base 10 numerals to base 2 numerals. Administer the following quiz, instructing students to convert the given numerals (base 10) to base 2: (a) 4; (b) 8; (c) 16; (d) 60; (e) 125.

Teaching Strategies

Begin the lesson by relating the "history" of the puzzle to the class.

W. W. Rouse Ball, in his book *Mathematical Recreations and Essays*, relates an interesting legend of the origin of a puzzle called the Tower of Hanoi. In the great temple at Benares, beneath the dome that marks the center of the world, rests a brass plate in which are fixed three diamond needles, each a cubit high and as thick as the body of a bee. During creation, God placed 64 gold disks of diminishing size on one of these needles, the largest disk at the base resting on the brass plate. This is the tower of Bramah.

According to the legend, the priests work day and night transferring the disks from one diamond needle to another according to the fixed laws of Bramah, which require that the priest on duty must *not* move *more than one disk at a time* and that he must place each disk on a needle so that *there is no smaller disk beneath it.* When the 64 disks have been transferred from the needle on which God placed them at creation to one of the other needles, tower, temple, and Brahmins alike will crumble into dust, and with a thunderclap the world will vanish.

The puzzle, which is sold commercially, can easily be made by each member of the class. Instruct students to cut out eight cardboard circles, each of different size. Have them punch three holes into a piece of thick cardboard so that the distance between the holes is greater than the outside radius of the largest disk. Next have them glue a dowel or pencil upright into each hole. In each disk, they should cut a hole at the center, wide enough for the dowel to fit through. Now they can place each disk on one of the dowels in order of size, with the largest one at the bottom. The arrangement of disks is called a *tower*.

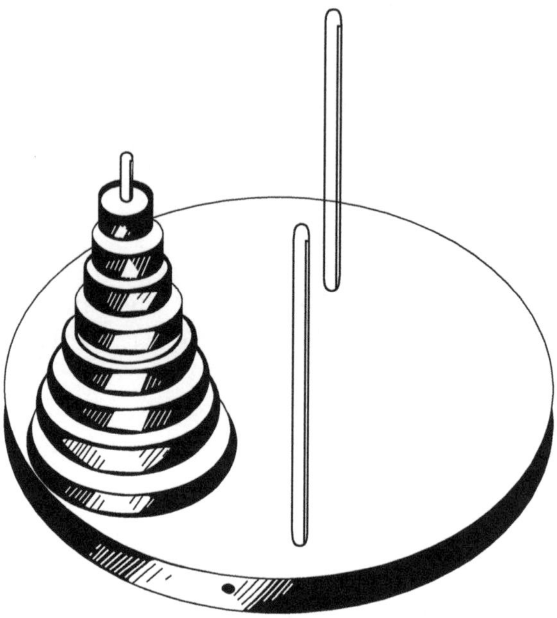

If you do not want to trouble yourself with cutting circles and gluing dowels into a board, you can make a simplified set by cutting eight squares of different sizes, and resting them on three plates instead of on dowels. In any case, be sure to observe the rules:

At the start, all the disks are placed on one post in order of size, the largest disk on the bottom. The puzzle involves shifting the disks, one at a time, from this post to another in such a way that a disk shall never rest on one smaller than itself. This should be done in the least possible number of moves. Remind students of the basic rules:

1. *Move only one disk at a time.*
2. *Never put a disk on top of a smaller disk.*

To familiarize students with the way the game works, demonstrate it first with only three disks. They should be able to transfer a tower of three disks in seven moves.

Now have them try it with four disks. To do this, seven moves are required to transfer the three top disks to one of the other two dowels. This frees the fourth disk which can then be moved to the vacant dowel. Seven more moves are now required to transfer the other three disks back on top of the fourth. Thus, the total number of moves required is 15.

When students consider the game with five disks, they must move the top four disks twice, once to free the bottom disk, and once to get them back on the bottom disk, after the bottom disk has been moved. Thus, moving five disks takes 31 moves: six disks, 63 moves. Ask the class how many moves are required to transfer seven disks. Eight disks?

As students begin to comprehend the challenge of the puzzle, an interesting mathematical problem will emerge: *What is the minimum number of moves required to shift a specific number of disks from one post to another?* To solve this problem, suggest that students denote the number of disks by n, the least number of moves required by $2^n - 1$. Therefore, if there are eight disks, the least number of moves is $2^8 - 1 = 256 - 1 = 255$.

Have students consider the Brahmins with their 64 disks of gold. How many moves will it take them? $2^{64} - 1 = 18,446,744,073,709,551,615$.

If the priests were to make one transfer every second and work 24 hours a day, 365 days a year, they would need more than *580 billion years* to perform the feat, assuming that they never made a mistake. How long would it take the priests to transfer half (or 32) of the disks? (4,294,967,296 sec. = 136 years.)

Now have the class consider the problem of moving eight disks, the Tower of Hanoi. Suggest that students number the disks; one to eight according to size, from the smallest to the largest. Also, have them number the moves from 1 to $225(2^8 - 1 = 225)$. As a class (or independent) project, they should write the number of each move in the binary scale. To discover which disk to transfer at each move, and where to place it, they can refer to the binary scale numeral that corresponds to that move. Then have them count the digits from the right until the first unit digit is reached. The number of digits counted tells which disk to move. For example, if the first 1 from the right is the third digit, then the third disk is moved. Now its placement must be determined. If there are no other digits to the left of the first 1, then the disk is placed on the dowel that has no disks on it. If there *are* other digits to the left of the first 1, students should count digits from the right again until they reach the second 1. The number of digits counted this time identifies a larger disk that was previously moved. Students must decide whether to place the disk they are moving on top of this larger disk or on the "empty" dowel. To decide which strategy to take, they should count the number of zeros between the first 1 from the right and the second 1 from the right. If there are no zeros between them, or if there is an even number of zeros between them, they should put the disk that they are moving onto the disk that the second 1 refers to. If the number of zeros between them is odd, they put the disk on the empty dowel.

The numbers 1 to 15, written in the binary scale are presented here, along with the instructions for the first 15 moves.

1	Move disk 1.
10	Move disk 2.
11	Place disk 1 on disk 2.
100	Move disk 3.
101	Place disk 1 not on disk 3.
110	Place disk 2 on disk 3.
111	Place disk 1 on disk 2.
1000	Move disk 4.
1001	Place disk 1 on disk 4.
1010	Place disk 2 not on disk 4.
1011	Place disk 1 on disk 2.
1100	Place disk 3 on disk 4.
1101	Place disk 1 not on disk 3.
1110	Place disk 2 on disk 3.
1111	Place disk 1 on disk 2.

Postassessment

To assess student progress have them complete the above table. Then have them make the first 25 moves on their model of the Tower of Hanoi.

References

Posamentier, A. S. and S. Krulik, *Strategy Games to Enhance Problem Solving Ability in Mathematics*, Hackensack, New Jersey: World Scientific Publishing, 2017.

Unit 7

What Day of the Week Was It?

This topic may be used for enrichment in a recreational spirit, as well as an interesting application of mathematics. Students will also enjoy seeing the relationship between astronomy and mod 7. In addition, students will be surprised to see how many factors have to be considered in this seemingly simple problem.

Performance Objectives

1. *Given any date, the student will determine the day of the week corresponding to this date.*
2. *Given any year, the student will determine the date of the Easter Sunday that year.*

Preassessment

The students must be familiar with the construction of the calendar.

Give the students a date in this year and ask them to indicate the day corresponding to this date. After they try this, ask them to do this for a date in the past. The students will be anxious to develop a rapid and accurate method for doing this.

Teaching Strategies

Start with a brief history of the calendar. The students will be most interested in knowing the development of the present day Gregorian Calendar and fascinated by how it was changed.

Discuss the relationship of the calendar to astronomy. Time can be measured only by observing the motions of bodies that move in unchanging cycles. The only motions of this nature are those of the celestial bodies. Hence, we owe to astronomy the establishment of a secure basis for the measurement of time by determining the lengths of the day, the month, and the year. A year is defined as the interval of time between two passages of the earth through the same point in its orbit in relation to the sun. This is the solar year. It is approximately 365.242216 mean solar days. The length of the year is not commensurable with the length of the day; the history of the calendar is the history of the attempts to adjust these incommensurable units in such a way as to obtain a simple and practical system.

The calendar story goes back to Romulus, the legendary founder of Rome, who introduced a year of 300 days divided into 10 months. His successor, Numa, added 2 months. This calendar was used for the following six and a half centuries until Julius Caesar introduced the Julian Calendar. If the year were indeed 365.25 days, the introduction of an additional day to 365 days once every 4 years, making the fourth year a leap year, would completely compensate for the discrepancy. The Julian Calendar spread abroad with other features of Roman culture, and was generally used until 1582.

The difficulty with this method of reckoning was that 365.25 was not 365.242216, and although it may seem an insignificant quantity, in hundreds of years it accumulates to a discrepancy of a considerable number of days. The Julian Year was somewhat too long and by 1582 the accumulated error amounted to 10 days.

Pope Gregory XIII tried to compensate for the error. Because the Vernal Equinox occurred on March 11 in 1582, he ordered that 10 days be suppressed from the calendar dates in that year so that the Vernal Equinox would fall on March 21 as it should. When he proclaimed the calendar reform, he formulated the rules regarding the leap years. The Gregorian Calendar has years (based on approximately 365.2425 days) divisible by four as leap years, unless they are divisible by 100 and not 400. Thus, 1700, 1800, 1900, 2100, ... are not leap years, but 2000 is.

In Great Britain and its colonies, the change of the Julian to the Gregorian Calendar was not made until 1752. In September of that year, 11 days were omitted. The day after September 2 was September 14. It is interesting to see a copy of the calendar for September 1752 taken from the almanac of Richard Saunders, Gent., published in London (Figure 1).

Mathematicians have pondered the question of the calendar and tried to develop ways of determining the days of any given date or holiday.

To develop a method for determining the day, the student should be aware that a calendar year (except for a leap year) is 52 weeks and one day long. If New Year's day in some year following a leap year occurs on a Sunday, the next New Year's will occur on Monday. The following New Year's day will occur on a Tuesday. The New Year's day of the leap year will occur on a Wednesday. Since there are 366 days in a leap year, the next New Year's day will occur on a Friday, and not on a Thursday. The regular sequence is interrupted every 4 years (except during years where numbers are evenly divisible by 100 but not evenly divisible by 400).

First develop a method to find the weekday for dates in the same year.

Suppose February 4 falls on Monday. On what day of the week will September 15 fall? Assuming that this calendar year is not a leap year, one need only:

1. Find the number of days between February 4 and September 15. First find that February 4 is the 35th day of the year and that September 15 is the 258th day of the year. (Table 1 expedites this.) The difference of 258 and 35 is the number of days, namely, 223.
2. Since there are 7 days in a week, divide 223 by 7. [$\frac{223}{7} = 31 + \text{remainder } 6.$]
3. The 6 indicates that the day on which September 15 falls is the sixth day after Monday, thus Sunday. In the case of a leap year, one day must be added after February 28 to account for February 29.

1752 September hath XIX Days this Year.

First Quarter, the 15th day at 2 afternoon.
Full Moon, the 23rd day at 1 afternoon.
Last Quarter, the 30th day at 2 afternoon.

M D	W D	Saints' Days Terms, &c.	Moon South	Moon Sets	Full Sea at Lond.	Aspects and Weather
1	f	Day br. 3.35	3 A 27	8 A 29	5 A 1	□ ♃ ☿
2	g	London burn.	4 26	9 11	5 38	Lofty winds

According to an act of Parliament passed in the 24th year of his Majesty's reign and in the year of our Lord 1751, the Old Style ceases here and the New takes its place; and consequently the next Day, which in the old account would have been the 3d is now to be called the 14th; so that all the intermediate nominal days from the 2d to the 14th are omitted or rather annihilated this Year; and the Month contains no more than 19 days, as the title at the head expresses.

M D	W D	Saints' Days Terms, &c.	Moon South	Moon Sets	Full Sea at Lond.	Aspects and Weather
14	e	Clock slo. 5 m.	5 15	9 47	6 27	HOLY ROOD D.
15	f	Day 12 h. 30 m.	6 3	10 31	7 18	and hasty
16	g		6 57	11 23	8 16	showers
17	A	15 S. AFT. TRIN.	7 37	12 19	9 7	
18	b		8 26	Morn.	10 22	More warm
19	c	Nat. V. Mary	9 12	1 22	11 21	and dry
20	d	EMBER WEEK	9 59	2 24	Morn.	weather
21	e	ST. MATTHEW	10 43	3 37	0 17	☌ ♀ ☿
22	f	Burchan	11 28	☾ rise	1 6	□ ♃ ♀
23	g	EQUAL D. & N.	Morn.	6 A 13	1 52	☌ ☉ ♂
24	A	16 S. AFT. TRIN.	0 16	6 37	2 39	☌ ☉ ☿
25	b		1 5	7 39	3 14	
26	c	Day 11 h. 52 m.	1 57	8 39	3 48	Rain or hail
27	d	EMBER WEEK	2 56	8 18	4 23	☌ ♂ ☿
28	e	Lambert bp.	3 47	9 3	5 6	now abouts
29	f	ST. MICHAEL	4 44	9 59	5 55	✶ ♄ ♀
30	g		5 43	11 2	6 58	

Figure 1.

A similar method for finding the weekday of the dates in the same year can be discussed as follows. Because January has 31 days, the same date in the subsequent month will be 3 days after that day in January; the same date in March will also be 3 days later than in January; in April, it will be 6 days later than in January. We can then construct a table of Index Numbers for the months that will adjust all dates to the corresponding dates in January:

January	0	April	6	July	6	October	0
February	3	May	1	August	2	November	3
March	3	June	4	September	5	December	5

(The Index Numbers are actually giving you the days between the months divided by 7 to get the excess days as in the previous method.)

Now you need only add the date to the Index Number of the month, divide by 7 and the remainder will indicate the day of the week.

Example: Consider the year 1925. January 1 was on a Thursday. Find March 12.

To do this, add $12 + 3 = 15$; divide $15/7 = 2$ remainder 1. This indicates Thursday. In leap years, an extra 1 has to be added for dates after February 29.

Table 1.

Date	1	2	3	4	5	6	7	8	9	10	11	12	13	14	15	16
January	1	2	3	4	5	6	7	8	9	10	11	12	13	14	15	16
February	32	33	34	35	36	37	38	39	40	41	42	43	44	45	46	47
March	60	61	62	63	64	65	66	67	68	69	70	71	72	73	74	75
April	91	92	93	94	95	96	97	98	99	100	101	102	103	104	105	106
May	121	122	123	124	125	126	127	128	129	130	131	132	133	134	135	136
June	152	153	154	155	156	157	158	159	160	161	162	163	164	165	166	167
July	182	183	184	185	186	187	188	189	190	191	192	193	194	195	196	197
August	213	214	215	216	217	218	219	220	221	222	223	224	225	226	227	228
September	244	245	246	247	248	249	250	251	252	253	254	255	256	257	258	259
October	274	275	276	277	278	279	280	281	282	283	284	285	286	287	288	289
November	305	306	307	308	309	310	311	312	313	314	315	316	317	318	319	320
December	335	336	337	338	339	340	341	342	343	344	345	346	347	348	349	350

Date	17	18	19	20	21	22	23	24	25	26	27	28	29	30	31
January	17	18	19	20	21	22	23	24	25	26	27	28	29	30	31
February	48	49	50	51	52	53	54	55	56	57	58	59			
March	76	77	78	79	80	81	82	83	84	85	86	87	88	89	90
April	107	108	109	110	111	112	113	114	115	116	117	118	119	120	
May	137	138	139	140	141	142	143	144	145	146	147	148	149	150	151
June	168	169	170	171	172	173	174	175	176	177	178	179	180	181	
July	198	199	200	201	202	203	204	205	206	207	208	209	210	211	212
August	229	230	231	232	233	234	235	236	237	238	239	240	241	242	243
September	260	261	262	263	264	265	266	267	268	269	270	271	272	273	
October	290	291	292	293	294	295	296	297	298	299	300	301	302	303	304
November	321	322	323	324	325	326	327	328	329	330	331	332	333	334	
December	351	352	353	354	355	356	357	358	359	360	361	362	363	364	365

Students will now want to find the day for a date for any given year. Point out that first one need know what day January 1 of the Year 1 fell and also make adjustments for leap years.

The day of the week on which January 1 of Year 1 fell can be determined as follows. Using a known day and date, we find the number of days that have elapsed since January 1 of the Year 1. Thus, since January 1, 1952, was Wednesday, in terms of the value of the solar year, the number of days since January 1 is $1951 \times 365.2425 = 712{,}588.1175$. Dividing by 7, we get 101,798 with a remainder of 2. The remainder indicates that 2 days should be counted from Wednesday. Since calculations refer to the past, the counting is done backwards, indicating that January 1 (in the Gregorian Calendar) fell on Monday.

One method for determining the day for any year suggests that dates in each century be treated separately. Knowing the weekday of the first day of that period, one could, in the same fashion as before, determine the excess days after that weekday (thus the day of the week that a given day would fall on for that century). For the years 1900–1999, the information needed is

1. The Index Numbers of the months (see earlier discussion).
2. January 1 of 1900 was Monday.
3. The number of years (thus giving the number of days over the 52 week cycles) that have elapsed since the 1st day of the year 1900.
4. The number of leap years (i.e., additional days) that have occurred since the beginning of the century.

Knowing this, we can ascertain how many days in that Monday-week cycle we need count.

Examples:

1. *May 9, 1914.* Add 9 (days into the month), 1 (Index Number of the month), 14 (Number of years since the beginning of the century), and 3 (number of leap years in that century thus far). $9+1+14+3 = 27$. Divide by 7, leaves 6, which is Saturday.
2. *August 16, 1937.* Add $16 + 2 + 37 + 9 = 64$. Divide by 7, leaves 1, which is Monday.

For the period, 1800–1899, the same procedure is followed except that January 1, 1800, was on Wednesday. For the period September 14, 1752, through 1799, the same procedure is followed except that the first day of that period would be Friday. For the period up to and including September 2,

1752, the same procedure is followed except that the whole year would be added and the number of the days would start with Friday.

Example: *May 13, 1240*

Add $13 + 1 + 1240 + 10 = 1264/7$, leaves 4, Monday.

There is another method for determining the day without having to consider separate periods.

Again we start by knowing the day of January 1 of the Year 1. We will not count the actual number of days that have elapsed since January 1 of Year 1, but count the number of excess days over weeks that have elapsed and to this number, add the number of days that have elapsed since January 1 of the given year. This total must be divided by 7, the remainder will indicate the number of days that must be counted for that week, thus the formula is, 1 (Monday) + the remainder of the division by 7 of (the number of years that have elapsed thus far + the number of days that have elapsed since January 1 of the given year + the number of leap years that have occurred since year 1) = the number of days of the week. The calculation of the number of leap years must take into account the fact that those years whose number ends with two zeros, which are not divisible by 400, are not leap years. Thus, from the total number of leap years, a certain number of leap years must be subtracted.

Example: *December 25, 1954*. $1 + 1953 + 488$ (leap years) -15 (century leap years $19 - 4$) $+ 358$ (number of days between January 1, 1954, and December 25, 1954) $= 2785$. Dividing by 7 gives remainder 6. Thus, December 25, 1954, fell on the sixth day of the week, Saturday.

Many other tables and mechanisms have been devised to solve the problem of determining days. The following are two nomograms devised for this.

The first (see Figure 2) consists of four scales and is to be used as follows:

1. With a straightedge, join the point on the first scale indicating the date with the proper month on the third scale. Mark the point of intersection with the second scale.
2. Join this point on the second scale with the point on the fourth scale indicating the proper century. Mark the point of intersection with the third scale.
3. Join this point with the point indicating the appropriate year on the first scale. The point of intersection with the second scale gives the desired

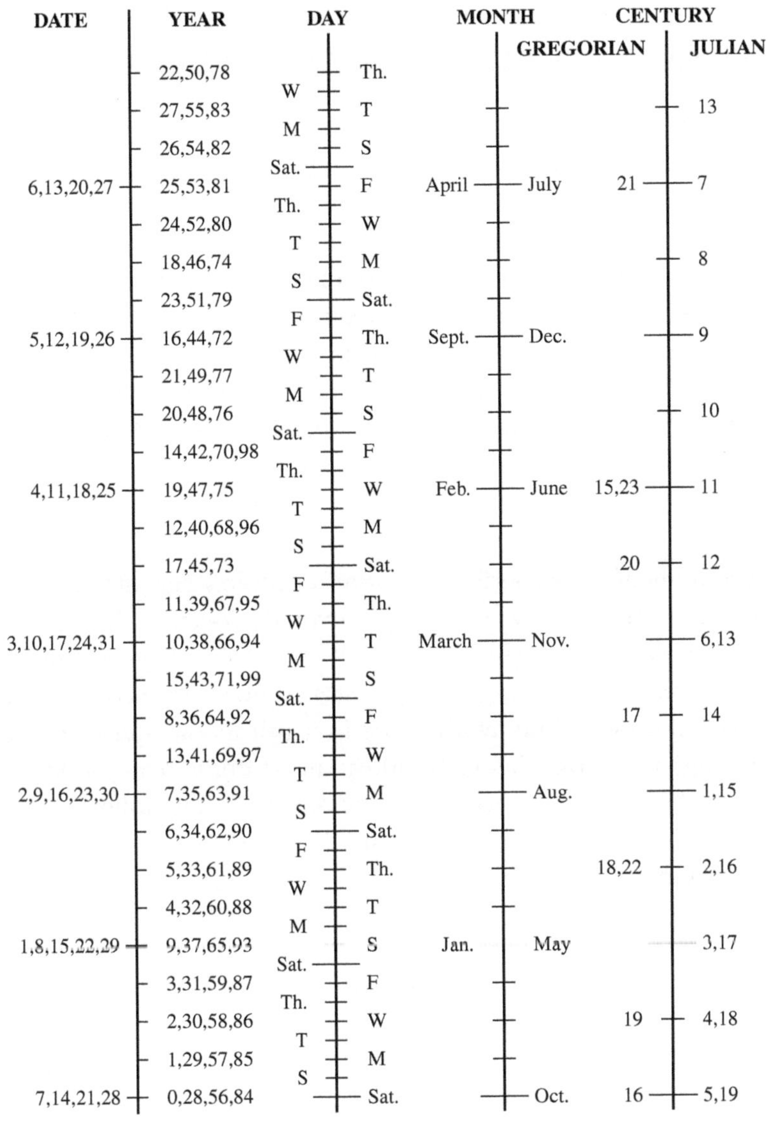

Figure 2. Perpetual calendar.

day of the week. (N.B. For the months of January and February, use the year diminished by 1.)

The second arrangement (see Figure 3) consists of three concentric rings intersected by seven radii. The procedure is:

1. Locate the date and the month on the outer ring; if they are two points, draw a line between them; if they coincide, draw a tangent.
2. Locate the century on the intermediate ring. Through this point, draw a line parallel to the line drawn, until it intersects the intermediate ring at another point. The point found will be a ring-radius intersection.
3. From the point just found, follow the radius to the inner ring, then locate the year. (If the month is January or February, use the preceding year.) Draw a line between these two points on the inner ring. (If they coincide on Saturday, then Saturday is the weekday sought.)
4. Now find the point where the Saturday radius cuts the inner ring, and through this Saturday point draw a line parallel to the line just drawn. The line will meet the inner ring at some radius–ring intersection. The weekday on this latter radius is the weekday of the date with which we began.

The problem of a perpetual calendar occupied the attention of many mathematicians and many of them devoted considerable attention to calculating the date of Easter Sunday. All church holidays fall on a definite date. The ecclesiastical rule regarding Easter is, however, rather complicated. Easter must fall on the Sunday after the first full moon that occurs after the Vernal (Spring) Equinox. Easter Sunday, therefore, is a movable Feast that may fall as early as March 22 or as late as April 25. The following procedure to find Easter Sunday in any year from 1900–1999 is based on a method developed by Gauss.

1. Find the remainder when the year is divided by 4. Call this remainder a.
2. Find the remainder when the year is divided by 7. Call this remainder b.
3. Find the remainder when the year is divided by 19. Multiply this remainder by 19, add 24, and again find the remainder when the total is divided by 30. Call this remainder c.
4. Now add $2a+4b+6c+3$. Divide this total by 7 and call the remainder d.

The sum of c and d will give the number of days after March 22 on which Easter Sunday will fall.

Example: Easter 1921

a. $\frac{21}{4}$ leaves 1

b. $\frac{21}{7}$ leaves 0

c. $\frac{21}{19}$ leaves 2; $[2(19)+24]/30$ leaves 2

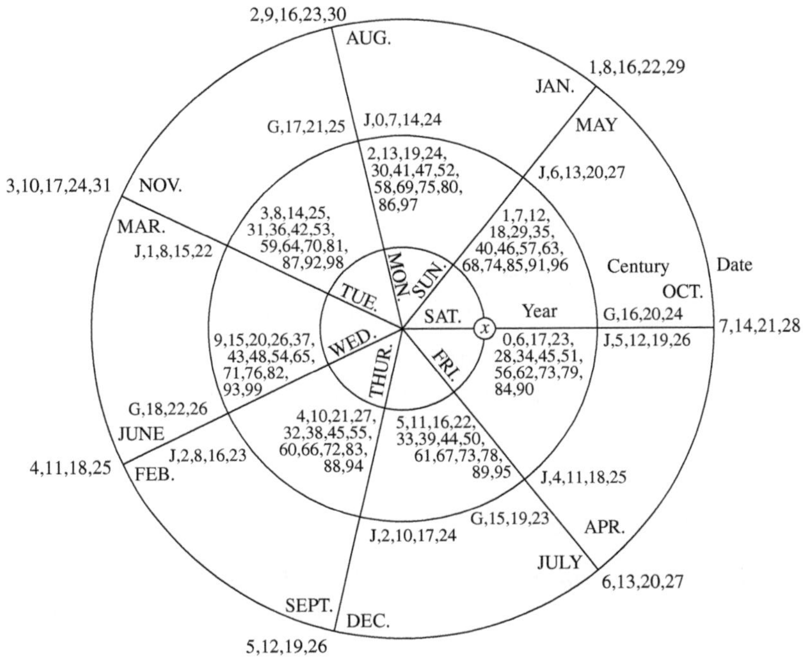

Figure 3. Perpetual calendar, radius–ring type.

d. $[2 + 0 + 12 + 3]/7$ leaves 3
$2 + 3 = 5$ days after March 22 = March 27

(The method above gives the date accurately except for the years 1954 and 1981. These years it gives a date exactly 1 week late, the correct Easters being April 18 and 19, respectively.)

Postassessment

1. Have the students determine the day of the week of their birth.
2. Have the students work out several given dates.

October 12, 1492	May 30, 1920	Christmas 1978
April 1, 1945	October 21, 1805	August 14, 1898
		July 4, 1776

3. Have the students find the dates of several Easter Sundays.
1944, 1969, 1950, 1978, 1930, 1977, 1929
4. George Washington was born February 11, 1732. Why do we celebrate it February 22?

Unit 8

Palindromic Numbers

This unit will define palindromic numbers and introduce some of their properties. The study of palindromic numbers is suited for any class: while it provides all students an approach for analyzing numbers and their relationships, certain aspects of this topic can be selected for the slow students (e.g., the reverse addition property) and more advanced properties can be investigated by the more adept students (e.g., modular palindromes).

Performance Objectives

1. *Students will state and analyze properties of palindromic numbers.*
2. *Students will construct new palindromes from any specified integer.*

Preassessment

Have students analyze the expression "Madam I'm Adam" and the words "rotator" and "reviver" and point out their peculiarity (they spell the same backward and forward). Indicate to them that such an expression is called a palindrome and that in mathematics, numbers having the same property like 343 and 59,695 are called palindromic numbers. Students can be asked to give their own examples of palindromic numbers and make a short list.

Teaching Strategies

After the students have compiled their lists, an analysis can be made of these numbers. Questions such as the following can be put to them to get a discussion going: Does a palindrome have an odd or even number of digits, or both? Are palindromic numbers prime, composite, or both? Is the square or cube of a palindromic number still a palindrome? Given a positive integer can a palindrome be constructed from some sort of operation on this integer? Students can attempt to answer these questions by testing their validity on the numbers from their lists or by seeking some new ones.

The students are ready at this point to study some of the following palindromic properties:

1. *Palindromic numbers contain both prime and composite numbers (e.g., 181 is a palindromic prime while 575 is a composite palindrome); however, a palindromic prime, with the exception of 11, must have an odd number of digits.*

   Proof of the latter: (by contradiction)

   Let p be a palindromic prime having an even number of digits. Let r be the sum of all digits in the even positions of the prime p and s be the sum of all the digits in the odd positions of the prime p. Since p is a palindrome with an even number of digits, the digits in odd positions duplicate the digits in even positions; therefore $s - r = 0$. But the test for divisibility by 11 states that a number is divisible by 11 if the difference between the sum of all digits in even positions and the sum of all digits in odd positions is 0 or a multiple of 11. Therefore, p has 11 as a factor and cannot be prime, a contradiction.

2. *All integers N, which yield palindromic squares, are not necessarily palindromes.* While there are infinitely many palindromes yielding palindromic squares (e.g., $22^2 = 484$ and $212^2 = 44{,}944$), there exist some non-palindromic integers whose squares are palindromes (e.g., $26^2 = 676$ and $836^2 = 698{,}896$) as well as some palindromic integers yielding nonpalindromic squares (e.g., $131^2 = 17{,}161$ and $232^2 = 53{,}824$).

   Repunits, numbers consisting entirely of 1s (by notation, R_k where k is the number of 1s) are palindromic numbers and produce palindromic squares when $1 \leq k \leq 9$: $R_2^2 = 121$: $R_3^2 = 12{,}321$ and in general $R_k^2 = 12 \ldots k \ldots 21$, where $k = 9$. However, when $k > 9$, the carrying in addition would lose the palindromic product (e.g., $R_{10}^2 = 12 \ldots 6790098 \ldots 21$).

   Square numbers are much richer in palindromes than randomly chosen integers.

3. *In general, numbers that yield palindromic cubes (some of which are prime and some composite) are palindromic in themselves.* The numbers N that yield palindromic cubes are as follows:

   a. $N = 1, 7, 11 (1^3 = 1, 7^3 = 343, 11^3 = 1{,}331)$.

   b. $N = 10^k + 1$ has a palindromic cube consisting of $k - 1$ zeros between each consecutive pair of 1,3,3,1: e.g., when $k = 1, N = 11$ and $N^3 = 1{,}331$; when $k = 2, N = 101$ and $N^3 = 1{,}030{,}301$; when $k = 3, N = 1{,}001$ and $N^3 = 1{,}003{,}003{,}001$, etc. Notice that when $k = 2m + 1$, $m > 0$, then N is divisible by 11 and hence is composite.

   c. N consisting of three 1s and any desired number (which must be even) of zeros is divisible by 3 and has palindromic cubes: e.g., $(111)^3 = 1{,}367{,}631$, $(10{,}101)^3 = 1{,}030{,}607{,}060{,}301$.

d. N = any palindromic arrangement of zeros and four 1's is a non-prime and has a palindromic cube, except when the same number of zeros appears in the three spaces between the 1's: e.g., $(11{,}011)^3 =$ 1,334,996,994,331, $(10{,}100{,}101)^3 =$ 1,030,331,909,339,091,330,301, whereas $(1{,}010{,}101)^3$ is not a palindrome.

The only $N < 2.8 \times 10^{14}$ that is not a palindrome yet yields a palindromic cube is $2{,}201^3 = 10{,}662{,}526{,}601$.

4. *Given any integer N we can often reach a palindrome by adding the number to its reversal (the number obtained by reversing the digits) and continuing the process until the palindrome is achieved.* For example, if $N = 798$, then $798 + 897 = 1{,}695$; $1{,}695 + 5{,}961 = 7{,}656$; $7{,}656 + 6{,}567 = 14{,}223$; $14{,}223 + 32{,}241 = 46{,}464$ (a palindrome). Whereas some numbers can reach a palindrome in only two steps (e.g., 75 and 48), there are others that reach it in 6 steps like 97, and still others like 89 and 98 that reach a palindrome in 24 steps. However, certain numbers like 196 when carried out to over 1000 steps still do not achieve palindromes, so this rule cannot be taken to hold for all integers but certainly for most since these cases are very rare. The rule does not hold in base two; the smallest counterexample is 10,110, which after 4 steps reaches the sum 10,*11*0,1*00*, after 8 steps it is 10,*111*0,1*000*, and after 12 steps it is 10*1,111*,01*0,000*. Every fourth step increases by one digit each of the two sequences italicized and it is seen that each of these increasing sums is not a palindrome. There are some generalities found in this process of reverse addition:

a. Different integers when subjected to this technique produce the same palindrome. For example, 554, 752, and 653 all produce the palindrome 11,011 in 3 steps. In general, all integers in which the corresponding digit pairs symmetrical to the middle have the same sum will produce the same palindrome in the same number of steps. (In this case, all the digit pairs add up to 9.) There are however different integers that produce the same palindrome in different numbers of steps. For example, 99 reaches 79,497 in 6 steps whereas 7,299 reaches it in 2 steps.

b. Two-digit numbers can be categorized according to the total sum of the two digits to ascertain the number of steps needed to produce a palindrome. It is obvious that if the sum of the digits is 9, only 1 step is needed; if their digit sum is 10 (e.g., 64 and 73), 2 steps are needed. Similar analyzes will lead the students to conclude that if their digit sum is $11, 12, 13, 14, 15, 16, 17$, or 18, a palindrome results after $1, 2, 2, 3, 4, 6, 24$, and 1 respectively. The students can be asked to perform an analysis of this type and put their results in table form.

This topic of palindromes can be investigated further by more adept students. Further investigation lies in the areas of multigrades with palindromic numbers as elements, special palindromic prime numbers such as primes with prime digits as elements, modular palindromes, and triangular and pentagonal numbers that are also palindromes.

Postassessment

1. Do the following numbers yield palindromic cubes: (a) 1,001,001; (b) 1,001,001,001; (c) 10,100,101; (d) 100,101.
2. Given the following two-digit numbers, indicate the number of steps required in reversal addition to reach a palindrome: (a) 56; (b) 26; (c) 91; (d) 96.
3. Perform the reverse addition technique on these integers and find other integers that will yield the same palindrome as these: (a) 174; (b) 8,699.

References

Posamentier, A. S. and B. Thaller, *Numbers Their Tales, Types and Treasures*, Amherst, New York: Prometheus Books, 2015.

Unit 9

The Fascinating Number Nine

This unit is intended to offer a recreational presentation of the many interesting properties of the number 9. A long-range goal of presenting these amusing number topics is to motivate further student investigation and insight into the properties of numbers.

Performance Objectives

1. *Students will demonstrate at least three properties of the number 9.*
2. *Students will provide an example of a short-cut calculation involving the number 9.*

Preassessment

Students should be familiar with the various field postulates and be reasonably adept with the operations of addition, subtraction, multiplication, and division. A knowledge of algebra is helpful but not essential.

Teaching Strategies

In presenting new ideas to a class, it is always best to build on what they already know. For example, ask students to multiply 53×99. Unsuspecting students will perform the calculation in the usual way. After their work has been completed, suggest the following:

Since

$$\begin{aligned} 99 &= 100 - 1 \\ 53 \times 99 &= 53\ (100 - 1) \\ &= 53\ (100) - 53(1) \\ &= 5{,}300 - 53 \\ &= 5{,}247 \end{aligned}$$

Now have them use this technique to multiply 42×999.

"Casting Out Nines" is a popular technique for checking calculations. For example, if students wish to check the addition $29 + 57 + 85 + 35 + 6 = 212$, they simply divide each number in the addition by 9 and retain only the remainder. Thus, they have $2, 3, 4, 8$, and 6, the sum of which is 23. (See following.)

$$\begin{array}{rcr} 29 & \rightarrow & 2 \\ 57 & \rightarrow & 3 \\ 85 & \rightarrow & 4 \\ 35 & \rightarrow & 8 \\ \underline{+\ 6} & \rightarrow & \underline{+6} \\ 212 & & 23 \end{array}$$

The remainder of $212 \div 9$ is 5. If the remainder of $212 \div 9$ is the same as the remainder of $23 \div 9$, then 212 could be the correct answer. In this case, where 5 is the remainder of both divisions by 9, 212 *could* be the correct answer. Students cannot be sure from this checking method whether in fact this answer is correct since a rearrangement of the digits, to say 221, would also yield the same remainder when divided by 9.

Division by 9 is not necessary to find the remainder. All one has to do is add the digits of the number (to be divided by 9) and, if the result is not a single digit number, repeatedly add the digits until a single digit remains. In the above example, the remainders are:

for 29: $2+9=11;\ 1+1=2$
for 57: $5+7=12;\ 1+2=3$
for 85: $8+5=13;\ 1+3=4$
for 35: $3+5=8$
for 6: 6

for the sum of $2+3+4+8+6=23;\ 2+3=5$
for 212: $2+1+2=5$

Students can use a similar procedure for other operations. For example, to check a multiplication operation: $239 \times 872 = 208{,}408$, they will find the remainders (when divided by 9) of each of the above numbers.

for 239: $2+3+9=14;\ 1+4=5$
for 872: $8+7+2=17;\ 1+7=8$
for the product $5 \times 8 = 40;\ 4+0=4$
for 208,408: $2+0+8+4+0+8=22;\ 2+2=4$

Stress to your classes that this is not a fool-proof check of a calculation, but simply one indication of possible correctness. Present this topic in a way that will have them begin to marvel at the interesting properties of the number 9.

Another unusual property of 9 occurs in the multiplication of 9 and any other number of 2 or more digits. Consider the example $65{,}437 \times 9$. An alternative to the usual algorithm is as follows:

1. Subtract the units digit of the multiplicand from 10.	$10-7=\boxed{3}$
2. Subtract each of the remaining digits from 9 and add the rest to the preceding number of the multiplicand (at right). For any two-digit sums, carry the tens digit to the next sum.	$9-3=6+7=1\,\boxed{3}$ $9-4=5+3+1=\boxed{9}$ $9-5=4+4=\boxed{8}$ $9-6=3+5=\boxed{8}$
3. Subtract 1 from the left-most digit of the multiplicand.	$6-1=\boxed{5}$
4. Now list the results in reverse order to get the desired product.	$\boxed{588{,}933}$

Although this method is somewhat cumbersome, it can set the groundwork for some rather interesting investigations into number theory.

To further intrigue your students with other fascinating properties of the number 9, have them multiply 12,345,679 by the first nine multiples of 9, and record their results:

$$12{,}345{,}679 \times 9 = 111{,}111{,}111$$

$$12{,}345{,}679 \times 18 = 222{,}222{,}222$$

$$12{,}345{,}679 \times 27 = 333{,}333{,}333$$

$$12{,}345{,}679 \times 36 = 444{,}444{,}444$$

$$12{,}345{,}679 \times 45 = 555{,}555{,}555$$

$$12{,}345{,}679 \times 54 = 666{,}666{,}666$$

$$12{,}345{,}679 \times 63 = 777{,}777{,}777$$

$$12{,}345{,}679 \times 72 = 888{,}888{,}888$$

$$12{,}345{,}679 \times 81 = 999{,}999{,}999$$

Students should realize that in the sequence of natural numbers (making up the above multiplicands) the number 8 was omitted. In other words, the number that is two less than the base, 10, is missing. Ask students how to extend this scheme to bases other than 10.

Now have them reverse the sequence of natural numbers including the 8, and multiply each by the first nine multiples of 9. The results will be astonishing:

$$987{,}654{,}321 \times 9 = 8\,888\,888\,889$$

$$987{,}654{,}321 \times 18 = 17\,777\,777\,778$$

$$987{,}654{,}321 \times 27 = 26\,666\,666\,667$$

$$987{,}654{,}321 \times 36 = 35\,555\,555\,556$$

$$987{,}654{,}321 \times 45 = 44\,444\,444\,445$$

$$987{,}654{,}321 \times 54 = 53\,333\,333\,334$$

$$987{,}654{,}321 \times 63 = 62\,222\,222\,223$$

$$987{,}654{,}321 \times 72 = 71\,111\,111\,112$$

$$987{,}654{,}321 \times 81 = 80\,000\,000\,001$$

Some other interesting properties of the number 9 are exhibited in the following. Have students discover them by carefully guiding them to the desired result. Stronger students should be encouraged to investigate these relationships and discover *why* they "work."

1.

$$9 \times 9 = 81$$
$$99 \times 99 = 9801$$
$$999 \times 999 = 998{,}001$$
$$9{,}999 \times 9{,}999 = 99{,}980{,}001$$
$$99{,}999 \times 99{,}999 = 9{,}999{,}800{,}001$$
$$999{,}999 \times 999{,}999 = 999{,}998{,}000{,}001$$
$$9{,}999{,}999 \times 9{,}999{,}999 = 99{,}999{,}980{,}000{,}001$$

2.

$$999{,}999 \times 2 = 1{,}999{,}998$$
$$999{,}999 \times 3 = 2{,}999{,}997$$
$$999{,}999 \times 4 = 2{,}999{,}996$$
$$999{,}999 \times 5 = 4{,}999{,}995$$
$$999{,}999 \times 6 = 4{,}999{,}994$$
$$999{,}999 \times 7 = 5{,}999{,}993$$
$$999{,}999 \times 8 = 6{,}999{,}992$$
$$999{,}999 \times 9 = 8{,}999{,}991$$

3.

$$1 \times 9 + 2 = 11$$
$$12 \times 9 + 3 = 111$$
$$123 \times 9 + 4 = 1111$$
$$1{,}234 \times 9 + 5 = 11{,}111$$
$$12{,}345 \times 9 + 6 = 111{,}111$$
$$123{,}456 \times 9 + 7 = 1{,}111{,}111$$
$$1{,}234{,}567 \times 9 + 8 = 11{,}111{,}111$$
$$12{,}345{,}678 \times 9 + 9 = 111{,}111{,}111$$

4.

$$9 \times 9 + 7 = 88$$
$$98 \times 9 + 6 = 888$$
$$987 \times 9 + 5 = 8{,}888$$
$$9876 \times 9 + 4 = 88{,}888$$
$$98765 \times 9 + 3 = 888{,}888$$
$$987654 \times 9 + 2 = 8{,}888{,}888$$
$$98{,}76{,}543 \times 9 + 1 = 88{,}888{,}888$$
$$98{,}765{,}432 \times 9 + 0 = 888{,}888{,}888$$

An interesting way to conclude this unit would be to offer your students a seemingly harmless challenge. That is, ask them to find an eight-digit number in which no digit appears more than once, and which, when multiplied by 9, yields a nine-digit number in which no digit appears more than once. Most of their attempts will fail. For example, $76{,}541{,}238 \times 9 = 688{,}871{,}142$, which has repeated 8s and 1s. Here are several correct numbers:

$$81{,}274{,}365 \times 9 = 731{,}469{,}285$$
$$72{,}645{,}831 \times 9 = 653{,}812{,}479$$
$$58{,}132{,}764 \times 9 = 523{,}194{,}876$$
$$76{,}125{,}483 \times 9 = 685{,}129{,}347$$

Postassessment

Ask students to:

1. Demonstrate three unusual properties of 9.
2. Show a short-cut for multiplying 547×99.
3. Explain now to "check" a multiplication calculation by "casting out nines."

References

Posamentier, A. S. and B. Thaller, *Numbers Their Tales, Types and Treasures*, Amherst, New York: Prometheus Books, 2015.

Posamentier, A. S. and I. Lehmann, *Mathematical Curiosities: A Treasure Grove of Unexpected Entertainments*, Amherst, New York Prometheus Book, 2014

Unit 10

Unusual Number Properties

The intention of this unit is to present a good supply of interesting number properties that can be best exhibited using a calculator.

Performance Objective

Students will investigate mathematical problems with the help of a calculator and then draw appropriate conclusions.

Preassessment

Students should be familiar with the basic functions of a calculator. The instrument required for this unit need only have the four basic operations.

Teaching Strategies

Perhaps one of the best ways to stimulate genuine excitement in mathematics is to demonstrate some short, simple, and dramatic mathematical phenomena. The following are some examples that should provide you with ample material with which to motivate your students toward some independent investigations.

Example 1

When 37 is multiplied by each of the first nine multiples of 3, an interesting result occurs. Let students discover this using their calculators.

$$37 \times 3 = 111$$

$$37 \times 6 = 222$$

$$37 \times 9 = 333$$

$$37 \times 12 = 444$$

$$37 \times 15 = 444$$

$$37 \times 18 = 666$$

$$37 \times 21 = 777$$

$$37 \times 24 = 888$$

$$37 \times 27 = 999$$

Example 2

When 142,857 is multiplied by $2, 3, 4, 5$, and 6, the products all use the same digits in the same order as in the original numbers, but each starting at a different point.

$$142{,}857 \times 2 = 285{,}714$$

$$142{,}857 \times 3 = 428{,}571$$

$$142{,}857 \times 4 = 571{,}428$$

$$142{,}857 \times 5 = 714{,}285$$

$$142{,}857 \times 6 = 857{,}142$$

When 142,857 is multiplied by 7, the product is 999,999. When 142,857 is multiplied by 8, the product is 1,142,856. If the millions digit is removed and added to the units digit $(142{,}856 + 1)$, the original number is formed. Have students investigate the product $142{,}857 \times 9$. What other patterns can be found involving products of 142,857?

A similar pattern occurs with the following products of 76,923: $1, 10, 9, 12, 3$, and 4; also $2, 5, 7, 11, 6$, and 8. Ask students to inspect the sum of the digits of each of the products obtained. They should discover a truly fascinating result! Ask if they can find other such relationships.

Example 3

The number 1,089 has many interesting properties. Have students consider the products of 1,089 and each of the first nine natural numbers.

$$1{,}089 \times 1 = 1{,}089$$

$$1{,}089 \times 2 = 2{,}178$$

$$1{,}089 \times 3 = 3{,}267$$

$$1{,}089 \times 4 = 4{,}356$$

$$1{,}089 \times 5 = 5{,}445$$

$$1{,}089 \times 6 = 6{,}534$$

$$1{,}089 \times 7 = 7{,}623$$

$$1{,}089 \times 8 = 8{,}712$$

$$1{,}089 \times 9 = 9{,}801$$

Tell students to notice the symmetry of the first two and last two columns of the products. Each column lists consecutive integers. Encourage students not just to establish and explain this unusual occurrence, but also to build on it. What makes 1,089 so unusual? What are the factors of 1,089? Why does 1,089 × 9 reverse the number 1,089? Does a similar scheme work for other numbers? These questions and others should begin to set the tone for further investigation. Naturally students' calculators will be an indispensable tool in their work. The calculators will permit students to see patterns rapidly and without the sidetracking often caused by cumbersome calculations.

Example 4

Some other interesting number patterns for students to generate are given as follows. Be sure to encourage students to extend the patterns produced and to try to discover why they exist.

$$1 \times 8 + 1 = 9$$

$$12 \times 8 + 2 = 98$$

$$123 \times 8 + 3 = 987$$

$$1{,}234 \times 8 + 4 = 9{,}876$$

$$12{,}345 \times 8 + 5 = 98{,}765$$

$$123{,}456 \times 8 + 6 = 987{,}654$$

$$1{,}234{,}567 \times 8 + 7 = 9{,}876{,}543$$

$$12{,}345{,}678 \times 8 + 8 = 98{,}765{,}432$$

$$123{,}456{,}789 \times 8 + 9 = 987{,}654{,}321$$

$$
\begin{aligned}
11 \times 11 &= 121 \\
111 \times 111 &= 12{,}321 \\
1{,}111 \times 1{,}111 &= 1{,}234{,}321 \\
11{,}111 \times 11{,}111 &= 123{,}454{,}321 \\
111{,}111 \times 111{,}111 &= 12{,}345{,}654{,}321 \\
1{,}111{,}111 \times 1{,}111{,}111 &= 1{,}234{,}567{,}654{,}321 \\
11{,}111{,}111 \times 11{,}111{,}111 &= 123{,}456{,}787{,}654{,}321 \\
111{,}111{,}111 \times 111{,}111{,}11 &= 12{,}345{,}678{,}987{,}654{,}321
\end{aligned}
$$

Example 5

Have your students compute the divisions indicated by each of the following fractions. Tell them to record their results.

$$
\begin{aligned}
\frac{1}{7} &= \overline{.142857} = \frac{142{,}857}{999{,}999} \\
\frac{2}{7} &= \overline{.285714} = \frac{285{,}714}{999{,}999} \\
\frac{3}{7} &= \overline{.428571} = \frac{428{,}571}{999{,}999} \\
\frac{4}{7} &= \overline{.571428} = \frac{571{,}428}{999{,}999} \\
\frac{5}{7} &= \overline{.714285} = \frac{714{,}285}{999{,}999} \\
\frac{6}{7} &= \overline{.857142} = \frac{857{,}142}{999{,}999}
\end{aligned}
$$

Students will notice the similar order of the repeating part along with the different starting points. Point out that the product of $7 \times .142857 = .999999$ (which is close to $7 \times \frac{1}{7} = 1$). Remind students that this is not the same as $7 \times \overline{.142857}$.

Some students may want to inspect this product in a nondecimal form:

$$
\begin{aligned}
7 \times 142{,}857 &= 999{,}999 \\
&= 999{,}000 + 999 \\
&= 1{,}000(142 + 857) + (142 + 857) \\
&= (142 + 857)(1{,}000 + 1) \\
&= 1{,}001(142 + 857) \\
&= 142{,}142 + 857{,}857
\end{aligned}
$$

A better insight into an investigation of these fractions would come after students have already calculated the following quotients:

$$\frac{1}{13} = \overline{.076923} = \frac{076923}{999{,}999}$$

$$\frac{3}{13} = \overline{.230769} = \frac{230{,}769}{999{,}999}$$

$$\frac{4}{13} = \overline{.307692} = \frac{307{,}692}{999{,}999}$$

$$\frac{9}{13} = \overline{.692307} = \frac{692{,}307}{999{,}999}$$

$$\frac{10}{13} = \overline{.769230} = \frac{769{,}230}{999{,}999}$$

$$\frac{12}{13} = \overline{.923076} = \frac{923{,}076}{999{,}999}$$

Once these have been fully discussed, students may wish to consider the remaining proper fractions with a denominator of 13. They should discover similar patterns and relationships.

The positive impact of the above examples will be lost if students are not immediately guided to investigate and extend their discoveries. While the calculator is the guiding tool for discovering new relationships, students' logical conjectures will come from a deeper investigation of the properties of numbers.

Postassessment

Have students complete the following exercises.

1. Multiply and add each of the following pairs of numbers:

   9,9
   24,3
   47,2
   497,2

   How do their sums compare with their products? (reverses)
2. Perform the indicated operations and justify the resulting patterns. Then extend the pattern and see if it holds true.

$$12{,}321 = \frac{333 \times 333}{1+2+3+2+1} = \frac{110{,}889}{9}$$
$$= 12{,}321$$

$$1{,}234{,}321 = \frac{4{,}444 \times 4{,}444}{1+2+3+4+3+2+1} = \frac{19{,}749{,}136}{16}$$
$$= 1{,}234{,}321$$

References

Posamentier, A. S. and B. Thaller, *Numbers Their Tales, Types and Treasures*, Amherst, New York: Prometheus Books, 2015.

Unit 11

Enrichment with a Calculator

A recommended beginning-of-the-term small group activity suitable for any level, geared to make students feel comfortable with one another, with the teacher, with a new group, or with the entire class, is a quick review of the calculator's memory and square root keys. This could be followed by a speed and accuracy "contest" involving arithmetic operations.

Most calculators are equipped with memory keys that make possible a variety of challenging activities. These keys provide an efficient way to save intermediate results. They can also store a number that needs to be used several times during the course of an exercise.

Example

Find the value of

$$5 \times 12 + 13 \times 16$$

Solution

Stress the order of operations by indicating that multiplication must be done before the addition so that

$$(5 \times 12) + (13 \times 16) = 60 + 208 = 268.$$

A longer range and more effective use of the calculator is as a problem-solving facilitator. Students who have difficulty solving problems are often faced with a double dilemma. They are unable to interpret the given problem into a solvable form, and they are unable to do the calculations necessary to compute an answer. Normally they cannot concentrate on problem solving until they can do the necessary calculations. However, by using the calculator, they can temporarily avoid the calculations pitfall and thereby concentrate on the key to successful problem solving: interpretation. When this aspect has been learned, a student can concentrate on mastering calculations as an essential ingredient in problem solving.

Performance Objective

Students will investigate mathematical problems with the help of a hand calculator and then draw appropriate conclusions.

Preassessment

Students should be familiar with the basic function of a calculator. The instrument required for this unit needs the four basic operations plus memory and square root keys.

For both practice and amusement, have students do the following on their calculators.

1. Compute:

$$2[60 - .243 + (12)(2{,}400)] - 1.$$

   To find what every man must pay, read your answer upside down.
2. Compute:

$$4{,}590.5864 + (568.3)(.007) - 1{,}379.26.$$

   Then turn your calculator upside down and after reading your answer, look inside your shoe.

   These two exercises should give your students a relaxed feeling about working with calculators. Now challenge your students to use the memory keys when finding the flaws in exercises such as these:
3. $15 \times 13 + 18 \times 32$
4. $-15 \times 13 + 226 - 81$
5. $\left(\frac{253}{13}\right) + (-23)$
6. $(335 - 281) \times (-81 + 37)$

Teaching Strategies

Begin the lesson with a simple but intriguing oddity. Have students consider the calendar for the month of May 2022. Tell them to make a square around *any* nine dates; one way to do so is shown in the following figure.

May 2022

S	M	T	W	T	F	S
1	2	3	4	5	6	7
8	9	10	11	12	13	14
15	16	17	18	19	20	21
22	23	24	25	26	27	28
29	30	31				

Next, students should add 8 to the smallest number in the square and multiply it by 9. In the above example, we have $(11 + 8) \times 9 = 171$. Then students can use their calculators to multiply the sum of the numbers in the middle row (or column) by 3 and find the same result, 171. Have your students try this for other selections of 9 numbers. You should have them realize that the sum of the numbers in the middle row or column multiplied by 3 is in fact the sum of the 9 numbers. Students can verify this easily with their calculators.

From this point, you have an excellent opportunity to investigate properties of the arithmetic mean, as such study will shed more light on this cute "calendar trick." Students should realize that the *middle number* of the square of 9 numbers is in fact the arithmetic mean of the selected numbers. The use of the calculator will relieve them of burdensome computations and permit them to focus all their attention on the mathematical concepts being discovered.

For their next number investigation, have your students select any three-digit number, say 538. Then have them enter it twice into their calculators, without pressing any operation buttons. Their display should show 538,538. Now have them divide by 7, then divide by 11, then divide by 13. Much to their surprise they will find their original number displayed. Immediately, student curiosity will arise. Ask them what single operation can be used to replace the 3 divisions. They should realize that a single division by $7 \times 11 \times 13 = 1{,}001$ was actually performed. Since $538 \times 1{,}001 = 538{,}538$, the puzzle is essentially solved; yet students may wish to try this scheme on their calculators using other numbers. This should strengthen their knowledge about numbers, especially 1,001, a rather significant number.

Students should now be motivated to try the following multiplications so that they can begin to appreciate (and predict) number patterns:

a. $3 \times 11 = 33$
$3 \times 111 = 333$
$3 \times 1111 = 3{,}333$
$3 \times 11111 = 33{,}333$

b. $4 \times 101 = 404$
$4 \times 10101 = 40{,}404$
$4 \times 1010101 = 4{,}040{,}404$
$4 \times 101010101 = 404{,}040{,}404$

c. $5 \times 1001 = 5{,}005$
$5 \times 110011 = 550{,}055$
$5 \times 11100111 = 55{,}500{,}555$

d. $65 \times 101 = 6{,}565$
$65 \times 10101 = 656{,}565$
$65 \times 1010101 = 65{,}656{,}565$

e. $65 \times 1001 = 65{,}065$
$65 \times 10001 = 650{,}065$
$65 \times 100001 = 6{,}500{,}065$
$65 \times 1001001 = 65{,}065{,}065$

f. $7 \times 11 = 77$
$7 \times 11 \times 101 = 7{,}777$
$7 \times 11 \times 10101 = 777{,}777$
$7 \times 111 \times 1001 = 777{,}777$

Now have students discover other ways of generating by multiplication: 777,777, 7,777,777, and 77,777,777. At this point, students ought to be interested enough to establish other number patterns from products.

Most of your students should now be ready to consider somewhat more sophisticated problems.

A *palindrome* is defined as "a word or verse reading the same backward or forward, e.g., *madam, I'm Adam.*" In mathematics, a number that reads the same in either direction is a palindrome. For example, have your students select any two-digit number and add to it the number whose digits are the reverse order of the original one. Now have them take the sum and add it to the number whose digits are in the reverse order. They should continue this process until a palindrome is formed. For example:

$$75 + 57 = 132$$
$$132 + 231 = 363, \text{a } \textit{palindrome}$$
$$79 + 97 = 176$$
$$176 + 671 = 847$$
$$847 + 748 = 1{,}595$$

$$1{,}595 + 5{,}951 = 7{,}546$$
$$7{,}546 + 6{,}457 = 14{,}003$$
$$14{,}003 + 30{,}041 = 44{,}044, \text{ a } \textit{palindrome}$$

No matter which original two-digit number is selected, a palindrome will eventually be formed. Using a calculator, the students will see various patterns arising, which should lead them to discover why this actually "works."

Encourage your students to carefully conjecture about other possible number relationships and then verify them using a calculator.

Postassessment

Tell your students to use their calculators to verify the following phenomenon for six different numbers. Then they should try to prove it.

1. Select any three-digit number in which the hundreds digit and the units digit are unequal. Then write the number whose digits are in the reverse order from the select number. Now subtract the smaller of these two numbers from the larger. Take the difference, reverse its digits, and add the "new" number to the original difference. What number do you always end up with? Why?
2. Evaluate $\frac{23.4\times17.6}{50\times8}$ to the nearest tenth.
3. Evaluate $\frac{2}{2-1}$ to the nearest hundredth.

Unit 12

Symmetric Multiplication

This unit shows how some numbers, because of their symmetry, can be multiplied easily through the use of "form multiplication."

Performance Objective

Given a form multiplication example, students will perform the multiplication using the technique described in this unit.

Preassessment

Have students multiply each of the following by conventional means:

a. 66,666 × 66,666   b. 2,222 × 2,222   c. 333 × 777

Teaching Strategies

After students have completed the above computations, they will probably welcome a more novel approach to these problems. Have them consider the following rhombic form approach.

$$\begin{array}{c} 66666 \\ \times\ 66666 \\ \hline 36 \\ 3636 \\ 363636 \\ 36363636 \\ 3636363636 \\ 36363636 \\ 3636 \\ 36 \\ \hline 4444355556 \end{array}$$

Students might wonder if this scheme works for other numbers of this type. Have them square 88,888 first by the conventional method and then by rhombic form multiplication. To do the latter, students should replace the "36s" in the previous example with "64s." Soon students will wonder how to extend this multiplication technique to squaring a repeated-digit number where the square of a digit is a *one-digit* number.

In squaring a number such as 2,222, students must write each partial product as 04.

$$\begin{array}{c} 2{,}222 \\ \times\ 2{,}222 \\ \hline 04 \\ 0404 \\ 040404 \\ 04040404 \\ 040404 \\ 0404 \\ 04 \\ \hline 4{,}937{,}284 \end{array}$$

At this juncture, students may be convinced that this will be true for all numbers of this type. Have them consider squaring an n-digit number $uuu \ldots uuu$, where $u^2 = 10s + t$ (or written in base 10 as st). This multiplication would require an nth order rhombic form (i.e., one that increases the number of sts by one in each of the first n rows, and then decreases by one st in each of the remaining $n - 1$ rows). The case where $n = 5$ is shown as follows.

$$\begin{array}{c}
uuuuu \\
\times\ uuuuu \\
\hline
st \\
stst \\
ststst \\
stststst \\
ststststst \\
stststst \\
ststst \\
stst \\
st \\
\hline
\end{array}$$

Students will be interested to notice that this multiplication technique can be further extended to finding the product of two *different* repeated-digit numbers. That is, if the product of $uuu \ldots u$ and $vvv \ldots v$ is sought, then the rhombic form of sts is formed, where $uv = 10s + t$. For example, the product of $8{,}888 \times 3{,}333 = 29{,}623{,}704$ or:

$$\begin{array}{c}
8888 \\
\times\ 3333 \\
\hline
24 \\
2424 \\
242424 \\
24242424 \\
242424 \\
2424 \\
24 \\
\hline
29623704
\end{array}$$

When students have mastered the rhombic form of multiplying repeated-digit numbers, you might want to show them another form of repeated-digit multiplication. A *triangular* form of multiplication is shown in the following. Notice that after students sum the triangular array, they must multiply by

6.

$$\begin{array}{r} 66{,}666 \\ \times\ 66{,}666 \\ \hline 6 \\ 666 \\ 66666 \\ 6666666 \\ 666666666 \\ \hline 740{,}725{,}926 \\ \times\ 6 \\ \hline 4{,}444{,}355{,}556 \end{array}$$

In general, to square an n digit repeated-digit number, sum the columns of the triangular array of the repeated digit (where there are n rows beginning with a single digit and increasing succeeding rows by two digits), and then multiply this sum by the repeated digit.

This multiplication technique may be extended to finding the product of two different repeated digit numbers. Have students form their own rule after considering the following example.

$$\begin{array}{r} 8{,}888 \\ \times\ 3{,}333 \\ \hline 8 \\ 888 \\ 88888 \\ 8888888 \\ \hline 9{,}874{,}568 \\ \times\ 3 \\ \hline 29{,}623{,}704 \end{array}$$

$$\begin{array}{r} 3{,}333 \\ \times\ 8{,}888 \\ \hline 3 \\ 333 \\ 33333 \\ 3333333 \\ \hline 3{,}702{,}963 \\ \times\ 8 \\ \hline 29{,}623{,}704 \end{array}$$

Notice that the number of lines of the triangular form equals the number of digits in each number being multiplied. The rest of the rule can be easily elicited from the students.

The only other variation in the multiplication of repeated-digit numbers is when the number of digits of each number being multiplied is not the same. Suppose an n digit number is being multiplied by an m digit number. Set up the triangular form as if both numbers had n digits (as was done earlier).

Then draw a diagonal to the right of the mth row and delete all the digits below the diagonal. The sum of the remaining digits is the desired product. The following example illustrates this procedure.

```
     44,444
      × 666
 ----------
         24
       2424
     242424
   24242424
 2424242424
   24242424
     242424
       2424
         24
 ----------
 29,599,704
```

Now have students multiply 66,666 × 444. They should get the same array of numbers to be summed as above. This implies that 44,444 × 666 = 66,666 × 444. After factoring, students should have no trouble justifying this equality.

Students should be encouraged to explain mathematically why the various form multiplications actually "work."

Postassessment

1. Have students compute each of the following using rhombic form multiplication:

   a. 22,222 × 77,777
   b. 9,999 × 9,999
   c. 444 × 333

2. Have students compute each of the following using triangular form multiplication:

   a. 555,555 × 555,555
   b. 7,777 × 4,444

Unit 13

Variations on a Theme — Multiplication

This unit will present unconventional methods for determining the product of two integers.

Performance Objective

Given two integers and a method of multiplication, students will compute the product.

Preassessment

Have students compute the product of 43 and 92 by *more than one method.*

Teaching Strategies

The preceding problem should serve as an excellent motivation for this unit. Most students will probably multiply the numbers correctly using the "conventional" method of multiplication shown as follows.

$$\begin{array}{r} 92 \\ \times\ 43 \\ \hline 276 \\ 368 \\ \hline 3{,}956 \end{array}$$

Before discussing other methods of multiplication, the teacher should show why the "conventional" multiplication algorithm works. It may be easily seen that

$$\begin{aligned} 43 \times 92 &= (40 + 3) \times 92 \\ &= 40 \times 92 + 3 \times 92 \\ &= 3{,}680 + 276 \\ &= 3{,}956 \end{aligned}$$

which is exactly what is done, although mechanically.

The Doubling Method

To multiply 43 by 92 construct the following columns of numbers, starting with 1 and 92, and double each number.

•	1	92
•	2	184
	4	368
•	8	736
	16	1,472
•	32	2,944

We stop at 32 because twice 32 is 64 which is larger than 43. We start with the last number in the first column and add the appropriate numbers such that the sum is 43. Hence, we choose $(32, 8, 2, 1)$.

Now we add the corresponding numbers in the second column.

$$\begin{array}{r} 92 \\ 184 \\ 736 \\ \underline{2{,}944} \\ 3{,}956 \end{array}$$

Therefore, $43 \times 92 = 3{,}956$. The reason this method works is illustrated in the following.

$$\begin{aligned} 43 \times 92 &= (32 + 8 + 2 + 1) \times 92 \\ &= (32 \times 92) + (8 \times 92) + (2 \times 92) + (1 \times 92) \\ &= 2{,}944 + 736 + 184 + 92 \\ &= 3{,}956 \end{aligned}$$

Russian Peasant Method

Again suppose we wish to multiply 43 and 92. Construct the following columns of numbers, starting with 43 and 92. In successive rows, halve the entries in the first column, rejecting the remainder of 1 when it occurs. In the second column, double each successive entry. This process continues until 1 appears in the first column.

*43	92
*21	184
10	368
* 5	736
2	1,472
* 1	2,944

Choose the numbers in the second column that corresponds to the "odd numbers" of the first column (those starred). Add those corresponding numbers of the second column and the result is the product of 43 and 92. That is, $92+184+736+2{,}944 = 3{,}956$. The proof that the Russian Peasant Method is always correct follows.

Assume "a" even: $a \cdot b = c$, where c is the desired result.

Assume "$\frac{1}{2}a^n$" odd: $\frac{1}{2}a - 2b = c$.

The next step in the method is:

$$\left[\frac{1}{2}\left(\frac{1}{2}a\right) - \frac{1}{2}\right] \cdot 4b = y.$$

Then using the distributive property:

$$\left[\frac{1}{4}a \cdot 4b\right] - \left[\frac{1}{2} \cdot 4b\right] = y.$$

Since $\frac{1}{4}a \cdot 4b = c$, then $c - 2b = y$.

Therefore, the new product, y, will be short of the correct answer, c, by $2b$ (which is the first desired number to be added since it is paired with an odd number, $\frac{1}{2}a$).

As the process continues, the "new products" will remain the same if ka (an entry in the first column) is even. If ka is odd and $ka \cdot mb = w$, the next product will decrease by mb (the number matches with the odd number). For example, $\left(\frac{1}{2}ka - \frac{1}{2}\right) \cdot 2mb = \left(\frac{1}{2}ka \cdot 2mb\right) - \left(\frac{1}{2} \cdot 2mb\right) = w - mb$.

Finally, when 1 appears in the first column:

$1 \cdot pb = Z$, with $pb = Z$.
$Z = c -$ all deductions (numbers referred to above matched with odd kas)
c (desired result) $= Z +$ all deductions.

A further consideration of the Russian Peasant Method for multiplication can be seen from the following illustration:

$$\begin{aligned}
*\ 43 \cdot 92 &= (21 \cdot 2 + 1)(92) = 21 \cdot 184 + 92 = 3{,}956 \\
*\ 21 \cdot 184 &= (10 \cdot 2 + 1)(184) = 10 \cdot 368 + 184 = 3{,}864 \\
10 \cdot 368 &= (5 \cdot 2 + 0)(368) = 5 \cdot 736 + 0 = 3{,}680 \\
*\ 5 \cdot 736 &= (2 \cdot 2 + 1)(736) = 2 \cdot 1{,}472 + 736 = 3{,}680 \\
2 \cdot 1{,}472 &= (1 \cdot 2 + 0)(1{,}472) = 1 \cdot 2{,}944 + 0 = 2{,}944 \\
*\ 1 \cdot 2{,}944 &= (0 \cdot 2 + 1)(2{,}944) = 0 + \frac{2{,}944}{3{,}956} = 2{,}944
\end{aligned}$$

Note that summing only those numbers in the second column whose corresponding entries in the first column are odd is justified by the above representation.

The teacher may wish to shed further light on this curiosity by presenting the binary nature of this multiplication.

$$\begin{aligned}
&(43)(92) \\
&= (1 \cdot 2^5 + 0 \cdot 2^4 + 1 \cdot 2^3 + 0 \cdot 2^2 + 1 \cdot 2^1 + 1 \cdot 2^0)(92) \\
&= 2^0 \cdot 92 + 2^1 \cdot 92 + 2^3 \cdot 92 + 2^5 \cdot 92 \\
&= 92 + 184 + 736 + 2{,}944 \\
&= 3{,}956
\end{aligned}$$

Other investigations should be encouraged by the students.

Lattice Multiplication

Once again consider the multiplication $43 \cdot 92$. To perform this method, a 2×2 array is constructed and diagonals are drawn as shown.

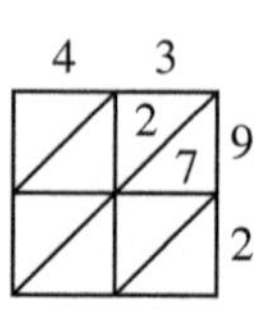

First multiply $3 \cdot 9 = 27$; the 2 is placed above the 7 as shown in the following.

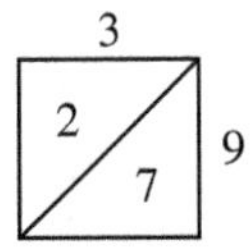

The next step is to multiply $4 \cdot 9 = 36$. Again, the 3 is placed above the 6 in the appropriate box.

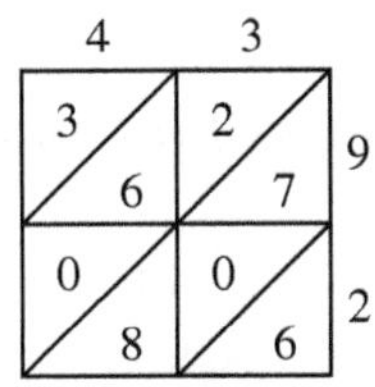

This process is continued, filling in the remainder of the square. Notice that $3 \cdot 2 = 6$ is recorded as 0/6.

Now that there are entries in all cells, add the numbers in the diagonal directions indicated, beginning at the lower right. The sums are circled.

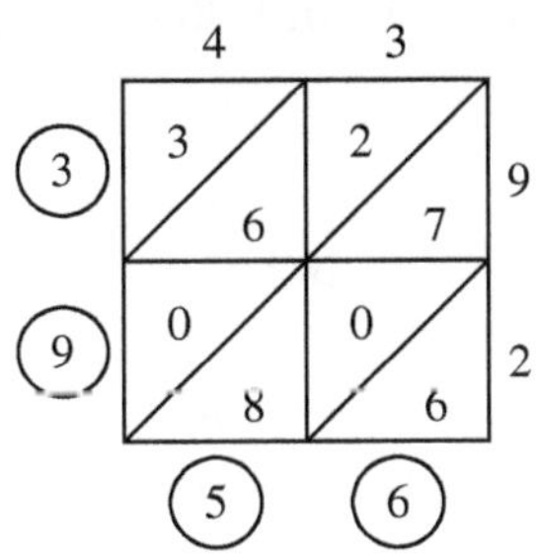

Notice that in the second addition, $8 + 0 + 7 = 15$, the 5 is recorded and the 1 is carried to the next diagonal addition. The correct answer (the product of $43 \cdot 92$) is then merely read from the circled numbers. That is, the answer is 3,956.

Trachtenberg System

The Trachtenberg System is a method for high-speed multiplication, division, addition, subtraction, and square root. There are numerous rules for these operations. Concern here will only be focused on the multiplication of two 2-digit numbers.

Again, suppose the product of 43 and 92 is desired.

Step 1. Multiply the two units digits

$$[3 \cdot 2 = 6] \qquad \begin{array}{r} 43 \\ \times\ 92 \\ \hline 6 \end{array}$$

Step 2. Cross-multiply and add mentally

$$[(9 \cdot 3) + (2 \cdot 4) = 27 + 8 = 35]$$

Place the 5 as shown and carry the 3 (as an addend to the next step)

$$\begin{array}{r} 43 \\ \times\ 92 \\ \hline 56 \end{array}$$

Step 3. Multiply the two tens digits and then add any number carried over from the previous step.

$$[9 \cdot 4 = 36 \text{ and } 36 + 3 = 39]$$

$$\begin{array}{r} 43 \\ \times\ 92 \\ \hline 3{,}956 \end{array}$$

The algebraic justification of this method is shown as follows:

Consider the two 2-digit numbers ab and mn (written in place-value form)

$$\begin{aligned} &(10a + b) \cdot (10m + n) \\ &= 10a \cdot 10m + 10a \cdot n + 10b \cdot m + bn \\ &= \underbrace{100am}_{\text{Step3}} + \underbrace{10an + 10bm}_{\text{Step 2}} + \underbrace{bn}_{\text{Step 1}} \end{aligned}$$

Another Method

A multiple of ten "squared" is an easy mental computation. This method of multiplication incorporates this idea. Have students consider the product of $m \cdot n$. They then choose x such that $x = 10p$ and $m < x < n$ with $x - m = a$ and $n - x = b$. Then $m \cdot n = (x - a)(x + b) = x^2 - ax + bx - ab$.

Consider the multiplication of $43 \cdot 92$.

Let $x = 60$.

$$\begin{aligned}\text{Then } 43 \cdot 92 &= (60 - 17)(60 + 32)\\ &= 3{,}600 + (-17 \cdot 60 + 60 \cdot 32) + (-17 \cdot 32)\\ &= 3{,}600 + (-1{,}020 + 1{,}920) + (-544)\\ &= 3{,}600 + 900 - 544\\ &= 3{,}956\end{aligned}$$

However, if the numbers are the same distance from the multiple of ten, the method is much faster. The middle term will be eliminated!

Suppose students wish to multiply $57 \cdot 63$. Then $57 \cdot 63 = (60 - 3)(60 + 3) = 60^2 - 3^2 = 3{,}600 - 9 = 3{,}591$.

Part of the skill in working with different multiplication methods involves selection of the most efficient method for the particular problem. This should be stressed with the class.

Other Methods

Now that students have been exposed to several methods for multiplication, the teacher should suggest that students do research and explore other methods for multiplication. These could then be presented to the class.

Postassessment

Multiply 52 by 76 using each of the four different methods.

References

Posamentier, A. S. and B. Thaller, *Numbers Their Tales, Types and Treasures*, Amherst, New York: Prometheus Books, 2015.

Unit 14

Ancient Egyptian Arithmetic

The study of a number system and its arithmetic is valuable to students on many levels. A student on a high level can delve into the mechanics of the system through an intricate comparison between that system and our own, and perhaps go off on a suitable tangent from there, such as an inspection of bases and still other number systems. For other students, it can serve as a reinforcement of basic arithmetic (multiplication and division) both of integers and of fractions since the students will want to check if this system really works. This unit introduces students to the ancient Egyptian numerical notation and to their system of multiplying and dividing.

Performance Objectives

1. *Given a multiplication problem, students will find the answer using an Egyptian method.*
2. *Given a division problem, students will find the answer using an Egyptian method.*
3. *Given a nonunit fraction, students will obtain its unit fraction decomposition.*

Preassessment

Students should be familiar with addition and multiplication of fractions, as well as with the distributive property of real numbers. A knowledge of bases may also be useful.

Teaching Strategies

An example of a simple grouping system can be seen in the Egyptian hieroglyphics. This numeral system is based on the number 10. The symbols used when representing their numbers on stone, papyrus, wood, and pottery were

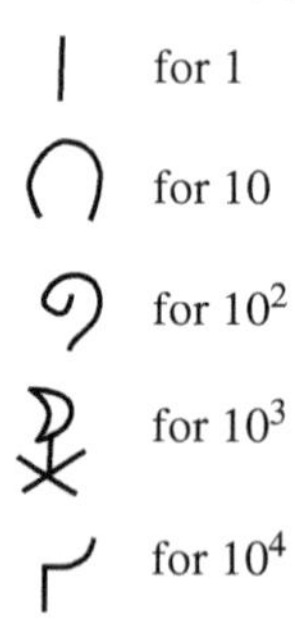

Therefore, 13,521 would be represented as

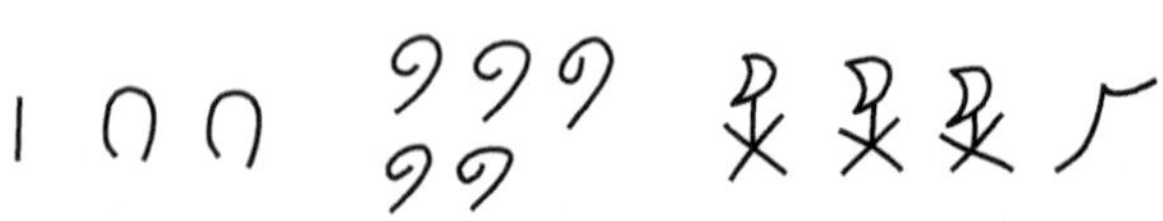

(Have students notice that the Egyptians wrote numbers from right to left.)

The Egyptians avoided a difficult multiplication or division method by using an easier (although at times longer) method. To multiply 14 by 27, they would have done the following:

	1	27
*	2	54
*	4	108
*	8	216
	16	432

To advance from any line to the next line, all the Egyptians had to do was double the number. Then, they picked out the numbers in the left-hand column that added up to 14 (the numbers with a *). By adding up the corresponding numbers in the right-hand column, they arrived at the answer: $54 + 108 + 216 = 378$. This is an application of the distributive property of multiplication over addition, for what the Egyptians did is equivalent to

$$27(14) = 27(2 + 4 + 8) = 54 + 108 + 216 = 378$$

Further justification for the method lies in the fact that any number can be expressed as the sum of powers of two. Investigate this process with your students to the extent you feel necessary.

The Egyptians performed division in a similar way. They viewed the problem $114 \div 6$ as 6 times whatever number equals 114.

1	6	*
2	12	*
4	24	
8	48	
16	96	*

Now since $114 = 6 + 12 + 96$, the Egyptians would have found that $114 = 6(1 + 2 + 16)$ or $6 \times 19 = 114$. The answer is 19.

While no problems could arise in the Egyptian method of multiplication, a slight one occurs with respect to their method of division. To call attention to this problem, ask your class to use the above method to solve $83 \div 16$.

1	16	*
2	32	
4	64	*
8	128	

Using $16 + 64 = 80$, the Egyptians were still missing 3. Since $1 \times 16 = 16$, they found that they needed fractions to complete this problem.

In the Egyptian number system, every fraction except $\frac{2}{3}$ was represented as the sum of "unit fractions," fractions whose numerators are 1. In this way, the Egyptians avoided some of the computational problems one encounters when working with fractions. Since their arithmetic was based on doubling, the only problem they had to deal with was how to change a fraction of the form $\frac{2}{n}$ to one of the form $\frac{1}{a} + \frac{1}{b} + \cdots$. They handled this problem with a table (found in the Rhind papyrus, dated approximately 1650 B.C.) which gives the decomposition of all fractions of the form $\frac{2}{n}$ for all odd n from 5 to 101. (Your class should be able to see why they considered only odd n.) A fraction such as $\frac{2}{37}$ was written as $\frac{1}{19} + \frac{1}{703}$ or, using the common notation for unit fractions, $\overline{19} + \overline{703}$ (This notation survives from the Egyptians who wrote a fraction such as $\frac{1}{4}$ as  and $\frac{1}{14}$ as  in hieroglyphics.) $\frac{2}{3}$ had its own symbol  and sometimes $\frac{1}{2}$ appeared as .

The need now arises to consider a rule that can be used to decompose a fraction of the form $\frac{2}{pq}$ (where p or q may be 1). Have the class consider $\frac{2}{pq} = \frac{1}{\frac{p(p+q)}{2}} + \frac{1}{\frac{q(p+q)}{2}}$. They can add the fractions on the right side of the equation together to prove this is true. Also have them notice that since pq is odd (since we only need a rule to decompose $\frac{2}{n}$ where n is odd, p and q are odd, so $p+q$ will be even, and therefore, $\frac{p+q}{2}$ will be an integer).

Students can decompose $\frac{2}{15}$ at least two ways. If they set $p = 3$ and $q = 5$, they will have $\frac{2}{15} = \frac{1}{\frac{3(8)}{2}} + \frac{1}{\frac{5(8)}{2}} = \frac{1}{12} + \frac{1}{20}$, or $\overline{12} + \overline{20}$. If they let $p = 1$ and $q = 5$, they will have $\frac{2}{15} = \frac{1}{\frac{1(16)}{2}} + \frac{1}{\frac{15(16)}{2}} = \frac{1}{8} + \frac{1}{120}$ or $\overline{8} + \overline{120}$.

It seems the Egyptians had other ways to decompose fractions so as to make the new denominator less complicated. For instance, we could also view $\frac{2}{15}$ as $\frac{4}{30}$.

Then we have $\frac{4}{30} = \frac{3}{30} + \frac{1}{30} = \overline{10} + \overline{30}$. Have students check these conversions.

Now, have students reconsider the earlier division problem $83 \div 16$.

1	16*
2	32
4	64*
8	128
$\overline{2}$	8
$\overline{4}$	4
$\overline{8}$	2*
$\overline{16}$	1*

By selecting a sum of 83 from the right column, they will arrive at the following answer: $1 + 4 + 8 + \overline{16} = 5 + 8 + \overline{16} = 5\frac{3}{16}$. Students should now be able to solve their arithmetic problems using methods developed by the ancient Egyptians.

Postassessment

1. Have students write each of the following numbers in hieroglyphics.

   a. 5,280 b. 23,057 c. $\frac{2}{25}$ d. $\frac{2}{35}$

2. Have students change each of the following fractions into two different unit fraction decompositions.

   a. $\frac{2}{27}$ b. $\frac{2}{45}$ c. $\frac{2}{99}$

3. Have students solve the following problems using Egyptian methods.

   a. 30×41 b. 25×137 c. $132 \div 11$ d. $101 \div 16$

Unit 15

Napier's Rods

Performance Objectives

1. *Students will construct a cardboard set of Napier's Rods.*
2. *Students will successfully perform multiplication examples using Napier's Rods.*

Preassessment

The only essential skill for this activity is the ability to do multiplication.

Teaching Strategies

Begin your presentation with a brief historical note about Napier's Rods. This multiplication "machine" was developed by John Napier (1550–1617), a Scotch mathematician, who was principally responsible for the development of logarithms. The device he developed consisted of flat wooden sticks with successive multiples of numbers 1–9 (see Figure 1).

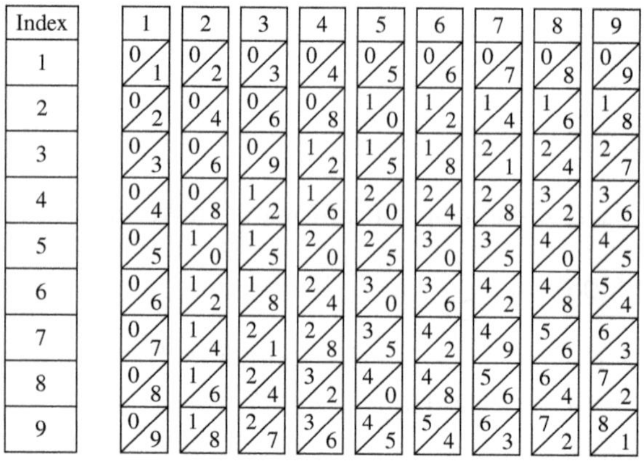

Figure 1.

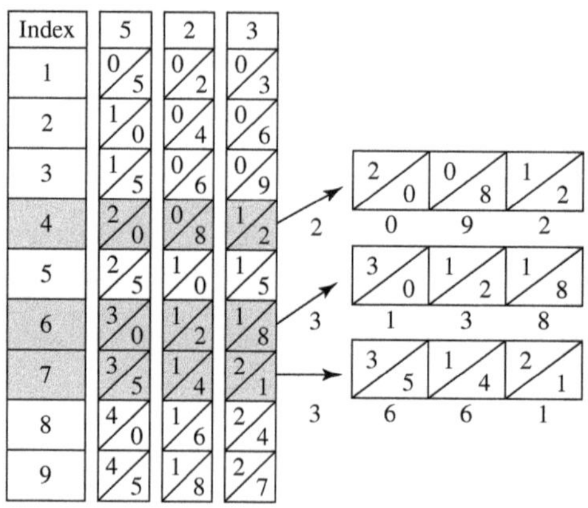

Figure 2.

Each student should be given an opportunity to construct his own set of Napier's Rods. Perhaps the best way to explain how to use Napier's Rods is to illustrate with an example using this device.

Consider the multiplication 523×467. Have students select the rods for 5, 2, and 3, and line them up adjacent to the Index Rod (see Figure 2).

Students must then select the appropriate rows from the index corresponding to the digits in the multiplier. In a diagonal direction, addition is done for each row (see Figure 2). The numbers thus obtained:

$$2{,}092 = 4 \times 523$$

$$3{,}138 = 6 \times 523$$

$$3{,}661 = 7 \times 523$$

are added after considering the appropriate place values of the digits from which they were generated.

$$\begin{aligned}467 &= 400 + 60 + 7\\(467)(523) &= (400)(523) + (60)(523) + (7)(523)\\(467)(523) &= 209{,}200\\&\ \ 31{,}380\\&\ \ \ \underline{3{,}661}\\&\ 244{,}241\end{aligned}$$

A careful discussion of this last step not only will insure a working knowledge of the computing "machine" by your students, but also should give them a thorough understanding as to *why* this technique "works."

1. 561×49
2. 308×275
3. $4{,}932 \times 7{,}655$

Postassessment

Have students use a set of Napier's Rods (which they have constructed) to multiply each of the following:

1. $361 \cdot 49$
2. $308 \cdot 275$
3. $4{,}932 \cdot 7{,}566$

References

Posamentier, A. S. and I. Lehmann, *Mathematical Amazement and Surprises*, Amherst, New York: Prometheus Books, 2009.

Unit 16

Unit Pricing

Performance Objectives

1. *Students will determine which of two given fractions is greater using the technique described on this card.*

2. *Students will determine which of two quantities of the same product has the more favorable price, given the quantity and the price for that quantity.*

Preassessment

The only skill students need for this activity is multiplication of whole numbers.

Teaching Strategies

Ask your students if they would rather buy a 32oz. jar of applesauce costing 30¢ or a 27 oz. jar costing 25¢. An organized thinker might translate the problem to one that asks which is larger, $\frac{30}{32}$ or $\frac{25}{27}$.

Students should realize that the fractions come from "price *per* ounce." The word *per* indicates division, so a fraction $\frac{\text{price}}{\text{ounces}}$ can be attained.

There are many ways in which two fractions can be compared, dividing the numerator by the denominator and comparing the resulting decimal, changing both fractions to equivalent fractions with the same denominator, and so on.

We shall now consider another method, which could easily be the most efficient method. Draw two arrows as indicated in the following. Then multiply as the arrows show, writing the product under the arrowheads.

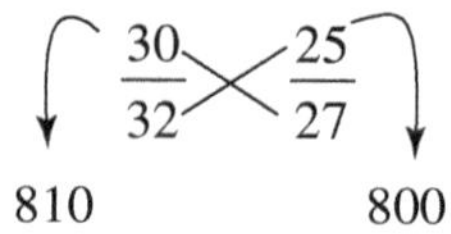

Simple inspection of these products indicates that 810 is the larger of the two; hence the fraction above the 810 (i.e., $\frac{30}{32}$) is the larger of the two fractions. Therefore, the unit price of the 27 ounce jar is lower and thus the better buy.

Before giving students problems involving comparison of unit prices, offer them some drill problems involving only comparison of fractions.

Postassessment

Select the larger of each for the following pairs of fractions.

1. $\frac{5}{6}, \frac{7}{8}$ 2. $\frac{8}{11}, \frac{17}{23}$

3. $\frac{7}{9}, \frac{4}{5}$

4. $\frac{13}{17}, \frac{19}{25}$

5. $\frac{11}{19}, \frac{5}{9}$

6. $\frac{7}{12}, \frac{18}{31}$

7. Which quantity has the lower unit price : a 7 oz. jar of mustard costing 11¢ or a 9 oz. jar of mustard costing 13¢?

Unit 17

Successive Discounts and Increases

This unit provides students with a simple technique for expressing various successive discounts and/or increases as one equivalent discount or increase. They may be rather fascinated by the case of solution that this method brings to a usually complicated consumer situation.

Performance Objectives

1. *Students will convert two or more successive discounts to one equivalent discount.*
2. *Students will convert two or more successive discounts and increases to one equivalent discount or increase.*

Preassessment

Use the following problem for diagnostic purposes as well as to motivate discussion.

Ernie is deciding where to buy a shirt. Barry's Bargain Store offers the shirt at a 30% discount off the list price. Cheap Charlie's Store usually offers the same shirt at a 20% discount off the same list price. However, today Cheap Charlie's Store is offering the shirt at 10% discount off the already discounted (20%) price. At which store will Ernie get a greater discount on the shirt today?

Teaching Strategies

Students may not immediately realize the difference in the discounts of the two stores mentioned in the problem above. Some students may feel that both stores offer the same discount. With your help they should begin to notice that, whereas Bargain Stores offers a 30% discount off the original list price, Cheap Charlie's Store only offers a 20% discount off the original list price, while the 10% discount is taken off the lower, already discounted price. Hence, Ernie gets a greater discount with the 30% discount.

At this juncture, your students may begin to wonder how large the actual difference is between the discounts offered by the two stores. You might then elicit from them that this quantitative comparison calls for finding a single discount equivalent to the two *successive discounts* of 20% and 10%.

Some students may suggest finding the required discounts by starting with a list price such as \$10.00. This would work well as 100 is the basis of percents. That is, a discount of 20% off \$10.00 yields a price of \$8.00. Then a 10% discount of \$8.00 yields a new price of \$7.20. Since \$7.20 may be obtained by a single discount of 28% off the original \$10.00, *successive discounts* of 20% and 10% are equivalent to a single discount of 28%. It is simple to compare this with the 30% discount from the original problem.

Students should consider a general method of converting any number of successive discounts to a single discount. Illustrate with two successive discounts percents of d_1 and d_2 operating on a price p.

Use the same procedure as before:

$p - \frac{pd_1}{100} = p\left(1 - \frac{d_1}{100}\right)$ represents the price after one discount has been computed; $\left[p\left(1 - \frac{d_1}{100}\right)\right] - \left[p\left(1 - \frac{d_1}{100}\right)\right]\left(\frac{d_2}{100}\right) = p\left(1 - \frac{d_1}{100}\right)\left(1 - \frac{d_2}{100}\right)$ represents the price after the second discount has been computed; $\left(1 - \frac{d_1}{100}\right)\left(1 - \frac{d_2}{100}\right)$ represents the percent that the new price is of the original price. Therefore, $1 - \left(1 - \frac{d_1}{100}\right)\left(1 - \frac{d_2}{100}\right)$ represents the discount taken off the original price in order to obtain the new price.

Hence, successive discounts of $d_1\%$ and $d_2\%$ are equivalent to a single discount percent of $1 - \left(1 - \frac{d_1}{100}\right)\left(1 - \frac{d_2}{100}\right)$. By translating this final algebraic expression into verbal form, the students should be able to establish the following simple technique for converting two successive discounts to a single equivalent discount:

1. Change each of the successive discounts to decimal fractions.
2. Subtract each of these decimal fractions from the whole (i.e., 1.00).

3. Multiply the results of Step 2.
4. Subtract the results of Step 3 from the whole (i.e., 1.00).
5. Change the result of Step 4 to percent form.

Applying this to the successive discounts of 20% and 10%, the students should show the following work:

1. 20% = .20 and 10% = .10
2. 1.00 − .20 = .80, and 1.00 − .10 = .90
3. (.80)(.90) = .72
4. 1.00 − .72 = .28
5. 28 = 28% (discount)

Students will notice that the rules above do not specify the number of successive discounts considered. This may prompt them to investigate the case where more than two successive discounts are to be converted to a single equivalent discount. Students should proceed in a manner similar to the one used above for two successive discounts. They should find that these same rules do, in fact, hold true in converting any number of successive discounts to a single equivalent discount.

A natural question to be expected at this juncture would probe the nature of *successive increases* or *successive decreases* and *increase.* An increase requires adding a percent of the price to the original price, while a discount requires subtracting a percent of the price from the original price. So the students could be expected to guess that the technique for converting successive increases or combinations of successive increases and discounts to a single increase or discount will be similar to the conversion technique used for successive discounts. Suggest they work out this technique.

Broadening the scope of this conversion technique to include increases as well as discounts will allow students to consider problems such as the following:

When the entrance price to a basketball game was decreased by 25%, the attendance at the game increased by 35%. What was the effect of these changes on the daily receipts?

In solving this problem, the student's work should resemble the following:

1. 25% = .25, and 35% = .35
2. 1.00 − .25 = .75 and 1.00 + .35 = 1.35
3. (.75)(1.35) = 1.0125
4. 1.0125 − 1.0000 = .0125
5. .0125 = 1.25% (increase)

You might wish to ask your students what they feel would be the net effect of a successive discount of 10% and increase of 10%. Students generally feel that these two changes counteract one another, leaving the original price unaltered. However, they should be encouraged to apply the conversion technique. The net effect of these two changes is in fact a discount of 1%.

The following chart of successive discounts and increases of the same percent should lead students to making an intelligent conjecture about the "break-even" point.

Successive changes							
% Discount	20	15	10	5	1	.5	.1
% Increase	20	15	10	5	1	.5	.1
Equivalent % discount change	4	2.25	1	.25	.01	.0025	.0001

Students may want to discover what combinations of successive discount and increase will leave the original price unaltered. One possible approach would be to use the conversion technique for a successive discount of $d\%$ and increase of $i\%$:

$$1 - \left(1 - \frac{d}{100}\right)\left(1 + \frac{i}{100}\right) = 0;$$

$$1 - \left(1 - \frac{d}{100} + \frac{i}{100} - \frac{di}{100^2}\right) = 0;$$

$$100d - 100i + di = 0;$$

$$d = \frac{100i}{100 + i} \text{ or } i = \frac{100d}{100 - d}$$

The following chart lists possible values for d and i.

d	0	9.0909	16.666	20	50	75	10	25
i	0	10	20	25	100	300	11.111	33.333

By now the students should have some insight into the topic of *successive percents*. The conversion technique introduced in this model is rather easy to remember as the basic steps merely call for subtraction (and/or addition), multiplication, and subtraction (and/or addition).

Postassessment

Have students try to solve several problems such as the following.

1. Alice wants to buy a dress whose list price is \$20. One store, which generally discounts its dresses 12%, is offering an additional 20% on its already discounted price. A nearby store offers the same dress at a single 32% discount. Which store offers the lower price?
2. When the price of a magazine was decreased by 15%, the sales increased by 20 %. How were the receipts affected by these changes?

References

Posamentier, A. S. *Students! Get Ready for the SAT I: Problem Solving Strategies and Practical Tests.* Thousand Oaks, CA: Corwin Press, 1996.

Posamentier, A. S. and S. Krulik. *Teachers! Prepare Your Students for the Mathematics for SAT I: Methods and Problem Solving Strategies.* Thousand Oaks, CA: Corwin, 1996.

Posamentier, A. S. and C. T. Salkind. *Challenging Problems in Algebra.* New York: Dover, 1996.

Posamentier, A. S. *Math Wonders to Inspire Teachers and Students.* Alexandria, VA: ASCD, 2003.

Posamentier, A.S. and C. Spreitzer, *The Mathematics of Everyday Life*, Amherst, New York: Prometheus Books, 2018.

Unit 18

Prime and Composite Factors of a Whole Number

This unit presents a different approach to the process of factoring a number. It allows students to find the complete set of all different factors of a composite whole number. At the same time, this unit helps students better understand the factorization process.

Performance Objectives

1. *Students will determine the total number of factors, prime and composite, of a given composite whole number.*
2. *Students will determine each of the elements of the set of prime and composite factors of this number.*
3. *Students will find the sum of all the elements of this set.*

Preassessment

Students should be familiar with the basic rules of divisibility and able to find the prime factorization of a number.

Teaching Strategies

To find the set of prime and composite factors of the given number, you should first find the prime factorization of the number and then determine all possible products of these factors.

To find the prime factorization of the number, the "peeling" technique may be used. For example, to find the prime factorization of 3,960, you should proceed as follows:

$$
\begin{array}{lll}
2\underline{)3{,}960} & 3{,}690 & = 2 \times 1{,}980 \\
2\underline{)1{,}980} & & = 2 \times 2 \times 990 \\
2\underline{)\ \ 990} & & = 2 \times 2 \times 2 \times 495 \\
3\underline{)\ \ 495} & & = 2 \times 2 \times 2 \times 3 \times 165 \\
3\underline{)\ \ 165} & & = 2 \times 2 \times 2 \times 3 \times 3 \times 55 \\
5\underline{)\ \ \ \ 55} & & = 2 \times 2 \times 2 \times 3 \times 3 \times 5 \times 11 \\
\qquad 11 & &
\end{array}
$$

The prime factorization of 3,960 is

$$2 \times 2 \times 2 \times 3 \times 3 \times 5 \times 11 = 2^3 \times 3^2 \times 5 \times 11$$

The total number of factors of a given number is determined by the product of the exponents (*each* increased by one) of the different factors in the prime factorization of the number expressed in exponential form.

Therefore, the total number of factors of 3,960 will be given by the product:

$$(3+1)(2+1)(1+1)(1+1) = 4 \times 3 \times 2 \times 2 = 48$$

To find each of the 48 factors, prepare the following self-explanatory table:

1	2	2^2	2^3
1	3	3^2	
1	5		
1	11		

Now have students multiply each number in the first row by each number in the second row:

1×1	1×2	1×2^2	1×2^3
1×3	2×3	$2^2 \times 3$	$2^3 \times 3$
1×3^2	2×3^2	$2^2 \times 3^2$	$2^3 \times 3^2$

Then each resulting product would be multiplied by each number in the third row:

$1 \times 1 \times 1$	$1 \times 1 \times 2$	$1 \times 1 \times 2^2$	$1 \times 1 \times 2^3$
$1 \times 1 \times 3$	$1 \times 2 \times 3$	$1 \times 2^2 \times 3$	$1 \times 2^3 \times 3$
$1 \times 1 \times 3^2$	$1 \times 2 \times 3^2$	$1 \times 2^2 \times 3^2$	$1 \times 2^3 \times 3^2$
$1 \times 1 \times 5$	$1 \times 2 \times 5$	$1 \times 2^2 \times 5$	$1 \times 2^3 \times 5$
$1 \times 3 \times 5$	$2 \times 3 \times 5$	$2^2 \times 3 \times 5$	$2^3 \times 3 \times 5$
$1 \times 3^2 \times 5$	$2 \times 3^2 \times 5$	$2^2 \times 3^2 \times 5$	$2^3 \times 3^2 \times 5$

You should continue this same process until all rows are exhausted. In the case of our example, 3,960, we finally have

$1 \times 1 \times 1 \times 1$	$1 \times 1 \times 1 \times 2$	$1 \times 1 \times 1 \times 2^2$	$1 \times 1 \times 1 \times 2^3$
$1 \times 1 \times 1 \times 3$	$1 \times 1 \times 2 \times 3$	$1 \times 1 \times 2^2 \times 3$	$1 \times 1 \times 2^3 \times 3$
$1 \times 1 \times 1 \times 3^2$	$1 \times 1 \times 2 \times 3^2$	$1 \times 1 \times 2^2 \times 3^2$	$1 \times 1 \times 2^3 \times 3^2$
$1 \times 1 \times 1 \times 5$	$1 \times 1 \times 2 \times 5$	$1 \times 1 \times 2^2 \times 5$	$1 \times 1 \times 2^3 \times 5$
$1 \times 1 \times 3 \times 5$	$1 \times 2 \times 3 \times 5$	$1 \times 2^2 \times 3 \times 5$	$1 \times 2^3 \times 3 \times 5$
$1 \times 1 \times 3^2 \times 5$	$1 \times 2 \times 3^2 \times 5$	$1 \times 2^2 \times 3^2 \times 5$	$1 \times 2^3 \times 3^2 \times 5$
$1 \times 1 \times 1 \times 11$	$1 \times 1 \times 2 \times 11$	$1 \times 1 \times 2^2 \times 11$	$1 \times 1 \times 2^3 \times 11$
$1 \times 1 \times 3 \times 11$	$1 \times 2 \times 3 \times 11$	$1 \times 2^2 \times 3 \times 11$	$1 \times 2^3 \times 3 \times 11$
$1 \times 1 \times 3^2 \times 11$	$1 \times 2 \times 3^2 \times 11$	$1 \times 2^2 \times 3^2 \times 11$	$1 \times 2^3 \times 3^2 \times 11$
$1 \times 1 \times 5 \times 11$	$1 \times 2 \times 5 \times 11$	$1 \times 2^2 \times 5 \times 11$	$1 \times 2^3 \times 5 \times 11$
$1 \times 3 \times 5 \times 11$	$2 \times 3 \times 5 \times 11$	$2^2 \times 3 \times 5 \times 11$	$2^3 \times 3 \times 5 \times 11$
$1 \times 3^2 \times 5 \times 11$	$2 \times 3^2 \times 5 \times 11$	$2^2 \times 3^2 \times 5 \times 11$	$2^3 \times 3^2 \times 5 \times 11$

However, students can obtain the same results in a simpler and quicker way: Find the divisors of each of the factors in the number's prime factorization when written in exponential form. In our example,

$$\underset{a}{2^3 \begin{cases} 2^1 = 2 \\ 2^2 = 4 \\ 2^3 = 8 \end{cases}} \quad \underset{b}{3^2 \begin{cases} 3^1 = 3 \\ 3^2 = 9 \end{cases}} \quad \underset{c}{5^1 \{5^1 = 5} \quad \underset{d}{11^1 \{11^1 = 11}$$

Have students prepare a table in which the first row is formed by number one and the numbers in *a* (see the following). Have pupils draw a line and multiply each number in b by each number above this line. Students will draw a new line and multiply the element in c by all the numbers above the second line. The process will continue until all divisors of each of the factors in the number's prime factorization are multiplied.

1	2	4	8	I
3	6	12	24	
9	18	36	72	II
5	10	20	40	
15	30	60	120	III
45	90	180	360	
11	22	44	88	
33	66	132	264	
99	198	396	792	
55	110	220	440	IV
165	330	660	1320	
495	990	1980	3960	

The table that contains the 48 factors we are looking for starts with number 1 and ends with our given number 3,960.

Part 1 is formed by number I and the factors in a. Part II is constituted by the products of each of the numbers in b and each of the numbers in I. Part III is made up by the products obtained when multiplying each of the numbers of c by each of the numbers in I and II. And finally, part IV is formed by multiplying each number of d by each of the numbers in I, II, and III. The table has $4 \times 12 = 48$ factors. All of the factors of 3,960 appear in this table: the prime as well as the composite factors of 3,960.

To find the sum of the factors of a number N, let us represent the prime factorization of N by $a^\alpha \cdot b^\beta \cdot c^\rho \cdot d^\theta$, such that $N = a^\alpha \cdot b^\beta \cdot c^\rho \cdot d^\theta$. The sum of all the factors of N will be given by the formula

$$s = \frac{a^{\alpha+1}1}{a-1} \cdot \frac{b^{\beta+1}1}{b-1} \cdot \frac{c^{\rho+1}1}{c-1} \cdot \frac{d^{\theta+1}1}{d-1}.$$

In our example, $N = 3{,}960, a = 2, b = 3, c = 5, d = 11, \alpha = 3, \beta = 2, \rho = 1, \theta = 1$ as $3{,}960 = 2^3 \times 3^2 \times 5 \times 11$. Therefore,

$$\begin{aligned} s &= \frac{2^4 - 1}{2-1} \cdot \frac{3^3 - 1}{3-1} \cdot \frac{5^2 - 1}{5-1} \cdot \frac{11^2 - 1}{11-1} \\ &= \frac{15}{1} \cdot \frac{26}{2} \cdot \frac{24}{4} \cdot \frac{120}{10} \end{aligned}$$

$$= 15 \times 13 \times 6 \times 12$$

$$= 14,040$$

Postassessment

1. Have students calculate the total number of factors and then find each of them (either prime or composite), for each of the following:
   a. 3,600 b. 540 c. 1,680 d. 25,725
2. Find the sum of all the factors in each of the cases above.

Unit 19

Prime Numeration System

This unit will present an unusual way to express numbers. Consideration of this "strange" numeration system should strengthen student understanding of a place value system as well as appreciation of prime factorization.

Performance Objectives

1. *Students will convert from the prime numeration system into base-ten numerals.*
2. *Students will convert numbers from base-ten into the prime numeration system.*

Preassessment

Students should know what a prime number is. Students should also be able to factor a base-ten numeral into its prime factors.

Teaching Strategies

To familiarize students with the prime numeration system, have students consider the following problems:

a. $5 \cdot 4 = 9$; b. $12 \cdot 24 = 36$; c. $8 \div 2 = 6$.

Initially students will be quite puzzled. After further inspection, those familiar with exponents will begin to conjecture along those lines. Yet, this system is probably quite different from any numeration system studied before.

In the prime numeration system, there is no base. The value of each place is a prime number. The first place (starting at the right) is the first prime, 2; the next place (to the left) is the next prime, 3. This continues with the consecutive prime numbers with each succeeding place (moving left) corresponding to the next consecutive prime. This can be shown by using a dash for each place and indicating its value below it.

$$\overline{29}\ \overline{23}\ \overline{19}\ \overline{17}\ \overline{13}\ \overline{11}\ \overline{7}\ \overline{5}\ \overline{3}\ \overline{2}$$

As with our base-ten system, this prime system continues indefinitely to the left.

To find the value of a number in base-ten, digits occupying each place are multiplied by their place value and then added. However, in this prime numeration system, the value of a number is obtained by taking each place value to the *power* of the number occupying that place and then *multiplying*. For example, the number 145_p (the subscript p will be used to indicate that the number is in the prime numeration system) equals $5^1 \cdot 3^4 \cdot 2^5 = 5 \cdot 81 \cdot 32 = 12{,}960$. Notice the exponents of the prime numbers $5, 3$, and 2 are $1, 4$, and 5, respectively. Have students practice converting from the prime numeration system to base-ten numeration. When they begin to feel somewhat comfortable with this work, have them consider the representation of 0 and 1. Have students express 0_p and 10_p as a base-ten numeral. Indicate to the students that by definition $2^0 = 1$. Elicit from students that representation of zero will be impossible in the prime numeration system.

To convert a base-ten numeral into the prime numeration system, a review of prime factorization is necessary. Explain to students that any whole number greater than one can be expressed as the product of prime factors in precisely one way (The Fundamental Theorem of Arithmetic). For example, 420 can be factored as follows: $7^1 \cdot 5^1 \cdot 3^1 \cdot 2^2$. Therefore, $420 = 1112_p$. Have students factor (a) 144, (b) 600, and (c) 1960 into their prime factors and represent their equivalents in the prime numeration system. Emphasize that exponents of the prime factors are the digits of the prime numeral.

Prime System	Base-Ten
0_p	$2^0 = 1$
1_p	$2^1 = 2$
2_p	$2^2 = 4$
3_p	$2^3 = 8$
4_p	$2^4 = 16$
5_p	$2^5 = 32$
6_p	$2^6 = 64$
7_p	$2^7 = 128$
8_p	$2^8 = 256$
9_p	$2^9 = 512$
10_p	$3^1 \cdot 2^0 = 3$
11_p	$3^1 \cdot 2^1 = 6$
12_p	$3^1 \cdot 2^2 = 12$
13_p	$3^1 \cdot 2^3 = 24$
14_p	$3^1 \cdot 2^4 = 48$
15_p	$3^1 \cdot 2^5 = 96$
16_p	$3^1 \cdot 2^6 = 192$
17_p	$3^1 \cdot 2^7 = 384$
18_p	$3^1 \cdot 2^8 = 768$
19_p	$3^1 \cdot 2^9 = 1536$
20_p	$3^2 \cdot 2^0 = 9$
21_p	$3^2 \cdot 2^1 = 18$
22_p	$3^2 \cdot 2^2 = 36$
23_p	$3^2 \cdot 2^3 = 72$
24_p	$3^2 \cdot 2^4 = 144$
25_p	$3^2 \cdot 2^5 = 288$
26_p	$3^2 \cdot 2^6 = 576$
27_p	$3^2 \cdot 2^7 = 1152$
28_p	$3^2 \cdot 2^8 = 2304$
29_p	$3^2 \cdot 2^9 = 4608$

When students have mastered this numeration system, challenge them with multiplication; $5_p \cdot 4_p$. Then $5_p \cdot 4_p$ may be rewritten as $2^5 \cdot 2^4 = 2^9 = 512$. Therefore, $5_p \cdot 4_p = 9_p$. Now have them consider $25_p \cdot 4_p = 3^2 \cdot 2^5 \cdot 2^4 = 3^2 \cdot 2^9 = 29_p$ (or 4,608). Other related exercises should be presented (e.g., $8_p \div 2_p$). Indicate that the operations of addition and subtraction would require conversion to base-ten numeration before actually adding or subtracting. These problems allow students to practice working with exponents in a new and unusual way.

The prime numeration system can be implemented to review the greatest common divisor and least common multiple of two numbers.

Suppose students were required to find the greatest common divisor of 18,720 and 3,150. They should change these two base-ten numerals to the prime numeration system to get $100{,}125_p$ and $1{,}221_p$. By listing the *smallest value of each place* to form a new number, they will obtain 121_p. Which is the greatest common divisor of the two numbers.

Now suppose students were faced with the problem of finding the least common multiple of 18,720 and 3,150. Having changed these two base-ten numerals to the prime numeration system to get $100{,}125_p$ and $1{,}221_p$, they must merely list the *largest value of each place* to get $101{,}225_p$, which is the least common multiple of the two numbers.

Students will enjoy applying the prime numeration system methods to other problems that require finding the greatest common divisor or the least common multiple of given numbers (more than two numbers may be considered at one time). The true value of these methods rests in the justification of these methods. Teachers should present these justifications as soon as students have mastered the techniques involved.

Now have students convert 0_p through 29_p into base-ten numerals and record their answers. Students will begin to see the base-ten numerals being generated in an unusual way.

Elicit from students other possible applications of the prime numeration system.

Postassessment

Students should:

1. Express each of the following numbers in the base-ten numeration system.
   a. 31_p b. 24_p c. 15_p d. 41_p e. 221_p f. 1234_p

2. Express each of the following base-ten numerals as a product of primes, and then in the prime numeration system.
   a. 50 b. 100 c. 125 d. 400 e. 1,000 f. 260 g. 350
3. Solve the following problems:
   a. $3_p \cdot 6_p$ b. $12_p \cdot 13_p$ c. $6_p \div 3_p$

Reference

Posamentier, A. S. and B. Thaller, *Numbers: Their Tales, Types, and Treasures*, Amherst, New York: Prometheus Books, 2015.

Unit 20

Repeating Decimal Expansions

The terms *never-ending* and *infinite* are often confusing to students. One of the first places they really come into contact with the concept is in the junior high school where they confront nonterminating decimal expansions. Students themselves realize the need for specific notation when they encounter the repeating decimals certain fractional forms produce. In this unit, students discover the patterns and seemingly inconsistent arithmetic procedures that occur with repeating decimals.

Performance Objectives

1. *Students will be able to determine which rational numbers will yield repeating decimal expansions and which will terminate.*
2. *Students will determine minimum length of a repeating cycle.*
3. *Students will be able to use decimal equivalents to find other repeating decimals.*
4. *Students will examine an alternate method to determine decimal expansions.*

Preassessment

Students should know how to change from the fractional form $\left(\frac{a}{b}\right)$ of a rational number to its decimal equivalent. They should be familiar with prime factorization.

Have students guess which of the following fractions would become repeating decimals: $\frac{1}{2}, \frac{1}{3}, \frac{1}{4}, \frac{1}{5}, \frac{1}{6}$. Have them work out these fraction-to-decimal conversions to check the accuracy of their guesses. Note also that a terminating decimal can be considered a repeating decimal with an infinite repetition of zeros.

Teaching Strategies

Begin by having students work with fractions of the form $\frac{1}{n}$. This will force them to focus their attention (and guesswork) on the denominator. If they have difficulty figuring out how to determine repeating decimals without actually performing the expansion, suggest they factor into primes each of the denominators to see if any patterns become apparent. They will quickly see that the decimal terminates if, and only if, the prime factors of the denominator are 2 s and/or 5 s. They can easily justify that when the numerator takes on a decimal point and a series of zeroes, it becomes a multiple of 10 (for division purposes only); since the only prime factors of 10 are 5 and 2, it is only division by these factors that will terminate the division process.

Challenge students to determine how many decimal places they must establish before a pattern becomes evident. In some cases, as in $\frac{1}{3}$, it will take two decimal places before the pattern is clear. In others, it will not be so simple. Ask students to find the repeating pattern of $\frac{1}{17}$. This decimal expansion has 16 places before any pattern establishes itself $\frac{1}{17} = .0588235294117647\overline{0588235294117647}$.

Some students may want to generalize and assume $\frac{1}{n}$ will have $(n-1)$ repeating digits. However, $\frac{1}{3} = .3$, one repeating digit, which quickly disproves their theory. However, by examining each of the expansions of $\frac{1}{n}$, students will see that each expansion has at most $(n-1)$ repeating digits. They should realize that each of the expansion digits comes from the remainder after the division process of the previous step. For each of the remainders, there are only $(n-1)$ choices. (The remainder cannot equal zero because then the process would terminate; it cannot equal n because then it would have been divisible once again.) If the remainder is the same as any previous remainder, students have found the repeating digits; if not they must continue until a remainder repeats. This will have to happen in, at most, $(n-1)$ steps. Therefore $\frac{1}{n}$, if repeating, will have at most $(n-1)$ repeating digits.

Students will also find it interesting to note that the repeating expansion for a number such as $\frac{1}{7}$ also yields the expansions for $\frac{2}{7}, \frac{3}{7}, \frac{4}{7}, \frac{5}{7}$, and $\frac{6}{7}$. The fraction $\frac{2}{7}$ can be rewritten in terms of $\frac{1}{7}$. Thus,

$$\frac{2}{7} = 2 \times \frac{1}{7} = 2 \times \overline{.142857} = \overline{.285714}.$$

By adding different repeating decimals, students will be able to find new repeating decimals. For example,

$$\begin{array}{rl} \frac{1}{3} = & \overline{.333333} \\ +\frac{1}{7} = & +\overline{.143857} \\ \hline \frac{10}{21} = & \overline{.476190} \end{array}$$

To find the general repeating digits for $\frac{1}{n}$ when $n = 21$, have students divide $\overline{.476190}$ by the numerator, 10, to get $\overline{.047619}$.

In working with repeating decimals and in performing arithmetic operations, students may come upon the fact that $1 = \overline{.9}$. This is a difficult concept for a junior high school student to grasp. The following proof, which they can perform for themselves, should clarify the situation.

$$\begin{array}{r} \frac{1}{3} = .\overline{3} \\ +\frac{2}{3} = .\overline{6} \\ \hline 1 = .\overline{9} \end{array}$$

Similarly, since

$$\begin{aligned} \frac{1}{9} \times 9 &= .\overline{1} \times 9 \\ \frac{9}{9} &= .\overline{9} \\ 1 &= .\overline{9} \end{aligned}$$

Students often concentrate and comprehend more thoroughly when they feel they are learning something new. The following method, which basically outlines the division process used to change from fractional to decimal form, gives students another means of finding the repeating decimal.

To find the decimal expansion of $\frac{3}{7}$, let $r_0 = \frac{3}{7}$, and multiply by 10 (comparable to bringing down the 0).

1. $\frac{3}{7} \times 10 = \frac{30}{7} = 4\frac{2}{7}$

Now let the 4 occupy the tenths place of the decimal and use $\frac{2}{7}$ as the new remainder, r_1. Repeat the process, using the fraction as the new remainder and retaining the whole as the decimal digit for the next place.

2. $\frac{2}{7} \times 10 = \frac{20}{7} = 2\frac{6}{7} \quad r_2 = \frac{6}{7}$ hundredths place = 2
3. $\frac{6}{7} \times 10 = \frac{60}{7} = 8\frac{4}{7} \quad r_3 = \frac{4}{7}$ thousandths place = 8
4. $\frac{4}{7} \times 10 = \frac{40}{7} = 5\frac{5}{7} \quad r_4 = \frac{5}{7}$ ten thousandths place = 5
5. $\frac{5}{7} \times 10 = \frac{50}{7} = 7\frac{1}{7} \quad r_5 = \frac{1}{7}$ hundred thousandths place = 7
6. $\frac{1}{7} \times 10 = \frac{10}{7} = 1\frac{3}{7} \quad r_6 = \frac{3}{7}$ millionths place = 1

Tell students to repeat the process until the remainder is the same as the one with which they began. In this case $r_6 = r_0 = \frac{3}{7}$, and the decimal expansion is $\frac{3}{7} = .\overline{428571}$.

A clear demonstration of this method is an excellent tool to help students better understand what is involved in the division process and why remainders are such a big factor in determining the length of the repetition.

Postassessment

Have students do the following:

1. Determine which of the following will terminate without actually finding their decimal expansions: $\frac{2}{9}, \frac{1}{8}, \frac{3}{13}, \frac{19}{20}$.
2. Determine the maximum number of digits in the repeating cycle of each: $\frac{1}{37}, \frac{4}{9}, \frac{3}{7}$.
3. Knowing that $\frac{1}{14} = .\overline{0714285}$, find the decimal expansion for $\frac{3}{14}$ without dividing.
4. Show that $.5 = .\overline{49}$. (*Hint*: think about the fractions $\frac{1}{3}$ and $\frac{1}{6}$.)
5. Using the alternate method described, evaluate $\frac{2}{9}$ as a repeating decimal.

Unit 21

Peculiarities of Perfect Repeating Decimals

This unit can be used as an interesting sidelight to the subject of decimals and fractions by showing "magical" properties of a certain class of numbers. These numbers are reciprocals of prime numbers whose decimal equivalents repeat after no less than P − 1 places, where P is the prime number. Such numbers are said to have *perfect repetends.* In any repeating decimal, the sequence that repeats is called the repetend. It would be advisable to use this unit after the preceding one.

Performance Objectives

1. *Students will test various examples of perfect repetends to verify specific principles.*
2. *Students will discover and reinforce important ideas about division, remainders, and decimal equivalents of fractions.*

Preassessment

Students should know that the decimal equivalents of some fractions have 0 remainders, while others have repeating periods of various lengths. They should begin by converting $\frac{1}{7}$ to a decimal.

Teaching Strategies

It should be noted that in converting $\frac{A}{P}$ to a decimal, the repetend can have no more than P − 1 places, because in dividing A by P there can be at most P − 1 different remainders, and as soon as a remainder appears for the second time, the same sequence will be repeated. Perfect repetends, as well as the sequence of P − 1 remainders that accompanies each one, have several interesting properties. Only the simplest of these will be discussed here, but a more thorough listing of the principles of repeating decimals appears in *Philosophy of Arithmetic* by Edward Brooks (Norwood Editions), pp. 460–485.

One of the simpler properties to explain is that multiples 1 to P − 1 of $\frac{1}{P}$ are cyclic variations of the repetend of $\frac{1}{P}$. After students find $\frac{1}{7} = \overline{.142857}$, students can multiply the decimal by $2, 3, 4, 5$, or 6, and get answers of $\overline{.285714}, \overline{.428571}, \overline{.571428}, \overline{.714285}, \overline{.857142}$, which are also decimal equivalents of $\frac{2}{7}, \frac{3}{7}, \frac{4}{7}, \frac{5}{7}, \frac{6}{7}$, respectively. Once this is understood, an easy way to find the multiples of $\frac{1}{7}$ is to find the last place first. For example, $4 \times \overline{.142857}$ ends in 8, so it must be $\overline{.571428}$. Where the period is longer, or where any digit appears more than once in the repetend, it may be necessary to find the last two or three digits first. An explanation of this cyclic variation is that in dividing P into 1, at the point where the remainder is A, the same sequence will begin as occurs when dividing P into A. Remember also that every possible $A(1 < AP)$ occurs as a remainder. Incidentally, when the repetend of $\frac{1}{P}$ is multiplied by P, the result is .999999. Some other perfect repetends are

$$\frac{1}{17} = .0588235294117647$$

$$\frac{1}{19} = .052631578947368421$$

$$\frac{1}{23} = .0434782608695652173913$$

The only others for P < 100 are $\frac{1}{29}, \frac{1}{47}, \frac{1}{59}, \frac{1}{61}, \frac{1}{97}$.

Another curiosity of these numbers is that if the repetend is divided into two equal shorter sequences, their sum is $\overline{.99999}$. A graphic illustration of this is shown as follows. The inner circle is the repetend of $\frac{1}{29}$, and the outer circle is the sequence of remainders occurring after each number on the inner circle. The figure has the following properties (as do similar figures for all perfect repetends):

1. Any two diametrically opposite terms of the repetend add up to 9.
2. Any two opposite remainders add up to 29.
3. To multiply the repetend by $a(1 < a < 29)$, find a in the circle of remainders and begin the new repetend with the decimal term following the one associated with a (clockwise).

Some students may not have the patience to test out these generalizations on this figure, but of course a similar figure can be made for any of the perfect repetends. Students may construct their own figures, starting only with the information that $\frac{1}{17}$ is such a number.

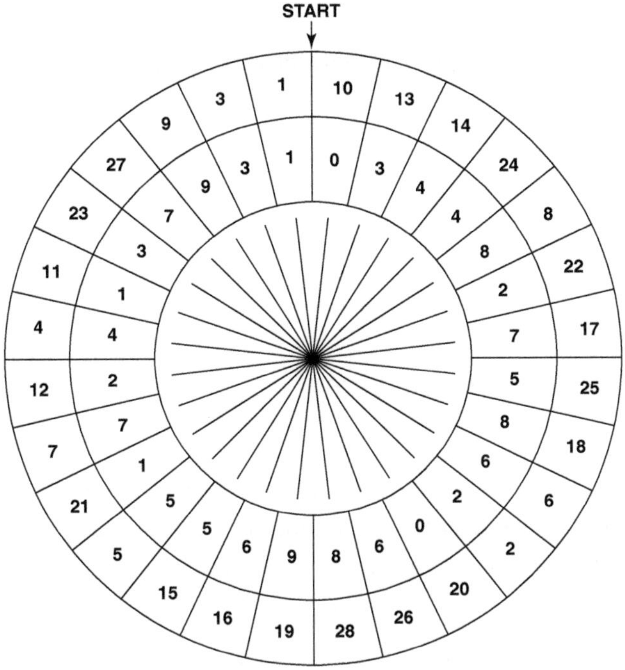

Here is one alternative to plowing straight ahead with division when generating a repetend: after dividing 19 into 1–5 places, we get a remainder of 3. $\frac{1}{19} = .05263\frac{3}{19}$.* But from this we know that $\frac{3}{19} = 3\left(.05263\frac{3}{19}\right) = .15789\frac{9}{19}$. So $\frac{1}{19} = .0526315789\frac{9}{19}$. But because we know that $\frac{1}{19}$ is a perfect repetend, 9 = first + tenth digit = second + eleventh = third + twelfth, etc., and we have generated all 18 digits.

This leads into a special property of the repetend of $\frac{1}{97}$, $\frac{1}{97} = .01\frac{3}{97} = .0103\frac{9}{97} = .010309\frac{27}{97} = .01030927\frac{81}{97}$. Unfortunately, 243 has 3 places, so this neat pattern changes, but we can still add powers of three in the following way to generate the repetend:

```
.0103092781
           243
             729
              2187
                6561 etc.
```

Students should be encouraged to discover other patterns of repetends.

---

* $\frac{1}{19} = .05263 + \frac{3}{19} \times 10^{-5}$; i.e., $\frac{3}{19}$ represents the 6th decimal place.

Postassessment

Have students generate any of the perfect repetends by using the rules shown here, then find multiples of the repetend. Explore with the class reasons why only primes have this peculiarity. For example, if $\frac{1}{14}$ had a perfect repetend, what happens to $\frac{2}{14}$ or $\frac{4}{14}$?

Unit 22

Patterns in Mathematics

This unit is designed for ninth-year students of mathematics. Parts of this unit could be used for enrichment of remedial classes in finding the patterns by observation alone.

Performance Objectives

1. *Students will find patterns by observation.*
2. *Students will find formulas for the patterns by trial and error.*
3. *Students will find the formulas for the patterns by discovering the rules for finding the constant and the coefficients of x and x^2.*

Preassessment

Challenge students to find succeeding numbers in the following patterns and formulas for the patterns:

a)

x	y
0	1
1	3
2	5
3	7
4	?
5	?

b)

x	y
0	1
1	4
2	7
3	10
4	?
5	?

c)

x	y
0	1
1	5
2	9
3	13
4	?
5	?

d)

x	y
0	3
1	5
2	7
3	9
4	?
5	?

Most students will be able to find the patterns and the formulas by trial and error. Have students fill in missing numbers, formulas, and note the differences between the successive ys. The completed charts will look like this. D denotes difference between successive ys:

a)

x	y	D
0	1	
1	3	2
2	5	2
3	7	2
4	9	2
5	11	2

$y = 2x + 1$

b)

x	y	D
0	1	
1	4	3
2	7	3
3	10	3
4	13	3
5	16	3

$y = 3x + 1$

c)

x	y	D
0	1	
1	5	4
2	9	4
3	13	4
4	17	4
5	21	4

$y = 4x + 1$

d)

x	y	D
0	3	
1	5	2
2	7	2
3	9	2
4	11	2
5	13	2

$y = 2x + 3$

Have students notice constants in each case. Do they observe any pattern? Perhaps they will notice that the constant is the value of y when x is zero. Draw their attention to the difference between the successive ys. Do they observe anything? Yes, the difference between the ys coefficient of x. Do several patterns of this type until students can quickly find the patterns and the formulas for the patterns.

Teaching Strategies

Give your students the following exercise and have them find the pattern and the formula if they can.

How many rectangles in all?

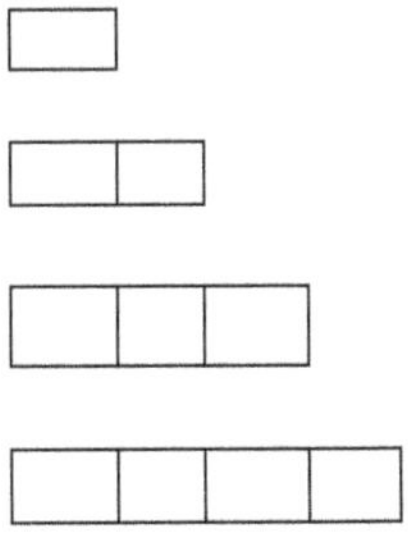

Complete the table

x No. of small rectangles	y Total no. of rectangles
0	
1	
2	
3	
4	
5	
6	

By observation of the rectangles, many will be able to find the pattern and fill in the table. Have them record the first difference. They will notice that it is not constant. Have them record the second difference. They will notice it is constant. Have them summarize their findings in a table. Perhaps some will find the formula also.

x No. of small rectangles	y Total no. of rectangles	D_1	D_2
0	0		
1	1	1	1
2	3	2	1
3	6	3	1
4	10	4	1
5	15	5	1
6	21	6	1

$$y = \frac{x^2}{2} + \frac{x}{2}$$

Have them do the same with the following pattern: What is the largest number of pieces you can make with x cuts?

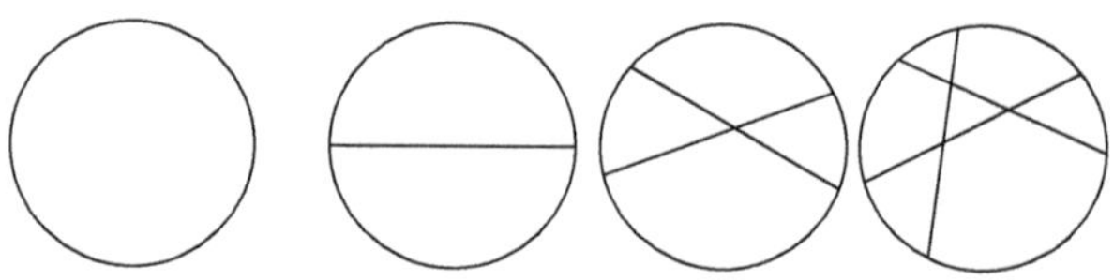

__cuts __pcs. __cuts __pcs. __cuts __pcs. __cuts __pcs.

Have them fill out the following:

x No. of cuts	y No. of pcs.	D_1	D_1
0	1		
1	2	1	
2	4	2	1
3	7	3	1
4	11	4	1
5	16	5	1

Notice the first difference. It is not constant. Notice the second difference. It is constant. Perhaps someone will be able to come up with the formula

$$y = \frac{x^2}{2} + \frac{x}{2} + 1.$$

Are there any patterns one can see for the values of the constants or the coefficients in the two preceding problems? Yes, the constant is the value of y when x is zero.

Let us examine the formula $ax^2 + bx + c = y$ and find the values of y for various values of x.

x	y	D_1	D_2
0	c		
1	$a + b + c$	$a + b$	
2	$4a + 2b + c$	$3a + b$	$2a$
3	$9a + 3b + c$	$5a + b$	$2a$
4	$16a + 4b + c$	$7a + b$	$2a$

Let us examine the pattern. As we found in the formulas, y is the constant when x is zero. The first difference is $a + b$, the sum of the coefficients of x^2 and x. The second difference is $2a$, twice the value of the coefficient of x^2. The value of the first difference when $x = 1$ is $a + b$. Since we know the value of a (it is one-half of the second difference), we can find the value of b by subtracting a from the first difference $(a + b)$. So if we reexamine the earlier pattern, we can derive the formula.

The constant is the value of y when x is zero. Therefore, the constant is 1. D_2 is $2a$. Since D_2 is 1, the value of a is $\frac{1}{2}$. D_1 is $a + b$. Since D_1 is 1, and a is $\frac{1}{2}$, the value of b is $\frac{1}{2}$. The formula therefore is

$$y = \frac{1}{2}x^2 + \frac{1}{2}x + 1.$$

Postassessment

Finish the tables and find the formulas for the following patterns by finding the first and second differences:

x	y	x	y	x	y	x	y
0	3	0	0	0	0	0	2
1	6	1	5	1	13	1	3
2	13	2	14	2	34	2	6
3	24	3	27	3	63	2	11
4		4		4		4	
5		5		5		5	

Unit 23

Googol and Googolplex

This unit presents a discussion of large numbers. It introduces students to the finite world of large numbers and the ease of expressing large numbers using scientific notation.

Performance Objectives

1. *Students will give examples of scientific notation as used in science and mathematics.*
2. *Given any number, students will convert it into scientific notation.*

Preassessment

Students should be able to solve the following problems:

1. Compute the following products. Find the solution without using a pencil.
   a. 10×63 b. $100 \times .05$ c. 1000×951.
2. Compute the following quotients.
   a. $470 \div 10$ b $4{,}862 \div 1000$ c. $46{,}000 \div 1000$
3. What is the largest number you can think of?

Teaching Strategies

You might want to tell the old story of two children engaged in a violent argument in the street. The argument stems from the fact that each child is trying to state a larger number than the other. Finally, they realize each one can state a larger number than the other.

Numbers are fun to play with and many interesting things can be done with them. However, all too often we forget what a number really is. The students should be asked, how large is one million? Can we visualize the sum of the first billion natural numbers? Why should we care about numbers of great magnitude, we never use them — or do we?

At this point, the teacher should state that scientists who use very large or very small numbers usually express these numbers in scientific notation.

To learn how to use this system of numeration, we will recall some patterns in mathematics. It is up to the teacher at this point to introduce scientific notation. You may wish to refer to any standard textbook for an appropriate development.

1. When a number is expressed as the product of a power of 10 and a number that is less than ten but greater than or equal to one ($1 \leq n < 10$), the number is said to be written in scientific notation.
2. A science teacher may be able to suggest very large or very small numbers that students have used or read about in their science classes. These can then be converted into scientific notation.

A general discussion should ensue on when large numbers are used (for example, grains of sand on a beach, stars in the sky, or in economics or science). Current newspaper and magazine articles abound with references to millions and billions. How many people have seen a million of anything? Most people do not have a clear idea of the size of a million.

To earn a million dollars, how long would you have to work at $100 a week? (almost 200 years). How many stars can you see naked eye on a clear night? No, not millions, but only about 3,500 (3.5×10^3 in scientific notation). A hundred sheets of paper make a stack about 5 mm thick or $\frac{1}{5}$ of an inch (25.4 mm equal an inch). A million sheets of paper would make a pile about 55 yards high, or roughly the height of a 12 story building. Suppose you are riding in a car at 65 mph. How long would it take you going constantly to travel one million miles? ($1\frac{3}{4}$ years).

Just how big is a billion? The 2021 federal budget called for more tax dollars than there were seconds since the birth of Christ. (*Note*: According to the ability of a particular class and allotted time, all large numbers should be converted to scientific notation by the student.) Students should be led to the fact that it is virtually impossible for the human mind to comprehend the enormity of a billion. Remember how high a pile of a million sheets of paper would be — a hundred sheets of paper make a stack about 5 mm thick (about $\frac{1}{5}$ of an inch). A billion sheets of paper would make a pile 31 miles high.

A car traveling nonstop at 100 mph would take 1,140 years to travel a billion miles. If you earn $100 a week, you would have to work 192,307 years to earn a billion dollars. (Some of these illustrations may be calculated on an electronic calculator.)

You should pose the following problem.

> John did me a favor the other day and I asked him what reward he would like. John, being very wise said, "Give me one penny for the first day, two pennies for the second day, four pennies for the third day, and likewise for sixty-four days, doubling the number of pennies for each successive day." How much money would I have to pay John?

Making a table gives students an opportunity to see numbers growing and the magnitude they can reach.

Number of Days	Number of Pennies
1	1
2	2
3	4
4	8
5	16
etc.	etc.
64	9,223,372,036,854,775,808

Now the sum of all the numbers in the second column is the number of pennies needed to pay John:

$$18{,}446{,}744{,}073{,}709{,}551{,}615$$

It is read: Eighteen quintillion, four hundred forty-six quadrillion, seven hundred forty-four trillion, seventy-three billion, seven hundred nine million, five hundred fifty-one thousand, six hundred fifteen. Students should be able to see that even though it is an enormous number it is not infinite, but finite. Have them express it approximately in scientific notation.

The next question that may arise at this point concerns the largest number that can be expressed by three digits. In ordinary notation, the answer is 999. What about 99^9? (Review of exponents reveals the fact that this means 99 multiplied by itself 8 times.) But if exponents are permitted, the answer is 9^{9^9}, that is, 9 with the exponent 9^9, or simply 9 with the exponent 387,420,489 (the product of 387,420,489 nines). If printed with 16 figures to an inch, it has been estimated, this huge number would fill 33 volumes of 800 pages each, printing 14,000 figures on a page. It has been estimated that this number is more than four million times as large as the number of electrons in the universe, but a *finite* number. (Ask students to find the largest three digit number that can be written with fours.)

Say the number of grains of sand at Coney Island is about 10^{20}. Students could be asked to devise a method for establishing this estimate. The number of electrons which pass through the filament of an ordinary light bulb in a minute equals the number of drops of water that flow over Niagara Falls in a hundred years. The reason for giving such examples of very large numbers to students is to emphasize that the elements of even very large sets can be counted.

Students may now ask what the largest number that has a name is. The term *googol* was coined to describe the figure 1 followed by a hundred zeros. Another term, *googolplex*, was invented for a still larger, but still finite, number consisting of a 1 followed by a googol of zeros. Thus, a googol times a googol would be a 1 with 200 zeros. Students who try to write a googolplex on the chalkboard or a sheet of paper will get some idea of the size of this very large but finite number (there would not be enough room to write it if you traveled to the farthest visible star, writing zeros all the way).

Astronomers find the light-year a very convenient unit of length in measuring great astronomical distances. The North Star is 47 light years away. What does this mean? Light travels at the rate of 186,000 miles per-second. In one year, light travels 6,000,000,000,000 (6×10^{12}) miles. This tremendous distance is called a light year. The nearest star is 4.4 light-years away, and the farthest known star is 1.4×10^9 light-years away. Have students consider this: It takes 47 light-years for light from Earth to reach the North Star. What would a person looking at the Earth from the North Star see today?

Postassessment

Have students complete the following:

1. The distance of the planet Pluto from the earth is approximately 4,700,000,000,000 miles. Express this answer in scientific notation (4.7×10^{12}).
2. The circumference of the earth at the equator is approximately 25,000 miles. Express this in scientific notation (2.5×10^4).

Unit 24

Mathematics of Life Insurance

This unit describes to students how insurance companies take into account probability and compound interest in calculating the net premium of life insurance.

Performance Objectives

1. *Students will use a compound interest formula to compute the value of money left in a bank for a given period at a given rate.*
2. *Students will compute the present value of money that increases to a given amount when left in a bank for a given period at a given rate.*
3. *Students will use appropriate probabilities and interest rate to calculate the net premium a life insurance policyholder must pay.*

Preassessment

Use the following problem for diagnostic purposes as well as to motivate the lesson. Out of 200,000 men alive at age 40, 199,100 lived at age 41. What is the probability that an insured man of age 40 will live at least one year? What is the probability that he will die within one year?

Teaching Strategies

By posing the above problem, students become aware of the applicability of probability theory to life insurance. These companies must be able to measure the risks against which people are buying the life insurance. To decide on the premiums, a life insurance company must know how many people are expected to die in any group. They do this by collecting data about the number of people who died in the past from each age group. Since the data is collected from a large number of events, the law of large numbers applies. This law states that *with a large number of experiments, the ratio of the number of successes to the number of trials gets very close to the theoretical probability.*

Life insurance companies construct mortality tables based on past deaths in order to predict the number of people who will die in each age group. Below is a portion of the Commissioners 1958 Standard Ordinary Mortality Table.

To construct this table, a sample of 10 million people was used. Their life span was recorded from birth till age 99. At each age level, the table records the number of people alive at the start of the year and the number of deaths that occurred during the year. Then the following ratio is computed:

$$\frac{\text{Number of deaths during year}}{\text{Number of people alive at start of year}}$$

This ratio is then converted to deaths per 1,000. The number of deaths per 1,000 is called the *death rate.* This death rate, as students will see, is crucial in computing the premium that policyholders will pay.

After this introduction, the teacher should ask the class: What is the probability that an 18-year-old will die if out of 6,509 18-year-olds at the beginning of the year, 11 died? The probability is $\frac{11}{6{,}509}$. However, life insurance companies prefer to transform this ratio into death rate per 1,000. The teacher should have the class change $\frac{11}{6{,}509}$ into $\frac{x}{1{,}000}$ by setting up the following proportion:

$$\frac{x}{1{,}000} = \frac{\text{Number dying during the year}}{\text{Number alive at start of the year}}$$

$x =$ death rate per 1,000.

Age	Number Living	Deaths Each Year	Deaths Per 1,000
0	10,000,000	70,800	7.08
1	9,929,200	17,475	1.76
2	9,911,725	15,066	1.52
3	9.896,659	14,449	1.46
4	9,882,210	13,835	1.40
10	9,805,870	11,865	1.21
11	9,794,005	12,047	1.23
12	9,781,958	12,325	1.26
13	9,769,633	12,896	1.32
18	9,698,230	16,390	1.69
25	9,575,636	18,481	1.93
30	9,480,358	20,193	2.13
42	9,173,375	38,253	4.17
43	9,135,122	41,382	4.53
44	9,093,740	44,741	4.92

Figure 1.

The answer to the above problem is

$$\frac{11}{6{,}509} = \frac{x}{1{,}000} \quad \text{or} \quad x = 1.69.$$

This means that 1.69 people out of the original 1,000 will have died by the end of the 18th year. The insurance company uses this information to calculate the premium it will charge a group of 18-year-olds. Suppose there were 1000 people aged 18 who insured themselves for \$1000 each for one year. How much would the company have to pay out at the end of the year? If 1.69 people die, the company will pay out \$1,690 ($1.69 \times 1{,}000 = 1{,}690$). Thus, how much must the company charge each of the 1,000 policyholders? (This does not take into account profit or operating expenses.) The \$1,690 divided evenly among 1,000 people equals \$1.69 per person.

In the previous discussion, students did not take into consideration the fact that money paid to the company earns interest during the year. So besides considering the death rate, the interest rate must also be taken into account when calculating the premium.

The teacher must now develop the concepts of compound interest. The teacher should ask the class how much money will be on deposit in a bank at the end of the year if one deposits \$100 at 5 percent interest. The answer is \$100 plus .05 (100) or 100×1.05 which is \$105. If the \$105 is kept in the bank another year, what what will it amount to? \$105 + .05(105) or \$100 × 1.05 × 1.05 or \$100 × $(1.05)^2$ which amounts to \$110.25. Have the students write the general formula using P = original principal, i = rate of interest per period, A = the amount of money at the end of the specified time, and n = the number of years the principal is on deposit. The formula is $A = P(1+i)^n$.

The teacher should now ask the students how much money they would have to deposit now in a bank whose rate of interest is 5 percent, if they wanted \$100 accumulated in one year from now. In the previous example, the students saw that \$100 grew to \$105 in one year's time. This information is used to set up a proportion: $\frac{x}{100} = \frac{100}{105} = .9524$, $x = 100(.9524) = \$95.24$.

How much would have to be deposited now to accumulate \$100 at the end of two years from now? $\frac{x}{100} = \frac{100}{110.25} = .9070, x = \$90.70$.

The students should now be able to derive a formula for calculating the present value from the formula for compound interest ($A = P(1+i)^n$).

This formula is $P = \frac{A}{(1+i)^n}$.

Your students will now return to the original problem of the life insurance company that has to pay out \$1,690 at the end of the year to the deceased 18-year-olds. What is the present value of \$1,690? In other words, how much must the insurance company collect at the beginning of the year so that it can pay out $1,690 at the end of the year? By using the present value formula, the students computed that for every $1 the company has to pay, it must collect $.9524 at the beginning of the year. If the company has to pay $1,690, then it has to collect $1,609.56 in total from its 1,000 18-year-olds ($1{,}690 \times .9524 = \$1{,}609.56$). Thus, each policyholder must contribute a premium of $\frac{\$1{,}609.56}{1{,}000} = 1.60956$ or about $1.61.

You may now pose another problem. Suppose another group of 1,000 people aged 25 bought policies for one year worth $1,000 apiece (the death benefit is $1,000). According to the mortality table, their death rate is 1.93, or 1.93 out of 1,000 25-year-olds die during their 25th year. What will the net premium be if the interest rate is 5 percent? Death rate per 1,000 at age $25 = 1.93$. Amount needed to pay claims $= (1.93 \times 1{,}000) =$ $1,930. Interest factor $=$ $.9524. Present value of claims due in 1 year ($\$1{,}930 \times .9524) = \$1{,}838.13$. Number of persons paying premium $= 1{,}000$. Net premium $\frac{\$1{,}838.13}{1{,}000} = 1.83813$ of $1.84. This process may be continued for additional years of insurance.

Postassessment

Calculate the net premium for a 2-year policy for a group of 1,000 all age 30, with interest at 5 percent. Death rate at 30 is 2.13 and death rate at 31 is 2.19.

Reference

Posamentier, A. S. and C. Spreitzer, *The Mathematics of Everyday Life*, Amherst, New York: Prometheus Books, 2015.

Unit 25

Geometric Dissections

Unlike Humpty Dumpty, dissected geometric figures can be put back together again. In fact, the primary purpose of dissections is to cut a plane rectilinear figure with straight lines in such a way that the resulting pieces can be reassembled into a desired figure. This unit will introduce the wide range of geometric dissections by emphasizing their mathematical as well as recreational value.

Performance Objectives

1. *Students will see familiar polygonal area formulas in a concrete and interrelated manner.*
2. *Students will transform certain polygonal figures into other polygonal figures of equal area through dissections.*

Preassessment

Present your students with the following problem: Given an equilateral triangle, dissect the triangle into four pieces, which can be put together to form a rectangle. One possible solution: construct the perpendicular bisector from C to point D on side $\overline{AB}$; from D draw a line segment to the midpoint of $\overline{BC}$; bisect $\angle$A extending the bisecting ray to point F on $\overline{CD}$. These four pieces will form a rectangle.

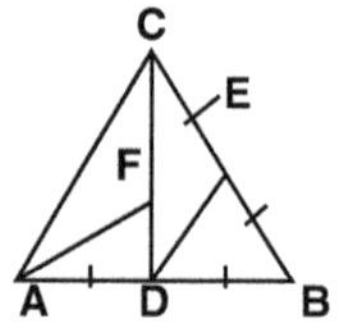

Figure 1.

Teaching Strategies

Begin discussion of dissections by demonstrating the area equality between a rectangle and a parallelogram with the same base. The dissection proceeds as follows. Using heavy paper or cardboard, construct a rectangle ABCD. Make a straight cut from vertex A to a point E on side $\overline{DC}$. Remove ΔADE placing side $\overline{AD}$ along side $\overline{BC}$ to form parallelogram ABE′E.

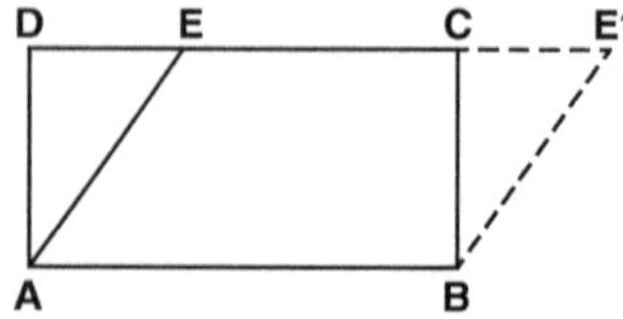

Figure 2.

In a similar manner, you can also demonstrate that a parallelogram and a trapezoid with the same base have equal areas. Consider any trapezoid; find the midpoint E of side $\overline{BC}$, and through E draw a line parallel to $\overline{AD}$ which intersects $\overline{AB}$ at X and $\overrightarrow{DC}$ at Y. Since ΔCEY and ΔBEX are congruent, the areas of trapezoid ABCD and parallelogram AXYD are equal.

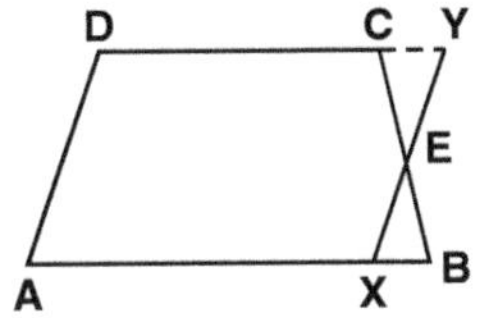

Figure 3.

The range of possible transformations of polygons into other polygons by means of dissections is vast. Janos Bolyai, one of the founders of non-Euclidean geometry, was the first to suggest that given any two polygons with equal area, either figure could be dissected a finite number of times such that upon rearrangement it would be congruent to the other. However, we are concerned with specific transformations that require a minimum number of dissections.

For example, you could consider the problem of dissecting a given acute triangle to form a rectangle. In Figure 4, first find the midpoints of sides $\overline{AC}$ and $\overline{BC}$ and connect these points to form $\overline{DE}$. From C construct a perpendicular to $\overline{DE}$ at X. Take ΔDXC and place it so that X is now at X′ and $\angle$DCX is adjacent to $\angle$CAB. Similarly move ΔEXC so that X is now at X″ and $\angle$ECX is adjacent to $\angle$CBA.

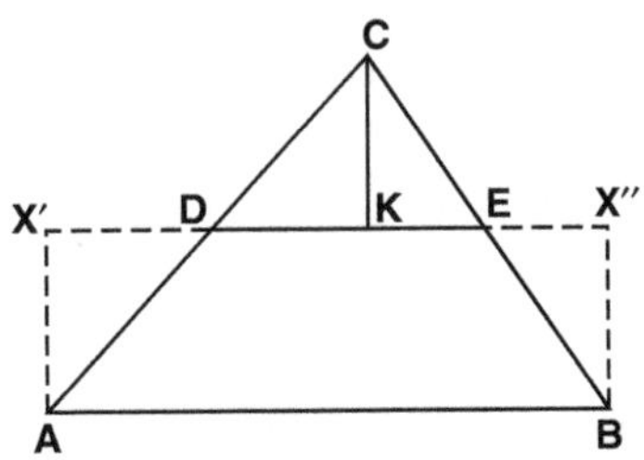

Figure 4.

To encourage students to begin solving dissection problems on their own, suggest that they carefully construct a 10 cm × 10 cm square as follows: Let AL = BG = 7, CT_1 = 3.1, AR_1 = 2.9, DN = 4.2; draw $\overline{AG}$ and $\overline{LN}$; on

$\overline{LN}$ let $LS_1 = 1.6$; on $\overline{AG}$ locate points R, S, K, and T such that $AR = 2.4$, $RS = 3.3$, $SK = 2.4$ and $KT = 3.3$; draw $\overline{RR_1}$, $\overline{SS_1}$, $\overline{KB}$, and $\overline{TT_1}$.

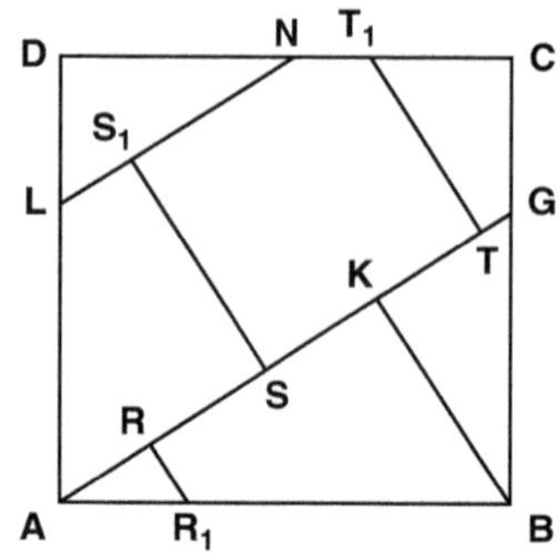

Figure 5.

After cutting, the student should have seven pieces. Using all of these pieces, students should attempt to form: (1) three squares of the same size and (2) an isosceles trapezoid.

A beautiful dissection is possible with three regular hexagons. Leaving the first hexagon uncut, dissect the second and third as shown in Figure 6. These 13 pieces can be combined to form a single hexagon.

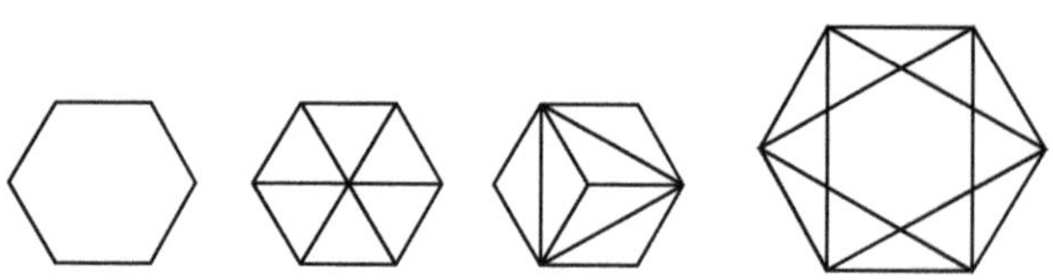

Figure 6.

This transformation can be considered in terms of rotations, and others in terms of reflections, translations, as well as rotations. You can then determine that a side of the larger hexagon is $\sqrt{3}$ times a side of the smaller hexagons. Since the area of the new hexagon is three times the area of each of the smaller hexagons, we have verified a significant relationship that holds between similar figures: the ratio of their areas is the square of the ratio between any two corresponding sides.

Postassessment

Students should complete the following exercises:

1. Demonstrate by dissection that a rectangle can be divided into two congruent trapezoids that each have one-half the area of the rectangle.

2. With the pieces from the dissected 10 cm × 10 cm square, form (1) a rectangle, and (2) a parallelogram.
3. Dissect the regular dodecagon that appears below into a square (cut along indicated lines).

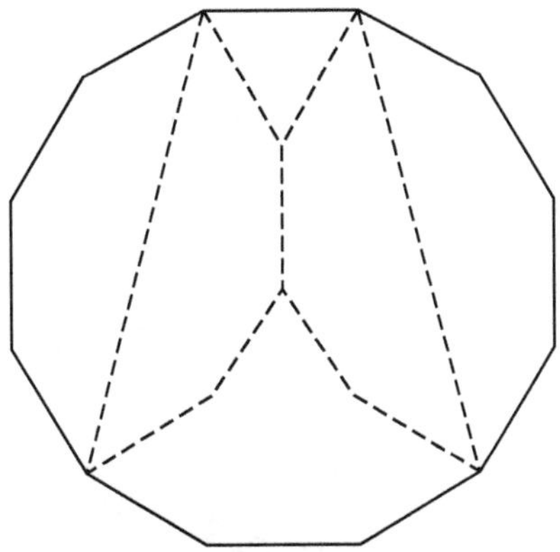

Unit 26

The Klein Bottle

This unit will provide students with an insight into one of the very fascinating topics of topology, the Klein bottle. They will be surprised to see a solid figure whose inside cannot be distinguished from its outside.

Performance Objectives

1. *Students will create a Klein bottle from a flat piece of paper.*
2. *Students will characterize a surface by certain topological properties.*
3. *Students will determine the Betti number of topological surfaces.*

Teaching Strategies

Before demonstrating how the above situation can be created, briefly discuss the one-sided topological figure, the Klein bottle. The Klein bottle was invented by Felix Klein, a German mathematician, in 1882. If we were to compare the Klein bottle to something realistic, we would use a flexible object, such as a cylinder with a hole cut through the surface. We would then stretch one end to make a wide base and the other end narrowed like the neck of a bottle. But we would have to bring these two-end circles together with their arrows running in opposite directions (see following diagram). Imagine the narrow end of the cylinder bent up, and plunged through the hole on the cylinder and joined to the wide base as in the following figure.

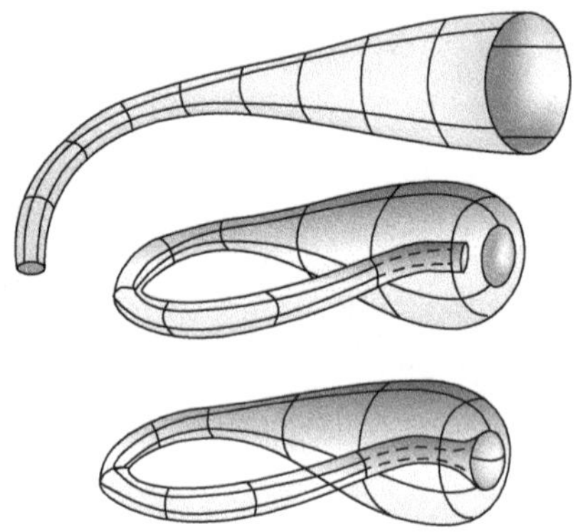

The hole on the surface of the cylinder should not be actually thought of as a hole but rather an intersection of surfaces covered by a continuation of the surface of the bottle.

Let us now return to the original problem. The situation can easily be visualized if we compare the sleeves of the jacket to the ends of the cylinder and one of the armholes to the hole in the cylinder. We have now created a figure that is topologically equivalent to the Klein bottle.

Once the students have a clear understanding as to what a Klein bottle appears to be, demonstrate how it can be created from a piece of paper. To construct a Klein bottle, what we are supposed to do with the flat piece of paper is to join the respective corners of the edges AB to A′B′, but we are also to join the remaining edges AB′ to A′B.

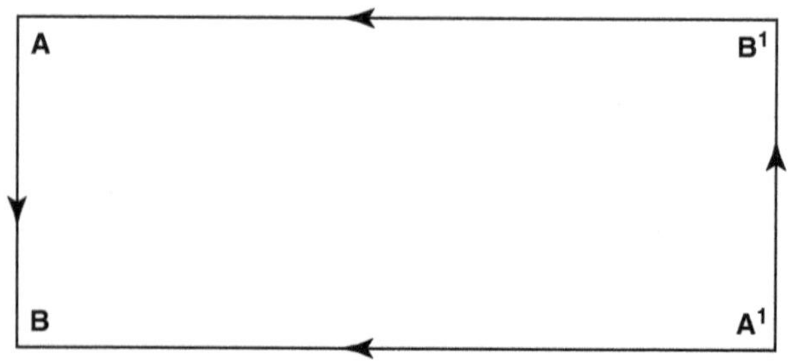

First create a cylinder by folding the sheet of paper in half and joining the open edges with a strip of tape. Cut a slot through the thickness of the paper nearest you about a quarter of the distance from the top. This will correspond to the "hole" in the surface of the cylinder. Fold the model in half and push the lower end through the slot. Join the edges as indicated by the arrows in the diagram. It is easily seen that this paper model is topologically identical to the Klein bottle created from the cylinder.

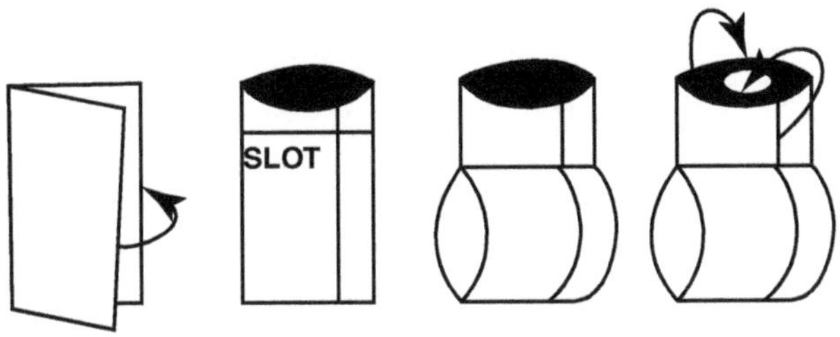

If we were now to examine the Klein bottle and try to distinguish the outside from the inside, and vice versa, we would find it impossible to do so. It would be evident that the surface is one-sided and edgeless, a notion very unusual to geometric figures.

Since it may be difficult to recognize a Klein bottle or any surface whose shape has been extremely distorted, it is necessary to be able to characterize each surface by simpler topological properties. Two of the properties have already been mentioned: number of edges and number of sides. A Klein bottle was found to be one-sided and have no edges. A third distinguishing feature of these surfaces is the *Betti number*. The Betti number is the maximum number of cross cuts (a simple cut with a pair of scissors that begins and ends on the edge) that can be made on a surface without dividing it into more than one piece. This means that a figure in the shape of a disk has a Betti number zero, since any cross cut will divide it into two pieces. On the other hand, the lateral surface of a cylinder has a Betti number of one.

Ask students why it would be difficult to determine the Betti number of a doughnut shaped figure or a Klein bottle using the cross cut method. Most students should realize that the problem here is that both of these topological figures contain no edges. Therefore, an alternate method using a *loop-cut* (it starts at any point on the surface, and returns to it without crossing itself, avoiding the edge entirely) provides another way of determining the Betti number. When using the loop-cut to determine the Betti number, we count the number of edges and say that the Betti number equals the number of loop-cuts we can make in a surface without dividing it into more pieces than there were edges. A doughnut-shaped figure requires two loop-cuts:

one horizontally and the other vertically so the Betti number is two. The Klein bottle also requires two loop-cuts as shown in the following diagram.

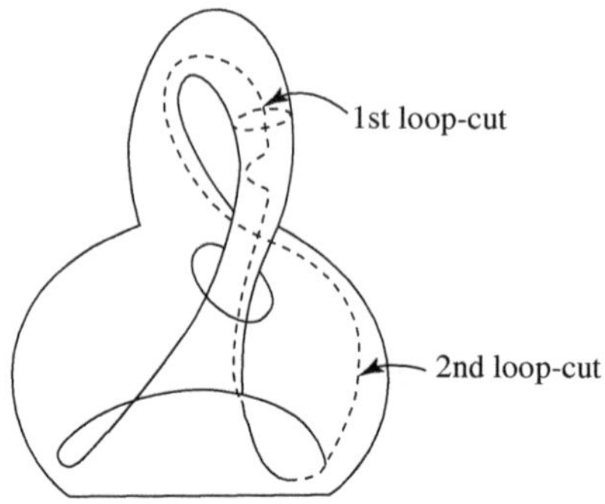

Postassessment

1. Have students determine the Betti number of the following surfaces:
   a. a tube
   b. a punctured tube
   c. a punctured sphere
2. Have students determine which figures would be created if you cut a Klein bottle in half.

Unit 27

The Four-Color Map Problem

Topology is a branch of mathematics related to geometry. Figures discussed may appear on plane surfaces or on three-dimensional surfaces. The topologist studies those properties of a figure that remain *after* the figure has been distorted or stretched according to a set of rules. A piece of string with its ends connected may take on the shape of a circle or a square. In going through this transformation, the order of the "points" along the string does not change. This retention of ordering has survived the distortion of shape, and is a property that attracts the interest of topologists.

Performance Objectives

1. *Students will state the Four-Color Map Problem.*
2. *Given a geographical map on a plane surface, the student will show, by example, that four colors are sufficient to successfully color the entire map.*

Preassessment

Students should know the meaning of common boundaries and common vertices as applied to geographical maps on a plane surface.

Teaching Strategies

Begin by indicating that this problem was only recently solved with the extensive aid of computers. Previously it was considered one of the famous unsolved problems of mathematics.

Have students analyze this fictional, geographic map (Figure 1) of eight different countries, and list all countries that have a common boundary with country H and countries that share a common vertex with region H. A map will be considered correctly colored when each country is completely colored and two countries that share a common boundary have different colors. Two countries sharing a common vertex may also share the same color. Have students color in several maps according to the rules for coloring as stated above (b/blue; r/red; y/yellow; g/green).

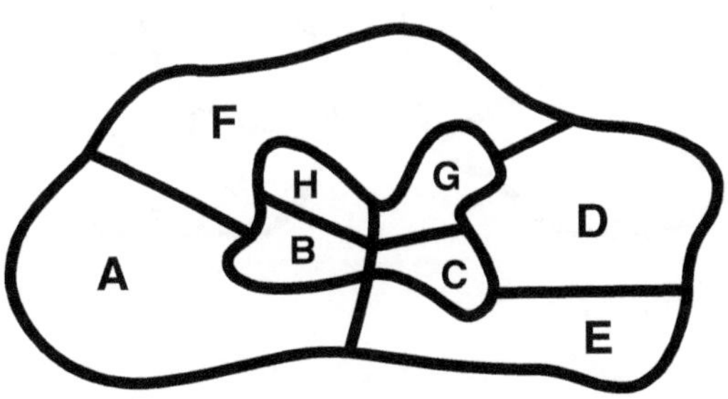

Figure 1.

This map (Figure 2) consists of two regions with one common boundary and therefore requires two colors to color correctly.

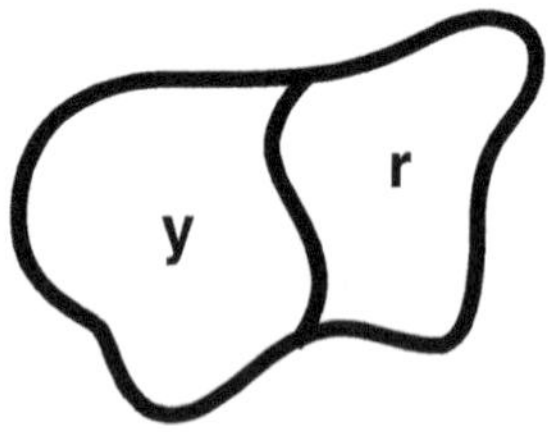

Figure 2.

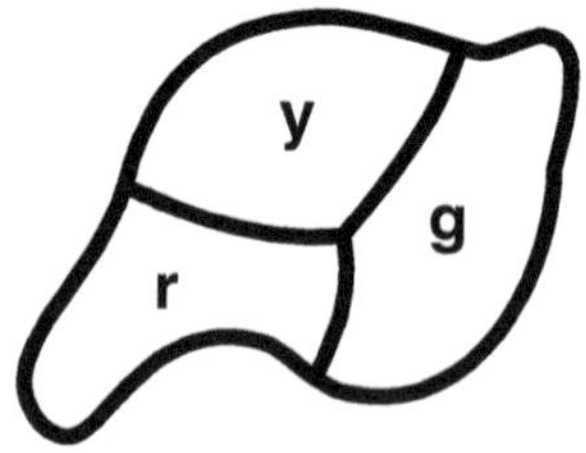

Figure 3.

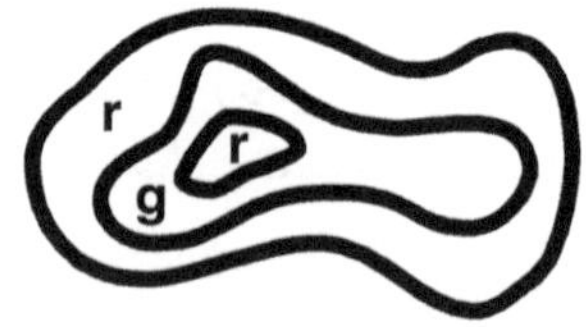

Figure 4.

This map (Figure 3) consists of three different regions and the students should conclude that three different colors are required to color it correctly. It seems as though a map with two regions requires two colors and a map with three regions requires three colors.

Ask the students if they can devise a map that has three different countries that will require less than three colors to color it. As an example see Figure 4.

Since the innermost country and the outermost country share no common boundary, they may share the color red and still retain their separate identity.

It seems reasonable to conclude that if a three-country map can be colored with less than three colors, a four-region map can be colored with less than four colors. Have the students create such a map.

Figure 5 has four regions and requires only two colors for correct coloring. Figure 6 also consists of four regions and requires three colors for correct coloring.

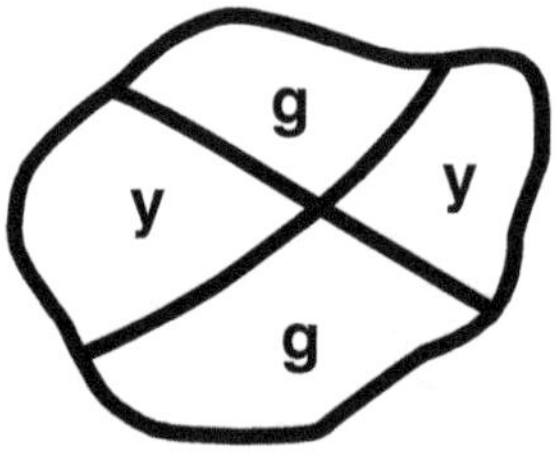

Figure 5.

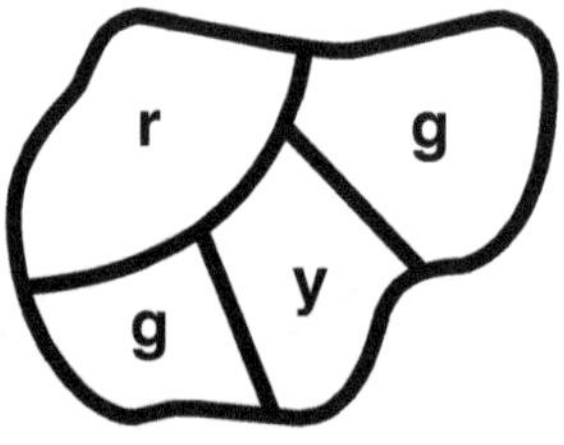

Figure 6.

Challenge students to devise a map that consists of four countries and requires exactly four colors for correct coloring. Before undertaking such a task, students should now realize that this map calls for each of the four countries to share a common boundary with the other three. Figure 7 is an example of this map.

Ask students to take the next logical step in this series of map-coloring problems. They should come up with the idea of coloring maps involving

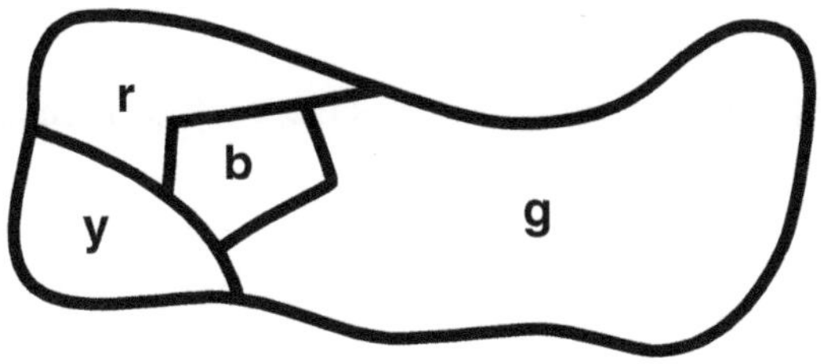

Figure 7.

five distinct regions. It will be possible to draw maps that have five regions and require two, three, or four colors to be colored correctly. The task of drawing a five-country map that *requires* five colors for correct coloring will be impossible. This curiosity can be generalized through further investigation and students should arrive at the idea that any map, on a plane surface, with any number of regions, can be successfully colored with four or fewer colors.

It is more satisfying to present the problem as a direct challenge in the following form: "Can you draw a geographic map, on a plane surface, with any number of regions, that requires five colors to be correctly colored?" This is the statement of the Four-Color Problem. Whereas the Three Famous Problems of antiquity have been proved to be impossible many years ago, this problem was only solved recently.

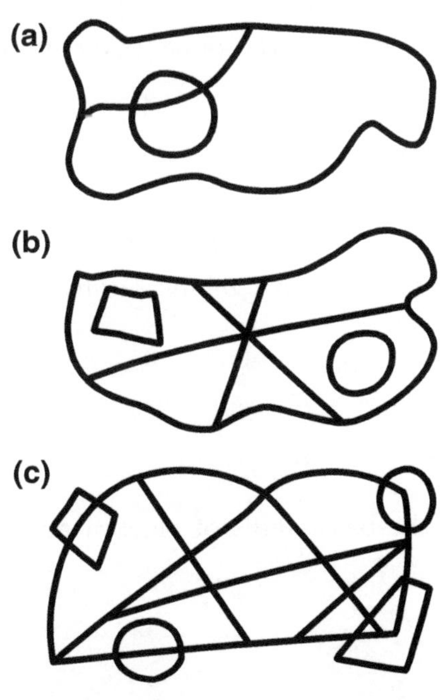

Postassessment

1. In a paragraph, using diagrams, describe what is meant by The Four-Color Problem of Topology.
2. Using the colors g/green, r/red, b/blue, y/yellow, show that it is possible to correctly color each of the following maps with four or fewer colors.
3. Draw a map that has an infinite number of regions but requires only two colors for correct coloring.

Reference

Appel, K. and W. Haken. The solution of the four-color-map problem. *Scientific American* **237**(4), 108–121, 1977.

Unit 28

Mathematics on a Bicycle

With the many variations of gears on the traditional ten-speed bicycle, there are lots of applications of mathematics. These ought to help students better understand their bicycles while at the same time reinforce their mathematics.

Performance Objectives

1. *Given the number of teeth (or sprockets) in the front and rear sprocket wheels, and diameter of the wheel, students will find gear ratios and distance traveled with each turn of the pedals. (New vocabulary will be developed.)*
2. *Students will be able to explain why pitch is important.*

Preassessment

Students should have the basic skills of algebra and be somewhat familiar with a bicycle.

Teaching Strategies

The adult bicycles that we shall consider have two wheels, front and rear cable brakes, gears — three, five, or ten — and are made of steel in its various alloys.

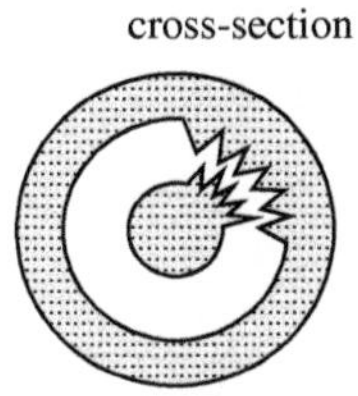

Let's examine first the differences in gearing between the three- and ten-speed bicycles, and in particular the mechanism of the ten-speed bicycle.

In a three-speed bicycle, the gearing mechanism is located within the rear hub (or axle). It is a clutch-type mechanism with pieces that interlock within the hub. It has constraints in that no ratio greater than the inside diameter of the rear hub can exist.

On a ten-speed bicycle, the back wheel has five sprocket wheels called a cluster, with the largest sprocket wheel closest to the spokes and then the rest gradually getting smaller. The gearing (i.e., the connection of sprocket wheels by a chain) is obtained by moving the chain from one sprocket to the other by means of a derailleur.

Lets examine closely the basic setup.

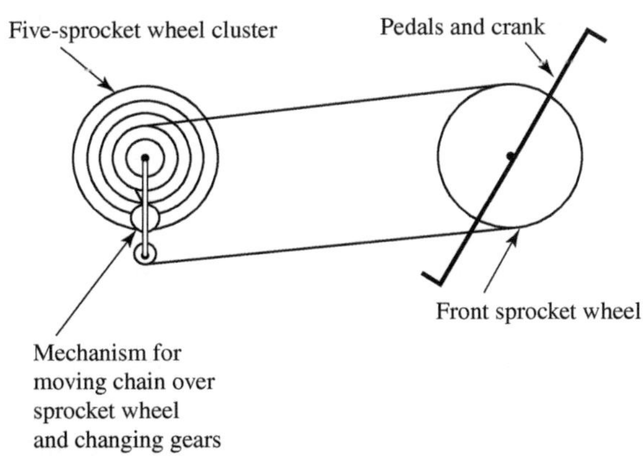

There exists a front and rear sprocket wheel with teeth set in gear by a connecting chain. The numbers of teeth on the front and rear sprocket wheels are important. Suppose the front sprocket wheel has 40 teeth and the rear sprocket wheel has 20 teeth; the ratio would then be $\frac{40}{20}$ or 2. This means that the rear sprocket wheel turns twice every time the front sprocket wheel turns once. But the rear sprocket wheel is attached to the bicycle wheel and a translation of energy again occurs depending on the diameter of the wheel. On a ten-speed bicycle, the diameter of the wheel (including the tire) is 27 inches. This arrangement is shown in the following diagram.

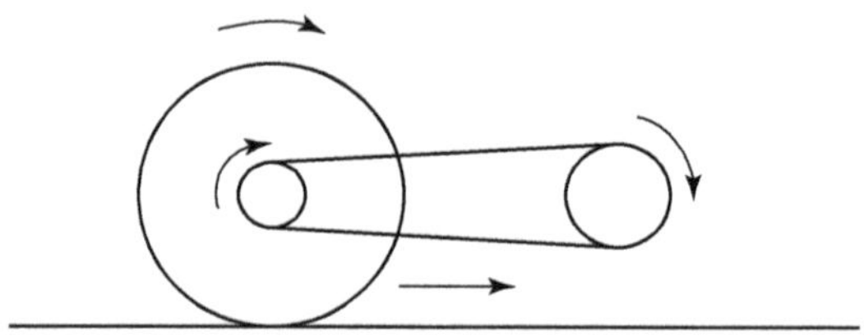

The relationship (when the bicycle wheel is included in the consideration) is: gear ratio $=$ ratio $\times$ diameter $= \frac{2}{1} = 27'' = 54$. The number generated here is usually between 36 and 108. It gives a comparison between gears and is useful for relating gear ratios to work performed.

For example, a rider using a sprocket wheel with 46 teeth in the front and a 16 tooth wheel in the rear along with a $27''$ wheel gets a rear ratio of $77.625 \approx 78$. Another rider using a 50-tooth front sprocket wheel and 16-tooth sprocket wheel in the rear gets a gear ratio of $84.375 \approx 84$, which would be harder to pedal than a 78 gear ratio.

Where does this extra difficulty in pedaling benefit the rider? If one multiplies the gear ratio obtained in the formula above by π, one gets the distance traveled forward with each turn of the pedals. Students should recall that Circumference $= \pi \times$ Diameter.

For example, the rider with the 78 gear ratio goes approximately 245 inches forward for each turn of the pedals, the rider with the 84 gear ratio goes approximately 264 inches forward for each complete turn of the pedals. Hence, the increase in work (increased difficulty in pedaling) is returned in greater distance per pedal revolution.

Now let us examine applications of various gearing ratio to the average rider.

Suppose Danny was riding comfortably on level ground in a 78 gear ratio and then he came to a rather steep hill. Should he switch to a higher or lower gear ratio?

Your reasoning should be as follows: If Danny switches to an 84 gear ratio, he will go 264 inches forward for each turn of the pedals. This requires a certain amount of work. To overcome the effects of gravity to get up the hill requires additional energy. So Danny would probably end up walking his bicycle. If Danny had switched to a lower gear, he would use less energy to turn the pedals and the additional energy required to climb the hill would make his gearing feel about the same as the 78 gear ratio. So the answer is to switch to a lower number gear ratio.

Danny will have to turn the pedal more revolutions to scale the hill; more than if he had chosen the 84 gear ratio and more than if he stayed in the 78 gear ratio. Remember, his gearing only feels like 78 because of the hill. This is the "trade-off" Danny made: more revolutions at a constant torque (angular force) instead of the same number of revolutions per distance with varying force.

The benefit of this trade-off is understood by comparing the human body to the engine. An engine works most effectively at a constant torque than a varying torque and compensates by changing gear ratios with changing speed and revolutions per minute.

A more concise description is given as follows. A car uses gears to overcome static friction and accelerate to operating speed while providing constant torque or less torque than overload to the engine. This is not similar to the bicycle because the human machine can overcome an increase in torque for the short period of time to accelerate the bicycle. When a bicycle is in motion, the only force needed to keep it in motion at constant speed on level ground is that required to overcome internal friction and wind resistance. This is the same for a car. Should a rider wish to accelerate quickly, he would want to turn the pedals as fast as possible. All machines (including the human machine) have optimum torquc capacity for this. There are two things that can happen to prevent the machine from reaching its maximum possible speed. First, if the torque is too high, it prevents rapid spin. This corresponds to a car in third gear trying to pass without a downshift. The engine doesn't have the power for rapid acceleration and can only accelerate slowly. The same is true for a rider trying to accelerate quickly in the harder gears; he is without the necessary power. Second is spin-out. This corresponds to when your car reaches 30 m.p.h. in first gear and cannot turn any faster even though there exists power for greater distance with each turn. An example is a manual shift car accelerating from a light in first without shifting. This compares on a bicycle to the rider turning the pedals as fast as possible but not at maximum force.

If the rider reaches maximum spin at maximum torque, he will reach maximum speed.

At this point you might want to have your students try some applications.

Model Problem. Max can turn a 68 gear ratio at 100 r.p.m. or a 72 gear ratio at 84 r.p.m. For maximum speed, which should Max choose? (These are the very considerations bicycle racers use in determining which gear to use final sprint.) Assume these speeds are constant for the duration of the sprint.

Solution. A 68 gear ratio times $\pi = 214$ inches per revolution (approximately). If Max spins 100 r.p.m., she is traveling at 21,400 inches/min. or 20.27 m.p.h.

A 72 gear ratio times $\pi = 226$ inches per revolution (approximately) at a rate of 84 r.p.m. would produce a speed of 18,984 inches/min. or 17.98 m.p.h. Therefore, Max would be better off sprinting in the 68 gear ratio.

As mentioned earlier, these torque and spin performance items are given careful attention by racers. A racer will carefully select his or her back sprocket wheel cluster depending on the course. A relatively flat course would necessitate a 13–18 tooth range in the rear sprocket wheel cluster with a 47 tooth inner front sprocket wheel and 50 tooth outer front sprocket wheel.

This is where the ten speeds come from. When the chain is on the 47 tooth sprocket wheel, there are five different gear ratios as the rear derailleur moves the chain through the five rear sprocket wheels. When the chain is on the 50 tooth sprocket wheel, there are again five different gear ratios.

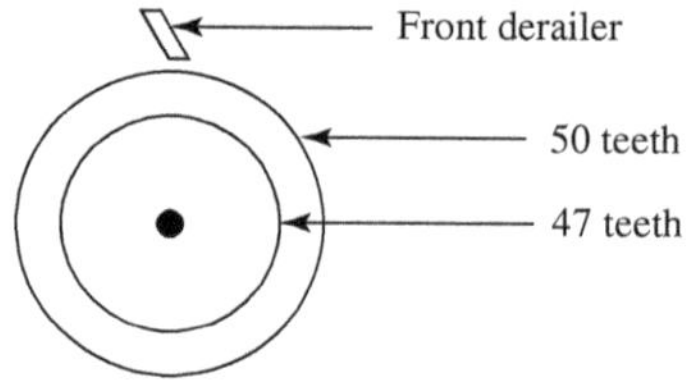

One other consideration a racer will make in selecting gears is inertia. You will notice that a 54 front sprocket wheel and 18-tooth rear sprocket wheel gives the same gear ratio as would a 48-tooth front sprocket wheel and a 16-tooth rear sprocket wheel; that is, $\frac{48}{16} = 3$. The rider will choose the $\frac{48}{16}$ because work is expended without return to accelerate through an angular acceleration a sprocket wheel of larger radius than a sprocket wheel of smaller radius due to inertia considerations. Since a 10″ radius sprocket wheel is the smallest to take the heavy shear forces, a 34-tooth sprocket wheel is the smallest available. We are currently using $\frac{1}{2}$ pitch (distance between teeth), an improvement over 1″ pitch to increase the number of ratios without letting sprocket wheels get too large. A well-made sprocket wheel would look like the following diagram, where most of the unnecessary mass is cut out.

Inertia $=$ M $\times$ distance from the axle of rotation squared. The smaller the distance, the smaller the inertia.

Thus, in selecting a ten-speed bicycle, remember that with each difference in price goes a difference in thought toward design, performance, and work required for riding.

As a final example, many inexpensive bicycles really only have 6–8 speeds because of duplication. Consider our previous example on inertia, where the choice was between a 48-tooth and 54-tooth front sprocket wheel. We saw duplication of the same gear ratio with a 16 and 18 rear sprocket wheel. This case occurs on many less expensive bicycles.

Postassessment

1. Lisa approaches a hill that raises whatever gear ratio she is in by 10. Lisa cannot pedal anything harder than a 62 gear ratio. If her three-speed bicycle has 48, 58, and 78 gear ratio, which should she use?
2. How far forward with each revolution of the pedals will a 78 gear ratio move a bicycle whose wheel radius is 27″?
3. Josh can spin a 72 gear ratio 80 r.p.m. and a 96 gear ratio 48 r.p.m. Which gives a greater velocity?

Reference

Posamentier, A. S. and C. Spreitzer, *The Mathematics of Everyday Life*, Amherst, New York: Prometheus Books, 2015.

Unit 29

Mathematics and Music

Students who are acquainted with operations on fractions but whose knowledge of music theory is limited will find a correlation between these fields.

Performance Objectives

1. *Students will demonstrate knowledge of certain formulas relating pitch of a note to properties of a string or an air column.*
2. *Students will know how to create Pythagoras' diatonic scale.*
3. *Students will show how Euclid proved that an octave is less than six whole tones.*

Preassessment

Obtain a stringed instrument such as a banjo, violin, or guitar. If these are unavailable, the science department can probably lend you a sonometer, which is a scientific instrument with strings used in experimentation.

Perform the following three demonstrations. In each case, have students determine whether the pitch becomes higher or lower.

1. Pluck a string, tighten it, then pluck it again.
2. Pluck a string. Then by pressing down on the middle of the string (fretting) cause only half the string to vibrate.
3. Using two strings of different diameters (thickness), pluck each one.

Teaching Strategies

Elicit from the students the following three facts:

1. As tension increases, pitch becomes higher.
2. As length decreases, pitch becomes higher.
3. As diameter decreases, pitch becomes higher.

At this point, explain that the above is grounded in mathematical formulas. However, these formulas use frequency, which is the number of vibrations of the string per second, rather than pitch. Since the pitch of a tone gets higher whenever the frequency increases, it will not really alter the formulas. They are

$$\underset{\left(\begin{array}{c}\text{strings are}\\ \text{same type}\end{array}\right)}{\frac{F_1^2}{F_2^2} = \frac{T_1}{T_2};} \qquad \underset{\left(\begin{array}{c}\text{tension}\\ \text{constant}\end{array}\right)}{\frac{F_1}{F_2} = \frac{L_2}{L_1};} \qquad \underset{\left(\begin{array}{c}\text{length and}\\ \text{tension constant}\end{array}\right)}{\frac{F_1}{F_2} = \sqrt{\frac{D_2}{D_1}}}$$

where F = frequency
T = tension
D = diameter of string

Have students try numerical examples: A string that vibrates at a frequency of 400 vps (vibrations per second) is 20 inches long. A second string of the same type is plucked in a similar manner. The tension being the same as in the first case.) If its frequency is 800 vps, how long is it? Have them solve $\frac{400}{800} = \frac{L_2}{20}$, concluding that the length of the second string is 10 inches. Another example could be the effect on the tension if the frequency of a string doubles. Elicit that the tension quadruples $\left(\frac{1^2}{2^2} = \frac{1}{4}\right)$.

Music and mathematics are also related to the creation of a scale. Pythagoras, familiar to most students for his work with the right triangle, produced a scale that could make beautiful melodies, but limited the combination of tones possible and the use of harmony.

Pythagoras felt that those tones, which were particularly pleasing, or *consonant*, were related to the numbers 1, 2, 3, and 4. He took several strings of the same length, letting the note C be the fundamental tone. If the sonometer is used, the teacher can demonstrate the basics of what Pythagoras did. This means that the string vibrates as a whole (see Figure 1(a)). To obtain the note C an octave higher, the string must vibrate in two parts, i.e., have twice the frequency (see Figure 1(b)). One can also accomplish the same thing by dividing a string into two parts of the ratio 1:2 (see Figure 1(c)).

In Diagram 3, vibrating $\overline{AD}$ and $\overline{DB}$ separately will have the same effect of producing two notes an octave apart. Thus, if C corresponds to the number 1, then C an octave higher would correspond to the number $\frac{2}{1}$ or 2. Pythagoras also added notes F and G corresponding to $\frac{4}{3}$ and $\frac{3}{2}$, respectively.

In Diagram 3, vibrating $\overline{AD}$ and $\overline{DB}$ separately will have the same effect of producing two notes an octave apart. Thus, if C corresponds to the number 1, then C an octave higher would correspond to the number $\frac{2}{1}$ or 2. Pythagoras also added notes F and G corresponding to $\frac{4}{3}$ and $\frac{3}{2}$, respectively.

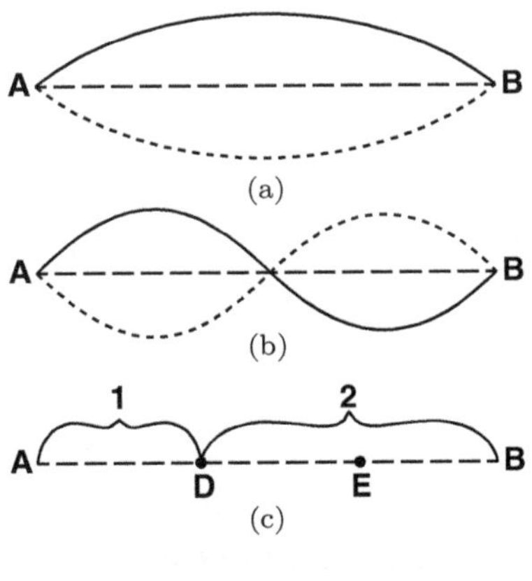

Figure 1.

Ask the students how the string can be divided into the ratio 3:2. Elicit that the string can be divided into five parts to obtain the result.

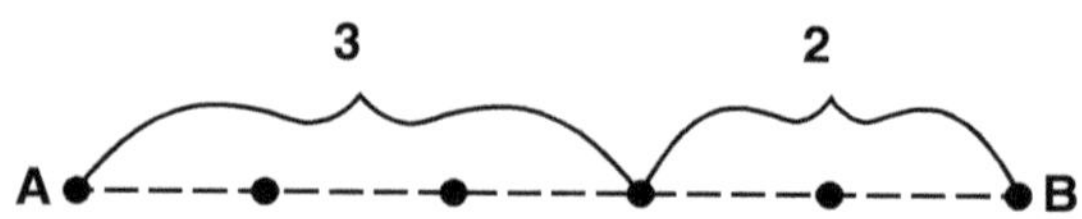

This is one reason why this note can be called a perfect fifth. To obtain a note corresponding to $\frac{4}{3}$, the string should be divided into seven parts as shown in Figure.

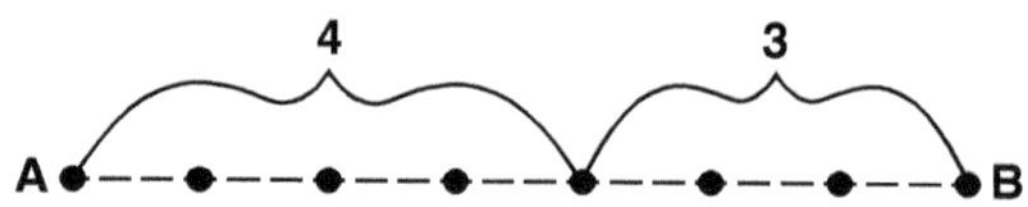

However, this note is called a fourth and not a seventh.

Pythagoras added to his scale $\frac{3}{2}$G or $\frac{3}{2} \cdot \frac{3}{2}\text{C} = \frac{9}{4}\text{C}$.

Since C $= 1$ and the octave C $= 2$, $\frac{9}{4}\text{C} = 2\frac{1}{4}\text{C}$ would not fit between these notes.

Challenge the students to find a tone "basically" the same that would fit between C and its octave. Have them recall that doubling or halving a frequency only changes a tone by an octave. Thus, instead of $\frac{9}{4}$, Pythagoras used $\frac{1}{2}$ of $\frac{9}{4}$ or $\frac{9}{8}$. By adding the third harmonic of each successive tone (i.e., multiplying by $\frac{3}{2}$), students will be able to obtain tones whose relative frequencies are $1, \frac{3}{2}, \frac{9}{8}, \frac{27}{16}, \frac{81}{64}, \frac{243}{128}$. Of course, some additional "halving" was done needed as in the case of $\frac{9}{8}$. Pythagoras' diatonic scale is thus obtained.

C	D	E	F	G	A	B	C	
1	$\frac{9}{8}$	$\frac{81}{64}$	$\frac{4}{3}$	$\frac{3}{2}$	$\frac{27}{16}$	$\frac{243}{128}$	2	(Relative Frequencies)

G, which occupies the fifth position, is considered a perfect fifth. This occurs whenever the ratio of the fifth to the first is $\frac{3}{2}$. Throughout the discussion, the fact that frequencies are in the same ratio as their lengths should be stressed.

Have students study the scale in its new form. Elicit that there is a constant ratio of $\frac{9}{8}$ between notes (except between E and F and between B and C where the ratio is $\frac{256}{243}$). It should also be noted that $\frac{9}{8}$ corresponds to a whole tone (W) while the other is called a semitone (S). Thus, the pattern obtained is as follows:

C		D		E		F		G		A		B		C
1	$\downarrow$	$\frac{9}{8}$	$\downarrow$	$\frac{81}{64}$	$\downarrow$	$\frac{4}{3}$	$\downarrow$	$\frac{3}{2}$	$\downarrow$	$\frac{27}{16}$	$\downarrow$	$\frac{243}{128}$	$\downarrow$	2
	W		W		S		W		W		W		S	

This is called a major scale.

However, there is some difficulty with harmony. When one sounds a tone on a musical instrument, it not only vibrates in one piece making the fundamental tone, but also in parts creating tones called overtones. The overtones have 2, 3, 4, and 5 times the frequency of the fundamental. The fifth overtone corresponds to 5 or $\frac{5}{4}$ if it is to be placed between 1 and 2 (recall continuous halving like $\frac{1}{2} \cdot \frac{1}{2}$ creates a similar tone). The closest tone on the Pythagorean scale is E of frequency $\frac{81}{64}$. When a C is played and then followed by an E, the ear expects to hear the same E just heard as an overtone of C. However, to the individual, the Pythagorean E can be quite disturbing. This disturbance is due to the fact that the two E's involved have only slightly different frequencies, one being $\frac{81}{64}$ and the other $\frac{5}{4}$ or $\frac{80}{64}$.

Postassessment

1. If the tension is constant and the length is increased, how is a string's pitch affected?
2. How does string tightening affect pitch?
3. Suppose C corresponds to $\frac{4}{5}$ instead of 1 in Pythagoras' scale. Find the relative frequencies of the next 8 notes of this major scale.

Reference

Posamentier, A. S. *The Pythagorean Theorem.* Amhearst, NY: Prometheus Books, 2010.

Unit 30

Mathematics in Nature

Performance Objective

Students will identify and explain where mathematics is found in nature in at least one situation.

Preassessment

A famous sequence of numbers (*The Fibonacci Numbers*) was the direct result of a problem posed by Leonardo of Pisa in his book *Liber Abaci* (1202) regarding the regeneration of rabbits. A brief review of this problem indicates that the total number of pairs of rabbits existing each month determined the sequence: $1, 1, 2, 3, 5, 8, 13, 21, 34, 55, 89, \ldots$

Fibonacci numbers have many interesting properties and have been found to occur in nature.

Teaching Strategies

Have students divide each number in the Fibonacci sequence by its right-side partner to see what sequence develops. They will get a series of fractions:

$$\frac{1}{1}, \frac{1}{2}, \frac{2}{3}, \frac{3}{5}, \frac{8}{13}, \frac{13}{21}, \frac{21}{34}, \frac{34}{55}, \frac{55}{89}, \ldots$$

Ask students if they can determine a relationship between these numbers and the leaves of a plant (have a plant on hand). From the standpoint of Fibonacci numbers, one may observe two items: (1) the number of leaves it takes to go (rotating about the stem) from any given leaf to the next one similarly placed (i.e., above it and in the same direction) on the stem; and (2) the number of revolutions as one follows the leaves in going from one leaf to another one similarly placed. In both cases, these numbers turn out to be the Fibonacci numbers.

In the case of leaf arrangement, the following notation is used: $\frac{3}{8}$ means that it takes three revolutions and eight leaves to arrive at the next leaf similarly placed. In general, if we let r equal the number of revolutions, and s equal the number of leaves, it takes to go from any given leaf to one similarly placed, then $\frac{r}{s}$ will be the *phyllotaxis* (the arrangement of leaves in plants). Have students look at Figure 1 and try to find the plant ratio. Draw a diagram on the board, and if possible, provide a live plant.

In this figure, the plant ratio is $\frac{5}{8}$.

The pine cone also presents a Fibonacci application. The bracts on the cone are considered to be modified leaves compressed into smaller space. Upon observation of the cone, one can notice two spirals, one to the left (clockwise) and the other to the right (counterclockwise). One spiral increases at a sharp angle, while the other spiral increases more gradually.

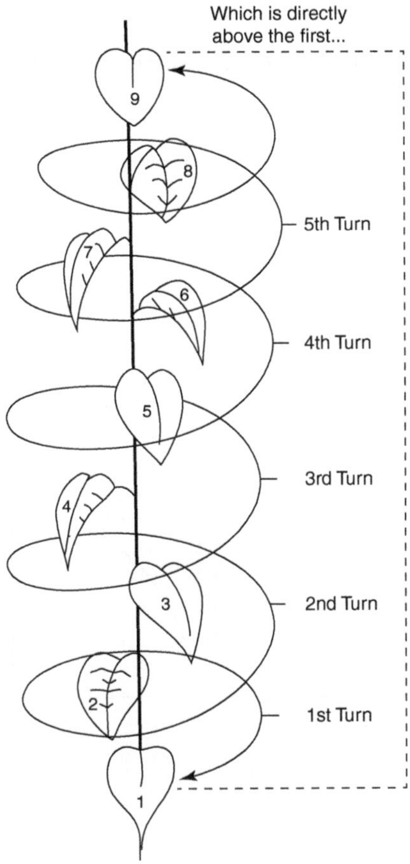

Figure 1.

Have students consider the steep spirals and count them as well as the spirals that increase gradually. Both numbers should be Fibonacci numbers. For example, a white pine cone has five clockwise spirals and eight counterclockwise spirals. Other pine cones may have different Fibonacci ratios. Later, have students examine the daisy to see where the Fibonacci ratios apply to it.

If we look closely at the ratios of consecutive Fibonacci numbers, we can approximate their decimal equivalents. Some are

1. $\frac{2}{3} = .666667$ 2. $\frac{3}{5} = .600000$

3. $\frac{89}{144} = .618056$ 4. $\frac{144}{233} = .618026$

Continuing in this manner, we approach what is known as the *golden ratio.* Point B in Figure 2 divides line $\overleftrightarrow{AC}$ into the golden ratio, $\frac{AB}{BC} = \frac{BC}{AC} \approx$ .618034.

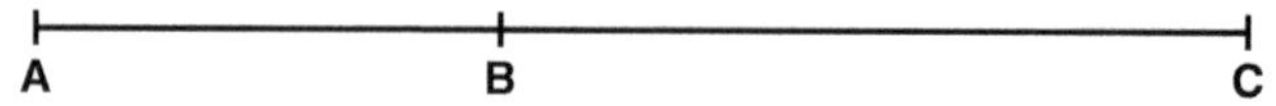

Figure 2.

Now consider the series of golden rectangles (Figures 3a and 3b), those whose dimensions are chosen so that the ratio of $\frac{\text{width}}{\text{length}}$ is the golden ratio $\left(\text{i.e., } \frac{W}{L} = \frac{L}{W+L}\right)$.

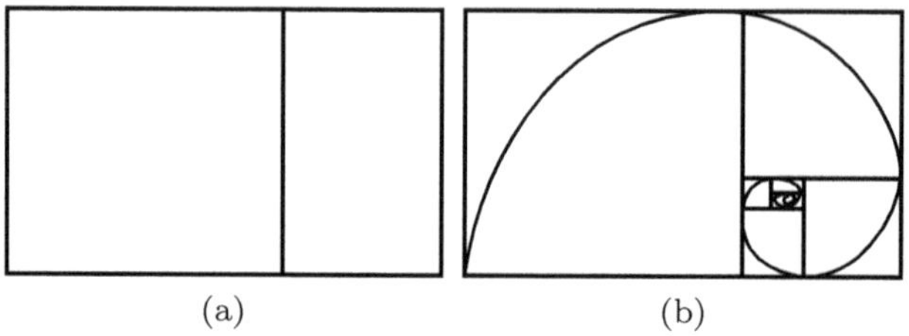

Figure 3.

If the rectangle (Figure 3a) is divided by a line into a square and a golden rectangle, and if we keep partitioning each new golden rectangle in the same way, we can construct a logarithmic spiral in the successive squares (Figure 3b). This type of curve is frequently found in the arrangements of seeds in flowers or in the shapes of seashells and snails. Bring in illustrations to show these spirals (Figure 4).

Figure 4.

For another example of mathematics in nature, students should consider the pineapple. Here there are three distinct spirals of hexagons: a group of *five* spirals winding gradually in one direction, a second group of 13 spirals winding more steeply in the same direction, and a third group of *eight* spirals winding in the opposite direction. Each group of spirals consists of a Fibonacci number. Each pair of spirals interacts to give Fibonacci numbers. Figure 5 shows a representation of the pineapple with the scales numbered in order. This order is determined by the distance (relative) each hexagon is from the bottom. That is, the lowest is numbered 0 and the next higher one is numbered 1. Note hexagon 42 is slightly higher than hexagon 37.

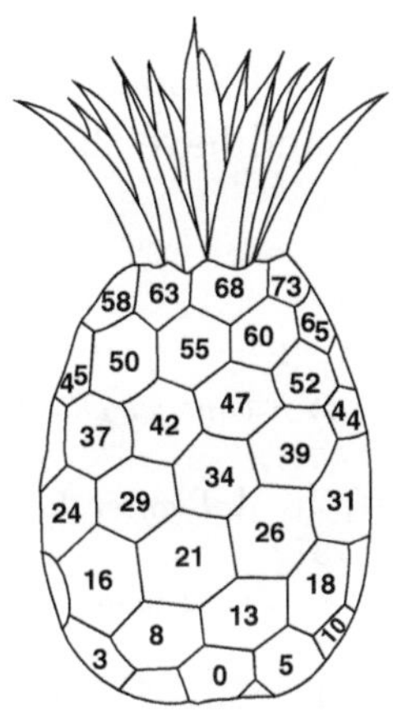

Figure 5.

See if students can note three distinct sets of spirals in Figure 5 that cross each other, starting at the bottom. One spiral is the $0, 5, 10$, etc., sequence, which increases at a slight angle. The second spiral is the $0, 13, 26$, etc., sequence, which increases at a steeper angle. The third spiral has the $0, 8, 16$, etc., sequence, which lies in the opposite direction from the other two. Have students figure out the common difference between the numbers in each sequence. In this case, the differences are $5, 8, 13$, all of which are Fibonacci numbers. Different pineapples may have different sequences.

In concluding this topic, consider briefly the regeneration of male bees. Male bees hatch from unfertilized eggs; female bees from fertilized eggs. The teacher should guide students in tracing the regeneration of the male bees. The following pattern develops:

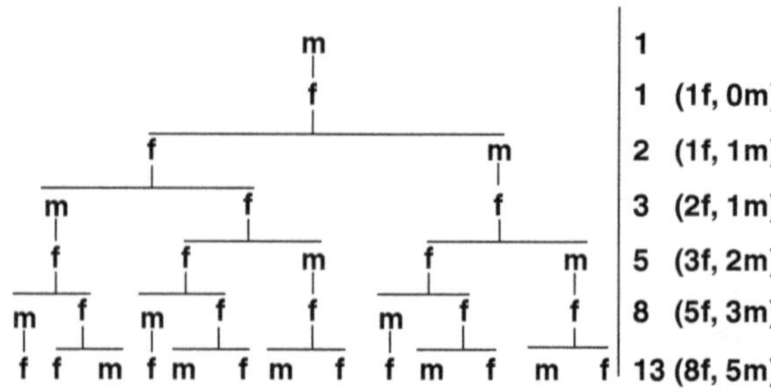

It should be obvious by now that this pattern is the Fibonacci sequence.

Postassessment

1. Ask students to explain two distinct ways mathematics manifests itself in nature.
2. Have students find examples of Fibonacci numbers in nature (other than those presented in this unit) and have them explain the manner in which the sequence is used.

References

Brother, U. Alfired. *An Introduction to Fibonacci Discovery.* San Jose, CA: The Fibonacci Association, 1965.

Bicknell, M. and Verner E. Hoggatt, Jr. *A Primer for the Fibonacci Numbers.* San Jose, CA: The Fibonacci Association, 1972.

Dunlap, Richard A. *The Golden Ratio and Fibonacci Numbers.* River Edge, NJ: World Scientific Publishing Co., 1997.

Posamentier, Alfred S. and Ingmar Lehmann. *The Fabulous Fibonacci Numbers.* Amherst, NY: Prometheus Books, 2007.

Unit 31

The Birthday Problem

Students are fascinated by problems that involve surprise or unpredictable outcomes. The "birthday problem" will engage them in the study of mathematical probability.

Performance Objective

In problems involving sequences of successive events, such as indications of birthdays, tossing of coins, drawing of cards, throwing of dice, students will calculate the probability that a specified outcome (a) occurs at least once, (b) fails to occur at all.

Preassessment

Ask the class what they think the probability is of two students in the class sharing the same birthday. The students will respond that the chances of this being true are remote. Surprise them by telling them that, in a class of 30 students, the probability of at least two students having the same birthday is approximately 0.68 (a probability of 1.00 indicates an absolute certainty). In a class of 35 students, this probability rises to about 0.80. Restate these probabilities in the language of "odds." Point out that the odds in favor of the desired outcome in the first instance are better than two to one and in the second about four to one. Copy and distribute a list of the 43 presidents of the United States, with the dates of their birth and death next to their names. Give students time to look for any dates the presidents had in common (two, Polk and Harding, were born on November 2; two, Fillmore and Taft, died on March 8; and three, Adams, Jefferson, and Monroe, died on July 4). Now take a class survey to determine if any students share the same birthday. If they do, this fact will reinforce the probability figures and will help convince them of the statistical plausibility. If they do not, indicate that no claim of absolute certainty was made.

Teaching Strategies

Review the following fundamental principles the students will need to know. Mathematical probabilities are stated as decimals between 0.00 and 1.00, and a probability of 0 (zero) means that a particular outcome is impossible, while a probability of 1 (one) means that a particular outcome is a certainty.

Each of the principles enumerated here may be illustrated by simple examples in the tossing of coins, the throwing of dice, the drawing of cards, etc. For example, the probability of throwing a total of 13 with a pair of ordinary dice is equal to zero, while the probability of throwing some number between 2 and 12 inclusive is equal to one. The probability of a desired outcome occurring can be calculated by forming a fraction whose numerator represents the number of "acceptable" or "successful" outcomes and whose denominator represents the sum of the "successful" outcomes and the "unsuccessful" outcomes or "failures." Symbolically, $P = \frac{S}{S+F}$ or $P = \frac{S}{T}$. P represents the probability of a particular event occurring, S the number of successful outcomes, F the number of failures, and T the total number of outcomes possible. Either form of this fraction may be converted to a decimal between zero and one, since the numerator may never exceed the denominator.

The students should also note that the probability of a desired outcome *failing* to occur would be equal to $\frac{F}{S+F}$. Since $\frac{S}{S+F} + \frac{F}{S+F} = \frac{S+F}{S+F} = 1$, it follows that $\frac{S}{S+F} = 1 - \frac{F}{S+F}$. The students should now state in words that the probability of a desired outcome *occurring* is equal numerically to 1.00 minus the probability of this outcome failing to occur. This statement enables them to complete the lesson.

Students should be familiar with a fundamental theorem of probability and it is presented here without proof: If the probability of an event is P_1, and if, after it has happened, the probability of a second event is P_2, then the probability that both events will happen is P_1P_2. Point out that this principle may be generalized to calculate the probability that a sequence of n events will occur, given that each preceding event has occurred, and that the result would be $P_1P_2P_3P_4 \ldots P_n$. For example, the students, by performing the following activities, will realize that the probability of drawing a spade from an ordinary deck of 52 cards is $\frac{13}{52}$ or 0.25 and the probability of *not* drawing a spade is $\frac{39}{52}$ or 0.75; where both probabilities refer to the drawing of a single card from the deck, they should note that $0.25 = 1.00 - 0.75$. The probability of tossing a head, followed by a tail, followed by another tail, where a "true" coin is tossed three times in succession, is $\frac{1}{2} \cdot \frac{1}{2} \cdot \frac{1}{2}$ or $\frac{1}{8}$, illustrating the use of the fundamental principle for dealing with successive events.

Back to the birthday problem. You might point out that it will be simpler to calculate the probability that *no* students in the class share the same birthday and then to subtract this result from 1.00, than to directly calculate the probability that at least two students in the class have the same birthday. Help the students to formulate the following representation of the probability that *no* students in the class share the same birthday:

$$\frac{365}{365} \cdot \frac{364}{365} \cdot \frac{363}{365} \cdot \frac{362}{365} \cdot \frac{361}{365} \cdot \frac{360}{365} \cdot \frac{359}{365} \cdot \frac{358}{365} \cdot \frac{357}{365} \cdot \frac{356}{365} \cdots .$$

There will be as many fractions in this product as there are students in the class. Note that this formulation is based on an ordinary year of 365 days. If any of your students has a February 29 birthday, use denominators of 366 and have the first fraction read $\frac{366}{366}$.

Explain that these fractions represent the probabilities that students questioned in sequence as to their birthdays would *not* name a day already mentioned by a preceding student. Point out the fundamental principle for calculating the probability of successive events — sequential questioning. They will be interested in learning how they can best perform the sequence of multiplications and divisions — the simplest approach is through the use of calculators.

The students will discover that the value of their product has decreased to about 0.32 when the number of factors has reached 30, and to about 0.20 when the number of factors has reached 35. Since these figures represent the probabilities that no students in the class shared the same birthday, they represent "failures," in the terminology of this problem. Using the principle of subtraction from 1.00, already mentioned, to arrive at the probabilities of "successes" — the probabilities that at least two students in the class share the same birthday — we arrive at 0.68 or 0.80 or some other high decimal, depending on the size of the class. When the number of people in a group reaches just 55, the probability of finding at least two with the same birthday reaches the astonishing value of 0.99!

Postassessment

Students who have met the performance objective will be able to answer these or similar probabilities correctly:

1. Represent the probability that, given a group of 15 people, at least two share the same birthday.

2. If a coin is tossed into the air five times, what is the probability that (a) none of the tosses will turn up heads and (b) at least one of the tosses will turn up heads?
3. If a card is drawn from an ordinary deck of 52 cards, examined and replaced, and if this is repeated four times, what is the probability that (a) at least one of the cards drawn is a spade, and (b) none of the cards drawn is an ace?

Unit 32

The Structure of the Number System

Performance Objective

Given any number, the student will identify it as belonging to the set of natural numbers, integers, rational numbers, real numbers, or complex numbers. The student will also convert any decimal representing a rational number to its equivalent fractional form and vice versa.

Preassessment

Assess students' ability with the following pretest, assuring them, of course, that this is a trial test and they will not be graded on it.

1. Identify the following numbers as belonging to the set of natural numbers, integers, rational numbers, real numbers, or complex numbers (name the "smallest" possible set in each case): -3, $\frac{5}{3}$, 17, $\sqrt{2}$, 3.14, $\frac{22}{7}$, $\sqrt{-9}$, $.\overline{4}$, $0.2133333\ldots, 2.71828\ldots, 0.121121112\ldots, .\overline{15}, -\frac{1}{4}, -\sqrt{16}$, etc.
2. Convert each of the following fractions to decimals: $\frac{3}{8}, \frac{7}{5}, \frac{2}{3}, \frac{7}{9}, \frac{5}{11}, \frac{5}{12}$.
3. Convert each of the following decimals to fractions: $0.875, .\overline{8}, .272727\ldots, 0.8333333\ldots$.

Students who do well on the pretest have already attained the performance objective. Give them a different assignment while you present this lesson to the rest of the class.

Teaching Strategies

Ask your students to solve a simple linear equation such as $3x + 5 = 11$. When the correct solution, $x = 2$, is offered, ask what kind of number this is. Students may use such terms as *whole number* or *positive number* in their responses. Explain that 2 is a *counting* number and that the set of counting numbers is known mathematically as the set of natural numbers. Elicit other illustrations of natural numbers and have the students describe the set by roster: $\text{N} = \{1, 2, 3, \ldots\}$. They should note that the elements of this set are ordered and infinite in number, and that the set possesses a first or smallest element, the number 1.

Proceed in a similar fashion to develop the concept of the set of integers. Modify the equation just explored by reversing the two constants: $3x + 11 = 5$. When they obtain $x = -2$ as the solution, the students will volunteer that this is a negative whole number or some other similar description. Introduce the term *integer* if it is not mentioned by the class. They will readily understand that this term is synonymous with the term *whole number* and that the natural numbers just studied are a subset of the set of integers. This can be illustrated by a Venn diagram that, at this stage of the lesson, consists of an inner circle labeled N to represent the set of natural numbers and an outer circle labeled I to represent the set of integers. This diagram will be built up as the lesson develops by the addition of three more circles, each completely encircling all of the preceding circles. Elicit several illustrations of integers and help your students to describe this set by roster: $\text{I} = \{\ldots, -3, -2, -1, 0, 1, 2, 3, \ldots\}$. They will note this set is also an infinite one and that it is ordered, but that it possesses *no* first element.

Now offer the equation $2x + 1 = 6$. When the answer, $\frac{5}{2}$, is forthcoming, students will recognize that this number is *not* a member of the set of natural numbers or the set of integers since it is not a whole number but, rather, a fraction. Point out that such numbers are formed by setting up the ratio of two integers, $\frac{a}{b}$, where the denominator, b, is not equal to zero (ask your students why not!). The term *rational number* derives naturally from the word *ratio.* Add the third circle to your Venn diagram, completely enclosing the previous two circles. Label the new circle Q for quotient.

Elicit numerous illustrations of rational numbers, including proper and improper fractions, both positive and negative. Students should be aware that the set of rational numbers is infinite and ordered, but that no roster can be prepared for this set. You might explain that the rational numbers are "everywhere dense" and that an infinitude of rationals exists between any two rational numbers.

We are now ready to examine the decimals. Students will generally exhibit some uncertainty as to which of these represent rational numbers. We must consider terminating decimals, nonterminating but repeating decimals, and nonterminating and nonrepeating decimals. Your students can obtain some clues by converting a few fractions such as $\frac{1}{8}, \frac{5}{9}$, and $\frac{1}{6}$ to decimals by dividing their numerators by their denominators. They will observe that the result in every case is either a terminating decimal or a nonterminating but repeating decimal. Students can show easily that every terminating decimal represents a rational number by simply writing each one as a decimal fraction.

Next, introduce nonterminating decimals. Students who believe that $0.\overline{3}$ represents a rational number can be challenged to write it in fractional form. Some may recognize this decimal as being equal to $\frac{1}{3}$. If so, challenge them with the decimal $0.\overline{5}$, which they obtained earlier themselves by converting $\frac{5}{9}$ to a decimal, or with $0.16666\ldots$ which they also obtained themselves by converting $\frac{1}{6}$ to a decimal. Ask them whether they could convert these decimals to fractions if they did *not* know the answers! Or, challenge them with the decimal $0.\overline{13}$, for which it is unlikely that they will know the result. If they need your help to convert such decimals to fractions, two illustrations will make the technique clear.

$$\begin{aligned}
\mathrm{N} &= 0.131313\cdots \\
&\text{Multiply by 100:} \\
100\,\mathrm{N} &= 13.131313\ldots \\
\mathrm{N} &= 0.131313\ldots \\
\hline
99\,\mathrm{N} &= 13 \text{ by subtraction} \\
\mathrm{N} &= \frac{13}{99} \\
\mathrm{N} &= .1666666\ldots \\
&\text{Multiply by 10:} \\
10\,\mathrm{N} &= 1.666666\cdots \\
\mathrm{N} &= 0.166666\cdots \\
\hline
9\,\mathrm{N} &= 1.5 \text{ by subtraction} \\
90\,\mathrm{N} &= 15 \\
\mathrm{N} &= \frac{15}{90} = \frac{1}{6}
\end{aligned}$$

Provide several illustrations for your students, including some with nonrepeating portions in the decimal before the repetend appears, as in the second example above. Help them to grasp the fact that such decimals represent rational numbers, even though their nonrepeating portions may be lengthy, as long as they are finite in length *and* they are followed by repetends of one or more digits.

Your students are now ready to consider the nonterminating, nonrepeating decimals. They are already familiar with some of these, notably $\pi = 3.14159\ldots$ and perhaps the square roots of some of the nonperfect squares. (Be certain they understand that such numbers as $\frac{22}{7}$ and 3.14 or 3.1416 are only *rational approximations* to the *irrational* number π.) Propose the equation $x^2 + 2 = 7$. Those who are familiar with the square root algorithm may be asked to work out $\sqrt{5}$ to a few decimal places to determine whether a pattern of repetition appears. They will discover, of course, that it does not, since $\sqrt{5} = 2.236\ldots$ and is irrational.

Students unfamiliar with the algorithm may refer to a table of square roots. They will discover that the only square roots that contain a repetand in their decimal representations are those of perfect squares. All other square roots are irrational numbers, since they are nonterminating, nonrepeating decimals. You may also wish to generalize this result to the nth roots of nonperfect nth powers.

Explain to your students that the set of the rational numbers together with the set of irrational numbers form the set of real numbers. Add a fourth circle to your Venn diagram, labeling it R, and have it completely enclose the three circles previously drawn. The students should realize that the sets of natural numbers, integers, and rational numbers are each proper subsets of the set of real numbers.

This development of the structure of the number system may be concluded with a brief treatment of the complex numbers. Ask your students to try to solve the equation $x^2 + 4 = 0$. Help them to see why answers such as $+2$ and -2 are incorrect. They should soon realize that no real number squared can equal -4, or any other negative number, for that matter. Explain that numbers that are not real are called "imaginary" and that the imaginary numbers and the real numbers together form the set known as the "complex numbers." You may wish to introduce the symbol $i = \sqrt{-1}$ so they can write a solution for their equation as $+2i$ and $-2i$. Complete the Venn diagram with the fifth and last circle, completely enclosing the other four circles, and showing the real numbers as a proper subset of the complex numbers, C.

Postassessment

Give students a test similar to the pretest. Compare each student's pre- and posttest answers to measure progress.

Unit 33

Excursions in Number Bases

Students learn early in their school careers that the base used in our everyday number system (the "decimal system") is the number 10. Later, they discover that other numbers can serve as bases for number systems. For example, numbers written in base 2 (the "binary system") are used extensively in computer work. This lesson will explore a variety of problems involving numbers written in many different positive integral bases.

Performance Objective

Students will solve a variety of numerical and algebraic problems involving numbers expressed as numerals in any positive integral base $b, b \geq 2$.

Preassessment

How far you go in this lesson will depend somewhat on the algebraic background of the class. Question students or evaluate their previous work to determine how well they understand the idea of place value in writing numerals, the meaning of zero and negative exponents, and the techniques for solving quadratic and higher degree equations.

Teaching Strategies

Review briefly the fact that decimal numerals are written by using a system of place values. Point out, for example, that in the numeral 356 the digit 3 represents 300 rather than merely a 3, the digit 5 represents 50 rather than merely a 5, and the digit 6 is a units digit and really does represent a 6. Briefly, $356 = 300 + 50 + 6 = 3(100) + 5(10) + 6 = 3(10)^2 + 5(10)^2 + 6(10)^0$. Likewise, $3,107 = 3(10)^3 + 1(10)^2 + 0(10)^1 + 7(10)^0$. Ask students for further illustrations. If necessary, review or teach at this point the meaning of the zero exponent and also of negative exponents, since these will be used later.

Explain that the use of the number 10 as a base is somewhat arbitrary, and students should note that other numbers can be used as bases. If the number 2 is used as a base, numbers are expressed as sums of powers of 2 rather than as sums of positive integral multiples of powers of 10, and the only digits used to represent numerals are 0 and 1. For example, the number 356 considered above is equal to $256+64+32+4 = 2^8+2^5+2^5+2^2 = 1(2)^8+0(2)^7+1(2)^6+1(2)^5+0(2)^4+0(2)^3+1(2)^2+0(2)^1+0(2)^0 = 101100100_{\text{two}}$, the subscript indicating the base. In base 3 (where the digits used to represent the numerals are 0,1, and 2) $356 = 243+81+27+3+2(1) = 1(3)^5+1(3)^4+1(3)^3+0(3)^2+1(3)^1+2(3)^0 = 111012_{\text{three}}$. In base 5 (where the digits used are $0, 1, 2, 3,$ and 4) $356 = 2(125) + 4(25) + 1(5) + 1(5) = 2(5)^3 + 4(5)^2+1(5)^1 + 1(5)^0 = 2411_{\text{five}}$. Subscripts should be written in words rather than in numerals to avoid any possible confusion. The class should note that when numerals are in base b, the only digits available for such representations are those from zero to $b-1$, and that if the value of b is greater than 10, new digits must be created to represent the numerals $10, 11, 12$, etc. Remind the class that numerals such as 2411_{five} should be read "two, four, one, one, base 5." Provide practice in writing and reading whole numbers in the numerals of bases other than base 10, according to class needs.

Next consider numbers other than integers. Help your students to see that 12.2_{ten} means $1(10)^1 + 2(10)^0 + 2(10)^{-1}$, since $10^0 = 1$ and $10^{-1} = \frac{1}{10}$, and that this number can be represented in the numerals of other bases just as integers can. For example, in base 5 we have $12.2_{\text{ten}} = 2(5)^1 + 2(5)^0 + 1(5)^{-1}$, since $\frac{1}{5} = \frac{2}{10}$, so $12.2_{\text{ten}} = 22.1_{\text{five}}$. Illustrate further with such problems as the conversion of 7.5_{ten} to base 2: $7.5_{\text{ten}} = 1(2)^2 + 1(2)^1 + 1(2)^0 + 1(2)^{-1} = 111.1_{\text{two}}$. Decimal numerals whose decimal parts are $.5\left(\frac{1}{2}\right), 25\left(\frac{1}{4}\right), 75\left(\frac{3}{4}\right), 125\left(\frac{1}{8}\right)$, etc., can be easily converted to base 2 numerals. For example, $8.75_{\text{ten}} = 1(2)^3 + 1(2)^{-1} + 1(2)^{-2}$, since $.75 = \frac{3}{4} = \frac{1}{2} + \frac{1}{4} = \frac{1}{21} + \frac{1}{22} = 2^{-1} + 2^{-2}$, so $8.75_{\text{ten}} = 1000.11_{\text{two}}$. Numbers can also be converted from numeral representations in one base to equivalent numeral representations in another base, where neither base is equal to 10. For example, 12.2_{four} can be represented in base 6 numerals as follows: $12.2_{\text{four}} = 1(4)^1+2(4)^0+2(4)^{-1} = 4+2+\frac{2}{4} = 6+\frac{3}{6} = 1(6)^1+0(6)^0+3(6)^{-1} = 10.3_{\text{six}}$ · In base 10, this is the numeral 6.5. Provide practice with these types of numerical problems according to the interests and abilities of your students.

The class is ready to consider algebraic problems next. Offer the following challenge: "In a certain base b, the number 52 is double the number 25. Find the value of b." The students should note that 52 (read "five, two") really represents the expression $5b + 2$, since $52_b = 5(b)^1 + 2(b)^0$. Accordingly, the problem states that $5b + 2 = 2(2b + 5)$. Solve for b to get $b = 8$. Checking shows that 52 eight $= 5(8) + 2 = 42_{\text{ten}}$ and $25_{\text{eight}} = 2(8) + 5 = 21_{\text{ten}}$ and $42 = 2(21)$. The above equation is only a linear one, but the following problem requires the use of a quadratic equation: "In what base b is the number represented by 132 twice the number represented by 33?" You have $132b = 1(b)^2 + 3(b)^1 + 2(b)^0$ and $33_b = 3(b)^1 + 3(b)^0$, so our equation becomes $b^2 + 3b + 2 = 2(3b + 3)$ or $b^2 - 3b - 4 = 0$. Solve for b in the usual fashion to obtain $b = -1$ (which must be rejected since the domain of b is positive), and $b = 4$, the only acceptable solution. Check: $132_{\text{four}} = 1(4)^2 + 3(4)^1 + 2(4)^0 = 1(16) + 3(4) + 2 = 30_{\text{ten}}$ and $33_{\text{four}} = 3(4)^1 + 3(4)^0 = 3(4) + 3 = 15_{\text{ten}}$; and 30 is twice 15. Offer students similar problems. If they have studied the solution of equations of degree higher than two, by synthetic division (since all results will be integral), include numbers whose representations in the bases being used involve more than three digits. For example: "In what base b is the number represented by the numeral 1213 triple the number represented by the numeral 221?" You have $1(b)^3 + 2(b)^2 + 1(b)^1 + 3(b)^0 = 3[2(b)^2 + 2(b)^1 + 1(b)^0]$ or $b^3 + 2b^2 + b + 3 = 3(2b^2 + 2b + 1)$, which simplifies to $b^3 - 4b^2 - 5b = 0$. Since this equation can be factored without resorting to synthetic division, solve it as follows: $b(b - 5)(b + 1) = 0$ and $b = 0, 5, -1$. As before, the only acceptable solution is the positive one, $b = 5$. Ask the class to check this result.

A final interesting algebraic application of number base problems is suggested by the following: "In base 10, the numeral 121 represents a number that is a perfect square. Does the numeral represent a perfect square in *any* other positive integral base?" Help your students investigate this problem as follows: $121_b = 1(b)^2 + 2(b)^1 + 1(b)^0 = b^2 + 2b + 1 = (b + 1)^2$. Surprise! The numeral 121 represents a perfect square in *any* positive integral base $b \geq 3$, and is the square of one more than the base number! Are there any other such numerals? Students may discover others by squaring such expressions as $b+2$ and $b+3$ to obtain the numerals 144 and 169. These perfect squares in base 10 are also perfect squares in any positive integral base containing the digits used in them ($b \geq 5$ and $b \geq 10$, respectively). It's not necessary for the coefficient of b to equal 1. If you square $2b + 1$, for example, you obtain $4b^2 + 4b + 1 = 441_b$, which will be a perfect square in any positive integral base $b \geq 5$. Invite your students to try to square expressions such as

$3b+1$, $2b+2, 4b+1$, etc., to obtain other perfect squares. Some may wish to continue this investigation into a search for perfect cubes, perfect fourth powers, etc. Help them to cube $b+1$, for example, to obtain b^3+3b^2+3b+1, indicating that the numeral 1331 is a perfect cube in any positive integral base $b \geq 4$ (in base 10,$1331 = 11^3$). As a matter of fact, 1331 is the cube of one more than the base number in each case! This study can be carried as far as the interest and ability of your class permits. Students familiar with the binomial theorem will find it convenient to use in expanding higher powers of such expressions as $b+1, 2b+1$, etc.

Postassessment

Students who have met the performance objective will be able to solve problems such as the following:

1. Represent the decimal numeral 78 as a numeral in base 5.
2. The number represented by the numeral 1000.1 in base 2 is represented by what numeral in base 8?
3. In a certain base b, the number represented by the numeral 54 is three times the number represented by the numeral 16. Find the value of b.
4. In a certain base b, the number represented by the numeral 231 is double the number represented by the numeral 113. Find the value of b.
5. In what bases would the numeral 100 represent a perfect square? In what bases would the numeral 1000 represent a perfect cube? Can you make a generalization of these results?

Unit 34

Raising Interest

Students are often confronted with advertisements by savings institutions offering attractive interest rates and frequent compounding of interest on deposits. Since most banks have a variety of programs, its valuable for potential depositors to understand how interest is calculated under each of the available options.

Performance Objective

Students will use the formula for compound interest to calculate the return on investments at any rate of interest, for any period of time, and for any commonly used frequency of compounding, including instantaneous (continuous) compounding. They will also determine which of two or more alternatives gives the best return over the same time period.

Preassessment

This lesson requires the ability to apply the laws of logarithms, so question students to be certain they're familiar with these laws. You should also determine the extent to which they are familiar with limits, since the class's background will help you determine how deeply you treat the concept of instantaneous compounding.

Teaching Strategies

Propose the following interesting problem: "In the year 1626, Peter Minuit bought Manhattan Island for the Dutch West India Company from the Native Americans for trinkets costing 60 Dutch guilders, or about \$24. Suppose Native Americans had been able to invest this \$24 at that time at an annual interest rate of 6%, and suppose further that this same interest rate had continued in effect all these years. How much money could the Present-day (let's say 1980) descendants of these Native Americans collect if (1) only simple interest were calculated, and (2) interest were compounded (a) annually, (b) quarterly and (c) continuously?" The answers to a, b, and c should surprise everyone!

Review briefly the formula for simple interest studied in lessons. The class will recall that simple interest is calculated by taking the product of the principle P, the annual interest rate r, and the time in years t. Accordingly, you have the formula $I = Prt$, and in the above problem $I = (24)(.06)(394) = \$567.36$ simple interest. Add this to the principal of \$24.00 to obtain the amount A of \$591.36 available at present. You have just used the formula for "amount," $A = P + Prt$.

With this relatively small sum in mind (for a return after 354 years!), turn to investigate the extent to which this return would have been improved if interest had been compounded annually instead of being calculated on only a simple basis. With a principle P, an annual rate of interest r, and a time $t = 1$, the amount A at the end of the first year is given by the formula $A_1 = P + Pr = P(1 + r)$. (The subscript indicates the year at the end of which interest is calculated.) Now $A_1 = P(1+r)$ becomes the principle at the beginning of the second year, upon which interest will be credited during the second year. Therefore, $A_2 = P(1+r) + P(1+r)r = P(1+r)(1+r) = P(1+r)^2$. Since the last expression represents the principle at the beginning of the third year, you have $A_3 = P(1+r)^2 + P(1+r)^2 r = P(1+r)^2(1+r) = P(1+r)^3$. By now, your students will see the emerging pattern and should be able to suggest the generalization for the amount after t years, $A_t = P(1+r)^t$.

Now try this formula on the \$24 investment made in 1626! Assuming annual compounding at 6% per annum, you have $A_{394} = 24(1+.06)^{394} =$ 224,244,683,837.57676. This means that the original \$24 is now worth almost \$22 billion! Most students are truly surprised by the huge difference between this figure and the figure \$533.76 obtained by computing simple interest.

Most banks now compound not annually, but quarterly, monthly, daily, or continuously, so next generalize the formula $A = P(1+r)^t$ to take into account compounding at more frequent intervals. Help your students observe that if interest is compounded semi-annually, the *periodic rate* would be only *one-half* the annual rate, but the number of periods would be *twice* the number of years: so $A = P\left(1+\frac{r}{2}\right)^{2t}$. Likewise, if interest is compounded quarterly $A = P\left(1+\frac{r}{4}\right)^{2t}$ · In general, if interest is compounded n times per year, you have $A = P\left(1+\frac{r}{n}\right)^{nt}$. This formula may be used for any finite value of n. Letting $n = 4$ in the problem yields $A = 24\left(1+\frac{.06}{4}\right)^{4(354)} = 24(1.015)^{1416} = 34,365,848,150$. The \$24 has now risen to about \$34 billion.

Students should note that changing the compounding from annually to quarterly increased the yield by about \$12 billion.

Students may now ask whether the yield can be increased indefinitely by simply increasing the frequency of compounding. A complete treatment of this question requires a thorough development of the concept of limits, but an informal, intuitive approach will suffice here. Have students first explore the simpler problem of an investment of \$1 at a nominal annual interest rate of 100% for a period of one year. This will give $A = 1\left(1+\frac{1.00}{n}\right)^n = \left(1+\frac{1.00}{n}\right)^n$. Ask the students to prepare a table of values for A for various common values of n, such as $n = 1$ (annual compounding), $n = 2$ (semiannual), $n = 4$ (quarterly), and $n = 12$ (monthly). They should note that the amount A does *not* rise astronomically as n increases, but rather rises slowly from \$2.00 ($n = 1$) to about \$2.60 ($n = 12$). Explain that the amount A would approach, but not quite reach, the value \$2.72.

(The extent to which you may wish to discuss the fact that $\lim_{n\to\infty} \left(1+\frac{1}{n}\right) n = e = 2.71828\ldots$ will depend on the backgrounds and abilities of your math students.)

Since investments generally don't draw 100% interest, you must next convert to a general interest rate or r. Setting $\frac{r}{n} = \frac{1}{k}$, you have $n = kr$, and $A = P\left(1 + \frac{r}{n}\right)^{nt}$ becomes $A = P\left(1 + \frac{1}{k}\right)^{krt} = P\left[\left(1 + \frac{1}{k}\right)^{k}\right]^{rt}$. Clearly, as n approaches infinity so does k, since r is finite, so the expression in brackets approaches the value e as a limit. You then have the formula $A = Pe^{rt}$ *for instantaneous compounding*, where r is the nominal annual rate of interest and t is the time in years.

Students might be interested in knowing that this formula is a special representation of the general "Law of Growth," which is usually written in the $N = N_0e^{rt}$ form, where N represents the final amount of a material whose initial amount was N_0. This law has applications in many other areas such as population growth (people, bacteria in a culture, etc.) and radioactive decay of elements (in which case it becomes the "Law of Decay," $N = N_0e^{-rt}$).

Completing the investment problem, using 2.72 as an approximation to e, you have $A = 24(2.72)^{.06(354)} = 40{,}780{,}708{,}190$.

Students can see that the "ultimate" return on a $24 investment (at a nominal annual interest rate of 6% for 354 years) is about $41 billion.

Students may now apply the formulas developed. Banks currently offer interest rates ranging from 5% to as much as 12% (usually for time deposits of 2 years or more) and compounding is commonly done quarterly, monthly, daily, or continuously. Students can work problems with varying principles, periodic rates, frequencies of compounding and time periods, and compare yields. They'll probably be surprised by what they learn!

Postassessment

Students who have met the performance objective will be able to answer questions such as these:

1. Banks offering 5% annual interest compounded quarterly claim that money doubles in 14 years. Is this claim accurate?
2. If you had $1,000 to invest for 2 years, would you get a greater return from a savings bank offering a 5% annual rate compounded quarterly or from a commercial bank offering a $4\frac{1}{2}$% annual rate compounded continuously?
3. Banks offering a 6% nominal annual rate compounded continuously on term savings of 2 years or longer claim that this rate is equivalent to an "effective annual rate" (the rate under annual compounding) of 6.27%. Prove that this is true, assuming a deposit of $500 (the usual minimum) for a period of 2 years.

Reference

Posamentier, A. S. and c. Sprietzer, *The Mathematics of Everyday Life*, Amherst, New York: Prometheus Books, 2018.

Unit 35

Reflexive, Symmetric, and Transitive Relations

In this lesson, students will have the opportunity to explore some properties of mathematical relations among numbers, geometric figures, sets, propositions, persons, places, and things.

Performance Objective

Students will identify a given relation as reflexive symmetric, transitive, or as an equivalence relation.

Preassessment

Ask students to describe what the mathematical term *relation* refers to. If you are not satisfied they understand the term, present some examples before beginning the lesson. You may wish to vary the relations you present students, according to their grade level and background in such areas as algebra, geometry, set theory, number theory, and logic.

Teaching Strategies

Begin with a consideration of a very simple relation such as "is equal to" for real numbers. From their previous math experience, students will recognize that any quantity a is equal to itself; that if a quantity a is equal to another quantity b, then b is also equal to a; and that if a quantity a is equal to another quantity b, and b is in turn equal to a third quantity c, then a is equal to c. Symbolically, we have $a = a, a = b \rightarrow b = a$, and $a = b$ and $b = c \rightarrow a = c$. The arrow is read "implies" as in ordinary symbolic logic. (Replace the arrow with the word if the class is unfamiliar with this notation.) Explain that when a quantity a has a given relation to itself (as in $a = a$) that relation is called *reflexive*. Further, when a quantity a has a given relation to another quantity b and this results in b having the same relation to a (as in $a = b \rightarrow b = a$) that relation is called *symmetric*. Add that when a quantity a has a given relation to another quantity b and b has the *same* relation to a third quantity c, and this results in a having that same relation to c (as in $a = b$ and $b = c \rightarrow a = c$) that relation is called *transitive*. A relation possessing all three of these properties is an *equivalence relation*.

Now invite students to examine some of the relations with which they are familiar from earlier work in mathematics. You've just established that "is equal to" is an equivalence relation. Follow up by considering the relations "is greater than" and "is less than" for real numbers. Your class will quickly discover that these relations are neither reflexive nor symmetric, but that they *are* transitive. An interesting variation is the relation "is not equal to" for real numbers. Although this relation is not reflexive, it is symmetric. Students may also think that this relation is transitive, but a simple counterexample will prove that it is not: $9 + 6 \neq 7 + 2$ and $7 + 2 \neq 11 + 4$, but $9 + 6 = 11 + 4$. So the relation "is not equal to" is not transitive.

Of course, none of the relations just considered is an equivalence relation. Have the class consider other relations, for example: "is a multiple of" (or "is divisible by") and "is a factor of" for integers. Both of these relations are reflexive and transitive, but neither is symmetric. Ask students to prove these facts algebraically. For the first relation, for example, they may write $a = kb$ and $b = mc$ where k and m are integers. Clearly, $a = 1a$ so a is divisible by a (reflexivity); $\frac{a}{b} = k$ since a is divisible by b but $\frac{b}{a} = \frac{1}{k}$, which is not an integer, so b is not divisible by a (no symmetry); $a = kb$ and $b = mc \rightarrow a = k(mc)$ or $\frac{a}{c} = km$, which is an integer, since the product of two integers is an integer (the set of integers being closed under multiplication), so a is divisible by c (transitivity).

Consider next some relations in geometry. First explore the relations "is congruent to" and "is similar to" for geometric figures. Students will have little difficulty recognizing that both of these are equivalence relations. Ask the class to examine each of these relations when it is negated. Each will then possess only the symmetric property.

The relations "is parallel to" and "is perpendicular to" are very interesting when applied to lines in a plane and to planes themselves. For example, for lines in a plane, "is parallel to" is symmetric *and* transitive but "is perpendicular to" is *only* symmetric. Ask your students why! They should recall such ideas from geometry as "lines parallel to the same line are parallel to each other" and "lines perpendicular to the same line are parallel to each other." These relations may also be negated as exercises.

Students who have some familiarity with set theory may explore the relations "is equal to" and "is equivalent to" as applied to sets. Since *equal sets* are containing identical elements, it's obvious that "is equal to" is an equivalence relation. *Equivalent* sets have the same number of elements (their elements can be placed into one-to-one correspondence with each other), but not necessarily identical ones. A little reflection reveals that "is equivalent to" is also an equivalence relation. Another interesting relation is "is the complement of" as applied to sets. The class should discover that this relation is symmetric, but that it is neither reflexive nor transitive. (If a is the complement of b and b is the complement of c, then a is not the complement of c, but rather $a = c$.)

An interesting relation from number theory is "is congruent to, modulo m" for integers. Students familiar with this concept should be able to prove easily that this is an equivalence relation, using simple algebra: $a = a(\text{mod}\ m)$ since $a - a = 0m$ (proving the reflexive property); $a \equiv b\ (\text{mod}\, m) \rightarrow b \equiv a(\text{mod} m)$ since $b - a = -(a - b) = -km$ (proving the symmetric property); $a \equiv b(\text{mod}\, m)$ and $b \equiv c(\text{mod}\, m) \rightarrow a = c(\text{mod}\, m)$ since $a - c = (a - b) + (b - c) = km + pm = (k + p)m$ (proving the transitive property).

Students familiar with symbolic logic may be invited to consider the relation "implies" for propositions (e.g., as designated by p, q, r). The alert student will recognize that this relation is reflexive, $p \rightarrow p$ (since any proposition implies itself) and transitive, $(p \rightarrow q) \wedge (q \rightarrow r) \rightarrow (p \rightarrow r)$ (since this can be proved to be a tautology by using a truth table) but that it is *not* symmetric, $(p \rightarrow q) \rightarrow (q \rightarrow p)$ is *false* (since the truth of a proposition does *not* guarantee the truth of its converse).

Now broaden the concept of relations from strictly mathematical settings to involve relations between persons, places, and things. Your class should find this amusing as well as instructive. Suggest a relation such as "is the father of: A little reflection reveals that this relation is not reflexive, not symmetric, and not transitive! It is obvious that a cannot be his own father (not reflexive); that if a is the father of b, then b is the son or daughter and not the father of a (not symmetric); and that if a is the father of b and b is the father of c, then a is the grandfather of c, not the father (not transitive)! Many similar relations may be considered, including "is the mother of," "is the brother of "(caution: *only* transitive, *not* symmetric, since b may be the sister of a), "is the sister of," "is the sibling of" (this one is symmetric), "is the spouse of," "is the ancestor of," "is the descendent of," "is taller than," and "weighs more than." Any of these relations may be explored in the negative sense as well as in the positive one. With respect to places, students may consider relations such as "is north of," "is west of" (caution: transitivity here is *not* necessarily true if places may be selected from anywhere on the globe rather than from merely a small area or only one country), "is at a higher altitude than," "is exactly one mile from" (symmetric only), and "is less than one mile from" (reflexive *and* symmetric). Relations among things may include "is above," "is older than," "costs as much as," and "costs more than," among others.

Postassessment

Students who have met the performance objective will be able to answer questions such as the following:

1. Identify each of the following relations as reflexive, symmetric, transitive, or an equivalence relation:

   a. "is supplementary to" for angles
   b. "is congruent to" for line segments
   c. "is a subset of" for sets
   d. "is a proper subset of" for sets
   e. "is equivalent to" for propositions
   f. "is wealthier than" for nations
   g. "is smaller than" for objects
   h. "is colder than" for places

2. Prove algebraically that the relation "is complementary to" for acute angles is symmetric but neither reflexive nor transitive.
3. Which of the following relations is reflexive and transitive but not symmetric?

   a. "is a positive integral power of" for real numbers
   b. "has the same area as" for triangles
   c. "is the converse of" for propositions
   d. "is younger than" for people

Unit 36

Bypassing an Inaccessible Region

This unit will present the problem of constructing a straight line through an inaccessible region using only straightedge and compasses, and without using the tools in or over this inaccessible region. This activity will provide an opportunity for students to exhibit creativity.

Performance Objectives

1. *Given a straight line segment with an endpoint on the boundary of an inaccessible region, students, using straightedge and compasses, will construct another straight line segment collinear with the given one and on the other side of the inaccessible region (an endpoint will be on the boundary of this region).*
2. *Given one point on either side of an inaccessible region, students, using only straightedge and compasses, will construct two collinear straight line segments, each having one given point as an endpoint and neither intersecting the inaccessible region.*

Preassessment

Students should be familiar with the basic geometric constructions using straightedge and compasses.

Teaching Strategies

To generate initial interest, begin this topic by developing a story about two countries that are separated by a mountain, and each of which wants to straight road and tunnel through the mountain. Neither country can decide how to dig the tunnel, so they both decide to construct a road on one side of the mountain at the point where the anticipated tunnel (the continuation of a straight road on the other side of the mountain) will emerge from the mountain. Using only straightedge and compasses, they seek to plot the path for this new road (Figure 1).

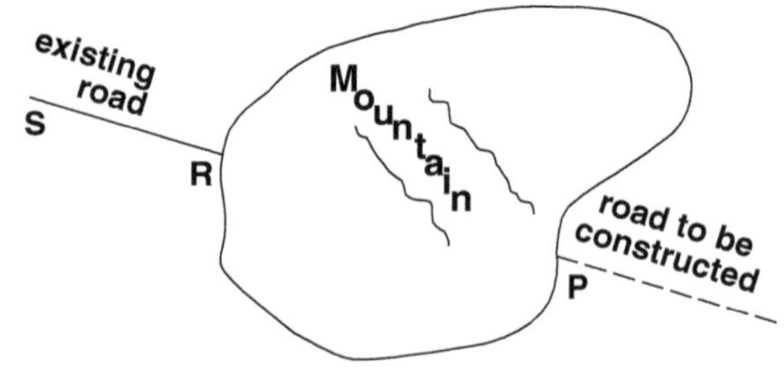

Figure 1.

Once students understand the problem, have them draw a diagram (maps) of this situation.

Students must construct the collinear "continuation of $\overline{SR}$" at point P (using straightedge and compasses) and never touch or go over the inaccessible region.

There are various ways to construct the collinear continuation of $\overline{SR}$ at P. One method is to erect a perpendicular (line ℓ) to $\overline{SR}$ at a convenient point N of $\overline{SR}$. Then at a convenient point M of line ℓ a perpendicular (line k) to ℓ is constructed (see Figure 2).

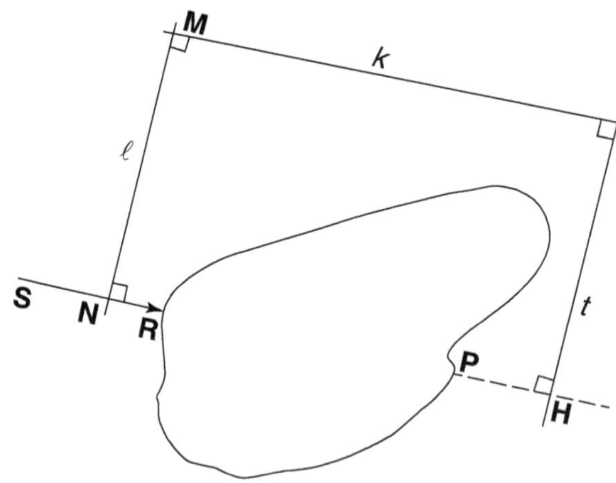

Figure 2.

At a convenient point G of line k, a perpendicular (line t) to line k is constructed. Point H is then obtained on line t, so that GH = MN. The line constructed perpendicular to line t at point H will be the required line through P and collinear with $\overline{SR}$. (Although P was collinear with $\overline{SR}$, it was virtually not needed for the construction.) The justification for this method is that a rectangle (minus part of a side) was actually constructed.

Another method for solving this problem involves replacing the above rectangle with an equilateral triangle, since angles of measure 60° are rather simple to construct. Figure 3 presents this method and ought to be rather self-explanatory.

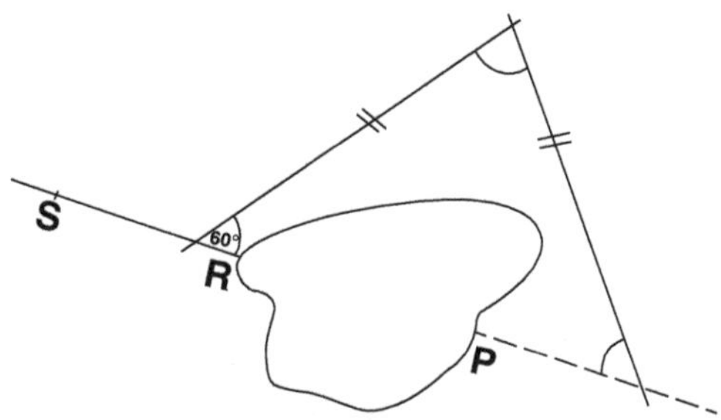

Figure 3.

The problem of constructing a straight line "through" an inaccessible region, when only the two endpoints (at either side of the region) are given, is a much more challenging problem. Naturally an appropriate story can be built around this situation.

To construct two collinear straight line segments at each of two points (P and Q) situated at opposite sides of an inaccessible region, by drawing any convenient line segment from point P and construct a perpendicular line to it at a convenient point R. This perpendicular should not intersect the inaccessible region (see Figure 4).

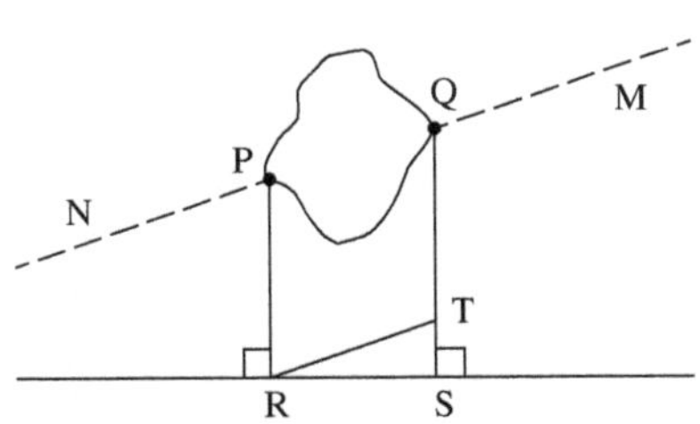

Figure 4.

Now construct a perpendicular from Q, to this last line drawn, intersecting it at S. Locate T on $\overleftrightarrow{QS}$ so that PR = QS. Draw $\overline{RT}$.

At P construct $\angle RPN \cong \angle PRT$, and at Q construct $\angle TQM \cong \angle QTR$. This completes the required construction, since $\overline{NP}$ and $\overline{QM}$ are extensions of side $\overline{PQ}$ of "parallelogram" PRTQ, and therefore are collinear.

There are many other methods of solving this problem. Many involve constructing similar triangles in order to then construct the two required lines. However, students select to approach this problem, they are apt to be led to a creative activity.

Postassessment

1. Have students construct a "continuation" of $\overline{SP}$ on the other side of the inaccessible region (using only straightedge and compass and not touching or going over the inaccessible region) (Figure 5).

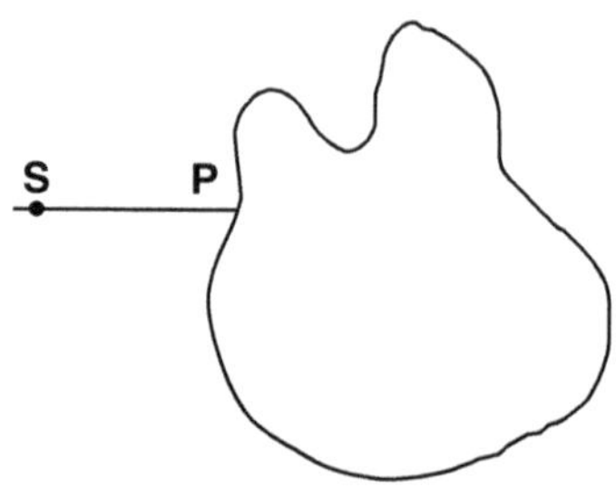

Figure 5.

2. Have students construct two collinear segments at opposite ends (P and Q) of an inaccessible region (using only straightedge and compasses and not touching or going over the inaccessible region). These post-assessment items become more challenging if original methods are sought (Figure 6).

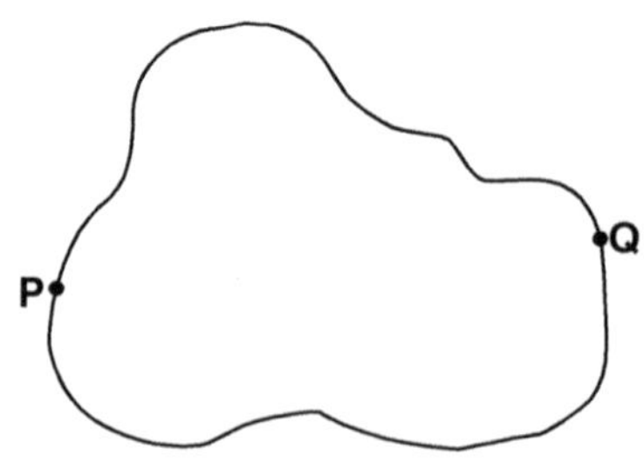

Figure 6.

This decision depends on the ability level of the class.

Unit 37

The Inaccessible Angle

Through a recreational application, this unit will provide the students the opportunity to use in novel ways various geometrical relationships they have learned. It also opens the door to a host of creative activities.

Performance Objective

Given an angle whose vertex is in an inaccessible region (hereafter referred to as an inaccessible angle), students will construct its angle bisector using straightedge and compasses.

Preassessment

Students should be familiar with the basic geometric constructions using straightedge and compasses.

See that students can properly bisect a given angle using only straightedge and compasses.

Teaching Strategies

After students have reviewed the basic geometric constructions, present them with the following situation:

Problem: Given an angle whose vertex is inaccessible (i.e., tell students that the vertex of the angle is in a region in which, and over which, a straightedge and compasses cannot be used), construct the angle bisector using only straightedge and compasses.

Most students' first attempts will probably be incorrect (see Figure 1). However, careful consideration of students' responses should serve as a guide to a correct solution. Students will eventually present some rather strange (and creative) solutions. All should be given careful attention.

To best exhibit the true source of creativity that this problem provides, three different solutions are presented.

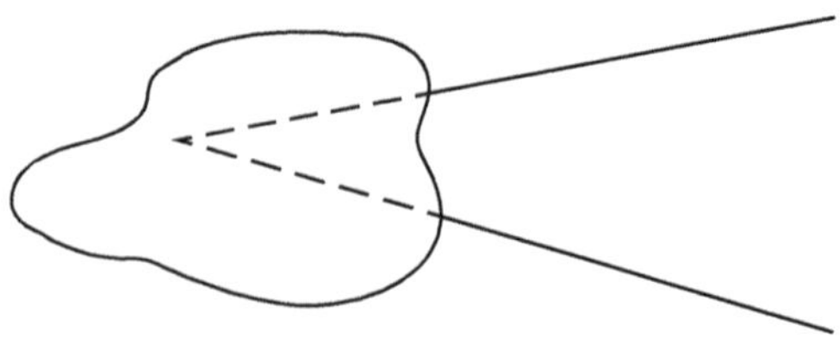

Figure 1.

Solution 1

Draw any line ℓ intersecting the rays of the inaccessible angle at points A and B. Label the inaccessible vertex P (Figure 2).

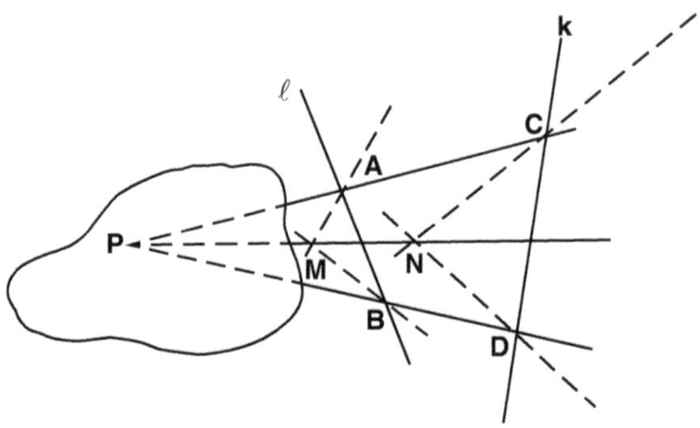

Figure 2.

Construct the bisectors of $\angle$PAB and $\angle$PBA, which then intersect at M. Remind students that since the angle bisectors of a triangle (here ΔAPB) are concurrent, the bisector of $\angle$P, which we are trying to construct, must contain point M.

In a similar way, draw any line k, intersecting the rays of the inaccessible angle at points C and D. Construct the bisectors of $\angle$PCD and $\angle$PDC, which intersect at N. Once again students should realize that, since the bisectors of a triangle (in this case ΔCPD) are concurrent, the bisector of $\angle$P must contain point N. Thus, it has been established that the required line must contain points M and N, and therefore by drawing $\overleftrightarrow{MN}$ the construction is completed.

Solution 2

Begin this method by constructing a line parallel to one of the rays of the inaccessible angle (see Figure 3). This can be done in any one of various ways.

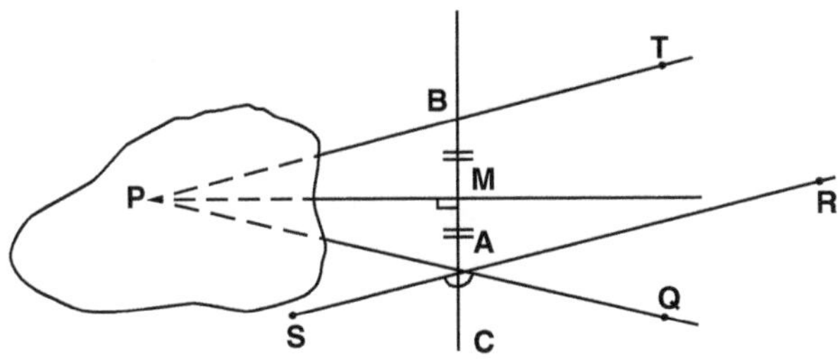

Figure 3.

In Figure 3, $\overleftrightarrow{RS}$ is parallel to $\overrightarrow{PT}$ (a ray of the inaccessible $\angle P$), and intersects $\overrightarrow{PQ}$ at point A. Construct the bisector of $\angle SAQ$, which will intersect $\overrightarrow{PT}$ at B. Since $\overleftrightarrow{SR} // \overrightarrow{PT}$, $\angle SAC \cong \angle PBA$. However, $\angle SAC \cong \angle CAQ \cong \angle PAB$. Therefore, $\angle PBA \cong \angle PAB$, thereby making ΔPAB isosceles. Since the perpendicular bisector of the base of an isosceles triangle also bisects the vertex angle, the perpendicular bisector of $\overline{AB}$ is the required angle bisector of the inaccessible angle ($\angle P$).

Solution 3

Start by constructing a line ($\overline{MN}$) parallel to one of the rays ($\overrightarrow{PT}$) of the inaccessible angle ($\angle P$), and intersecting the other ray at point A (see Figure 4).

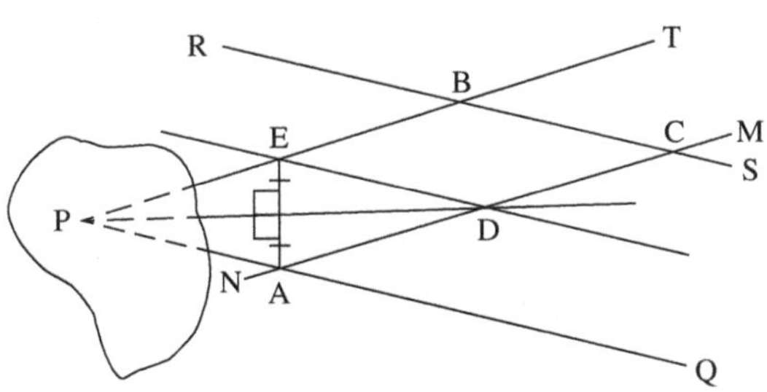

Figure 4.

Then construct a line ($\overleftrightarrow{RS}$) parallel to the other ray ($\overrightarrow{PQ}$) of the inaccessible angle intersecting $\overrightarrow{PT}$ and $\overrightarrow{MN}$ at points B and C, respectively. With a pair of compasses, mark off a segment, $\overline{AD}$, on $\overrightarrow{AC}$ of the same length as $\overline{BC}$.

Through D, construct $\overleftrightarrow{DE} \,//\, \overrightarrow{PQ}$, where E is on $\overrightarrow{PT}$. It can now be easily shown that ED = AD (since EBCD is a parallelogram and ED = BC. Since PEDA is a parallelogram with two adjacent sides congruent ($\overline{ED} \cong \overline{AD}$), it is a rhombus. Thus, the diagonal $\overline{PD}$ is the bisector of the inaccessible angle. $\overline{PD}$ can be constructed simply by bisecting ∠EDA or constructing the perpendicular bisector of $\overline{EA}$.

After presenting these solutions to your students, other solutions created by the students should follow directly. Free thinking should be encouraged to promote greater creativity.

Postassessment

Present students with an inaccessible angle and ask them to bisect it.

Unit 38

Triangle Constructions

Often teachers will justify the basic triangle congruence postulates by showing that unique triangles can be constructed with such given data as the lengths of three sides of a triangle or perhaps the lengths of two sides and the measure of the included angle. This unit will extend this usually elementary discussion of triangle constructions to some rather interesting problems.

Performance Objective

Given the measures of three parts of a triangle (which determine a triangle) students will analyze and construct the required triangle with straightedge and compasses.

Preassessment

Students should be familiar with the basic geometric constructions normally taught in the high school geometry course.

Teaching Strategies

To begin to familiarize students with this topic, have them construct the triangle, where the measures of two angles and the length of the included side are given (Figure 1).

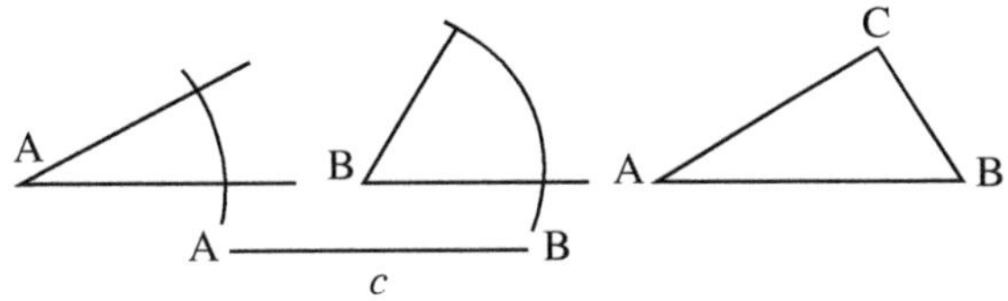

Figure 1.

Students will draw a line and mark off the length of $\overline{AB}$ (sometimes referred to by c, the length of the side opposite $\angle C$). By constructing angles A and B at either end of $\overline{AB}$, they eventually find that they have constructed a *unique* ΔABC.

Surely, if the students were given the measures of the three angles of a triangle, each student would probably construct a triangle of different size (although all should be the same shape). Yet if the students were given the lengths of the three sides of a triangle, they would all construct triangles congruent to one another. At this point, students should realize that certain data will determine a *unique* triangle while other data will not. A student is thus limited in investigating such cases where the measures of only sides and angles are provided. Students will want to consider other parts of triangles as well. Present the following problem:

> Construct a triangle given the lengths of two sides and the length of an altitude to one of these sides.

We shall write this problem as $[a, b, h_a]$, where h_a is the length of the altitude to side a.

To do this construction, take the point H_a on any line and erect a perpendicular H_aA (using the usual straightedge and compasses method) of length h_a (Figure 2). With arc (A, b) (*Note*: This ordered pair symbol is merely a short way of referring to the circle with center A and radius b) intersect the base line at C, then with arc (C, a), intersect this base line again at B and B′. The *two* solutions are ΔABC and ΔAB′C, each of which has the given $\{a, b, h_a\}$. Further inspection of this solution will indicate that

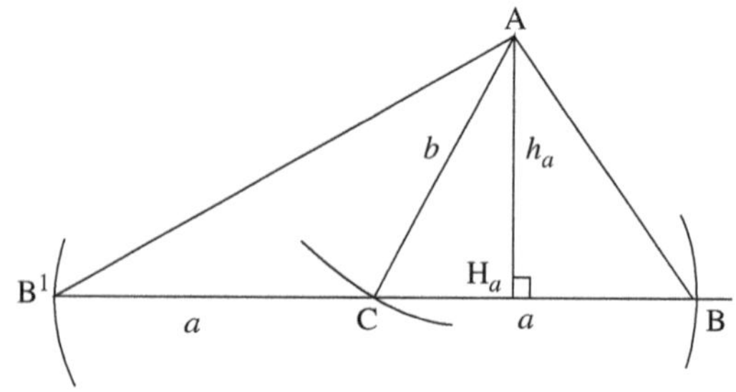

Figure 2.

$b > h_a$ was a necessary condition, and that if $b = h_a$ there would have only been *one* solution.

A much simpler problem is to construct a triangle given $\{a, b, h_c\}$. Here students begin in much the same way. On any line, erect at H_c a perpendicular length h_c. At C, the other extremity of h_c, draw (C, a) and (C, b). Their points of intersection with the original base line will determine points B and A, respectively (Figure 3). Once again, a discussion of uniqueness should follow.

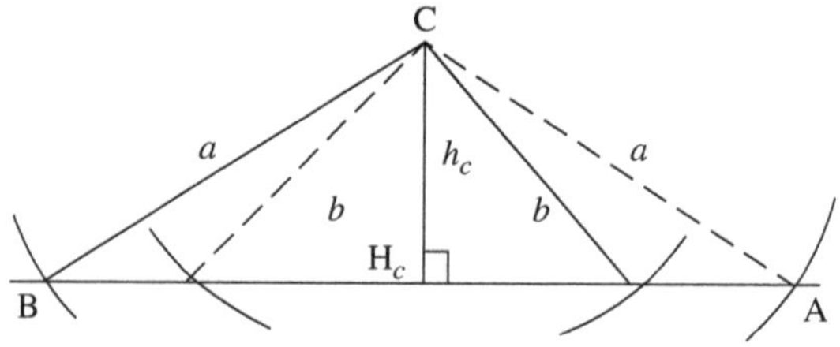

Figure 3.

The figure above should help in this discussion.

Some triangle constructions require a good deal more analysis before actually beginning the construction. An example of such a problem is to construct the triangle given the lengths of its three medians $\{m_a, m_b, m_c\}$.

One approach for analyzing this problem is to consider the finished product, ΔABC.

The objective here is to be able to construct one of the many triangles shown in the above figure by various elementary methods. By extending m_a (the median to side a), one third of its length to a point D and then drawing $\overline{BD}$ and $\overline{CD}$, we have created a triangle, ΔBGD, which is easily constructible. Since the medians of a triangle trisect each other, we know that BG $= \frac{2}{3}m_b$. Since GM$_a$ = DM$_a$ $= \frac{1}{3}m_a$ (by construction), and BM$_a$ = CM$_a$, we may conclude that BGCD is a parallelogram. Therefore, BD = GC $= \frac{2}{3}m_c$. It is then rather simple to construct ΔBGD, since its sides are each two thirds the length of one of the given medians (lengths easily obtained). After constructing ΔBGD, the students should be able to complete the required construction by (1) extending BG one half its length to point M_b, (2) extending $\overline{\text{DG}}$ its own length to A, and (3) extending $\overline{\text{BM}_a}$ its own length to C (where M$_a$ is the midpoint of $\overline{\text{DG}}$). The required triangle is then obtained by drawing $\overline{\text{AM}_b}$ to intersect $\overline{\text{BM}_a}$ at point C, and drawing $\overline{\text{AB}}$ (Figure 4).

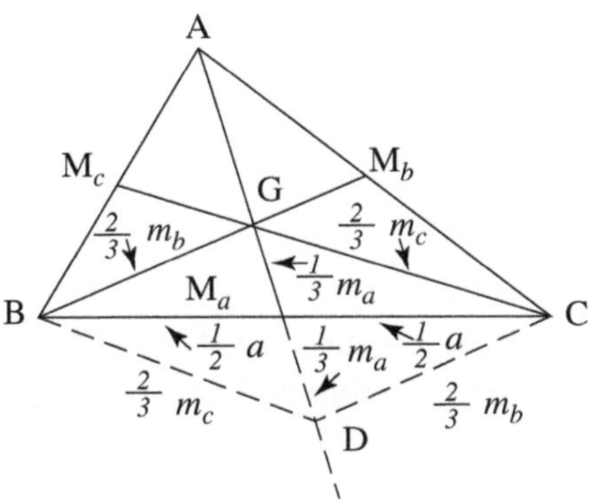

Figure 4.

Not only does this problem review for the students many important concepts from elementary geometry, but it also provides an excellent opportunity for students to practice "reverse" reasoning in analyzing the problem.

For additional practice, have students construct ΔABC given $\{a, h_b, m_c\}$. Once again have students first inspect the desired triangle. They should note that ΔCBH$_b$ can be easily constructed by erecting at H$_c$ a perpendicular to $\overleftrightarrow{\text{AC}}$ of length h_b. At its extremity, B, draw (B, a) to intersect $\overleftrightarrow{\text{AC}}$ at C, to complete ΔCBH$_b$. Further inspection of the above figure suggests that ΔCDB may also be constructed. Construct a line $\overleftrightarrow{\text{DB}}$ parallel to $\overleftrightarrow{\text{AC}}$ and intersecting arc (C, $2m_c$) at D. To find A, construct a line $\overleftrightarrow{\text{AD}}$ parallel to $\overleftrightarrow{\text{CB}}$ and intersecting $\overleftrightarrow{\text{CH}b}$ at A. Since ADBC is a parallelogram, $\overleftrightarrow{\text{CD}}$ bisects

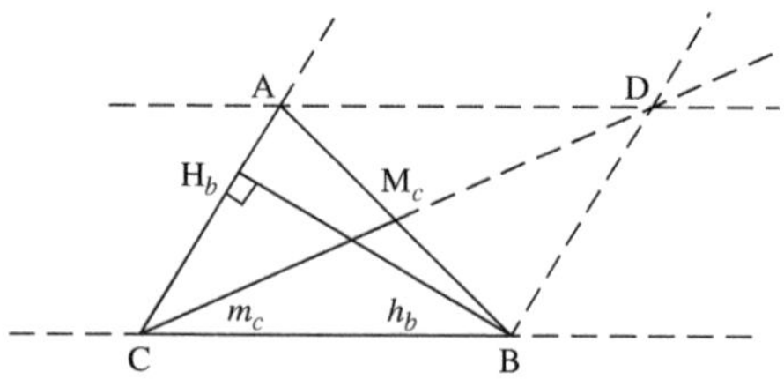

Figure 5.

$\overleftrightarrow{AB}$ at M$_c$, and $CM_c = \frac{1}{2}CD = m_c$ (Figure 5). Thus, the problem is analyzed in a rather reverse fashion, and then the required triangle is constructed.

When you consider the measures of other parts of a triangle such as angle bisectors, radius of the inscribed circle, radius of the circumscribed circle, and the semi-perimeter (as well as the measures of parts considered earlier in this unit), then there exist 179 possible triangle construction problems, where each consists of measures of three of these parts of a triangle. While some may be rather simple (e.g., $\{a, b, c\}$), others are somewhat more difficult (e.g., $\{h_a, h_b, h_c\}$).

This type of construction problem may very well serve as a springboard for a more careful study of this topic as well as various other geometric construction problems. A recently published book that contains much more on this topic (including the *complete* list of the 179 triangle constructions!) as well as a variety of other stimulating geometric construction topics (e.g., a review of the basic constructions, a variety of applications, and circle constructions) is Posamentier, A. S., *Advanced Euclidean Geometry: Excursions for Secondary Teachers and Students.* Hoboken, NJ: John Wiley & Sons, 2002.

Postassessment

Have students construct triangles given:

1. $\{a, b, m_a\}$
2. $\{a, h_b, t_c\}$
3. $\{a, h_b, h_c\}$
4. $\{h_a, m_a, t_a\}$
5. $\{h_a, h_b, h_c\}$

N.B. t_a is the length of the angle bisector of $\angle$A.

Reference

Posamentier, A. S. and I. Lehmann, *The Secrets of Triangles: A Mathematical Journey, Amherst*, New York: Prometheus Books, 2012.

Unit 39

The Criterion of Constructibility

This unit will develop a criterion of constructibility for the Euclidean tools, straightedge and compasses.

Performance Objectives

1. *Students will state the criterion of constructibility.*
2. *Students will represent algebraic expressions geometrically (in terms of given lengths).*

Preassessment

Ask students to represent geometrically AB + CD and AB − CD, given AB and CD.

Teaching Strategies

Most students should have been able to successfully do the above problem. Now let AB = a and CD = b.

$$\overset{a}{A\underline{\hspace{5em}}B} \qquad \overset{b}{C\underline{\hspace{5em}}D}$$

The next logical concern would be to represent the *product* of two given line segments. Here, however, a segment of unit length must be introduced.

To construct ab, two cases should be considered: (I) where $a > 1$ and $b > 1$, and (II) where $a < 1$ and $b < 1$.

In the first case (I), students would construct the following figure. Note $\overline{MN} // \overline{AB}$, and $\angle C$ is any convenient angle.

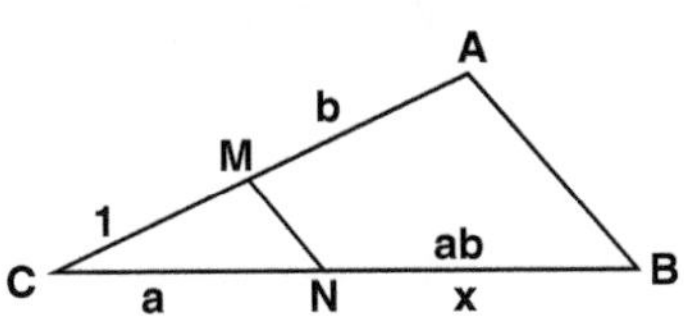

Since $\overline{MN}//\overline{AB}$, $\frac{x}{a} = \frac{b}{1}$, and $x = ab$; thus $\overline{NB}$ is the segment of desired length (i.e., ab). Note that $ab > a$, and $ab > b$, which is expected if $a > 1$ and $b > 1$.

In the second case (II), students would proceed in the same way as in case (I), however since $a < 1$ and $b < 1$, it should be made clear that geometrically $ab < a$, and $ab < b$.

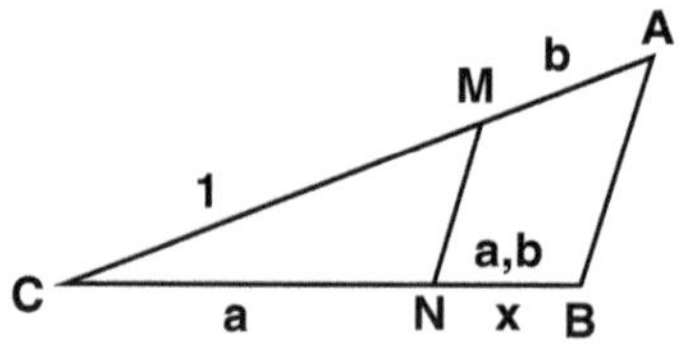

Students should now be challenged to discover a similar scheme for constructing a line segment that can represent the quotient of two given line segments, a and b. Again two cases should be considered:

Case I: $a < b < 1$. Once again have students construct the figure above, where $\overline{MN}//\overline{AB}$.

In this case, either $a < \frac{a}{b} < b < 1$ or $a < b < \frac{a}{b} < 1$.

Students should be encouraged to verify this.

Case II: $b < a \leq 1$. Proceed as above to construct the following figure.

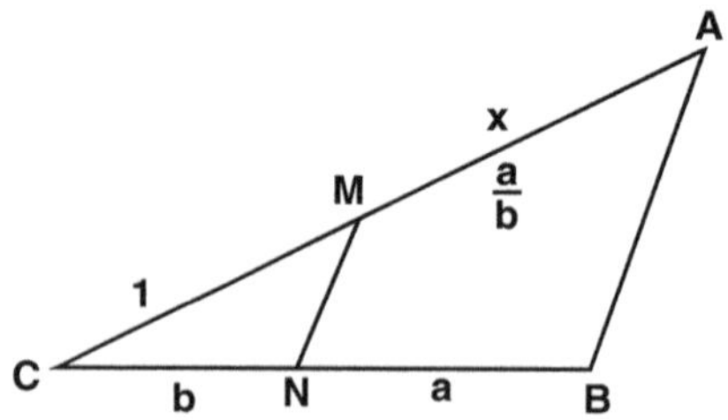

Here $b < a < \frac{a}{b}$.

Until now all segments considered were of positive length. Students will now become curious about using segments of negative length to represent products and quotients.

To consider segments of negative length, we must introduce number line axes, horizontal and oblique. To find ab, locate A on the horizontal axis so that OA $= a$, and locate B on the oblique axis so that OB $= b$. Draw the line through the 1 on the oblique axis and A. Through B draw a line parallel to the first line, intersecting the horizontal axis in a point C. Thus, OC $= ab$.

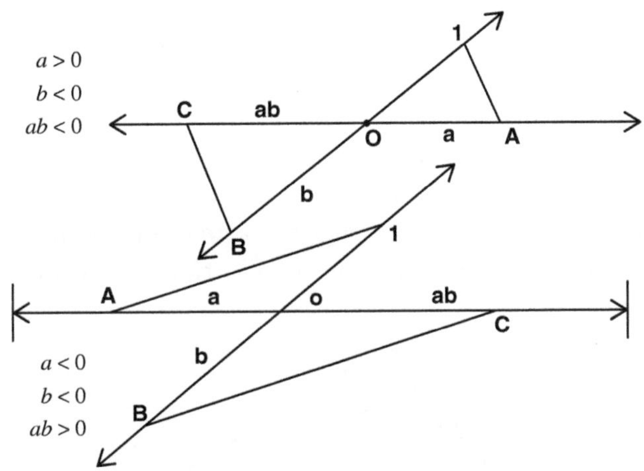

Students will notice that a and b were marked off on different axes and that in each case the product, ab, was appropriately less than or greater than zero.

As before, we shall find a quotient by considering division as the inverse of multiplication. To find $\frac{a}{b}$, we find x such that $bx = a.a > 0$, and $b > 0$, then $\frac{a}{b} > 0$.

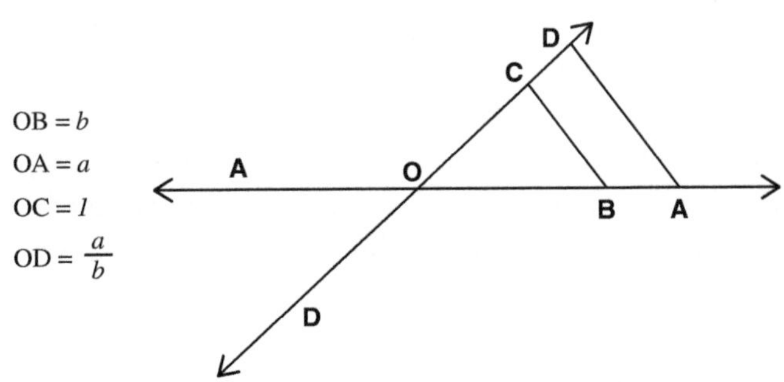

Similarly, when $a < 0$ and $b > 0$, it follows that $\frac{a}{b} < 0$.

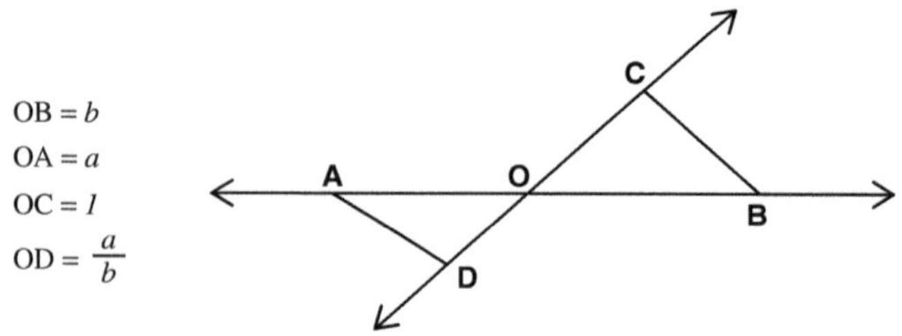

Have students consider division by zero. That is, where is B if $b = 0$? What happens to OD?

The only remaining operation for which a geometric representation is needed is square root extraction. Here students merely construct a semicircle on $1 + a$ (where $\sqrt{a}$ is sought). Then at the common endpoint, B, of segment 1 and a, erect a perpendicular to intersect the semicircle at D. Thus, $\text{BD} = \sqrt{a}$. Students should be able to apply mean proportional theorems to prove this.

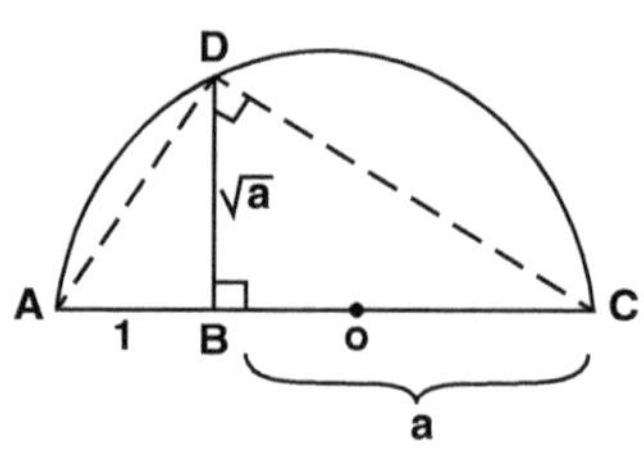

The solution to a construction problem may be expressed as a root of an equation. For example, consider the problem of duplication of a cube. We must find the edge of a cube whose volume is twice that of a given cube. That is, we must find x, when $x^3 = 2$.

If we can obtain the solution by a finite number of applications of the operations $+, -, \times, \div$, and $\sqrt{}$, using given segments and a unit length, then the construction is possible.

Conversely, if the construction is possible, then we can obtain it by a finite number of applications of addition, subtraction, multiplication, division, and extractions of square root, using the given segments and an arbitrary unit of length. We know that the straight lines and circles that we construct are determined either by given segments or those obtained from the intersections of two straight lines, a straight line and a circle, or two circles. To show the converse above, we must show that these intersections can be obtained from the coefficients of the equations by a finite number of applications of the operations of addition, subtraction, multiplication, division, and extraction of square root.

Two straight lines

$$y = mx + b$$
$$y = m'x + b' \quad m \neq m'$$

have as their point of intersection the point (x, y) with

$$x = \frac{b - b'}{m - m'} \quad y = \frac{mb' - m'b}{m - m'}$$

These relationships are obtained from the equations by applying the above operations. An equation for a circle with radius r and center (c, d) is $(x - c)^2 + (y - d)^2 = r^2$. To find the intersection of the circle with the line $y = mx + b$, we can substitute for y in the equation for the circle

$$(x - c)^2 + (mx + b - d)^2 = r^2$$

This forms a quadratic equation in x. Since the solution of the quadratic $ax^2 + bx + c = 0$ is

$$x = \frac{-b \pm \sqrt{b^2 - 4ac}}{2a}$$

we know that the quadratic $(x-c)^2+(mx+b-d)^2 = r^2$ has a root that can be obtained from the known constants by applying the above five operations.

The intersection of two circles is the same as the intersection of one circle with the common chord. Thus, this case can be reduced to finding the intersection of a circle and a line.

Criterion of Constructibility: A proposed geometric construction is possible with straightedge and compass alone if and only if the numbers that define algebraically the required geometric elements can be derived from those defining the given elements by a finite number of rational operations and extractions of square root.

Postassessment

1. Restate and explain the Criterion of Constructibility.
2. Given lengths $a, b, 1$ construct a line segment of length $\sqrt{\frac{ab}{a+b}}$.

Reference

Posamentier, A. S. *Advanced Euclidean Geometry: Excursions for Secondary Teachers and Students.* Hoboken, NJ: John Wiley & Sons, 2002.

Unit 40

Constructing Radical Lengths

Often students ask how a line of length $\sqrt{2}$ can be constructed. This activity will address itself to this question as well as find the length of other radical segments.

Performance Objective

Students will construct a segment of a given radical length after being given a unit length.

Preassessment

Students should be able to apply the Pythagorean theorem and be familiar with the basics of geometric constructions with straightedge and compasses.

Teaching Strategies

Ask students to construct a triangle with one side of length $\sqrt{2}$ (be sure to tell them to select a convenient unit length). In all likelihood, they will draw an isosceles right triangle with a leg of length 1. By the Pythagorean theorem, they will find that the hypotenuse has length $\sqrt{2}$.

Now have them construct a right triangle using this hypotenuse and have the other leg be of unit length. This newly formed right triangle has a hypotenuse of length $\sqrt{3}$. Students should easily discover that fact using the Pythagorean theorem.

By repeating this process, students will generate, in sequence, radicals of integers, i.e., $\sqrt{2}, \sqrt{3}, \sqrt{4}, \sqrt{5}, \ldots$ Figure 1, frequently referred to as a "radical spiral," shows this process.

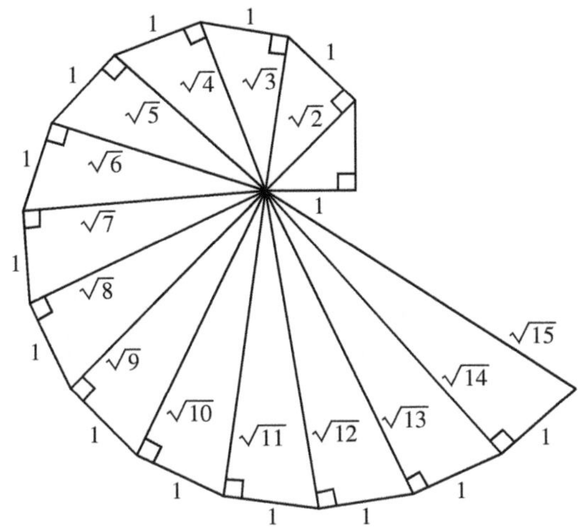

Figure 1.

A student might ask if there is a more expedient method for constructing $\sqrt{15}$, rather than generating a radical spiral up to $\sqrt{15}$.

Lead students to recall one of the "mean proportional" theorems. In following Figure 2, CD is the mean proportional between AD and BD.

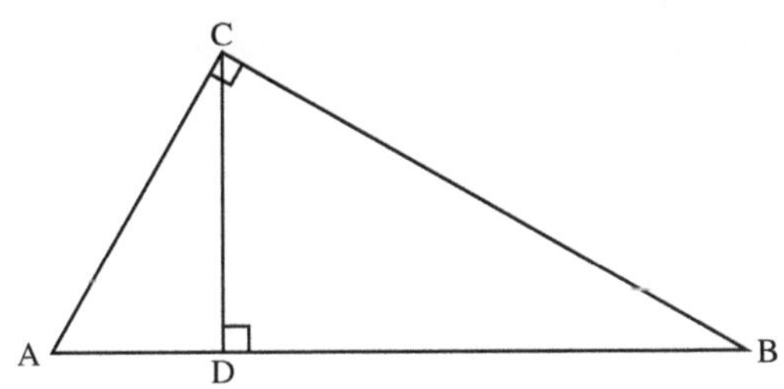

Figure 2.

That is, $\frac{AD}{CD} = \frac{CD}{BD}$, or $(CD)^2 = (AD)(BD)$, which implies that $CD = \sqrt{(AD)(BD)}$.

This relationship will help them construct a segment of length $\sqrt{15}$ in one construction. All they need to do is to construct the above figure and let $AD = 1$ and $BD = 15$; then $CD = \sqrt{(1)(15)} = \sqrt{15}$.

What they should do is draw a segment of length 16 and partition it into two segments of lengths 1 and 15. At the partitioning point, have them erect

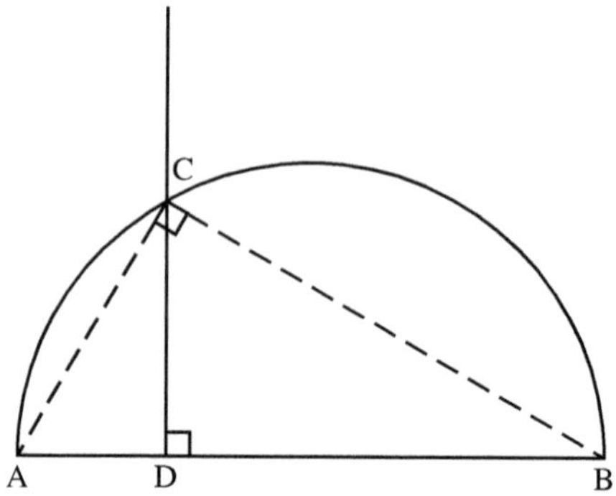

Figure 3.

a perpendicular to this segment. The intersection of the perpendicular and the semicircle having the segment of length 16 as diameter determines the other endpoint of the perpendicular segment of length $\sqrt{15}$ (see Figure 3).

The dashed lines are merely needed to justify the construction.

Postassessment

1. Construct a radical spiral up to $\sqrt{18}$.
2. Construct a segment of length $\sqrt{18}$ using a given unit length. Do *not* construct a radical spiral here.

Unit 41

Constructing a Pentagon

Performance Objective

Students will construct a regular pentagon given the length of the radius of the circumscribed circle.

Preassessment

Students should be familiar with the properties of regular polygons.

Teaching Strategies

Begin your lesson by having the class consider a regular decagon whose radius is 1 (Figure 1).

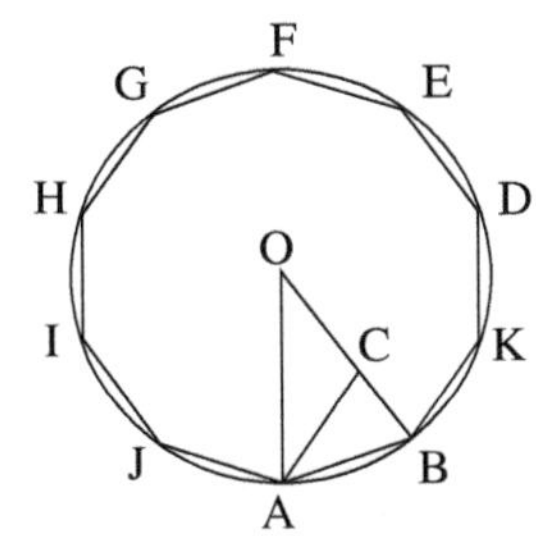

Figure 1.

With center O, draw $\overline{OA}$ and $\overline{OB}$ to form isosceles ΔAOB. The class should easily see that m$\angle$AOB $= 36°$ (i.e., $\frac{360}{10} = 36$). Therefore m$\angle$OAB $=$ m$\angle$OBA $= 72°$.

Isolate ΔAOB for clarity (Figure 2).

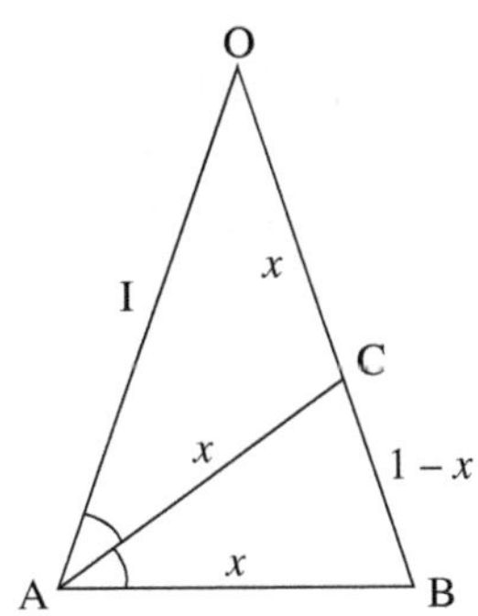

Figure 2.

Draw angle bisector $\overline{AC}$. Therefore, $m\angle$OAC $= 36°$, making ΔOCA isosceles. Similarly, ΔCAB is isosceles. Moreover, $\Delta AOB \sim \Delta BAC$. If we let $x =$ OC, then CB $= 1 - x$; also CA $= x =$ AB.

From the similarity, students should obtain the proportion: $\frac{1}{x} = \frac{x}{1-x}$, which leads to the equation $x^2 + x - 1 = 0$. This equation has two roots, one of which has geometric significance: $x = \frac{\sqrt{5}-1}{2}$.

Now have students consider the construction of this value of x. At any point A of a line, erect a perpendicular of length $1 = \mathrm{OA}$, and construct the unit circle (i.e., the circle with radius of length 1) tangent to that line at A. On the line, make $\mathrm{AP} = 2$ and then draw $\overline{\mathrm{OP}}$. By using the Pythagorean theorem, students should establish that $\mathrm{OP} = \sqrt{5}$ and $\mathrm{PQ} = \mathrm{OP} - \mathrm{OQ} = \sqrt{5} - 1$. Finally the perpendicular bisector of $\overline{\mathrm{PQ}}$ gives us $\mathrm{QR} = QR = \frac{\sqrt{5}-1}{2} = x$ (Figure 3).

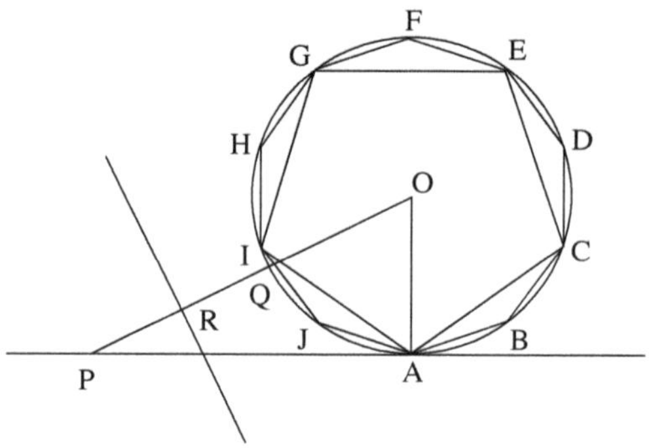

Figure 3.

Students should now mark consecutive segments of x on the original unit circle. When done accurately, this value of x will give exactly 10 arcs on the circle. After students have constructed the decagon, they ought to realize that by joining the alternate vertices of the decagon, the desired pentagon is formed.

Postassessment

Have students construct a regular pentagon given a specific unit length.

For an unusual method for constructing a (almost) regular pentagon, see *101 Plus Great Ideas for Introducing Key Concepts in Mathematics* by Alfred S. Posamentier and Herbert A. Hauptman (Corwin Press, 2006).

Reference

Posamentier, A. S. and I. Lehmann. *The Glorious Golden Ratio.* Amherst, NY: Prometheus Books, 2012.

Unit 42

Investigating the Isosceles Triangle Fallacy

This unit offers an opportunity to consider fully the Isosceles Triangle Fallacy. This fallacy can be used to reinforce the concept of betweenness.

Performance Objectives

1. *Students will exhibit the Isosceles Triangle Fallacy.*
2. *Students will indicate the "error" in the Isosceles Triangle Fallacy and prove their conjecture.*

Preassessment

Students should be familiar with the various methods for proving triangles congruent, as well as angle measurement in a circle.

Teaching Strategies

Begin the discussion by challenging your students to draw any *scalene* triangle on the chalkboard, which you will then prove isosceles.

To prove the scalene ΔABC isosceles, draw the bisector of $\angle C$ and the perpendicular bisector of $\overline{AB}$. From their point of intersection, G, draw perpendiculars to $\overline{AC}$ and $\overline{CB}$, meeting them at points D and F, respectively.

There are four possibilities for the above description for various scalene triangles. Figure 1, where $\overline{CG}$ and $\overline{GE}$ meet inside the triangle:

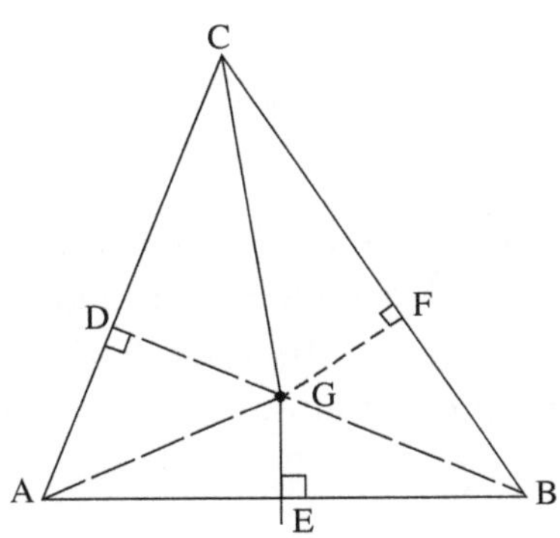

Figure 1.

Figure 2, where $\overline{CG}$ and $\overline{GE}$ meet on $\overline{AB}$:

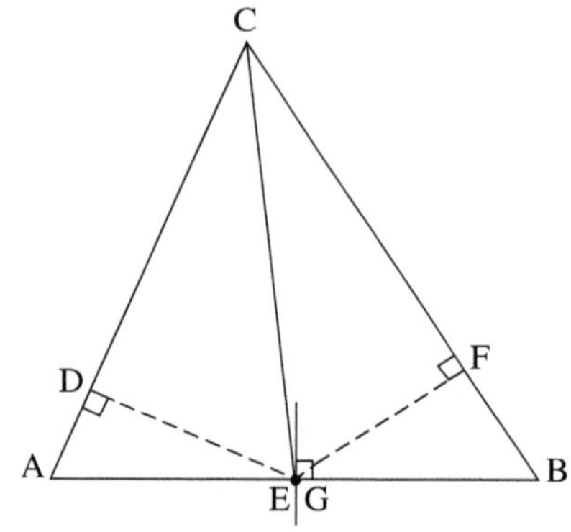

Figure 2.

Figure 3, where $\overline{CG}$ and $\overline{GE}$ meet outside the triangle, but the perpendiculars $\overline{GD}$ and $\overline{GF}$ fall on $\overline{AC}$ and $\overline{CB}$:

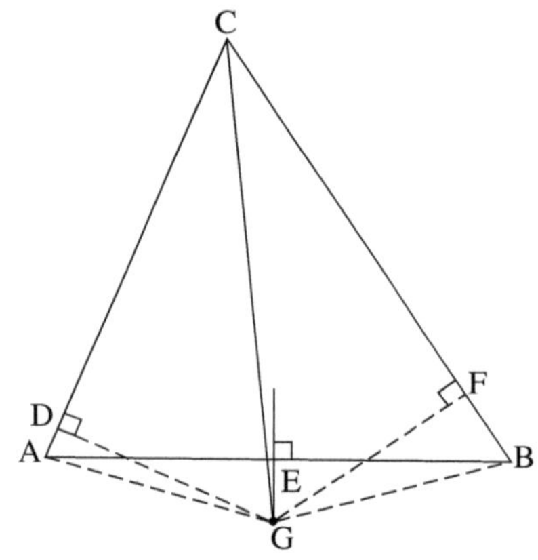

Figure 3.

Figure 4, where $\overline{CG}$ and $\overline{GE}$ meet outside the triangle, but the perpendiculars $\overline{GD}$ and $\overline{GF}$ meet $\overline{CA}$ and $\overline{CG}$ outside the triangle.

The "proof" of the fallacy can be done with any of the above figures. Have students follow the "proof" on any (or all) of these figures.

Given:

ABC is scalene.

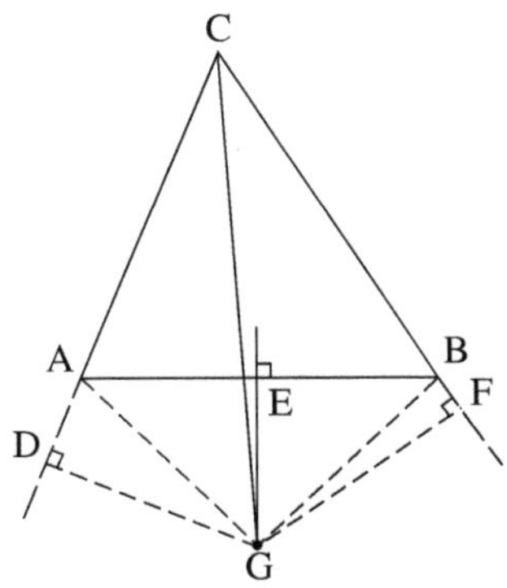

Figure 4.

Prove:

AC = BC (or ΔABC is isosceles)

Proof: Since $\angle$AGG $\cong$ $\angle$BCG and *rt* $\angle$CDG $\cong$ *rt* $\angle$CFG, ΔCDG $\cong$ ΔCFG (SAA). Therefore DG = FG and CD = CF $\cdot$ AG = BG (a point on the perpendicular bisector of a line segment is equidistant from the endpoints of the line segment) and $\angle$ADG and $\angle$BFG are right angles, ΔDAG $\cong$ ΔFBG (H.L.). Therefore DA = FB. It then follows that AC = BC (by addition in Figures 1, 2, and 3; and by subtraction in Figure 4).

At this point, students will be quite disturbed. They will wonder where the error was committed that permitted this fallacy to occur. Some students will be clever enough to attack the figures. By rigorous construction, students will find a subtle error in the figures:

a. The point G *must* be outside the triangle.
b. When perpendiculars meet the sides of the triangle, one will meet a side *between* the vertices, while the other will not.

Some discussion of Euclid's neglect of the concept of betweenness should follow. However, the beauty of this particular fallacy is the powerful proof of items *a* and *b* (above), which indicate the *error* of the fallacy.

Begin by considering the circumcircle of ΔABC. The bisector of $\angle$ACB must contain the midpoint, M, of $\widehat{AB}$ (since $\angle$ACM and $\angle$BCM are congruent

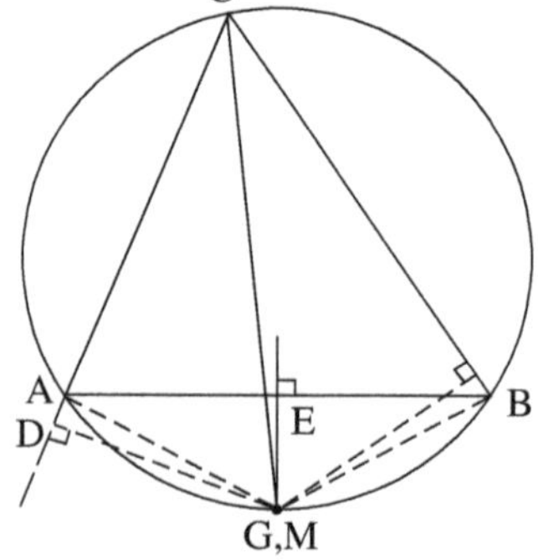

Figure 5.

inscribed angles). The perpendicular bisector of $\overline{AB}$ must bisect $\widehat{AB}$, and therefore pass through M. Thus, the bisector of $\angle ACB$ and the perpendicular bisector of $\overline{AB}$ intersect *outside* the triangle at M (or G). This eliminates the possibilities of Figures 1 and 2.

Now have students consider inscribed quadrilateral ACBG. Since the opposite angles of an inscribed (or cyclic) quadrilateral are supplementary, $m\angle CAG + m\angle CBG = 180$. If $\angle CAG$ and $\angle CBG$ are right angles, then $\overline{CG}$ would be a diameter and ΔABC would be isosceles. Therefore, since ΔABC is scalene, $\angle CAG$ and $\angle CBG$ are not right angles. In this case, one must be acute and the other obtuse. Suppose $\angle CBG$ is acute and $\angle CAG$ is obtuse. Then in ΔCGB the altitude on $\overline{CB}$ must be *inside* the triangle, while in obtuse ΔCAG, the altitude on $\overline{AC}$ must be *outside the triangle.* (*This is usually readily accepted by students but can be easily proved.*) The fact that one and *only one* of the perpendiculars intersects a side of the triangle *between* the vertices destroys the fallacious "proof." It is important that the teacher stress the importance of the concept of betweenness in geometry.

Postassessment

Have students:

1. "Prove" that any given scalene triangle is isosceles.
2. Indicate (and prove) where the "proof" in question 1 is fallacious.
3. Discuss the concept of betweenness in terms of its significance in geometry.

Reference

Posamentier, A. S. *Advanced Euclidean Geometry.* New York, John Wiley Publishing, 2002.

Posamentier, A. S. *Math Wonders to Inspire Teachers and Students.* Alexandria, VA: Association for Supervision and Curriculum Development, 2003.
Posamentier, A. S. and I. Lehmann. *The Secrets of Triangles.* Amherst, NY: Prometheus Books, 2012.

Unit 43

The Equiangular Point

This unit will develop interesting geometric relationships from an unusual geometric configuration. The topic is appropriate for any student who has mastered most of the high school geometry course.

Performance Objectives

1. *Students will define the equiangular point of an acute triangle.*
2. *Students will locate the equiangular point of an acute triangle.*
3. *Students will state at least three properties of the figure used to locate the equiangular point of an acute triangle.*

Preassessment

Before attempting to present this unit to your classes, review with them angle measurement of a circle and the basic properties of congruence and similarity.

Teaching Strategies

Begin your presentation by challenging students with the following problem:

Given: Acute ΔABC. Furthermore, ΔACD and ΔABF are equilateral (Figure 1).
Prove: DB = CF

Although this problem uses only the most elementary concepts of the high school geometry course, students tend to find it somewhat challenging. What seems to be most perplexing is the selection of the proper pair of

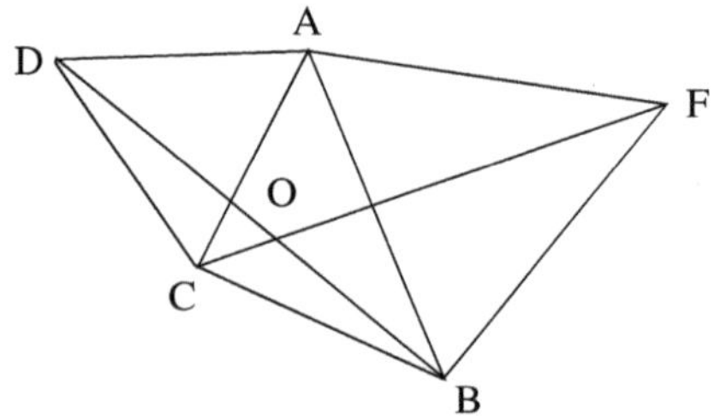

Figure 1.

triangles to prove congruent. If, after a few minutes, students do not find these, tell them to name triangles that use the required segments $\overline{DB}$ and $\overline{CF}$ as sides. Soon they will realize that they must prove $\Delta CAF \cong \Delta DAB$.

Next arises the problem of *how* to prove these triangles congruent. Lead students to realize that overlapping triangles usually share a common element. Here the common element is $\angle CAB$. Since ΔACD and ΔABF are equilateral, $m\angle DAC = 60°$, $m\angle FAB = 60°$, and $m\angle DAB = m\angle FAC$ (addition). Since ΔACD is equilateral, $AD = AC$, and since ΔABF is equilateral, $AB = AF$. Therefore, $\Delta CAF \cong \Delta DAB$ (S.A.S), and thus $DB = CF$.

Once students have fully understood this proof, have them consider a third equilateral triangle, ΔBCE, drawn on side $\overline{BC}$. Ask them to compare the length of $\overline{AE}$ to that of $\overline{DB}$ and $\overline{CF}$ (Figure 2).

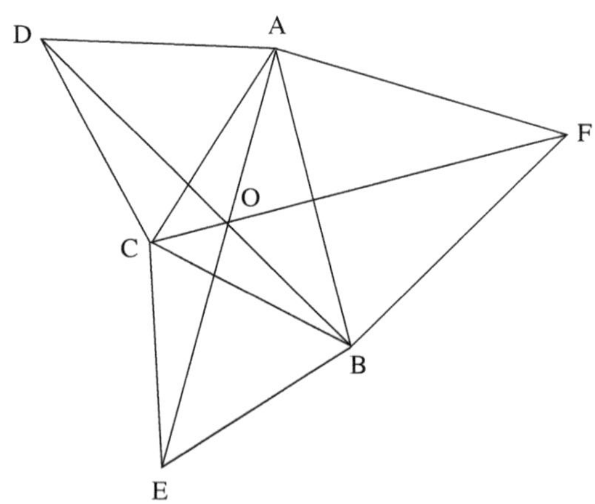

Figure 2.

Most students will realize that all three segments have the same length. A proof of this is done in the same way as the previous one. That is, simply have them prove that $\Delta CAE \cong \Delta CDB$ to get $AE = DB$.

The fact that AE = DB = CF is quite astonishing when we bear in mind that ΔABC was *any* acute triangle. A number of equally surprising results can now be established from this basis. Present each of these separately, but once each has been proved, carefully relate it to the previously established facts.

1. *The line segments* $\overline{AE}, \overline{DB}$, *and* $\overline{CF}$ *are concurrent.*

Proof: Consider the circumcircles of the three equilateral triangles, ΔACD, ΔABF, and ΔBCE.

Let K, L, and M be the centers of these circles (see Figure 3).

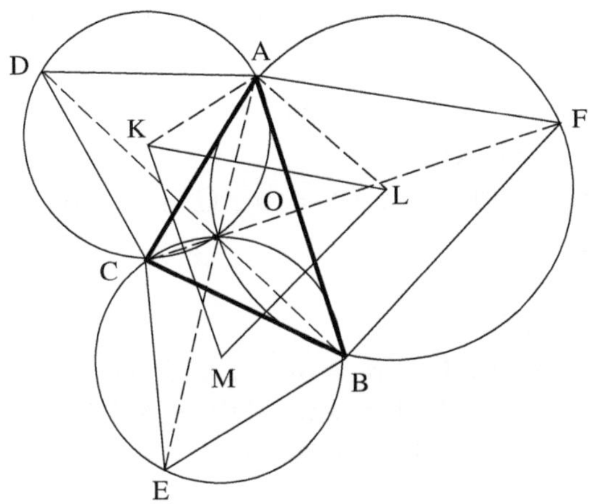

Figure 3.

Circles K and L meet at points O and A. Since $m\widehat{ADC} = 240°$, and we know that $m\angle AOC = \frac{1}{2}(m\widehat{ADC})$, $m\angle \widehat{AOC} = 120°$, similarly, $m\angle AOB = \frac{1}{2}(m\widehat{ADC}) = 120°$. Therefore, $m\angle COB = 120°$, since a complete revolution = 360°.

Since $m\widehat{CEB} = 240°$, $\angle COB$ is an inscribed angle and point O must lie on circle M. Therefore, we can see that the three circles are concurrent, intersecting at point O.

Now have students join point O with points A, B, C, D, E, and F. $m\angle DOA = m\angle AOF = m\ \angle FOB = 60°$, and therefore $\overleftrightarrow{DOB}$. Similarly $\overleftrightarrow{COF}$ and $\overleftrightarrow{AOE}$.

Thus, it has been proved that $\overline{AE}, \overline{CF}$, and $\overline{DB}$ are concurrent, intersecting at point O (which is also the point of intersection of circles K, L, and M).

Now ask the class to determine the point in ΔABC at which the three sides subtend congruent angles. They should quickly recall that we just proved that m$\angle$AOB = m$\angle$AOC = m$\angle$BOC = 120°. Thus, the point — called the *equiangular* point of a triangle — at which the sides of ΔABC subtend congruent angles is point O.

2. *The circumcenters* K, L, *and* M *of the three equilateral triangles* ΔACD, ΔABF, *and* ΔBCE, *respectively, determine another equilateral triangle.*

Proof: Before beginning this proof, review briefly with students the relationship among the sides of a 30-60-90 triangle.

Have your students consider equilateral ΔDAC. AK is 2/3 of the altitude (or median), so we obtain the proportion AC:AK = $\sqrt{3}$:1.

Similarly, in equilateral ΔAFB, AF:AL = $\sqrt{3}$:1.
Therefore, AC:AK = AF:AL.
m$\angle$KAC = m$\angle$LAF = 30°, m$\angle$CAL = m$\angle$CAL (reflexive), and m$\angle$KAL = m$\angle$CAF (addition).
Therefore, ΔKAL $\sim$ ΔCAF.
Thus, CF:KL = CA:AK = $\sqrt{3}$:1
Similarly, we may prove DB:KM = $\sqrt{3}$:1, and AE:ML = $\sqrt{3}$:1. Therefore, DB:KM = AE:ML = CF:KL. But since DB = AE = CF, as proved earlier, we obtain KM = ML = KL. Therefore, ΔKML is equilateral.

As a concluding challenge to your class, ask them to discover other relationships in Figure 3.

Postassessment

To test student comprehension of this lesson, give them the following exercises.

1. Define *the equiangular point* of an acute triangle.
2. Draw any acute triangle. Using straightedge and compasses, locate the equiangular point of the triangle.
3. State three properties found in Figure 3 above.

Reference

Posamentier, A. S. *Advanced Euclidean Geometry.* New York, NY: John Wiley Publishing, 2002.

Posamentier, A. S. and I. Lehmann, *The Secrets Of Triangles: A Mathematical Journal,* Amherst, New York: Prometheus Books, 2012.

Unit 44

The Minimum-Distance Point of a Triangle

This unit will develop a search for the point in a triangle the sum of whose distances to the vertices is a minimum.

Performance Objectives

1. *Students will prove that the sum of the distances to the sides of an equilateral triangle from an interior point is constant.*
2. *Students will locate the minimum distance point of a triangle with no angle of measure* 120° *or greater.*

Preassessment

Students should be familiar with basic concepts of geometric inequalities.

Ask students to find the position of a point of a quadrilateral, the sum of whose distances to the vertices is a minimum.

Teaching Strategies

Begin the discussion by having students consider the location of the point in the interior of a given quadrilateral, the sum of whose distances from the vertices is the smallest possible (from here on we shall refer to such a point as the *minimum-distance point*).

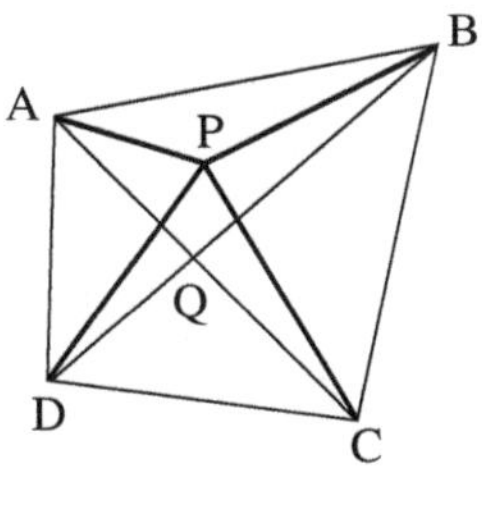

Figure 1.

You can expect most students to guess that the point of intersection of the diagonals (point Q in Figure 1) would be this minimum-distance point.

Although the conjecture is a clever one, try to elicit a justification (proof) for this point selection.

Let students select any point P(not at Q) in the interior of quadrilateral ABCD (Figure 1). $PA + PC > QA + QC$ (since the sum of the lengths of two sides of a triangle is greater than the length of the third). Similarly, $PB + PD > QB + QD$. By addition, $PA + PB + PC + PD > QA + QB + QC + QD$, which shows that the sum of the distances from the point of intersection of the diagonals of a quadrilateral to the vertices is less than the sum of the distances from *any other* interior point of the quadrilateral to the vertices.

The next logical concern of students is usually, "What is the minimum-distance point in a triangle?" Before tackling this question, it is useful to first consider another interesting theorem that will later help students develop the minimum-distance point of a triangle.

Once again ask students to use their intuition and reason inductively. Have them construct a large equilateral triangle. Then have them select any interior point and carefully measure its distances from the three *sides* of the equilateral triangle. After students have recorded the sum of these three distances, ask them to repeat the procedure three times, each time with a different interior point. Accurate measurements should yield equal distance sums for each point selected. Consequently, students ought to be able to draw the following conclusion: *The sum of the distances from any point in the interior of an equilateral triangle to the sides of the triangle is constant.* Two proofs of this interesting finding are provided here.

Method I:

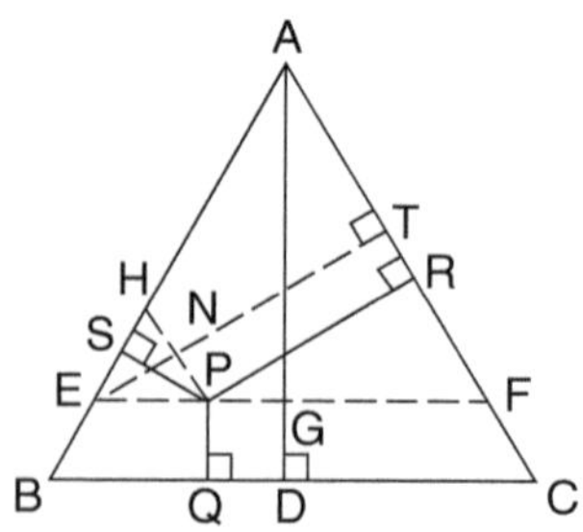

Figure 2.

In equilateral $\Delta ABC, \overline{PR} \perp \overline{AC}, \overline{PQ} \perp \overline{BC}, \overline{PS} \perp \overline{AB}$, and $\overline{AD} \perp \overline{BC}$ (Figure 2).
Draw a line through P parallel to $\overline{BC}$ meeting $\overline{AD}, \overline{AB}$, and $\overline{AC}$ at G, E, and F, respectively.
$PQ = GD$.
Draw $\overline{ET} \perp \overline{AC}$. Since ΔAEF is equilateral, $\overline{AG} \cong \overline{ET}$ (all the altitudes of an equilateral triangle are congruent).
Draw $\overline{PH} // \overline{AC}$ meeting $\overline{ET}$ at N. $\overline{NT} \cong \overline{PR}$.
Since ΔEHP is equilateral, altitudes $\overline{PS}$ and $\overline{EN}$ are congruent.
Therefore, we have shown that $PS+PR = ET = AG$. Since $PQ = GD, PS+PR+PQ = AG+GD = AD$, a constant for the given triangle.

Method II:

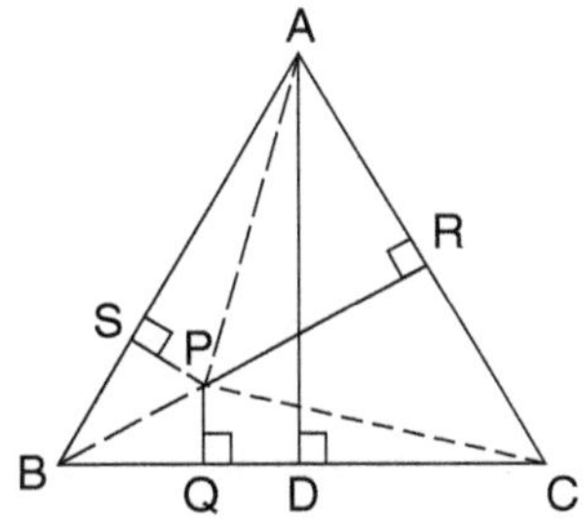

Figure 3.

In equilateral $\Delta ABC, \overline{PR} \perp \overline{AC}, \overline{PQ} \perp \overline{BC}, \overline{PS} \perp \overline{AB}$, and $\overline{AD} \perp \overline{BC}$, Draw $\overline{PA}, \overline{PB}$, and $\overline{PC}$ (Figure 3).

$$\begin{aligned} &\text{The area of } \Delta ABC \\ &\quad = \text{area of } \Delta APB + \text{area of } \Delta BPC + \text{area of } \Delta CPA \\ &\quad = \frac{1}{2}(AB)(PS) + \frac{1}{2}(BC)(PQ) + \frac{1}{2}(AC)(PR). \end{aligned}$$

Since $AB = BC = AC$, the area of $\Delta ABC = \frac{1}{2}(BC)[PS + PQ + PR]$. However, the area of $\Delta ABC = \frac{1}{2}(BC)(AD)$, therefore, $PS+PQ+PR = AD$, a constant for the given triangle.

Students are now ready to consider the original problem: to find the minimum-distance point of a triangle. We shall consider a scalene triangle with no angle having a measure greater than 120°.

Students, realizing the apparent need for symmetry in this problem, may suggest selecting the point at which the sides subtend congruent angles. If they are to accept this conjecture, they must prove it.

We shall therefore prove that *the point in the interior of a triangle (with no angle greater than 120°), at which the sides subtend congruent angles, is the minimum-distance point of the triangle.*

Proof:

In Figure 4, let M be the point in the interior of ΔABC, where m$\angle$AMB = m$\angle$BMC = m$\angle$AMC = 120°.

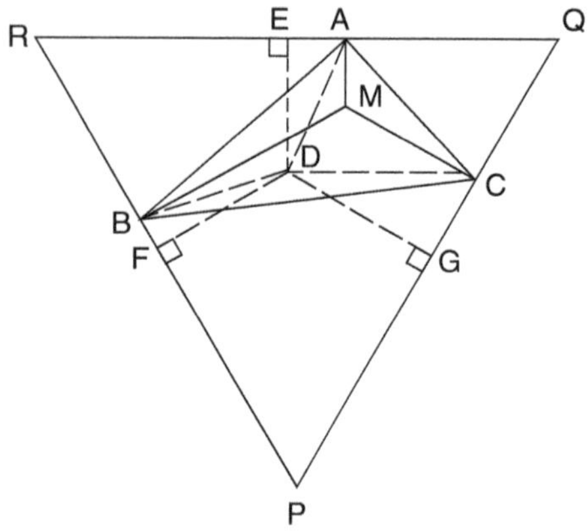

Figure 4.

Draw lines through A, B, and C that are perpendicular to $\overline{\text{AM}}$, $\overline{\text{BM}}$, and $\overline{\text{CM}}$, respectively.

These lines meet to form equilateral ΔPQR. (To prove ΔPQR is equilateral, notice that each angle has measure 60°. This can be shown by considering, for example, quadrilateral AMBR. Since m$\angle$RAM = m$\angle$RBM = 90°, and m$\angle$AMB = 120°, it follows that m$\angle$ARB = 60°.) Let D be *any other* point in the interior of ΔABC. We must show that the sum of the distances from M to the vertices is less than the sum of the distances from D to the vertices. From the theorem we proved above, MA + MB + MC = DE + DF + DG (where $\overline{\text{DE}}, \overline{\text{DF}}$, and $\overline{\text{DG}}$ are the perpendiculars to $\overline{\text{REQ}}, \overline{\text{RBP}}$, and $\overline{\text{QGP}}$, respectively).

But DE + DF + DG < DA + DB + DC. (The shortest distance from an external point to a line is the length of the perpendicular segment from the point to the line.)

By substitution:

$$\text{MA} + \text{MB} + \text{MC} < \text{DA} + \text{DB} + \text{DC}.$$

Now having proved the theorem, students may wonder why we chose to restrict our discussion to triangles with angles of measure less than 120°. Let them try to construct the point M in an obtuse triangle with one angle of measure 150°. The reason for our restriction should become obvious.

Postassessment

To test student comprehension of the above exercises, ask them to:

1. Prove that the sum of the distances to the sides of an equilateral triangle from an interior point is constant.
2. Locate the minimum-distance point of a triangle with no angle of measure greater than 120°.
3. Locate the minimum-distance point of a quadrilateral.

Reference

Posamentier, A. S. *Advanced Euclidean Geometry.* New York, NY: John Wiley Publishing, 2002.

Posamentier, A. S. and I. Lehmann, *The Secrets Of Triangles: A Mathematical Journal,* Amherst, New York: Prometheus Books, 2012.

Unit 45

The Isosceles Triangle Revisited

Early in the high school geometry course, students perform many practice proofs using isosceles triangles. One such proof involves proving that the angle bisectors of the base angles of an isosceles triangle are congruent. Although this is a rather simple proof, its converse is exceedingly difficult — perhaps among the most difficult statements to prove in Euclidean geometry. This unit presents several methods by which students can prove the statement.

Performance Objective

Students will prove that if two angle bisectors of a triangle are congruent, then the triangle is isosceles.

Preassessment

Students should have had practice with geometric proofs, including indirect proofs.

Teaching Strategies

Begin your presentation by asking students to prove:

The angle bisectors of the base angles of an isosceles triangle are congruent.

You may wish to start them uniformly:

Given: Isosceles ΔABC, with $AB = AC$, $\overline{BF}$ and $\overline{CE}$ are angle bisectors.
Prove: $\overline{BF} \cong \overline{CE}$

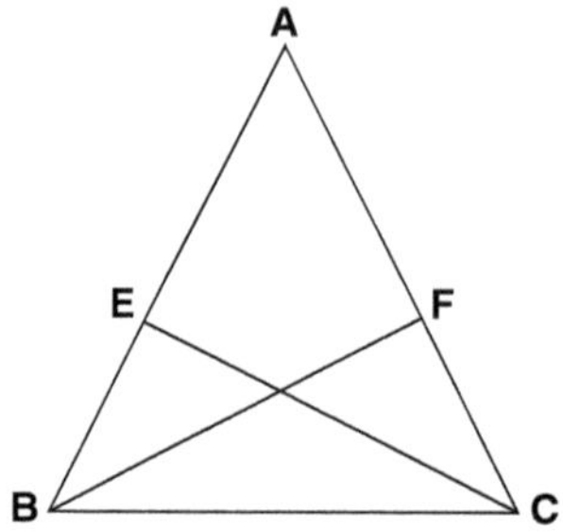

Proof: $m\angle FBC = \frac{1}{2}m\angle ABC$, and $m\angle ECB = \frac{1}{2}m\angle ACB$.

Since $m\angle ABC = m\angle ACB$ (base angles of an isosceles triangle), $m\angle FBC = m\angle ECB$.

Since $\overline{BC} \cong \overline{BC}$, $\Delta FBC \cong \Delta ECB$(ASA). Therefore, $\overline{BF} \cong \overline{CE}$.

When students have completed this proof, ask them to state the converse of the statement just proved.

If two angle bisectors of a triangle are congruent, then the triangle is isosceles.

Challenge students to prove this new statement. It is highly unlikely that your students will be able to prove this statement in a short time, so you may wish to show them some of the following proofs. They will be quite astonished that the converse of a rather simply proved statement is so difficult to prove. Each of the following proofs is quite instructional and merits special attention.

Given: $\overline{AE}$ and $\overline{BD}$ are angle bisectors of ΔABC, $\overline{AE} \cong \overline{BD}$.
Prove: ΔABC is isosceles.

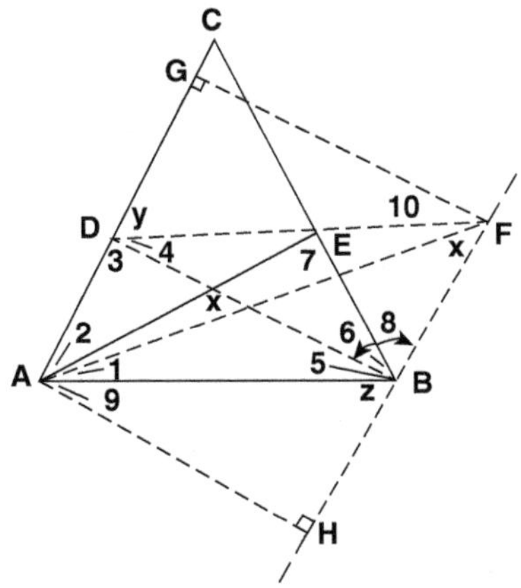

Proof:

Draw $\angle$DBF $\cong$ $\angle$AEB so that $\overline{\text{BF}} \cong \overline{\text{BE}}$.
Draw $\overline{\text{DF}}$.
Also draw $\overline{\text{FG}} \perp \overline{\text{AC}}$ and $\overline{\text{AH}} \perp \overline{\text{FH}}$.
By hypothesis, $\overline{\text{AE}} \cong \overline{\text{DB}}, \overline{\text{FB}} \cong \overline{\text{EB}}$.
And $\angle 8 \cong \angle 7$.
Therefore, ΔAEB $\cong$ ΔDBF (SAS) DF = AB and m$\angle 1$ = m$\angle 4$.
m$\angle x$ = m$\angle 2$ + m$\angle 3$ (exterior angles) of a triangle)
m$\angle x$ = m$\angle 1$ + m$\angle 3$ (substitution)
m$\angle x$ = m$\angle 4$ + m$\angle 3$ (substitution)
m$\angle x$ = m$\angle 7$ + m$\angle 6$ (exterior angles of a triangle)
m$\angle x$ = m$\angle 7$ + m$\angle 5$ (substitution)
m$\angle x$ = m$\angle 8$ + m$\angle 5$ (substitution)
Therefore, m$\angle 4$ + m$\angle 3$ = m$\angle 8$ + m$\angle 5$ (transitivity).
Thus, m$\angle z$ = m$\angle y$.
Right ΔFDG $\cong$ right ΔABH(SAA), DG = BH, and FG = AH.
Right ΔAFG $\cong$ right ΔFAH(HL) and AG = FH.
Therefore, GFHA is a parallelogram.
Also, m$\angle 9$ = m$\angle 10$ (from ΔABH and ΔFDG).
m$\angle$DAB = m$\angle$DFB (subtraction)
m$\angle$DFB = m$\angle$EBA (from ΔDBF and ΔAEB)
Therefore, m$\angle$DAB = mΔEBA (transitivity), and ΔABC is isosceles.

The following proofs of this theorem are *indirect proofs* and may deserve special introduction.

Given: $\overline{\text{BF}}$ and $\overline{\text{CE}}$ are angle bisectors of ΔABC $\cdot$ $\overline{\text{BF}} \cong \overline{\text{CE}}$
Prove: $\triangle$ABC is isosceles.

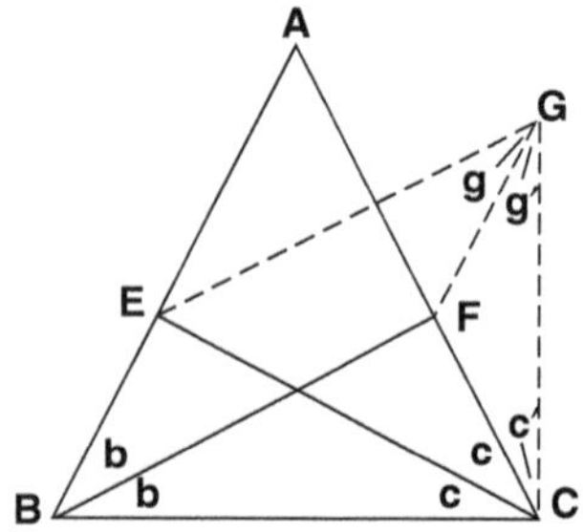

Indirect Proof I:

Assume ΔABC is *not* isosceles.
Let m$\angle$ABC > m$\angle$ACB.
$\overline{\text{BF}} \cong \overline{\text{CE}}$ (hypothesis) $\overline{\text{BC}} \cong \overline{\text{BC}}$.
m$\angle$ABC > m$\angle$ACB (assumption) $\overline{\text{CF}} > \overline{\text{BE}}$.
Through F, construct $\overline{\text{GF}}$ parallel to $\overline{\text{EB}}$.
Through E, construct $\overline{\text{GE}}$ parallel to $\overline{\text{BF}}$.
BFGE is a parallelogram.
$\overline{\text{BF}} \cong \overline{\text{EG}}, \overline{\text{EG}} \cong \overline{\text{CE}}$, ΔGEC is isosceles. $\text{m}\angle(g + g') = \text{m}\angle(c + c')$.
But $\text{m}\angle g = \text{m}\angle b$ $\text{m}\angle(b + g') = \text{m}\angle(c + c')$.
Therefore, $\text{m}\angle g' < \text{m}\angle c'$, since $\text{m}\angle b > \text{m}\angle c$.
In ΔGFC, we have CF < GF.
But GF = BE.
Thus, CF < BE.

The assumption of the inequality of m$\angle$ABC and m$\angle$ACB leads to two contradictory results, CF < BE and CF > BE. Therefore, ΔABC is isosceles.

A second indirect proof follows:

Given: $\overline{\text{BE}}$ and $\overline{\text{DC}}$ are angle bisectors of $\Delta\text{ABC} \cdot \overline{\text{BE}} \cong \overline{\text{DC}}$.
Prove: ΔABC is isosceles.

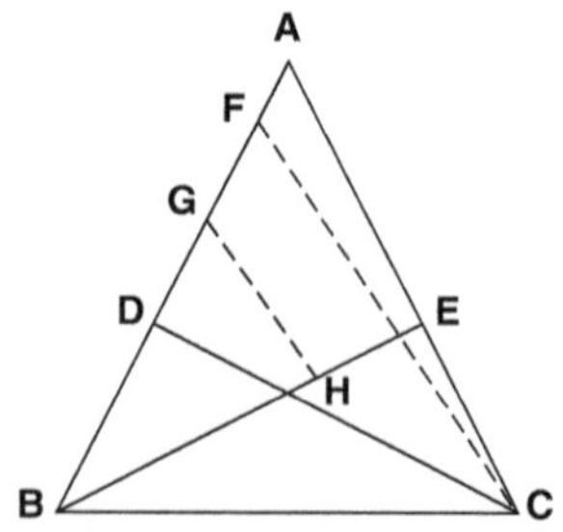

Indirect Proof II:

In ΔABC, the bisectors of angles ABC and ACB have equal measures (i.e., BE = DC).
Assume that m$\angle$ABC < m$\angle$ACB; then m$\angle$ABE < m$\angle$ACD.
We then draw $\angle$FCD congruent to $\angle$ABE. Note that we may take F between B and A without loss of generality. In ΔFBC, FB > FC. (If the measures of two angles of a triangle are not equal, then the measures of the sides opposite these angles are also unequal, the side with the greater measure being opposite the angle with the greater measure.)
Choose a point G so that $\overline{BG} \cong \overline{FC}$.
Then draw $\overline{GH} // \overline{FC}$.
Therefore, $\angle$BGH $\cong$ $\angle$BFC (corresponding angles), and ΔBGH $\cong$ ΔCFD (ASA).
Then it follows that BH = DC.

Since BH < BE, this contradicts the hypothesis that the angle bisectors are equal. A similar argument will show that it is impossible to have m$\angle$ACB < m$\angle$ABC.

It then follows that m$\angle$ACB = m$\angle$ABC and that ΔABC is isosceles.

Postassessment

Have students prove that if two angle bisectors of a triangle are congruent, then the triangle is isosceles.

Reference

Posamentier, A. S. and Charles T. Salkind. *Challenging Problems in Geometry.* New York: Dover, 1996.
Posamentier, A. S. and I. Lehmann, *The Secrets Of Triangles: A Mathematical Journal*, Amherst, New York: Prometheus Books, 2012.

Unit 46

Reflective Properties of the Plane

Performance Objective

Given a line and two points on one side of a line, students will determine the shortest combined path from one point to the line and then to the second point.

Preassessment

Using the following illustration, ask students to locate the precise point on the cushion $\overline{PQ}$ of the "billiard table" that ball A must hit in order to then hit ball B (assume no "English" on the ball) (Figure 1).

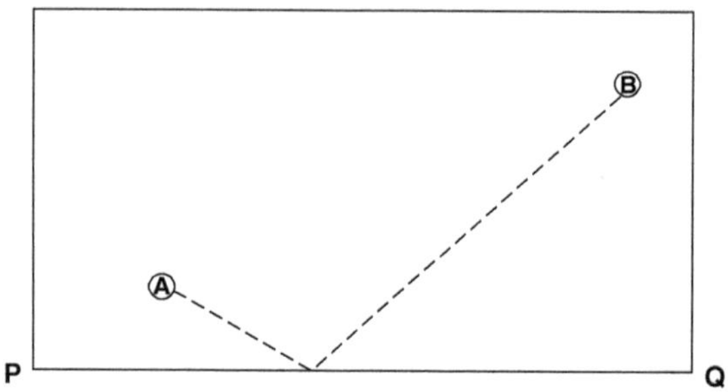

Figure 1.

The disagreement as to where to hit the ball will develop enough interest to motivate the topic of the properties of reflection.

Teaching Strategies

Have the class try to prove the following property: "A ray of light will make equal angles with a mirror before and after reflecting off a mirror." (This theorem can be easily proved after considering the following proof.)

To find the *shortest path* from point A to line m then to point B in Figure 2, consider the perpendicular from A to line m (meeting line m at point C). Let D be the point on $\overrightarrow{AC}$ such that $\overline{AC} \cong \overline{CD}$. Point D is called the *reflected image* of point A in line m.

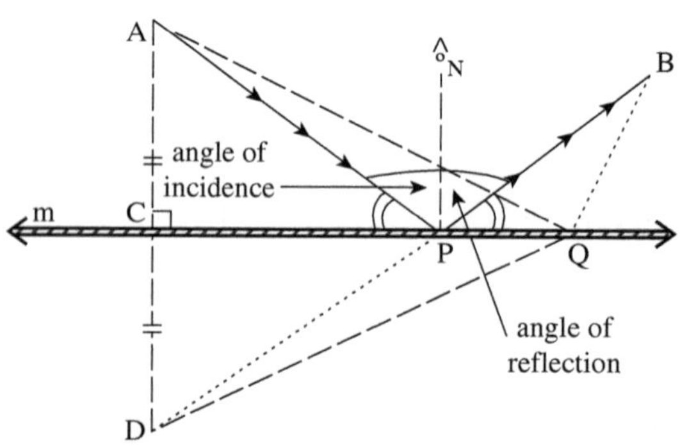

Figure 2.

The point of intersection of $\overline{BD}$ and line m determines point P, the required point in the original problem. However, what must be shown now is that AP + PB is less than *any other* path from A to line m (say at point Q), and then to B.

Students might be more comfortable stating this as a "formal proof":

Given: Points A and B are on the same side of line $m\,\overleftrightarrow{ACD} \perp \overleftrightarrow{CPQ}$, where Q is any point on $\overleftrightarrow{CP}$ (other than P).
$\overleftrightarrow{DPB}$
$\overline{AC} \cong \overline{CD}$

Prove: $AP + PB < AQ + QB$

Outline of proof: Because line m is the perpendicular bisector of $\overline{ACD}$, $\overline{AP} \cong \overline{DP}$ and $\overline{AQ} \cong \overline{QD}$. In ΔDQB, $BD < BQ + QD$ (triangle inequality). Since $BD = DP + PB$, $AP + PB < AQ + BQ$.

You can now show the class that since $\angle BPQ \cong \angle CPD$, and $\angle APC \cong \angle CPD$, that $\angle APC \cong \angle BPQ$. If $\overleftrightarrow{PN} \perp$ line m, then $\angle APN$, the angle of incidence, is congruent to $\angle BPN$, the angle of reflection.

Have students apply the properties of reflection to the billiard table problem. A billiard ball will rebound off a cushion as a ray of light "bounces" off a mirror. Thus, if a ball is at position A (Figure 3) and the player desires to hit it off cushion $\overline{PQ}$ to position B, he can aim his shot at the point on $\overline{PQ}$ where he would see B (if a mirror were placed along $\overline{PQ}$).

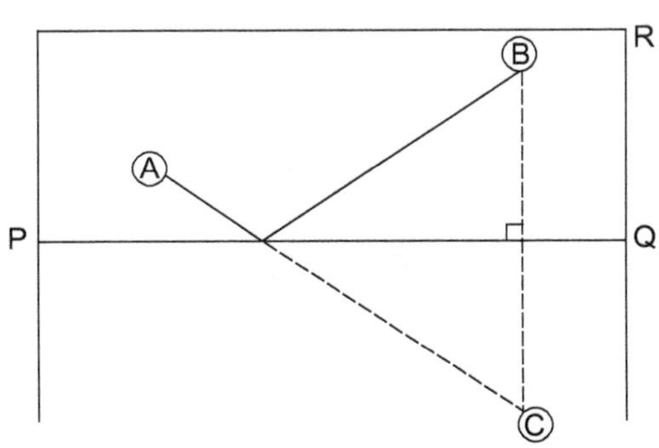

Figure 3.

Now have students consider the problem of hitting two cushions ($\overline{PQ}$ then $\overline{QR}$) before hitting B.

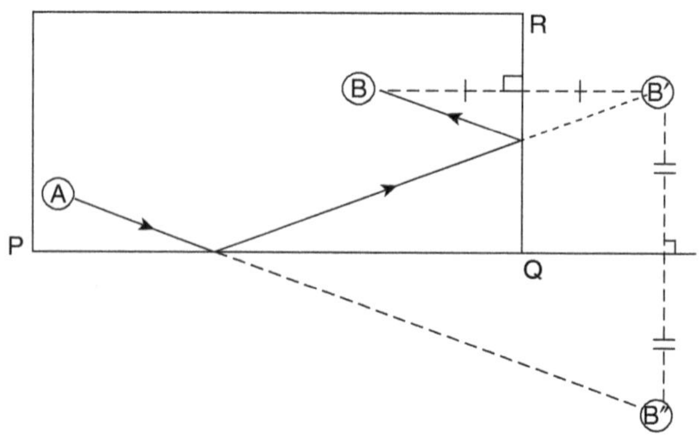

Figure 4.

Consider the reflected image of B in $\overline{QR}$; call it B′ (Figure 4). Now they merely have to consider the problem of where to hit the ball from A off cushion $\overleftrightarrow{PQ}$ so that it will roll toward B′. To do this, have them take the reflected image of B′ in $\overleftrightarrow{PQ}$ then the intersection of the line connecting A and the reflected image of B′ (call it B″) and $\overleftrightarrow{PQ}$ is the point to aim for to make this two-cushion shot. Students may envision this point as the reflection of the ball at B in the mirror placed along $\overleftrightarrow{PQ}$ that they would see as the reflection in the mirror placed along $\overline{QR}$.

A motivated class may now wish to look beyond the double reflection to where the angle between the two planes is always fixed as a right angle.

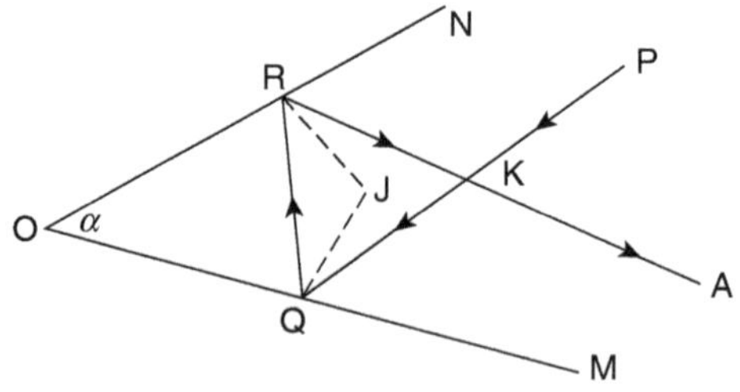

Figure 5.

Two mirrors with a fixed dihedral angle between them are called *angle mirrors*. Have the class prove that if an observer shines a light into the angle mirrors so that the ray reflects off one mirror, then the other, the final reflected ray will form an angle with the original ray, which is double the dihedral angle between the two mirrors. In other words:

Given: Mirrors OM and ON, let $\angle NOM = \alpha$ also a light ray originating at point P aimed at point Q reflecting off OM onto $\overrightarrow{ON}$ then to A.
Prove: $m\angle PKR = 2\alpha$ (Figure 5)
Outline of proof: Draw the normals (the perpendiculars) to the planes (the mirrors) at the points of incidence in each plane, i.e., $\overleftrightarrow{QJ} \perp \overleftrightarrow{OM}$, at point Q, and draw $\overleftrightarrow{RJ} \perp \overleftrightarrow{ON}$, at point R (the point of incidence in mirror $\overleftrightarrow{ON}$). Then, by the property of reflection, $\overleftrightarrow{OJ}$ and $\overleftrightarrow{RJ}$ bisect $\angle PQR$ and $\angle QRA$, respectively. Then $m\angle PKR = m\angle KQR + m\angle KRQ$ (exterior angle of a triangle theorem). Then $m\angle PKR = 2(m\angle JQR + m\angle QRJ)$ and $m\angle PKR = 2(180 - m\angle RJQ)$. But, both $\angle JRO$ and $\angle JQO$ are right angles, so $m\angle ROQ = 180 - m\angle RJQ$ (the sum of the measures of the interior angles of a quadrilateral is 360). By substituting, $m\angle PKR = 2(m\angle ROQ) = 2\alpha$.

One of the applications of angle mirrors is that when the dihedral angle is 45°, the ray will be reflected 90°. Such a pair of mirrors is often called an "optical square," because it is used to determine perpendicular lines of sight.

To show how the optical square is used, station one student at each of the three points O, A, and B, so that the three points will define a triangle (Figure 6). Using the optical square, they will be able to determine where the perpendicular from B to $\overleftrightarrow{AO}$ meets $\overleftrightarrow{AO}$. Have the student standing at O look at the student standing at A. Have another student, P, holding the optical square move along the line of sight from O to A until (at some point m) the student at O is able to see the student at B in the angle mirror. The point m is the foot of the altitude from B to $\overleftrightarrow{OA}$.

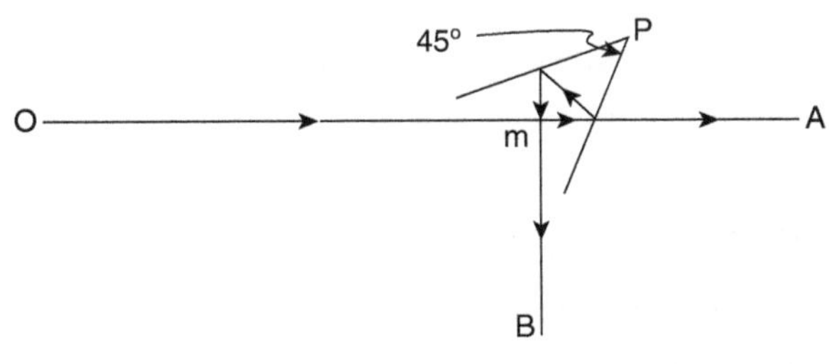

Figure 6.

Postassessment

Using the property of reflection, have students prove that the height of a flagpole is

$$x = \frac{h \cdot BC}{AB}$$

where x is the height of the flagpole and h is the height of an observer.

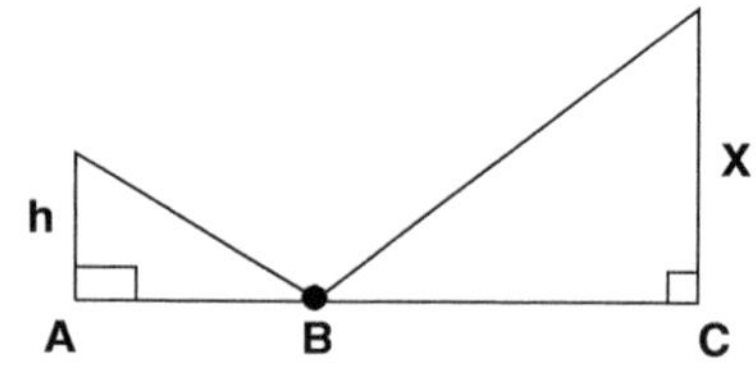

Unit 47

Finding the Length of a Cevian of a Triangle

This unit presents a method for finding the length of *any* line segment joining a vertex of a triangle with any point on the opposite side. Such a line segment is called a cevian, after Giovanni Ceva who developed a theorem about the concurrency of such line segments. This technique is particularly useful to students since it fills a void in many curricula. Students are usually taught methods for finding the lengths of special cevians such as altitudes and some medians. However, by using Stewart's theorem (named after Matthew Stewart who published it in 1745), students will now be able to find the length of *any* cevian of a triangle.

Performance Objectives

1. *Students will find the length of a specific cevian of a given triangle, all of whose side (and segment) lengths are known.*
2. *Students will use a special formula to find the length of an angle bisector of a triangle, given the lengths of its sides.*

Preassessment

Students should have mastered most of the standard high school geometry course. For review purposes, have students work the following problem:

> In a triangle whose sides have lengths 13, 14, and 15, what is the length of the altitude to the side of length 14?

Teaching Strategies

One of the major skills needed to develop Stewart's theorem is a working knowledge of the Pythagorean theorem. The problem stated above requires this skill.

After students have drawn the diagram required by this problem, they will immediately see two right triangles.

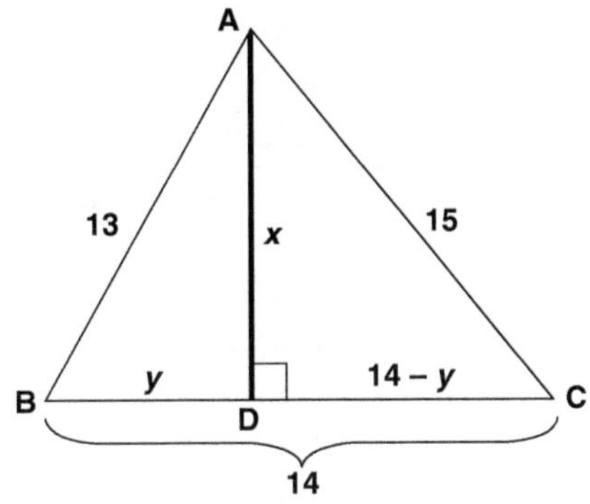

Figure 1.

To Figure 1, they will apply the Pythagorean theorem twice, once to ΔACD and a second time to ΔABD.

$$\begin{aligned}
\text{For } \Delta\text{ACD:}\quad & x^2 + (14-y)^2 = 225\\
\text{For } \Delta\text{ABD:}\quad & x^2 + y^2 = 169\\
\text{By subtraction:}\quad & (14-y)^2 - y^2 = 56\\
& 196 - 28y + y^2 - y^2 = 56\\
& y = 5\\
& \text{and then } x = 12.
\end{aligned}$$

Thus, students will see two right triangles with integral length sides: $5, 12, 13$ and $9, 12, 15$.

Now challenge your students to find the length of the angle bisector from vertex A of ΔABC. After a short time, their frustration will be evident. At this juncture, have the class stop working and discuss with them Stewart's theorem.

Stewart's theorem:

In Figure 2, the theorem states that:

$$a^2n + b^2m = c(d^2 + mn).$$

By this theorem, d may be found if a, b, m, and n are known. The proof of this most useful theorem follows.

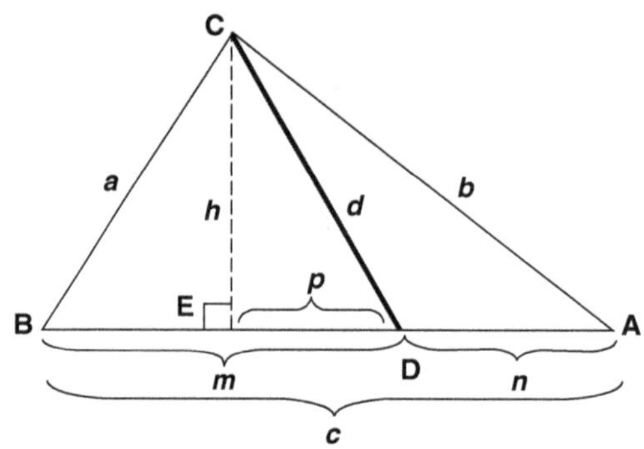

Figure 2.

Proof:

In ΔABC, let BC $= a$, AC $= b$, AB $= c$, CD $= d$. Point D divides $\overline{\text{AB}}$ into two segments; BD $= m$ and DA $= n$. Draw the altitude CE $= h$ and let ED $= p$.

To proceed with the proof of Stewart's theorem, we first derive two necessary formulas. The first one is applicable to ΔCBD. We apply the Pythagorean theorem to ΔCEB to obtain

$$(\text{CB})^2 = (\text{CE})^2 + (\text{BE})^2.$$

$$\text{Since BE} = m - p, a^2 = h^2 + (m - p)^2. \qquad \text{((I))}$$

However, by applying the Pythagorean theorem to ΔCED, we have $(\text{CD})^2 = (\text{CE})^2 + (\text{ED})^2$, or $h^2 = d^2 - p^2$. Replacing h^2 in equation (I), we obtain

$$a^2 = d^2 - p^2 + (m - p)^2,$$

$$a^2 = d^2 - p^2 + m^2 - 2mp + p^2.$$

$$\text{Thus, } a^2 = d^2 + m^2 - 2mp. \qquad \text{((II))}$$

A similar argument is applicable to ΔCDA. Applying the Pythagorean theorem to ΔCEA, we find that

$$(\mathrm{CA})^2 = (\mathrm{CE})^2 + (\mathrm{EA})^2.$$

$$\text{Since EA} = (n+p), b^2 = h^2 + (n+p)^2. \tag{III}$$

However, $h^2 = d^2 - p^2$ substitute for h^2 in (III) as follows:

$$b^2 = d^2 - p^2 + (n+p)^2,$$

$$b^2 = d^2 - p^2 + n^2 + 2np + p^2.$$

$$\text{Thus, } b^2 = d^2 + n^2 + 2np. \tag{IV}$$

Equations (II) and (IV) give us the formulas we need. Now multiply equation (II) by n to get

$$a^2n = d^2n + m^2n - 2mnp, \tag{V}$$

and multiply equation (IV) by m to get

$$b^2m = d^2m + n^2m + 2mnp. \tag{VI}$$

Adding (V) and (VI), we have

$$a^2n + b^2m = d^2n + d^2m + m^2n + n^2m$$
$$+ 2mnp - 2mnp.$$

Therefore, $a^2n + b^2m = d^2(n+m) + mn(m+n)$.

Since $m + n = c$, we have $a^2n + b^2m = d^2c + mnc$, or

$$a^2n + b^2m = c(d^2 + mn).$$

Your students should now be ready to find the length of the median from vertex A of ΔABC, where AB $= 13$, BC $= 14$, and AC $= 15$ (Figure 3). All they need to do is apply Stewart's theorem as follows:

$$c^2n + b^2m = a(d^2 + mn)$$

However, since $\overline{\mathrm{AD}}$ is a median $m = n$.

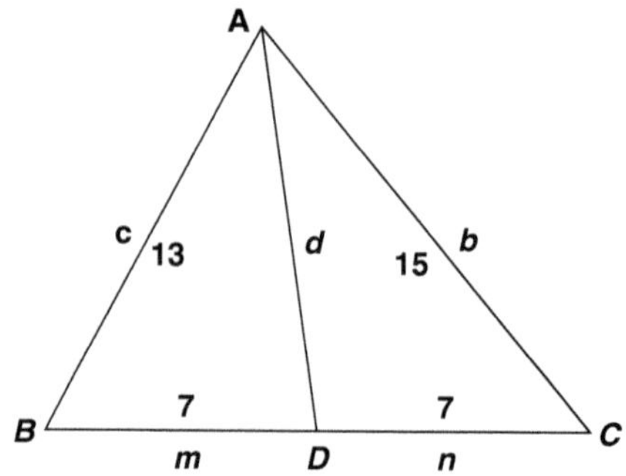

Figure 3.

Substituting in the above formula:

$$13^2(7) + 15^2(7) = 14(d^2 + 49).$$

Therefore, $d = 2\sqrt{37}$.

To find the length of an angle bisector of a triangle, Stewart's theorem leads to a very concise relationship, which students will find easy to use.

Have students consider $\triangle$ABC with angle bisector $\overline{\text{AD}}$.
By Stewart's theorem, we obtain the following relationship:

$$c^2n + b^2m = a(t_a^2 + mn), \text{ or}$$

$$t_a^2 + mn = \frac{c^2n + b^2m}{a}, \text{ or}$$

as illustrated by Figure 4.

But, $\frac{c}{b} = \frac{m}{n}$. (The bisector of an angle of a triangle divides the opposite side into segments whose measures are proportional to the measures of the other two sides of the triangle. The converse is also true.) Therefore, $cn = bm$.

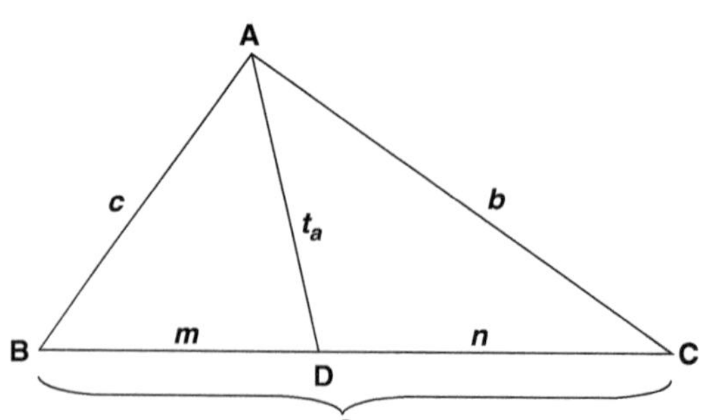

Figure 4.

Substituting in the above equation,

$$t_a^2 + mn = \frac{cbm + cbn}{m+n} = \frac{cb(m+n)}{m+n} = cb.$$

Hence, $t_a^2 = cb - mn$.

At this juncture, your students should be able to find the length of *any* cevian of a triangle. For reinforcement, present problems involving angle bisectors and medians before going on to other types of cevians.

Postassessment

Have students complete the following exercises:

1. Find the length of an altitude drawn to the longest side of a triangle whose sides have lengths 10, 12, and 14.
2. Find the length of a median drawn to the longest side of a triangle whose sides have lengths 10, 12, and 14.
3. Find the length of an angle bisector drawn to the longest side of a triangle whose sides have lengths 10, 12, and 14.
4. In ΔPQR, if $\text{PR} = 7, \text{PQ} = 8, \text{RS} = 4$, and $\text{SQ} = 5$, find PS when S is on $\overline{\text{RQ}}$.

Reference

Posamentier, A. S. *Advanced Euclidean Geometry: Excursions for Secondary Teachers and Students.* New York, NY: John Wiley Publishing, 2002.

Unit 48

A Surprising Challenge

This activity will alert geometry students to the fact that what may appear easy may actually be quite difficult.

Performance Objective

Given a geometric problem of the kind posed here, students will properly analyze it and solve it.

Preassessment

Students should be able to handle geometric proofs with relative ease before attempting this unit. The problem posed here is quite difficult to prove, yet easy to state. It ought to be within the reach of a slightly above-average student of high school geometry.

Teaching Strategies

The geometric problem you are about to pose to your students appears to be quite simple and certainly innocent.

Problem: ΔABC is isosceles $(CA = CB) \cdot m\angle ABD = 60°, m\angle BAE = 50°$, and $m\angle C = 20°$. Find the measure of $\angle EDB$.

Students should be given a fair amount of time to grapple with this problem. Immediately, they will find the measures of most of the angles in the diagram. However, they will soon realize that this problem was not as simple as they first imagined, since they will be most likely unable to solve the problem. At this point, you can begin your discussion of a solution of this problem.

Students will be quick to realize that auxiliary lines are necessary in order to solve this problem. Suggest that they draw $\overline{DG} // \overline{AB}$, where G is on $\overline{CB}$ (Figure 1). Then draw $\overline{AG}$ intersecting $\overline{BD}$ at F. The last segment to be drawn is $\overline{EF}$ (see Figure 2).

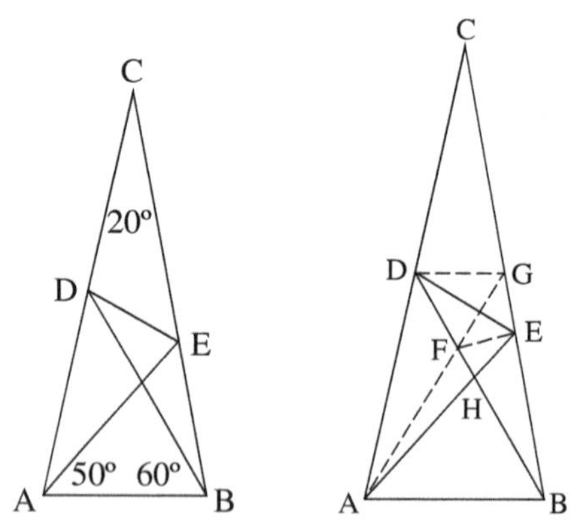

Figure 1. Figure 2.

Students should be able to prove that $\angle ABD \cong \angle BAG$. Then $m\angle AGD = m\angle BAG = 60°$ (alternate interior angles of parallel lines). Thus, $m\angle AFB$ must be 60° and ΔAFB is equilateral, and AB = FB.

Since $m\angle EAB = 50°$, and $m\angle ABE = 80°$, $m\angle AEB = 50°$ making ΔABE isosceles and AB = EB. Therefore, FB = EB (transitivity), and ΔEFB is isosceles.

Since $m\angle EBF = 20°$, $m\angle BEF = m\angle BFE = 80°$. As $m\angle DFG = 60°$, $m\angle GFE = 40° \cdot GE = EF$ (equal length sides of an isosceles triangle), and $DF = DG$ (sides of an equilateral triangle). Thus, DGEF is a kite, i.e., two isosceles triangles externally sharing a common base. $\overline{DE}$ bisects $\angle GDF$ (property of a kite), therefore $m\angle EDB = 30°$.

Another possible method of solution follows:

In isosceles ΔABC, $m\angle ACB = 20°$, $m\angle CAB = 80°$, $m\angle ABD = 60°$, and $m\angle EAB = 50°$.

Draw $\overline{BF}$ so that $m\angle ABF = 20°$; then draw $\overline{FE}$ (Figure 3). In ΔABE, m AEB = 50° (sum of measures of the angles of a triangle is 180°) therefore ΔABE is isosceles and AB = EB. (I)

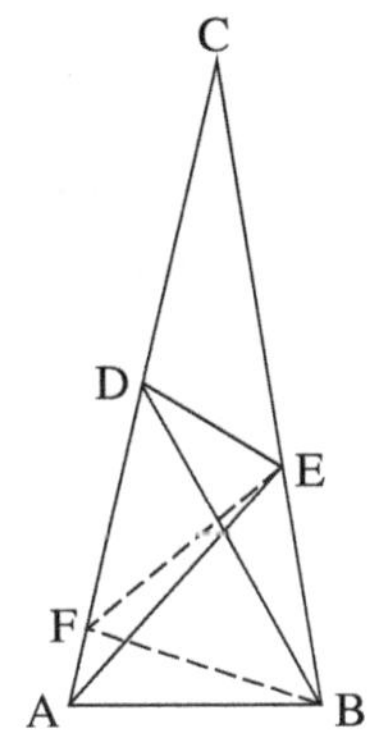

Figure 3.

Similarly, ΔFAB is isosceles, since $m\angle AFB = m\angle FAB = 80°$.
Thus, AB = FB. (II)
From (I) and (II), EB = FB. Since $m\angle FBE = 60°$, ΔFBE is equilateral and EB = FB = FE. (III)
Now, in ΔDFB, $m\angle FDB = 40°$, and $m\angle FBD = m\angle ABD - m\angle ABF = 60° - 20° = 40°$. Thus, ΔDFB is isosceles and FD = FB. (IV)

It then follows from (III) and (IV) that FE = FD, making ΔFDE isosceles, and m$\angle$FDE = m$\angle$FED. Since m$\angle$AFB = 80° and m$\angle$EFB = 60°, then m$\angle$AFE, the exterior angle of isosceles ΔFDE, equals 140°, by addition. It follows that m$\angle$ADE = 70°.

Therefore, m$\angle$EDB = m$\angle$ADE − m$\angle$FDB = 70° − 40° = 30°.

There are various other methods for solving this problem. One source for seven solutions of this problem is *Challenging Problems in Geometry* by A. S. Posamentier and C. T. Salkind, pp. 149–154 (New York: Dover, 1996).

Postassessment

Have students discover another solution for the above problem.

Unit 49

Making Discoveries in Mathematics

This activity is intended to permit the student to make discoveries based on observation, and then propose a conclusion.

Performance Objective

Faced with a mathematical pattern students will state their discoveries and propose conclusion.

Teaching Strategies

This activity will be comprised of a series of mathematical mini-activities, each of which will require the student to discover a pattern or relationship and then state his or her conclusion.

1. Select any two consecutive square numbers (e.g., 4 and 9). Give one prime between these two numbers. Repeat this for 10 other pairs of consecutive square numbers. Now try to find a pair of consecutive square numbers that do *not* have a prime number between them. What conclusion can you draw from this experiment?

2. Select any even integer greater than 2. Now express this even integer as the sum of exactly two prime numbers. For example, $8 = 3 + 5$, and $18 = 7+11$. Repeat this for at least 25 even integers before you draw any conclusions.
3. Draw *any* triangle. Using a protractor carefully trisect each of the angles of the triangle. Locate the points of intersection of the adjacent angle trisectors as illustrated in Figure 1.

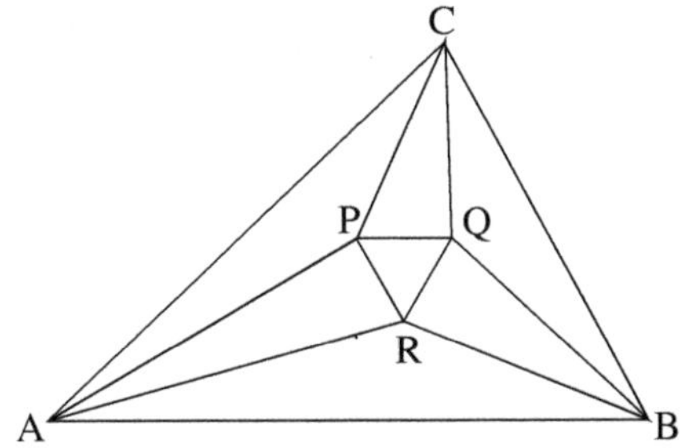

Figure 1.

Join these three points and inspect the triangle formed. Repeat this construction for at least six different triangles before drawing any conclusion.
4. Draw *any* parallelogram. Construct an equilateral triangle externally on two adjacent sides as shown in Figure 2. Then join the two remote vertices of the equilateral triangles to each other and to the farthest vertex of the parallelogram. What kind of triangle is formed? Before stating a conclu sion, repeat this experiment with at least six different parallelograms.

Students ought to be able to actually prove this last example. However, the other three examples should not be attempted; 1 and 2 have never been proved, and 3 is extremely difficult to prove.*

You might try to replicate these experiments with others similar to these. It is important for students to learn to trust their intuition in mathematics and be able to draw correct inductive conclusions.

*Two proofs of this theorem can be found in *Challenging Problems in Geometry* by A.S. Posamentier and C. T. Salkind (New York: Dover, 1996).

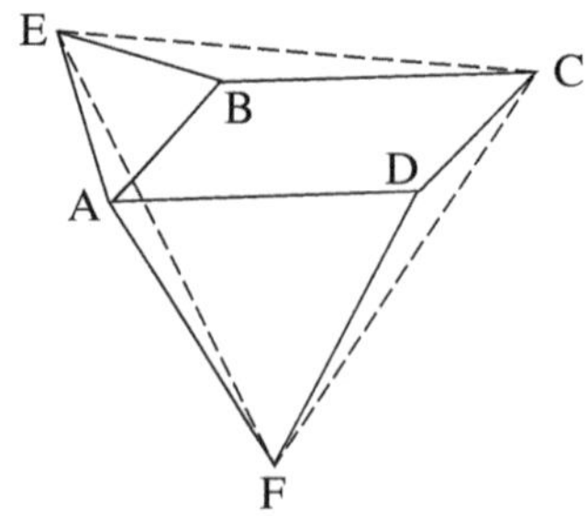

Figure 2.

Postassessment

Ask students to find the sum of the first $1, 2, 3, 4, 5, 6, \ldots$ 15 odd integers and list the 15 different sums. Then have the students state a logical conclusion.

Unit 50

Tessellations

Performance Objectives

1. *Given a regular polygon, each student will determine whether it will tessellate a plane.*
2. *Given a combination of various regular polygons, each student will determine whether they will tessellate a plane.*

Preassessment

Before beginning this lesson, explain to students that when polygons are fitted together to cover a plane with no spaces between them and no overlapping, the pattern is called a *tessellation*. (Mention the pattern of tiles on bathroom floors as one of the more common tessellations.) A tessellation made up entirely of congruent regular polygons that meet so that no vertex of one polygon lies on a side of another is referred to as a regular tessellation. Explain further that a network of equilateral triangles, a checkerboard pattern of squares, and a hexagonal pattern are the *only* tessellations of regular polygons that exist.

Teaching Strategies

To mathematically show why the previous three patterns are the only tessellations that fit the description, ask the class to suppose m regular polygons are required to fill the space around one point (where the vertex of the angles of the polygon is situated). If they assume that each regular polygon has n sides, each interior angle of the polygon is equal to $\frac{(n-2)180°}{n}$. Therefore, $\frac{m(n-2)180°}{n} = 360°$ and $(m-2)(n-2) = 4$.

Considering the nature of the problem, both integers m and n are greater than 2. If $m = 3$, then $n = 6$. If $m > 3$, then $n < 6$ and since $n > 2$ only the values $n = 3$, $n = 4$, and $n = 5$ need be considered. If $n = 3$, then $m = 6$; if $n = 4$, then $m = 4$. If $n = 5, m$ is nonintegral; therefore, the only solutions are $m = 3$, $n = 6$; $m = 4$, $n = 4$; $m = 6$, $n = 6$, $n = 3$. Have students suggest more convenient ways of symbolizing these tessellations $(6^3, 4^4, 3^6)$. Use the following diagrams to show that no other regular polygon has an interior angle that will divide 360° (see Figures 1 and 2).

Through further investigation, the range of tessellations may be expanded. Tessellations can also be formed by fitting together two or more kinds of regular polygons, vertex to vertex, in such a way that the same polygons, in the same cyclic order, surround each vertex. These are called *semiregular* tessellations in which there can be no fewer than three and no more than six polygons at any vertex.

Ask students to consider a ternary arrangement (three polygons share one point as vertex). Because the sum of the angles around any vertex must be 360°, a ternary arrangement of polygons of n_1, n_2, sides, respectively, will be possible only if

$$\left(\frac{n_1-2}{n} + \frac{n_2-2}{n_2} + \frac{n_3-2}{n_3}\right) 180° = 360°.$$

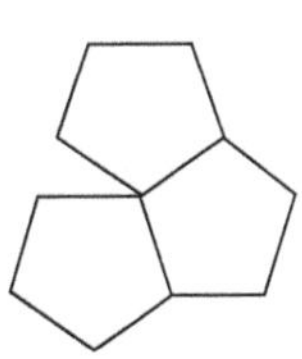
Figure 1.

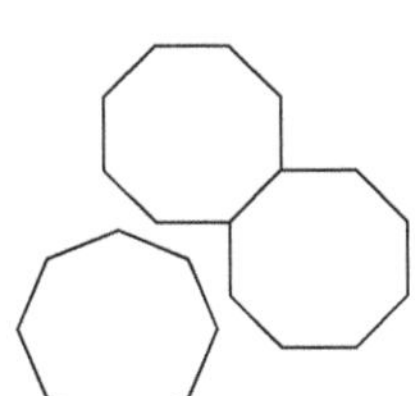
Figure 2.

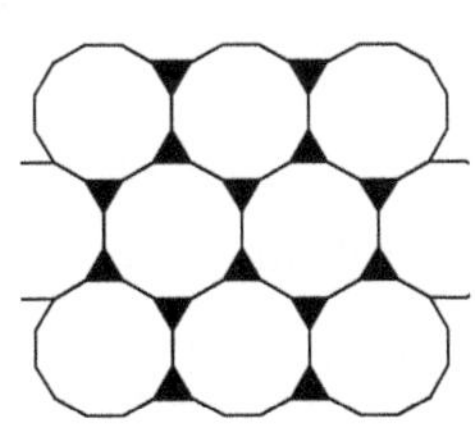
Figure 3.

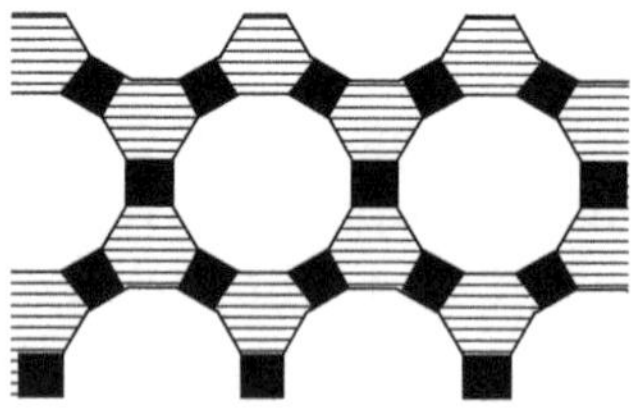

Figure 4.

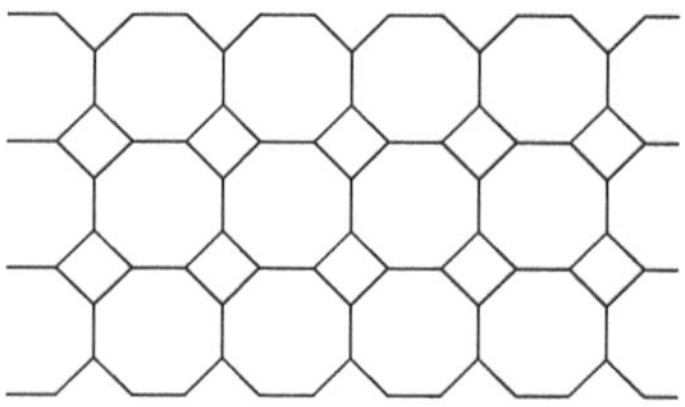

Figure 5.

Figure 6.

Figure 7.

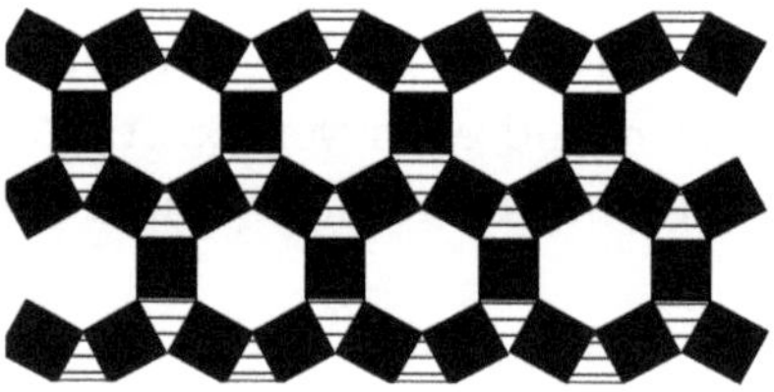

Solution 13

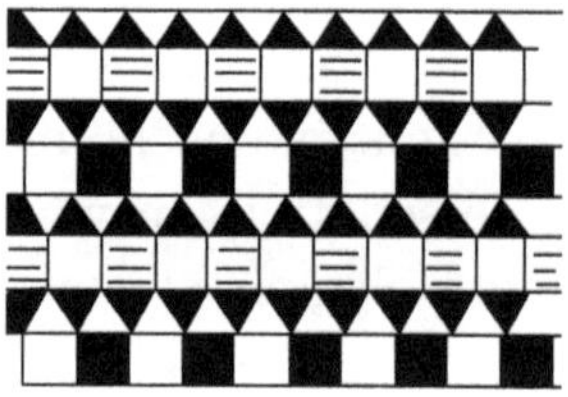

Solution 15

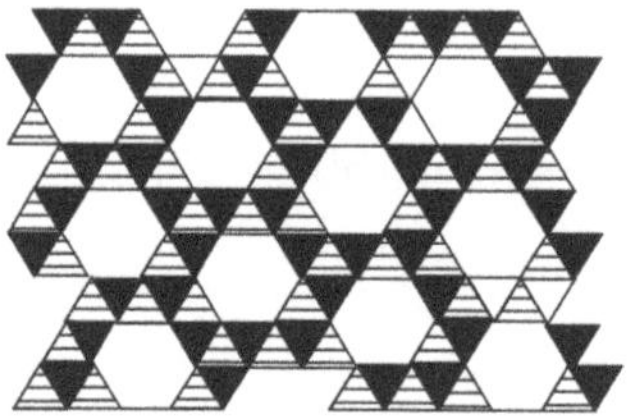

Solution 16

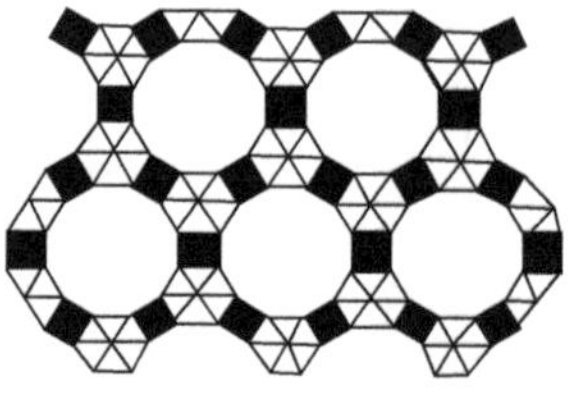

Solutions 11 and 17

From this, we obtain

$$\left(\frac{n_1}{n_1} - \frac{2}{n_1} + \frac{n_2}{n_2} - \frac{2}{n_2} + \frac{n_3}{n_3} - \frac{2}{n_3}\right) 180^\circ = 360^\circ.$$

$$1 + 1 + 1 - 2\left(\frac{1}{n_1} + \frac{1}{n_2} + \frac{1}{n_3}\right) = 2.$$

Therefore, $\frac{1}{n_1} + \frac{1}{n_2} + \frac{1}{n_3} = \frac{1}{2}$.

In a similar way, students can find the following conditions for other possible arrangements.

$$\frac{1}{n_1} + \frac{1}{n_2} + \frac{1}{n_3} + \frac{1}{n_4} = 1$$

$$\frac{1}{n_1} + \frac{1}{n_2} + \frac{1}{n_3} + \frac{1}{n_4} + \frac{1}{n_5} = \frac{3}{2}$$

$$\frac{1}{n_1} + \frac{1}{n_2} + \frac{1}{n_3} + \frac{1}{n_4} + \frac{1}{n_5} + \frac{1}{n_6} = 2$$

Following are the 17 possible integer solutions that need be considered (Table 1).

(Solutions 10, 14, and 17 have already been discussed. Solutions 1, 2, 3, 4, 6, and 9 each can be formed at a single vertex, but they cannot be extended to cover the whole plane.) They are made up of different combinations of triangles, squares, hexagons, octagons, and dodecagons.

Any of the remaining solutions can be used as the only type of arrangement in a design covering a whole plane except solution 11, which must be used in conjunction with others, e.g., 5 or 15.

Have the class consider what happens in solution 5. Here two dodecagons and a triangle meet at a vertex. The extended figure can be formed by

Table 1.

No.	n_1	n_2	n_3	n_4	n_5	n_6
1	3	7	42			
2	3	8	24			
3	3	9	18			
4	3	10	15			
5	3	12	12			
6	4	5	20			
7	4	6	12			
8	4	8	8			
9	5	5	10			
10	6	6	6			
11	3	3	4	12		
12	3	3	6	6		
13	3	4	4	6		
14	4	4	4	4		
15	3	3	3	4	4	
16	3	3	3	3	6	
17	3	3	3	3	3	3

juxtaposing dodecagons as in Figure 3. The remaining spaces form the triangles.

Solution 7, composed of dodecagons, hexagons, and squares, one at each vertex, gives a more complicated pattern (Figure 4).

A juxtaposition of octagons (Figure 5) forms solution 8. The empty spaces provide the areas needed for the squares.

Two different patterns can be obtained from Solution 12 by the juxtaposition of hexagons. In one, the hexagons have edges in common; in the other they have only vertices in common (Figures 6 and 7). The empty spaces form triangles or diamond shapes composed of pairs of triangles.

Call on individual students to determine and draw the patterns for the remaining solutions.

Postassessment

1. Which of the following regular polygons will tessellate a plane: (a) a square; (b) a pentagon; (c) an octagon; (d) a hexagon.
2. Which of the following combinations of regular polygons will tessellate a plane: (a) an octagon and a square; (b) a pentagon and a decagon; (c) a hexagon and a triangle.

Unit 51

Introducing the Pythagorean Theorem

This unit is intended for students taking the regular geometry course.

Performance Objective

Given appropriate measures, the student will use the Pythagorean theorem to solve geometric problems.

In addition, it is expected that student appreciation for the Pythagorean theorem will increase.

Preassessment

Have your students answer the following question:

Can a circular table top with a diameter of 9 feet fit through a rectangular door whose dimensions are 6 feet wide and 8 feet high?

Teaching Strategies

Students will immediately realize that the table top can possibly only fit through the door if it is tilted. Thereupon they will find a need for determining the length of the diagonal of this $6'$ by $8'$ rectangle. This is where you ought to introduce the Pythagorean theorem. There are over 360 proofs of the Pythagorean theorem available (see Elisha S. Loomis, *The Pythagorean Proposition*, National Council of Teachers of Mathematics, Washington, DC, 1968). A teacher may select the proof that she feels would be most interesting and intelligible for her particular class. Some proofs rely heavily on algebra while others are purely geometric.

Once the Pythagorean theorem has been proved, the student is ready to apply his knowledge of the theorem to some problems. Surely, he can now find that the length of the diagonal of the door (of the original problem) is 10 feet, and hence conclude that the table top would certainly fit through. There are many other "practical" problems that may be used to offer further application of the Pythagorean theorem. For example, suppose your students wanted to find the diameter of a pipe. All they would have to do is place a carpenter's measuring square as shown in the figure. Then by measuring the length of x, the diameter would merely be $4.828\,x$. Students should of course be asked to analyze this. The broken lines in the diagram will help in the justification of this situation. Applying the Pythagorean theorem to the right triangle shown:

$$\mathrm{R}^2 + \mathrm{R}^2 = (\mathrm{R} + x)^2, \quad \text{or} \quad \mathrm{R} = x(1 + \sqrt{2}).$$

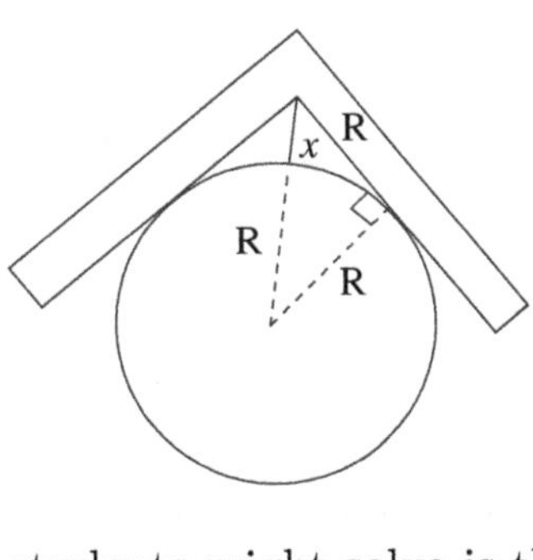

Another problem the students might solve is that of finding the original diameter of broken plate where only a segment of the circle remains. Once again the following diagram depicts the situation: The lengths AB and CD are measurable, and the broken lines are provided only for a discussion of

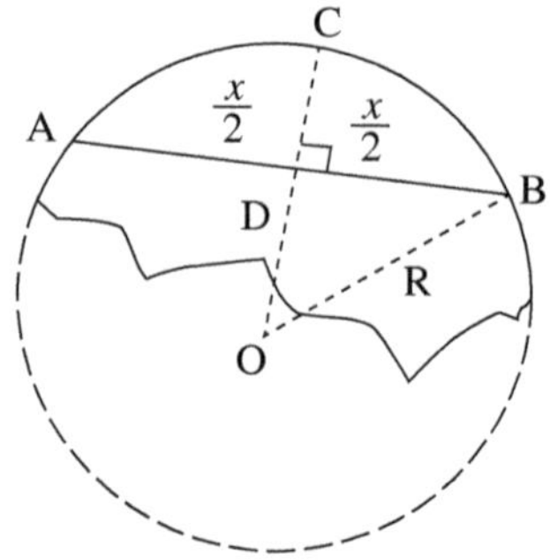

the solution. Let AB $= x$, CD $= y$ and OB (the radius to be found) = R. Thus, OD $=$ R $-$ y and by the Pythagorean theorem (in ΔODB) $(\mathrm{R}-\mathrm{y})^2 + \frac{x^2}{4} = \mathrm{R}^2$, and then $\mathrm{R} = \frac{y}{2} + \frac{x^2}{8y}$ so that the diameter (in terms of the measurable lengths x and y) is $y + \frac{x^2}{4y}$.

From a strictly geometric point of view, there are some rather interesting relationships which may be proved by applying the Pythagorean theorem. You may wish to present some of these to your class as further application of this theorem.

1. If E is any point on altitude $\overline{\mathrm{AD}}$, then $(\mathrm{AC})^2 - (\mathrm{CE})^2 = (\mathrm{AB})^2 - (\mathrm{EB})^2$.

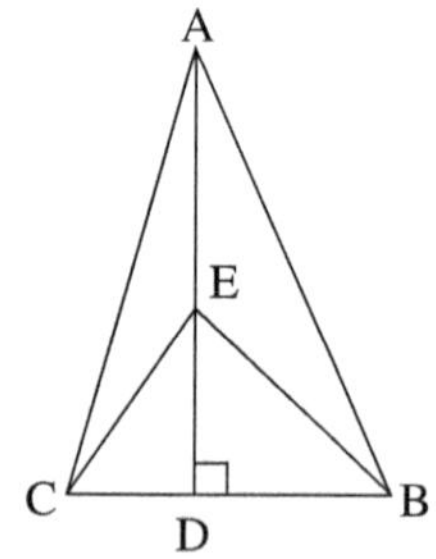

2. If medians $\overline{\mathrm{AD}}$ and $\overline{\mathrm{BE}}$ of ΔABC are perpendicular, then AB $= \sqrt{\frac{(\mathrm{AC})^2+(\mathrm{BC})^2}{5}}$.
3. If from any point inside a triangle, perpendiculars are drawn to the sides of the triangle, the sum of the squares of the measures of every other segment of the sides so formed equals the sum of the squares of the remaining three segments. That is in ΔABC in the following figure:

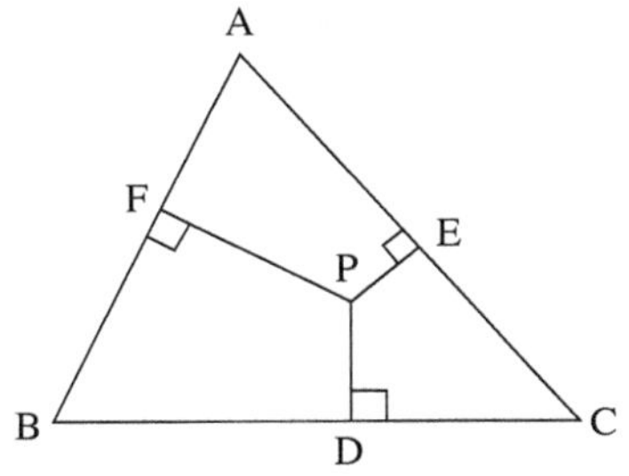

4. If $\overline{AD}, \overline{BE}$, and $\overline{CF}$ are medians of ΔABC,
   a. then $\frac{3}{4}[(AB)^2 + (BC)^2 + (CA)^2] = (AD)^2 + (BE)^2 + (CF)^2$.
   b. then $5(AB)^2 = 4(AE)^2 + (BE)^2$, if $m\angle C = 90$.

The complete solutions of these problems and many other more challenging problems can be found in *Challenging Problems in Geometry* by Alfred S. Posamentier and Charles T. Salkind (New York: Dover, 1996).

Once students have a fair command of this celebrated theorem, they are ready to consider a generalization of it.

To this juncture, the students have considered the Pythagorean theorem as $a^2 + b^2 = c^2$, where a and b represented the *lengths* of the legs of a right triangle and c represented the *length* of its hypotenuse. However, this statement could also be interpreted to mean the following: "The sum of the *areas* of the squares on the legs of a right triangle equals the *area* of the square on the hypotenuse."

For the following right triangle, $\mathscr{A}S_a + \mathscr{A}S_b = \mathscr{A}S_c$ ($\mathscr{A}$ represents "area of").

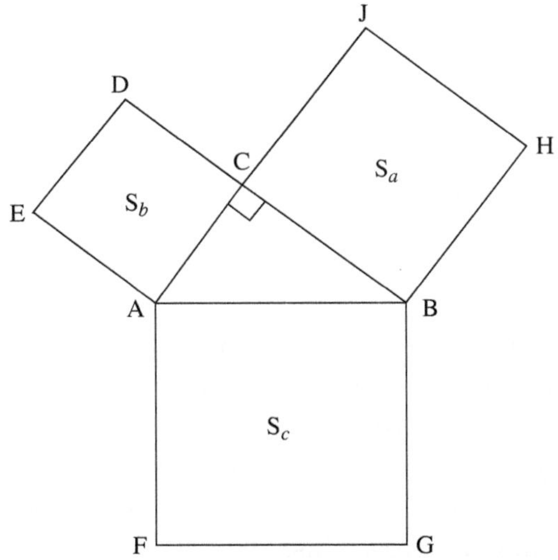

Now have students replace these squares with semicircles with diameters $\overline{BC}$, $\overline{AC}$, and $\overline{AB}$, or have them replace the squares with any similar polygons so that the corresponding sides are on the sides of ΔABC.

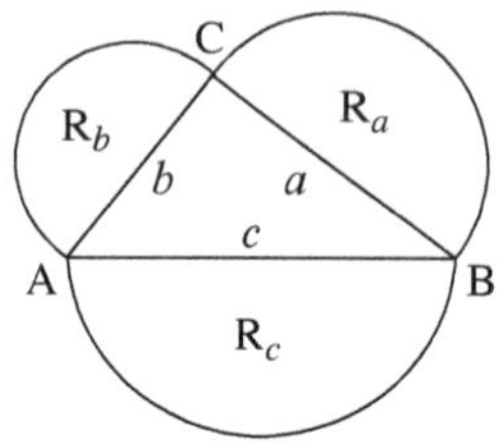

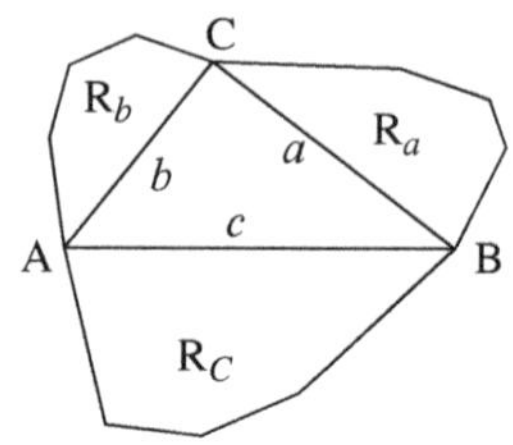

From a basic area relationship,

$$\frac{\mathscr{A}R_a}{\mathscr{A}R_c} = \frac{a^2}{c^2} \quad \text{and} \quad \frac{R_b}{R_e} = \frac{b^2}{c^2}$$

Then $\dfrac{\mathscr{A}R_a + \mathscr{A}R_b}{\mathscr{A}R_c} = \dfrac{a^2+b^2}{c^2}$.

However, by the Pythagorean theorem, $a^2 + b^2 = c^2$ so that $\frac{\mathscr{A}R_a+\mathscr{A}R_b}{\mathscr{A}R_c} = 1$, and $\mathscr{A}R_a + \mathscr{A}R_b = \mathscr{A}R_c$. The interesting significance of this extension of the Pythagorean theorem ought to be highlighted. Then have students pose other extensions.

Before leaving the geometric discussion of the Pythagorean theorem, you might wish to show students how the converse of this theorem can be used to determine whether an angle of a triangle is acute, right or obtuse, given the lengths of the sides of the triangle.

That is,

if $a^2 + b^2 = c^2, \angle C$ is a right angle;
if $a^2 = b^2 > c^2, \angle C$ is acute;
if $a^2 + b^2 < c^2, \angle C$ is obtuse.

These relationships should prove to be quite fascinating and useful to the students.

Having considered the Pythagorean theorem from a geometric standpoint, it should be interesting to consider this theorem from a number theoretic point of view. A *Pythagorean Triple*, written as (a, b, c), is a set of three positive integers a, b, and c, where $a^2 + b^2 = c^2$. For any Pythagorean Triple (a, b, c) and any positive integer $k, (ka, kb, kc)$ is also a triple. Your students should be able to prove this.

A *Primitive Pythagorean Triple* is a Pythagorean Triple whose first two members are *relatively prime*, one even and the other odd. Introduce this: where $a^2+b^2=c^2$ (and m and n are natural numbers, and $m>n$), $a=m^2-n^2$, $b=2mn$ and $c=m^2+n^2$. (For a development of these relationships, see W. Sierpinski, *Pythagorean Triangles*. New York: Yeshiva University Press, 1962.) After setting up a table such as the following, students will begin to conjecture about properties of m and n that generate specific types of Pythagorean Triples. Students will also begin to group different types of Pythagorean Triples.

m	n	m^2-n^2	$2mn$	m^2+n^2
2	1	3	4	5
3	2	5	12	13
4	1	15	8	17
4	3	7	24	25
5	4	9	40	41
3	1	8	6	10
5	2	21	20	29

Some questions to anticipate are: What must be true about m and n in order for (a,b,c) to be a Primitive Pythagorean Triple? Can c of this triple even? Why must the even member of a Primitive Pythagorean Triple be divisible by 4? What must be true about m and n in order that the third member of a Primitive Pythagorean Triple exceed one of the other members by 1? Why is one side of a Primitive Pythagorean Triple always divisible by 5? And why is the product of the three members of any Pythagorean Triple divisible by 60?

In a short, while students will begin to probe the parity of numbers and the relationships of Pythagorean Triples. This genuine interest brought about by a rather elementary and superficial introduction to a topic in number theory may be the beginning of a student's investigation into a heretofore unfamiliar field.

Thus, a study of the Pythagorean theorem has a wide range of possibilities for interesting your students. You must take the initiative of introducing these variations on the theme. If this is properly done, your students will carry these endeavors further.

Postassessment

1. Have students explain why the following "Proof Without Words" shows how some paper folding can prove the Pythagorean theorem.

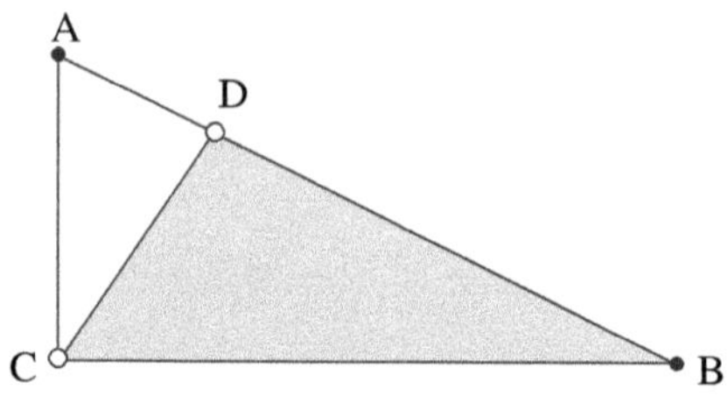

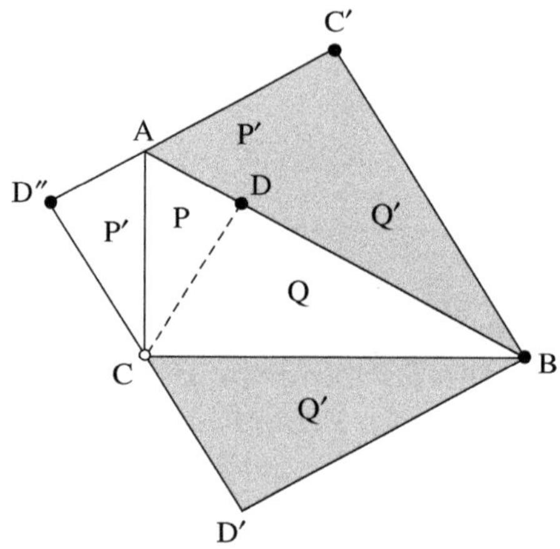

2. Have students show why one member of a primitive Pythagorean triple must always be even.
3. Have student show why the product of a primitive Pythagorean triple is always a multiple of 60.

References

Posamentier, A. S. *Math Wonders To Inspire Teachers and Students.* Alexandria, VA: Association for Supervision and Curriculum Development, 2003.

Posamentier, A. S., J. H. Banks, and R.L. Bannister. *Geometry, Its Elements and Structure.* 2nd edn. New York: Dover, 2014.

Posamentier, A. S. *The Pythagorean Theorem.* Amherst, NY: Prometheus Books, 2010.

Unit 52

Trisection Revisited

Performance Objectives

1. *Students will trisect a given angle using any of the four methods presented.*
2. *Students will prove the four methods of trisection.*

Preassessment

Students should have a working knowledge of algebra. They should also have mastered constructions commonly taught in high school geometry and proofs of those constructions.

Teaching Strategies

After demonstrating and proving the following construction, discuss why it is not really a solution to the ancient problem of trisecting an angle using only Euclidean tools.

Given $\angle AOB_0$ with $m\angle AOB_0 = x$.
Construct $\angle AOB_n$ such that $m\angle AOB_n = 2x/3$ (Figure 1).

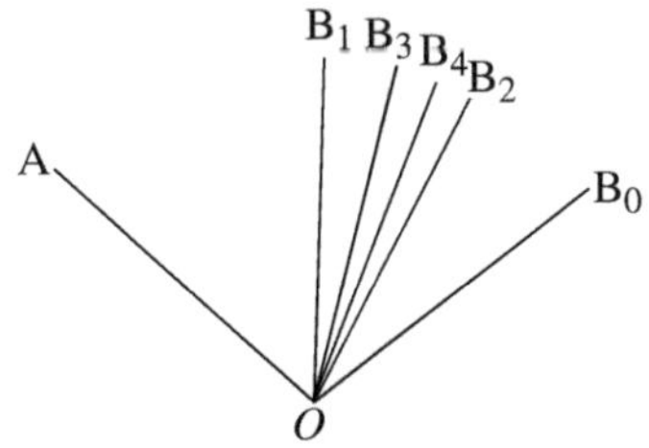

Figure 1.

Construction and Proof:

1. Construct $\overline{OB_1}$, the bisector of $\angle AOB_0$ then $m\angle AOB_1 = x - \frac{1}{2}x$.
2. Construct $\overline{OB_2}$, the bisector of $\angle B_1OB_0$ then $m\angle AOB_2 = x - \frac{1}{2}x + \frac{1}{4}x$.
3. Construct $\overline{OB_3}$, the bisector of $\angle B_1OB_2$ then $m\angle AOB_3 = x - \frac{1}{2}x + \frac{1}{4}x - \frac{1}{8}x$.

4. Construct $\overline{\text{OB}_4}$, the bisector of $\angle \text{B}_3\text{OB}_2$ then $m\angle\text{AOB}_4 = x - \frac{1}{2}x + \frac{1}{4}x - \frac{1}{8}x + \frac{1}{16}x$.
5. Continuing in this fashion, we will reach $m\angle\text{AOB}_n = x + \frac{1}{2}x + \frac{1}{4}x - \frac{1}{8}x + \cdots \pm \left(\frac{1}{2}\right)^n x$.
   Then we multiply by $\left(\frac{1}{2}\right)$ to obtain $\left(\frac{1}{2}\right) m\angle\text{AOB}_n = \frac{1}{2}x - \frac{1}{4}x + \frac{1}{8}x - \frac{1}{16}x + \cdots \pm \left(\frac{1}{2}\right)^{n+1} x$.
   Now we add the second equation to the first to get

$$\left(\frac{3}{2}\right) m\angle\text{AOB}_n = x \pm \left(\frac{1}{2}\right)^{n+1} x$$

$$m\angle\text{AOB}_n = \frac{2x}{3}\left[1 \pm \left(\frac{1}{2}\right)^{n+1}\right].$$

6. Now we observe that as n increases to infinity (which corresponds to carrying out an *infinite number* of construction operations), the term $\left(\frac{1}{2}\right)^{n+1}$ approaches zero. Then $m\angle\text{AOB}_n$ approaches $\frac{2x}{3}$.

The second construction adds to the Euclidean tools a strange looking device called a *tomahawk* (first published by Bergery in the third edition of *Geometrie Appliquee al' Industrie*, Metz, 1835).

To construct a tomahawk, start with a line segment $\overline{\text{RS}}$ trisected at U and T. Draw a semicircle about U with radius $\overline{\text{UT}}$ and draw $\overline{\text{TX}}$ perpendicular to $\overline{\text{RS}}$ Complete the instrument as shown in Figure 2.

To trisect any $\angle$AOB, simply place the implement on the angle so that S falls on $\overline{\text{OB}}$, $\overline{\text{TX}}$ passes through vertex O, and the semicircle is tangent to $\overline{\text{AO}}$ at some point, say D. Then, since we may easily show that $\Delta\text{DOU} \cong \Delta\text{TOU} \cong \Delta\text{TOS}$, we have $m\angle\text{DOU} = m\angle\text{TOU} = m\angle\text{TOS} = \frac{1}{3}m\angle\text{AOB}$.

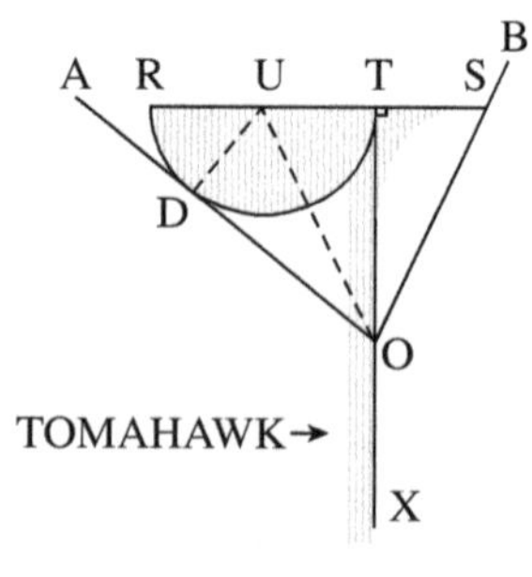

Figure 2.

The third construction is implied by a theorem given by Archimedes. In it, we use a straightedge on which a line segment has been marked. This extension of Euclidean tools makes possible an *insertion principle trisection.*

To demonstrate the insertion principle to students, have them try the following problem using Euclidean tools (see Figure 3).

Given $\overline{MN}$ with curves q and n (such that the smallest distance between q and n is $\leq$ MN) and point O not on q or n.
Construct a line through O that intersects q and n at M_1 and N_1, respectively, so that $M_1N_1 = MN$.

Except for certain special cases, this problem is impossible using only Euclidean tools. Now have students mark a line segment on their straightedges whose measure is equal to MN. It is now a simple matter to adjust the marked straightedge until it describes a line through O with the distance between the two intersections equal to MN.

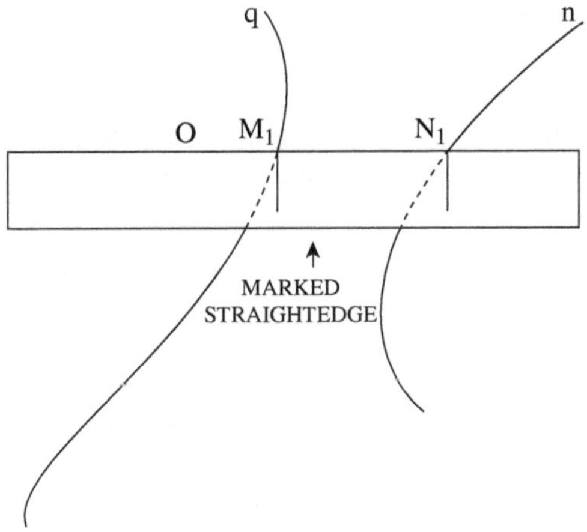

Figure 3.

Now students are ready for the insertion principle trisection (Figure 4).

Given Circle O with central $\angle$AOB.
Construct $\angle$ADB so that $m\angle ADB = \frac{1}{3}m\angle AOB$.

Construction:

1. Draw $\overrightarrow{AC}$.
2. Mark AO on a straightedge.

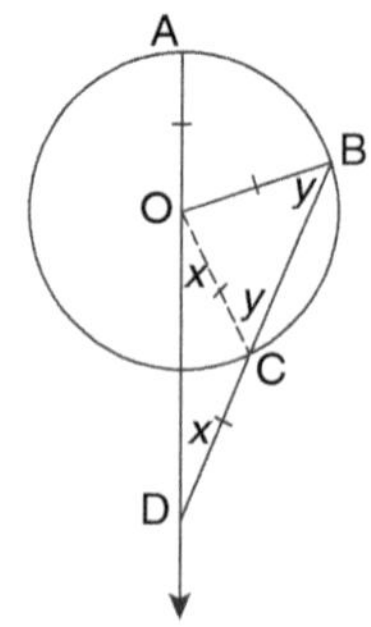

Figure 4.

3. Using the insertion principle, draw $\overline{\text{BD}}$ such that D is on $\overrightarrow{\text{AO}}$ and $\overline{\text{BD}}$ intersects circle O at C with $\overline{\text{AO}} \cong \overline{\text{CD}}$. Then $m\angle\text{ADB} = \frac{1}{3}m\angle\text{AOB}$.

Proof:

1. Draw $\overline{\text{OC}}$.
2. By construction, $\overline{\text{AO}} \cong \overline{\text{BO}} \cong \overline{\text{CO}} \cong \overline{\text{DC}}$ (since the first three are radii of circle O and the last was constructed to be congruent to $\overline{\text{AO}}$).
3. ΔOCD and ΔBOC are isosceles.
4. Therefore, $m\angle\text{DOC} = m\angle\text{CDO} = x$ and $m\angle\text{OCB} = m\angle\text{OBC} = y$.
5. Since $\angle$OCB is the exterior angle of $\Delta\text{OCD}, m\angle\text{OCB} = m\angle\text{DOC} + m\angle\text{CDO} = 2x$ or $y = 2x$.
6. Similarly since $\angle$AOB is the exterior angle of $\Delta\text{OBD}, m\angle\text{AOB} = m\angle\text{ADB} + m\angle\text{OBD} = x + 2x = 3x$.
7. Thus, $m\angle\text{ADB} = \frac{1}{3}m\angle\text{AOB}$.

Ceva's method of trisection (the last of this unit) utilizes a device that consists of four hinged straightedges. In the diagram of Ceva's linkage (Figure 5), points C, D, E, and O are pivots such that the figure CDEO is a rhombus. To trisect a given angle A′O′B′ one must first draw the circle about the vertex O′ with radius equal to the length of a side of the rhombus CDEO. Ceva's instrument is then placed on the angle so that O and O′ coincide. It is then adjusted until $\overrightarrow{\text{DC}}$ and $\overrightarrow{\text{DE}}$ go through the points where $\overline{\text{OA}'}$ and $\overline{\text{OB}'}$ intersect the circle, points A and B, respectively. Then

$$m\angle\text{AOF} = m\angle\text{FOG} = m\angle\text{GOB} = \left(\frac{1}{3}\right) m\angle\text{AOB}.$$

The proof uses the rhombus CDEO to obtain $m\angle\text{ACG} = m\angle\text{COE} = m\angle\text{FEB} = m\angle\text{CDE} = x$. Then $m\angle\text{FOG} = x$. Noting that points C and E are on the circle, we have $\angle$ACG and $\angle$FEB inscribed in the circle. Then

$m\angle ACG = x = \frac{1}{2}m\angle AOG$ and $m\angle FEB = x = \frac{1}{2}m\angle FOB$, which gives us $2x = m\angle AOG$ and $2x = m\angle FOB$. Clearly then $m\angle AOF = m\angle FOG = m\angle GOB = \frac{1}{3}m\angle AOB$.

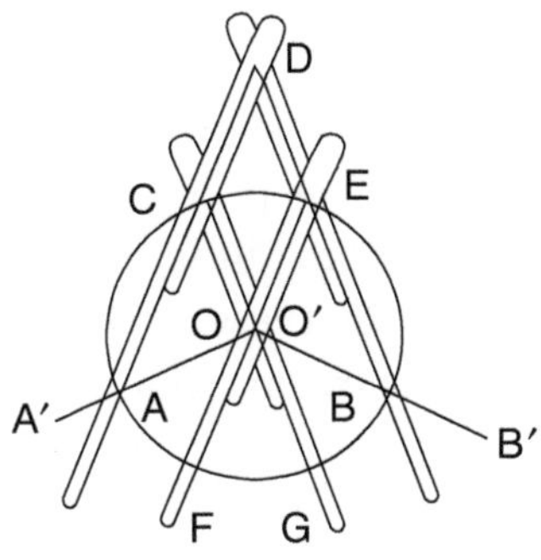

Figure 5.

Postassessment

1. Prove the *tomahawk* trisection method valid.
2. Trisect any given arbitrary angle using any two of the methods presented in this unit.

Unit 53

Proving Lines Concurrent

This unit will present the student with a theorem that is quite useful in some cases when proving lines concurrent.

Performance Objective

Given appropriate problems, students will apply Ceva's theorem to prove lines concurrent.

Preassessment

Have students try to prove any of the following:

1. Prove that the medians of a triangle are concurrent.
2. Prove that the angle bisectors of a triangle are concurrent.
3. Prove that the altitudes of a triangle are concurrent.

Teaching Strategies

An above-average geometry student should, given enough time, be able to prove some of these theorems. The proofs they would normally attempt (synthetically) are among the more difficult in the high school geometry course. Challenging students with these rather difficult problems sets the stage for the introduction of a theorem, that will permit these problems to be done quite easily.

This theorem, first published in 1678 by the Italian mathematician Giovanni Ceva, is stated as follows:

Three lines drawn from the vertices A, B, and C of ΔABC meeting the opposite sides in points L, M, and N, respectively, are concurrent if and only if

$$\frac{\text{AN}}{\text{NB}} \cdot \frac{\text{BL}}{\text{LC}} \cdot \frac{\text{CM}}{\text{MA}} = 1.$$

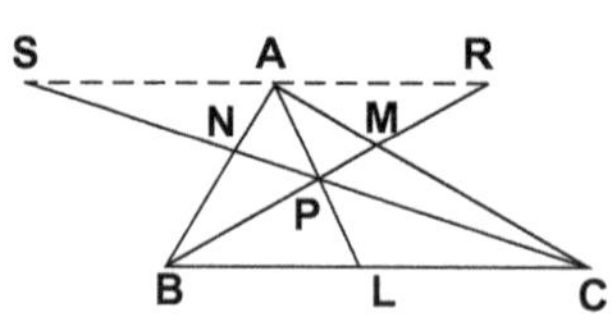

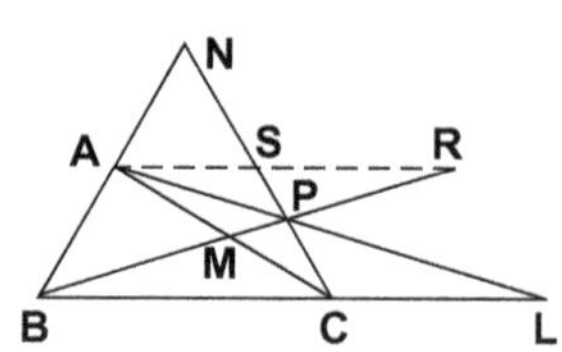

Note: There are two cases: The three lines meeting inside or outside the given triangle.

Before applying this theorem to the problems posed earlier, it might be wise to prove the theorem.

Given: ΔABC, with N on $\overleftrightarrow{\text{AB}}$, M on $\overleftrightarrow{\text{AC}}$, and L on $\overleftrightarrow{\text{BC}}$, also $\overleftrightarrow{\text{AL}}$, $\overleftrightarrow{\text{BM}}$,$\overleftrightarrow{\text{CN}}$ are concurrent at P.

Prove: $\frac{\text{AN}}{\text{NB}} \cdot \frac{\text{BL}}{\text{LC}} \cdot \frac{\text{CM}}{\text{MA}} = 1$

Proof: Draw a line through A, parallel to $\overleftrightarrow{\text{BC}}$ meeting $\overleftrightarrow{\text{CP}}$ at S and $\overleftrightarrow{\text{BP}}$ at R.

$$\Delta\text{AMR} \sim \Delta\text{CMB}.$$

Therefore, $\frac{\text{AM}}{\text{MC}} = \frac{\text{AR}}{\text{CB}}$. (I)

ΔBNC ∼ ΔANS

Therefore, $\frac{\text{BN}}{\text{NA}} = \frac{\text{CB}}{\text{SA}}$. (II)

$\Delta\text{CLP} \sim \Delta\text{SAP}$

Therefore, $\frac{\text{CL}}{\text{SA}} = \frac{\text{LP}}{\text{AP}}$. (III)

$\Delta\text{BLP} \sim \Delta\text{RAP}$

Therefore, $\frac{\text{BL}}{\text{RA}} = \frac{\text{LP}}{\text{AP}}$. (IV)

From (III) and (IV), we get $\frac{\text{CL}}{\text{SA}} = \frac{\text{BL}}{\text{RA}}$ or $\frac{\text{CL}}{\text{BL}} = \frac{\text{SA}}{\text{RA}}$. (V)

Now multiplying (I),(II), and (V) yields

$$\frac{\text{AM}}{\text{MC}} \cdot \frac{\text{BN}}{\text{NA}} \cdot \frac{\text{CL}}{\text{BL}} = \frac{\text{AR}}{\text{CB}} \cdot \frac{\text{CB}}{\text{SA}} \cdot \frac{\text{SA}}{\text{RA}} = 1.$$

Since Ceva's theorem is biconditional, it is necessary to prove the converse of the implication we have just proved.

Given: ΔABC with N on $\overleftrightarrow{\text{AB}}$ M on $\overleftrightarrow{\text{AC}}$ and L on $\overleftrightarrow{\text{BC}}$ also

$$\frac{\text{BL}}{\text{LC}} \cdot \frac{\text{CM}}{\text{MA}} \cdot \frac{\text{AN}}{\text{NB}} = 1.$$

Prove: $\overleftrightarrow{\text{AL}}$, $\overleftrightarrow{\text{BM}}$ and $\overleftrightarrow{\text{CN}}$ are concurrent.

Proof: Let $\overleftrightarrow{\text{BM}}$ and $\overleftrightarrow{\text{AL}}$ meet at P. Let $\overleftrightarrow{\text{CP}}$ meet AB at N′. Since $\overleftrightarrow{\text{AL}}$, $\overleftrightarrow{\text{BM}}$, $\overleftrightarrow{\text{CN}}$ are concurrent, by the part of Ceva's theorem we have already proved, we get:

$$\frac{\text{BL}}{\text{LC}} \cdot \frac{\text{CM}}{\text{MA}} \cdot \frac{\text{AN}'}{\text{N}'\text{B}} = 1.$$

However, $\frac{\text{BL}}{\text{LC}} \cdot \frac{\text{CM}}{\text{MA}} \cdot \frac{\text{AN}}{\text{NB}} = 1$ (given).

Therefore, $\frac{AN'}{N'B} = \frac{AN}{NB}$, so that N and N′ must coincide.

Thus, the three lines are concurrent.

Students should now be ready to apply Ceva's theorem to the three problems posed earlier.

1. Prove that the medians of a triangle are concurrent.
   Proof: In $\Delta ABC, \overline{AL}, \overline{BM}$, and $\overline{CN}$ are medians (see following figure below).

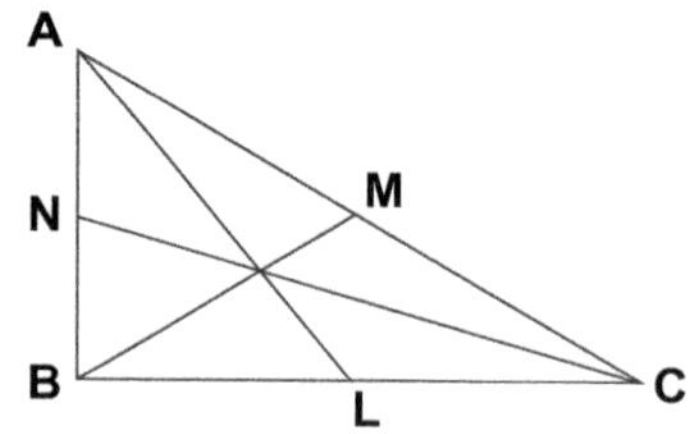

   Therefore, AN = NB, BL = LC, and CM = MA.
   By multiplication (AN)(BL)(MC) = (NB)(LC)(MA), or $\frac{AN}{NB} \cdot \frac{BL}{LC} \cdot \frac{CM}{MA} = 1$.
   Thus, by Ceva's theorem, $\overleftrightarrow{AL}$, $\overleftrightarrow{BM}$ and $\overleftrightarrow{CN}$ are concurrent.
2. Prove that the angle bisectors of a triangle are concurrent.
   Proof: In $\Delta ABC, \overline{AL}, \overline{BM}$, and $\overline{CN}$ are interior angle bisectors (see following figure).

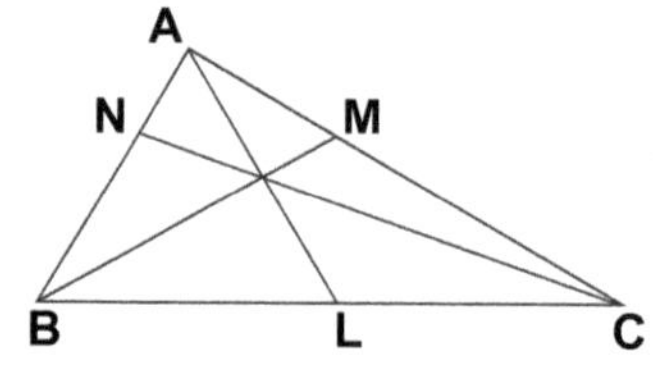

   Since an angle bisector of a triangle partitions the opposite side into segments proportional to the two remaining sides of the triangle, it follows that

$$\frac{AN}{NB} = \frac{AC}{BC}, \frac{BL}{LC} = \frac{AB}{AC}, \text{ and } \frac{CM}{MA} = \frac{BC}{AB}.$$

   Then by multiplying

$$\frac{AN}{NB} \cdot \frac{BL}{LC} \cdot \frac{CM}{MA} = \frac{AC}{BC} \cdot \frac{AB}{AC} \cdot \frac{BC}{AB} = 1.$$

   Thus by Ceva's theorem the three angle bisectors are concurrent.

3. Prove that the altitudes of a triangle are concurrent.
   Proof: In $\Delta ABC, \overline{AL}, \overline{BM}$, and $\overline{CN}$ are altitudes (see following figure).

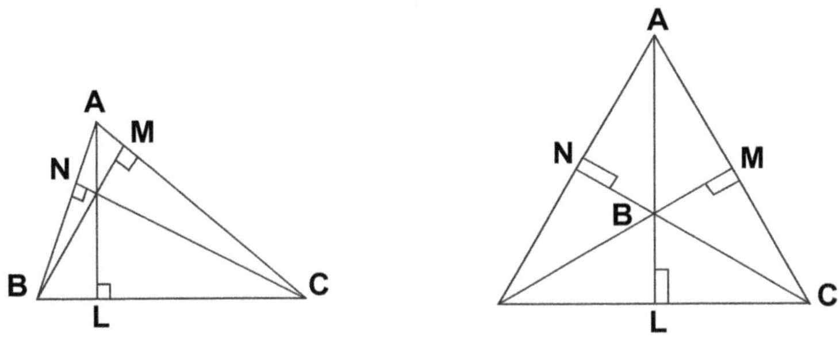

$$\Delta ANC \sim \Delta AMB, \quad \text{and} \quad \frac{AN}{MA} = \frac{AC}{AB}$$

$$\Delta BLA \sim \Delta BNC, \quad \text{and} \quad \frac{BL}{NB} = \frac{AB}{BC}$$

$$\Delta CMB \sim \Delta CLA, \quad \text{and} \quad \frac{CM}{LC} = \frac{BC}{AC}$$

By multiplying these three fractions, we get

$$\frac{AN}{MA} \cdot \frac{BL}{NB} \cdot \frac{CM}{LC} = \frac{AC}{AB} \cdot \frac{AB}{AC} \cdot \frac{BC}{AC} = 1.$$

Thus, by Ceva's theorem, the altitudes are concurrent.

These are some of the simpler applications of Ceva's theorem. One source for finding more applications of Ceva's theorem is *Challenging Problems in Geometry* by A. S. Posamentier and C. T. Salkind (New York: Dover, 1996).

Postassessment

1. Have students use Ceva's theorem to prove that when ΔABC has points P, Q, and R on sides $\overline{AB}, \overline{AC}$, and $\overline{BC}$, respectively, and when $\frac{AQ}{QC} = \frac{BR}{RC} = 2$, and $AP = PB$, it follows that $\overline{AR}, \overline{BQ}$, and $\overline{CP}$ are concurrent.
2. ΔABC cuts a circle at points $E, E'D, D'F, F'$ (see following figure). Prove that if $\overline{AD}, \overline{BF}$, and $\overline{CE}$ are concurrent, then $\overline{AD'}, \overline{BF'}$, and $\overline{CE'}$ are also concurrent.

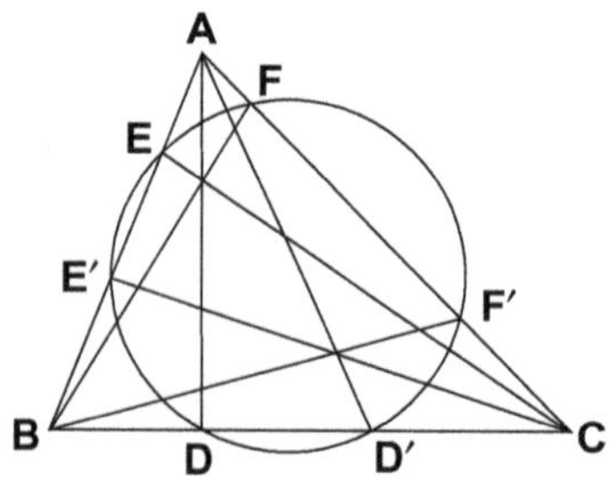

References

Posamentier, A. S. *Advanced Euclidean Geometry: Excursions for Secondary Teachers and Students.* New York, NY: John Wiley Publishing, 2002. (Contains lots of applications of Ceva's theorem.)
Posamentier, A. S. *The Secrets of Triangles.* Amherst, NY: Prometheus Books, 2012.

Unit 54

Squares

This unit will strengthen students' skills at proving quadrilaterals to be squares in addition to revisiting the topic of concurrency.

Performance Objective

Students will explain a method for proving concurrency.

Preassessment

Students should be familiar with the various properties of a square and should have had some experience in proving quadrilaterals to be squares.

Teaching Strategies

Have students construct a square externally on each side of a given parallelogram (see following figure). Have them locate the center of each square by drawing the diagonals. Ask the class what figure they believe will result by joining the centers of consecutive squares. Natural curiosity should motivate them to try to prove that PQRS is a square.

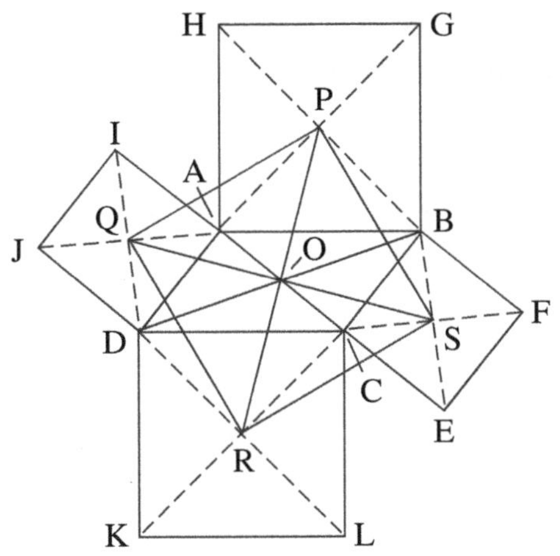

Proof:

ABCD is a parallelogram.

Points P, Q, R, and S are the centers of the four squares ABGH, DAIJ, DCLK, and CBFE, respectively.

PA = DR and AQ = QD (each is one-half a diagonal).

∠ADC is supplementary to ∠DAB, and ∠IAH is supplementary to ∠DAB (since ∠IAD and ∠HAB are right angles). Therefore, ∠ADC ≅ ∠IAH.

Since m∠RDC = m∠QDA = m∠HAP = m∠QAI = 45°, ∠RDQ ≅ ∠QAP.

Thus, ΔRDQ ≅ ΔPAQ (SAS) and QR = QP.

In a similar fashion, it may be proved that QP = PS and PS = RS.

Therefore, PQRS is a rhombus.

Since ΔRDQ ≅ ΔPAQ, ∠DQR ≅ ∠AQP; therefore, PQR ≅ ∠DQA (by addition).

Since ∠DQA ≅ right angle, ∠PQR ≅ right angle, and PQRS is a square.

Careful drawing of the figure above should indicate that the diagonals of square PQRS and the diagonals of parallelogram ABCD are concurrent. This proof deserves special attention since it illustrates an all too often neglected skill: proving concurrency.

To prove that the diagonals of square PQRS are concurrent with the diagonals of parallelogram ABCD, we must prove that a diagonal of the square and a diagonal of the parallelogram bisect each other. In other words, we prove that the diagonals of the square and the diagonals of the parallelogram all share the same midpoint (i.e., point O).

BAC ≅ ∠ACD, and m∠PAB = m∠RCD = 45, therefore, ∠PAC ≅ ∠RCA.

Since $\angle AOP \cong \angle COR$ and $AP = CR$, $\Delta AOP \cong \Delta COR(SAA)$.
Thus, $AO = CO$, and $PO = RO$.

Since $\overline{DB}$ passes through the midpoint of $\overline{AC}$ (diagonals bisect each other) and similarly, $\overline{QS}$ passes through the midpoint of $\overline{PR}$ and since $\overline{AC}$ and $\overline{PR}$ share the same midpoint (i.e., O), we have shown that $\overline{AC}$, $\overline{PR}$, $\overline{DB}$, and $\overline{QS}$ are concurrent (i.e., all pass through point O).

Postassessment

Ask students to explain a method for proving lines concurrent. It is expected that they will explain the method used in this lesson.

Unit 55

Proving Points Collinear

This unit will present the student with a theorem that is quite useful in certain cases, when proving points collinear.

Performance Objective

Given appropriate problems students will apply Menelaus' theorem to prove points collinear.

Preassessment

Have students try to prove that the interior angle bisectors of two angles of a nonisosceles triangle and the exterior angle bisector of the third angle meet the opposite sides in three collinear points.

Teaching Strategies

The average student of high school geometry is not properly trained or equipped to prove points collinear. Thus, in most cases, you will find the preassessment problem beyond student ability. However, this unit will provide you with sufficient student interest to introduce a theorem that will provide a simple solution.

This theorem, originally credited to Menelaus of Alexandria (about 100 A.D.), is particularly useful in proving points collinear. It states that:

Points P, Q, and R on sides $\overline{AC}$, $\overline{AB}$, and $\overline{BC}$ of ΔABC are collinear if and only if

$$\frac{AQ}{QB} \cdot \frac{BR}{RC} \cdot \frac{CP}{PA} = 1.$$

This is a two part (biconditional) proof.

Part I to prove $\frac{AQ}{QB} \cdot \frac{BR}{RC} \cdot \frac{CP}{PA} = 1$.

Proof: Points P, Q, and R are collinear. Consider the line through C, parallel to $\overline{AB}$, and meeting $\overline{PQR}$ at D.

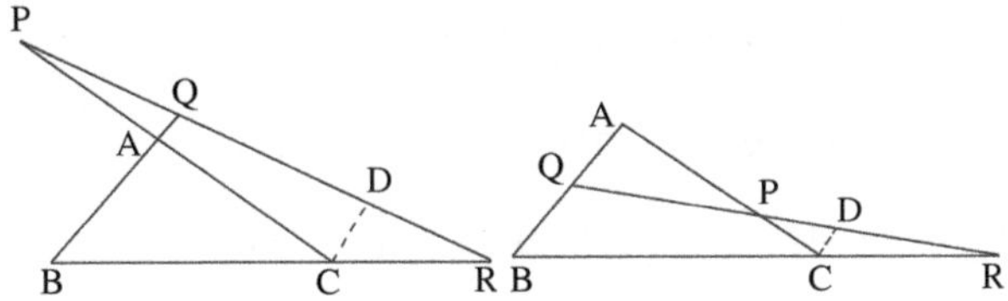

Since $\Delta DCR \sim \Delta QBR$, $\frac{DC}{QB} = \frac{RC}{BR}$ or $DC = \frac{(QB)(RC)}{BR}$. (a)

Similarly, since $\Delta PDC \sim \Delta PQA$, $\frac{DC}{AQ} = \frac{CP}{PA}$, or $DC = \frac{(AQ)(CP)}{PA}$ (b)

From (a) and (b) : $\frac{(QB)(RC)}{BR} = \frac{(AQ)(CP)}{PA}$.

Therefore (QB)(RC)(PA) = (AQ)(CP)(BR), which indicates that $\frac{AQ}{QB} \cdot \frac{BR}{RC} \cdot \frac{CP}{PA} = 1$.

Part II involves proving the converse of the implication proved in Part I, since this theorem is bi-conditional.

Proof: In the figures above, let the line through R and Q meet $\overleftrightarrow{AB}$ at P′. Then by the theorem just proved $\frac{AQ}{QB} \cdot \frac{BR}{RC} \cdot \frac{CP'}{P'A} = 1$.

However, by hypothesis,

$$\frac{AQ}{QB} \cdot \frac{BR}{RC} \cdot \frac{CP}{PA} = 1.$$

Therefore, $\frac{CP'}{P'A} = \frac{CP}{PA}$ and P and P′ must coincide.

At this point, students should be ready to apply Menelaus' theorem to the problem presented in the preassessment.

Given: ΔABC where $\overline{\text{BM}}$ and $\overline{\text{CN}}$ are interior angle bisectors and $\overline{\text{AL}}$ bisects the exterior angle at A.
Prove: N, M, and L are collinear.

Have students recall the important proportionality theorem about the angle bisector of a triangle.

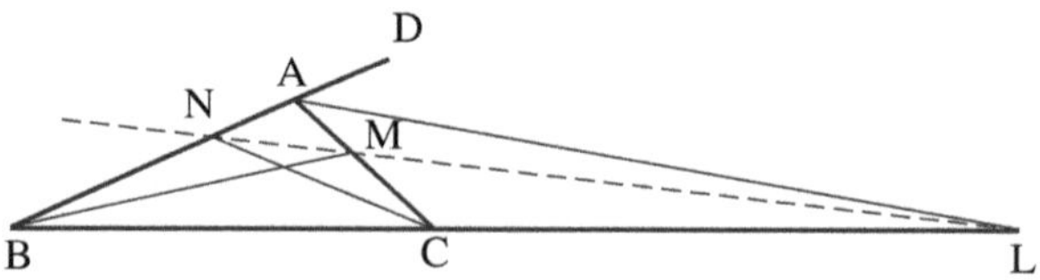

Proof: Since $\overline{\text{BM}}$ bisects $\angle$ABC, $\frac{\text{AM}}{\text{MC}} = \frac{\text{AB}}{\text{BC}}$.

Since $\overline{\text{CN}}$ bisects $\angle$ACB, $\frac{\text{BN}}{\text{NA}} = \frac{\text{BC}}{\text{AC}}$.

Since $\overline{\text{AL}}$ bisects the exterior angle at A, $\frac{\text{CL}}{\text{BL}} = \frac{\text{AC}}{\text{AB}}$.

Therefore, by multiplication:

$$\frac{\text{AM}}{\text{MC}} \cdot \frac{\text{BN}}{\text{NA}} \cdot \frac{\text{CL}}{\text{BL}} = \frac{\text{AB}}{\text{BC}} \cdot \frac{\text{BC}}{\text{AC}} \cdot \frac{\text{AC}}{\text{AB}} = 1.$$

Thus, by Menelaus' theorem N, M, and L must be collinear.

To provide further practice by applying this useful theorem, have students consider the following problem.

Prove that if tangents to the circumcircle of ΔABC, at A, B, and C, meet sides $\overleftrightarrow{\text{BC}}$, $\overleftrightarrow{\text{AC}}$, and $\overleftrightarrow{\text{AB}}$ at points P, Q, and R, respectively, then points P, Q, and R are collinear.

Proof: Since $m\angle\text{BAC} = \frac{1}{2}m\widehat{\text{BC}} = m\angle\text{QBC}, \Delta\text{ABQ} \sim \Delta\text{BCQ}$ and $\frac{\text{AQ}}{\text{BQ}} = \frac{\text{BA}}{\text{BC}}$

or $\dfrac{(\text{AQ})^2}{(\text{BQ})^2} = \dfrac{(\text{BA})^2}{(\text{BC})^2}$. (I)

However, $(\text{BQ})^2 = (\text{AQ})(\text{CQ})$. (II)

Substituting (II) into (I) yields $\dfrac{\text{AQ}}{\text{CQ}} = \dfrac{(\text{BA})^2}{(\text{BC})^2}$. (III)

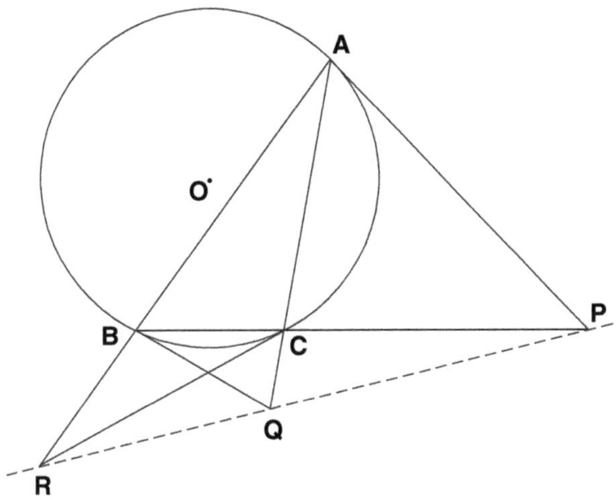

Similarly, $m\angle BCR = \frac{1}{2}m\widehat{BC} = m\angle BAC$; therefore,

$$\Delta CRB \sim \Delta ARC \text{ and } \frac{CR}{AR} = \frac{CR}{AC}, \text{ or } \frac{(CR)^2}{(AR)^2} = \frac{(BC)^2}{(AC)^2}. \quad (IV)$$

However, $(CR)^2 = (AR)(RB)$. (V)

Substituting (V) into (IV) yields

$$\frac{RB}{AR} = \frac{(BC)^2}{(AC)^2}. \quad (VI)$$

Students should now be asked to use the same scheme to prove $\Delta CAP \sim \Delta ABP$ and in a similar manner obtain

$$\frac{PC}{BP} = \frac{(AC)^2}{(BA)^2}. \quad (VII)$$

Now multiplying these proportions [i.e., (III), (VI), and (VII)] yields

$$\frac{AQ}{CQ} \cdot \frac{RB}{AR} \cdot \frac{PC}{BP} = \frac{(BA)^2}{(BC)^2} \cdot \frac{(BC)^2}{(AC)^2} \cdot \frac{(AC)^2}{(BA)^2} = 1.$$

Thus, by Menelaus' theorem, P, Q, and R are collinear.

Postassessment

Have students use Menelaus' theorem to prove that the exterior angle bisectors of any nonisosceles triangle meet the opposite sides in three collinear points. The following figure should be useful.

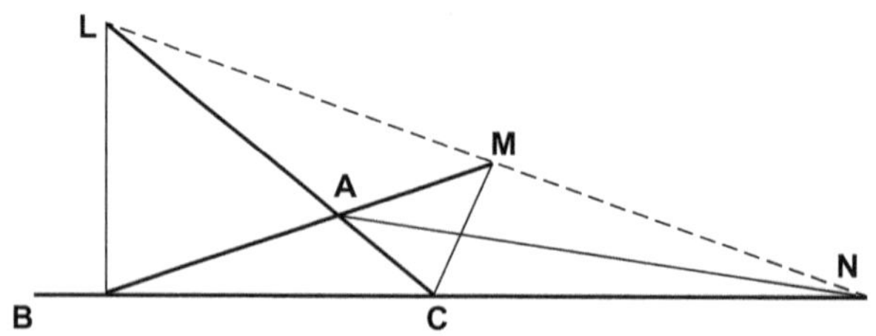

References

Posamentier, A. S. and C. T. Salkind. *Challenging Problems in Geometry.* New York: Dover, 1996.

Posamentier, A. S. A*dvanced Euclidean Geometry: Excursions for Secondary Teachers and Students.* New York, NY: John Wiley Publishing, 2002.

Posamentier, A. S. *The Secrets of Triangles.* Amherst, NY: Prometheus Books, 2012.

Unit 56

Angle Measurement with a Circle

This unit presents a rather unusual method for developing the theorems on angle measurement with a circle, normally considered in the tenth-grade geometry course.

Performance Objectives

1. *Given appropriate materials, students will generate the various angle measurement theorems in the manner developed in this unit.*
2. *Given problems that require the use of the theorems discussed in this unit, students will be able to solve them successfully.*

Preassessment

Students should be familiar with an inscribed angle and the relationship of its measure to that of its intercepted arc.

Teaching Strategies

In addition to using the usual classroom materials, you should prepare the following:

1. A piece of cardboard with two dark-colored pieces of string attached, forming an angle of convenient size.
2. A cardboard circle with an inscribed angle congruent to the "string angle."

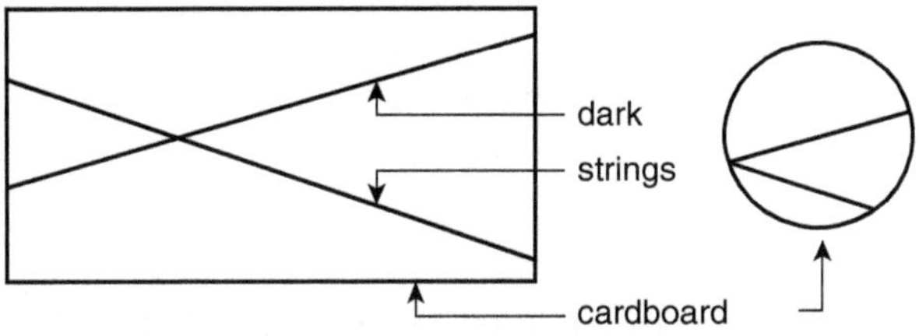

Naturally, it would be best if each student could prepare his or her own set of these materials in order to perform the following activities individually.

Refresh your students' memories about the relationship of an inscribed angle and its intercepted arc. Have them place the circle under the strings so that the two angles coincide:

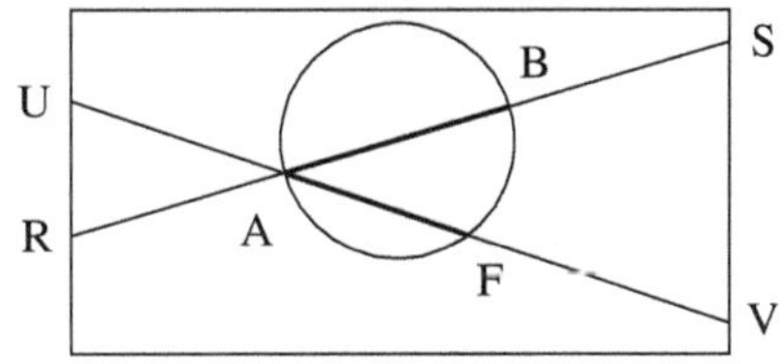

$$m\ \angle BAF = \frac{1}{2} m\ \widehat{BF}$$

Now have students slide the circle to the position illustrated in the following, where the rays of $\angle BAF$ are respectively parallel to the rays of the "string angle," $\angle NMQ$, and where the circle is tangent to $\overleftrightarrow{UQV}$ at M.

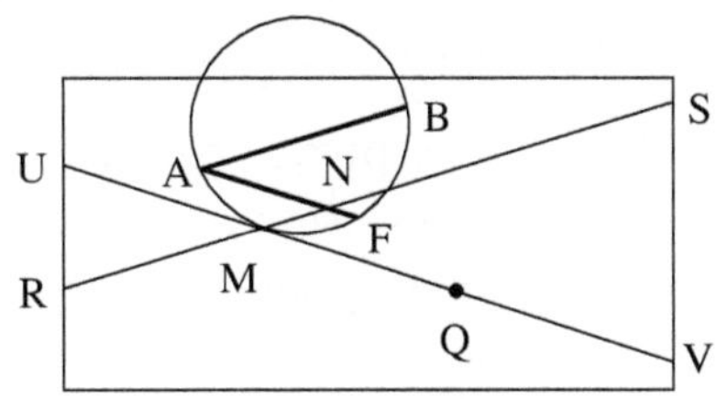

Students should realize that $m\widehat{FM} = m\widehat{AM}$, and $m\widehat{AM} = m\widehat{BN}$ (due to the parallel lines). Therefore $m\widehat{FM} = m\widehat{BN}$. Since $m\angle NMQ = m\angle BAF$, and $m\angle BAF = \frac{1}{2}m\widehat{BF} = \frac{1}{2}(m\widehat{BN} + m\widehat{NF}) = \frac{1}{2}(m\widehat{FM} + m\widehat{NF}) = \frac{1}{2}(m\widehat{MN}m\angle NMQ = \frac{1}{2}m\widehat{MN}$. This proves the theorem that *the measure of an angle formed by a tangent and a chord of a circle is one half the measure of its intercepted arc.*

Now have your students slide the circle to a position where the vertex of the string angle is on a $\overline{AF}$ and where $\overline{AB}//\overline{RS}$.

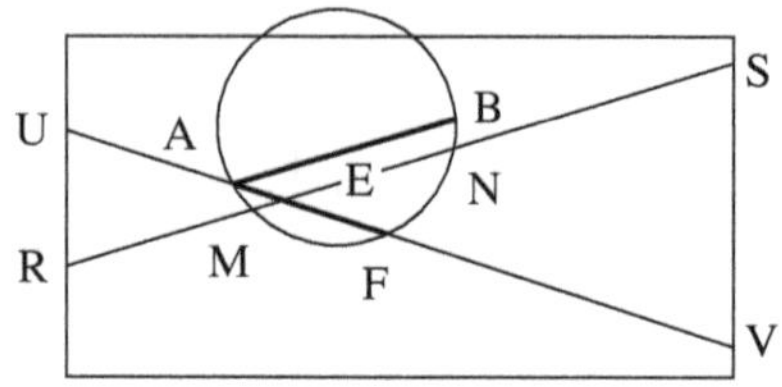

Once again because parallel lines exist here $(\overleftrightarrow{AB}//\overleftrightarrow{MN}) = m = \widehat{AM} = m\widehat{BN})$, and $m\angle BAF = m\angle NEF$. The students should now see that $m\angle BAF = \frac{1}{2}m\widehat{BF} = \frac{1}{2}(m\widehat{BN} + m\widehat{NF}) = \frac{1}{2}(m\widehat{AM} + m\widehat{NF})$. They may then conclude that $m\angle NEF = \frac{1}{2}(m\widehat{AM} + m\widehat{NF})$. This proves the theorem that *the measure of an angle formed by two chords intersecting in a point in the interior of a circle is one half the sum of the measures of the arcs intercepted by the angle and its vertical angle.*

To consider the next type of angle, have your students slide the circle to the position illustrated below, where the string angle now appears as an angle formed by two secants.

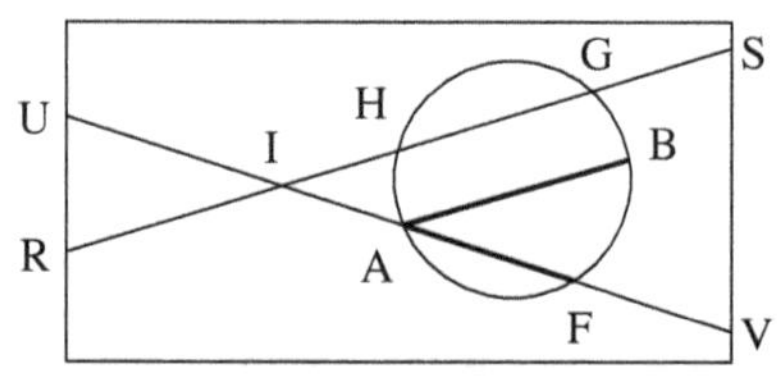

In this new position $\overline{AB}//\overleftrightarrow{GI}$ and $\overline{AF}$ is in $\overleftrightarrow{IF}$. Because $\overline{AB}//\overleftrightarrow{GI}$ $m\widehat{BG} = m\widehat{HA}$, and $m\angle BAF = m\angle GIF$ Have students once again follow the reasoning that $m\angle BAF = \frac{1}{2}m\widehat{BF} = \frac{1}{2}(m\widehat{BF} + m\widehat{BG} - m\widehat{BG}) = \frac{1}{2}(m\widehat{BF} + m\widehat{BG} - m\widehat{HA}) = \frac{1}{2}(m\widehat{GBF} - m\widehat{HA})$. They may then conclude that $m\angle GIF = \frac{1}{2}(m\widehat{GBF} - m\widehat{HA})$, which proves the theorem that *the measure of an angle formed by two secants intersecting in a point in the exterior of a circle is equal to one half the difference of the measures of the intercepted arcs.*

The next position of the circle will enable students to consider an angle formed by a tangent and a secant intersecting in the exterior of a circle.

Have students slide the circle to the position as indicated in the following illustration.

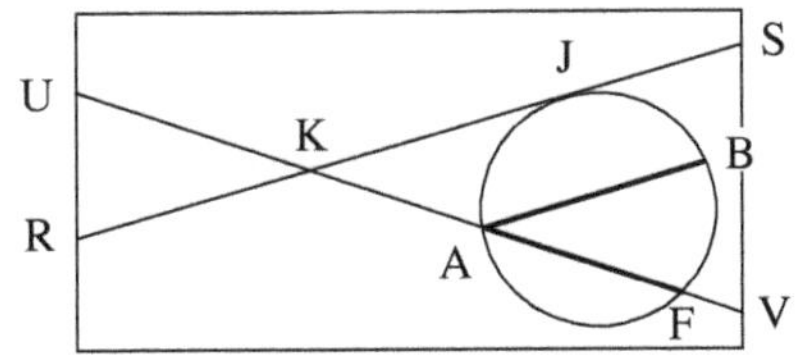

Here $\overline{AB}//\overleftrightarrow{KJS}$, $\overline{AF}$ is in $\overleftrightarrow{KV}$ and the circle is tangent to $\overline{KS}$ at J. Because $\overline{AB}//\overline{KJS}, m\widehat{JA} = m\widehat{JB}$ and $m\angle BAF = m\angle JKF$.

By now students should be able to produce the following without much difficulty: $m\angle BAF = \frac{1}{2}m\widehat{BF} = \frac{1}{2}(m\widehat{BF} + m\widehat{JB} - m\widehat{JB}) = \frac{1}{2}(m\widehat{BF} + m\widehat{JB} - m\widehat{JA} = \frac{1}{2}(m\widehat{JBF} - m\widehat{JA})$. They should then conclude that $\angle JKF = \frac{1}{2}(m\widehat{JBF} - m\widehat{JA})$, which proves the theorem that *the measure of an angle formed by a secant and a tangent to a circle intersecting in a point exterior to the circle is equal to one half the difference of the measures of the intercepted arcs.*

The last type of angle to be considered is an angle formed by two tangents. To form this angle, the circle should be positioned tangent to each of the two strings so that each string is parallel to one of the rays of the angle in the circle.

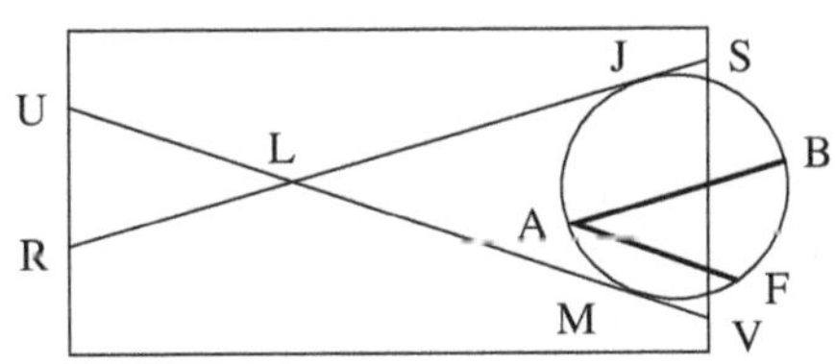

With the circle in the above position $\overline{AB}//\overleftrightarrow{LJS}$ and $\overline{AB}//\overleftrightarrow{LMV}$. Students should now be able to complete this proof independently. They should reason that $m\widehat{JB} = m\widehat{JA}$ and $m\widehat{MF} = m\widehat{MA}$; also $m\angle BAF = m\angle JLM$. Hence, $m\angle BAF = \frac{1}{2}m\widehat{BF} = \frac{1}{2}(m\widehat{BF} + m\widehat{JB} + m\widehat{MF} - m\widehat{JB} - m\widehat{MF}) = \frac{1}{2}(m\widehat{BF} + m\widehat{JB} + m\widehat{MF} - m\widehat{JA} - m\widehat{MA}) = \frac{1}{2}m\widehat{JBM} - \frac{1}{2}m\widehat{JAM}$. Thus, $m\angle JLM = \frac{1}{2}(m\widehat{JBM} - m\widehat{JAM})$, which proves the theorem that *the measure of an angle formed by two tangents is equal to one half the difference of the measures of the intercepted arcs.*

To summarize this presentation, have students realize that (1) the measure of an angle whose vertex is on the circle is one half the measure of the intercepted arc, (2) the measure of an angle whose vertex is *inside* the circle is one half the *sum* of the measures of the intercepted arcs, and (3) the measure of an angle whose vertex is *outside* the circle is one half the *difference* of the measures of the intercepted arcs.

As an alternative method for using this technique with your classes, see *Geometry, Its Elements and Structure*, 2nd edn., by A. S. Posamentier, J. H. Banks, and R. L. Bannister (Dover publications, 2014), pp. 396–402.

Postassessment

Have students redevelop some of the above theorems using methods presented in this unit.

Reference

Posamentier, A. S. and H. A. Hauptman. 100+ *Great Ideas for Introducing Key Concepts in Mathematics*. Thousand Oaks, CA: Corwin Press, 2006.

Unit 57

Trisecting a Circle

To partition a circle into two regions of equal area is a rather simple matter. However, to partition a circle into *three* regions of equal area is a more interesting problem. In this unit, students will investigate various methods of accomplishing this.

Performance Objective

Students will be able to partition a circle into three regions of equal area.

Preassessment

Students should be able to perform some simple geometric constructions using a straightedge and compass. They should also be familiar with the Pythagorean theorem and the formula for the area of a circle.

Teaching Strategies

Ask students to partition a circle into two regions of equal area (see Figure 1). The obvious solution is for them merely to draw the diameter of the given circle. Now ask students to partition a circle into three regions of equal area (hereafter referred to as "trisecting a circle"). This, too, should cause no problem as students will realize that they must merely construct (using straightedge and compasses) three adjacent angles of measure 120°.

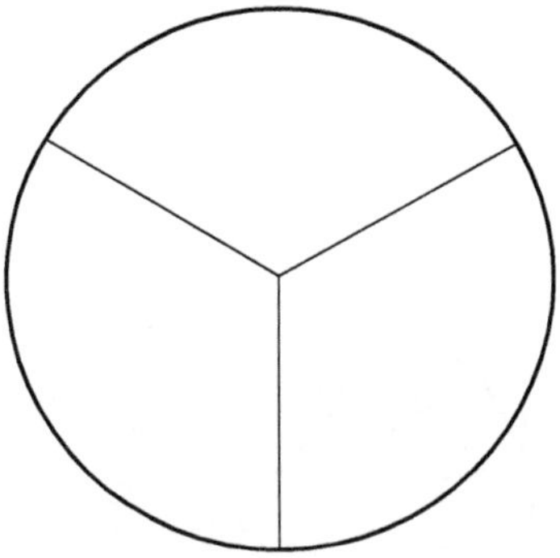

Figure 1.

To construct this trisection, they simply mark off six equal arcs along the circle with the compasses open to the radius of the circle. You may wish to justify this construction by referring to the inscribed hexagon, similarly constructed.

If you now ask students for another method of trisecting a circle, you will find them experimenting with another symmetry about the center. Ultimately this experimentation should lead to a consideration of two concentric circles, each concentric with the given circle. The problem then is to determine the lengths of the radii of the two circles.

Suppose students first find the radius, x, of a circle whose area is $\frac{1}{3}$ that of a given circle of radius r. Then $\pi x^2 = \frac{1}{3}\pi r^2$, which yields $x = \frac{r}{\sqrt{3}} = \frac{r\sqrt{3}}{3}$. In a similar way, they can find the radius, y, of a circle whose area is $\frac{2}{3}$ that of a given circle of radius r. That is, $\pi y^2 = \frac{2}{3}\pi r^2$, which yields $y = \frac{r\sqrt{2}}{\sqrt{3}} = \frac{r\sqrt{6}}{3}$.

Now that the lengths have been established, the only problem remaining is to do the actual construction. Have students begin with a circle of radius r. To construct x, rewrite $x = \frac{r\sqrt{3}}{3}$ as $\frac{x}{\sqrt{3}} = \frac{r}{3}$. Then mark off the lengths r and 3 on a convenient line segment (see Figure 2).

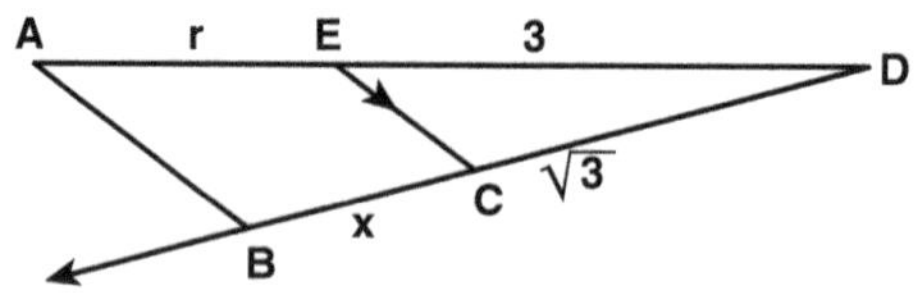

Figure 2.

With any convenient angle, have students mark off a length $\sqrt{3}$ along this newly drawn ray. To construct a line segment of length $\sqrt{3}$, students may use any convenient method. For example, the radical spiral may be used (Figure 3).

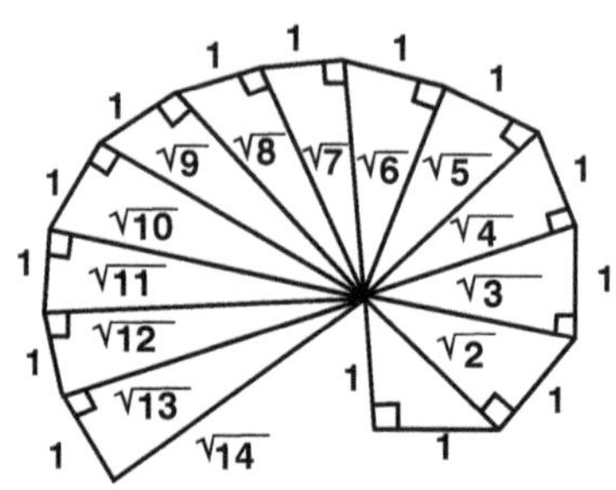

Figure 3.

Another method for constructing a line segment of length $\sqrt{3}$ would involve setting up a diagram as shown in Figure 4.

Once this length ($\sqrt{3}$) has been marked off along $\overline{\text{DCB}}$ (see Figure 2), students can construct a line through A parallel to $\overline{\text{EC}}$ to meet $\overleftrightarrow{\text{DC}}$ at B. Using proportions, they can establish that $x = \text{BC} = \frac{r\sqrt{3}}{3}$.

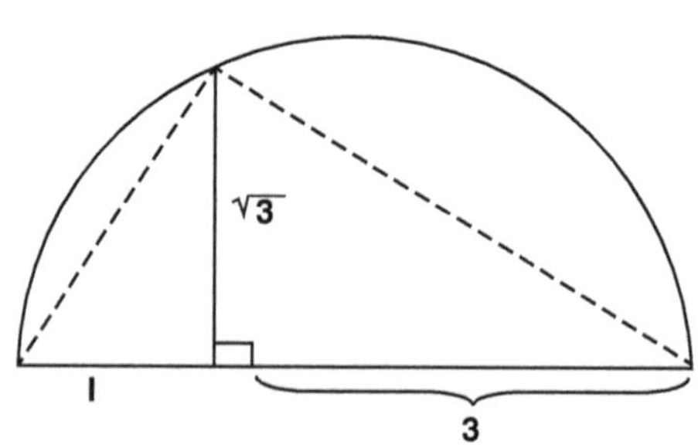

Figure 4.

Thus, tell students to draw a circle of radius x concentric with the given circle (Figure 5). The smaller circle has an area $\frac{1}{3}$ that of the large circle. To complete the trisection students should construct a circle, radius y, concentric with the given circle.

The area of the circle of radius y must be $\frac{2}{3}$ the area of the circle radius r. Therefore, $\pi y^2 = \frac{2}{3}\pi r^2$, and $y = \frac{r\sqrt{2}}{\sqrt{3}} = \frac{r\sqrt{6}}{3}$. Have students construct y in a manner similar to the construction of x and then draw the circle concentric with the others (see the dotted-line circle in Figure 5).

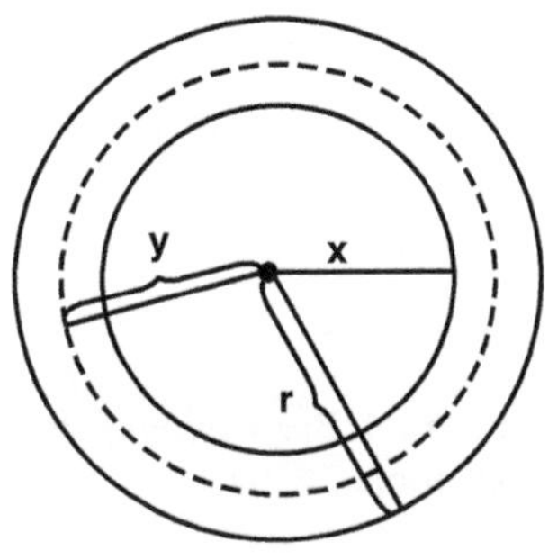

Figure 5.

The resulting figure shows a trisected circle.

A more intriguing trisection of a circle involves a rather unusual partitioning.

In Figure 6, the diameter of the given circle is trisected at points C and D. Four semicircles are then drawn as shown in the figure. Each of the two shaded regions is $\frac{1}{3}$ the area of the given circle. Therefore, the nonshaded region must also be $\frac{1}{3}$ the area of the circle and thus the circle is trisected.

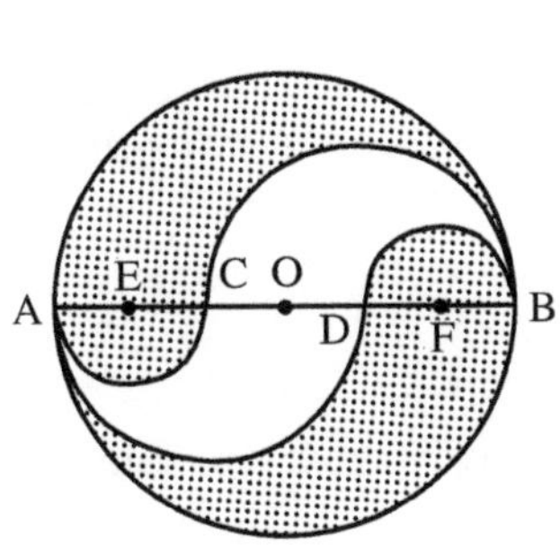

Figure 6.

To prove this trisection valid, students need to show that one of the shaded regions has an area $\frac{1}{3}$ that of the original circle. The area of the "upper" shaded region = area semicircle AB − area semi-circle BC + area semicircle AC. Let AE = r, AO = $3r$, and BD = $2r$. Therefore, the area of the "upper" shaded region $= \frac{1}{2}\pi(3r)^2 - \frac{1}{2}\pi(2r)^2 + \frac{1}{2}\pi r^2 = \frac{9\pi r^2}{2} - \frac{4\pi r^2}{2} + \frac{\pi r^2}{2} = 3\pi r^2$. However, the area of the original circle to be trisected $= \pi(3r)^2 = 9\pi r^2$. Thus, the area of each shaded region is $\frac{1}{3}$ the area of the original circle, which is then trisected.

Postassessment

Give students a circle and ask them to partition it into three regions of equal area.

Reference

Posamentier, A. S. and I. Lehmann. π: *A Biography of the World's Most Mysterious Number*. Amherst, NY: Prometheus Books, 2004.

Unit 58

Ptolemy's Theorem

This unit will offer the student a very powerful theorem about cyclic (inscribed) quadrilaterals.

Performance Objective

Given appropriate problems, students will apply Ptolemy's theorem to successfully solve the problem.

Preassessment

Present students with an isosceles trapezoid with bases of length 6 and 8 and legs of length 5. Ask them to find the length of a diagonal of the trapezoid.

Teaching Strategies

Students who are familiar with the Pythagorean theorem should be able to solve this problem with two applications of this theorem. However, most students, after being shown this method, will certainly welcome a less tedious method of solution. This is when you introduce Ptolemy's theorem.

> *Ptolemy's theorem*: In a cyclic (inscribed) quadrilateral, the product of the lengths of the diagonals is equal to the sum of the products of the lengths of the pairs of opposite sides.

Before proving this theorem, be sure students understand the statement of the theorem and understand what a cyclic quadrilateral is. Some of the more popular theorems about cyclic quadrilaterals ought to be reviewed here. Examples of noncyclic quadrilaterals should also be given, so that students better appreciate cyclic quadrilaterals.

Proof: Consider quadrilateral ABCD inscribed in circle O. Draw a line through A to meet $\overrightarrow{\mathrm{CD}}$ at P, so that $m\angle\mathrm{BAC} = m\angle\mathrm{DAP}$.

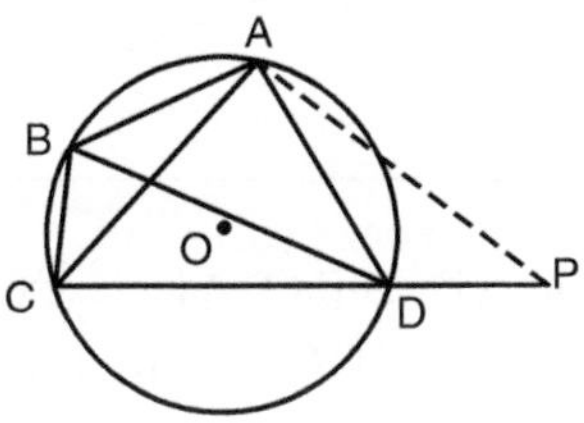

Since quadrilateral ABCD is cyclic, $\angle\mathrm{ABC}$ is supplementary to $\angle\mathrm{ADC}$. However, $\angle\mathrm{ADP}$ is supplementary to $\angle\mathrm{ADC}$. Therefore, $m\angle\mathrm{ABC} = m\angle\mathrm{ADP}$. We can then prove $\Delta\mathrm{BAC} \sim \Delta\mathrm{DAP}$, and $\frac{\mathrm{AB}}{\mathrm{AD}} = \frac{\mathrm{BC}}{\mathrm{DP}}$, or $\mathrm{DP} = \frac{(\mathrm{AD})(\mathrm{BC})}{\mathrm{AB}}$. Since $m\angle\mathrm{BAC} = m\angle\mathrm{DAP}, m\angle\mathrm{BAD} = m\angle\mathrm{CAP}$. Since $\Delta\mathrm{BAC} \sim \Delta\mathrm{DAP}, \frac{\mathrm{AB}}{\mathrm{AD}} = \frac{\mathrm{AC}}{\mathrm{AP}}$. Therefore, $\Delta\mathrm{ABD} \sim \Delta\mathrm{ACP}$, then $\frac{\mathrm{BP}}{\mathrm{CP}} = \frac{\mathrm{AB}}{\mathrm{AC}}$, or $\mathrm{CP} = \frac{(\mathrm{AC})(\mathrm{BD})}{\mathrm{AB}}$ But, $\mathrm{CP} = \mathrm{CD} + \mathrm{DP}$.

By substitution $\frac{(\mathrm{AC})(\mathrm{BD})}{\mathrm{AB}} = \mathrm{CD} + \frac{(\mathrm{AD})(\mathrm{BC})}{\mathrm{AB}}$.

Now simplifying this expression gives us the desired result:

$$(\mathrm{AC})(\mathrm{BD}) = (\mathrm{AB})(\mathrm{CD}) + (\mathrm{AD})(\mathrm{BC})$$

which is Ptolemy's theorem.

Show students how Ptolemy's theorem may be used to solve the preassessment problem. Since an isosceles trapezoid is a cyclic quadrilateral, Ptolemy's theorem may be used to get $d^2 = (6)(8) + (5)(5) = 73$.

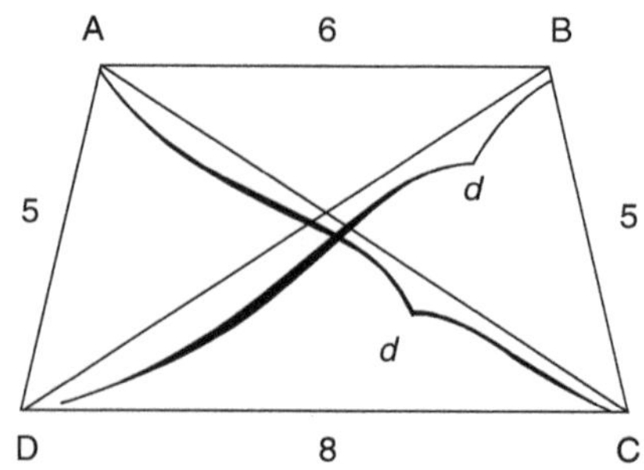

Therefore, the length of a diagonal (d) is $\sqrt{73}$.

Students are often curious if a "new" theorem is consistent with theorems they learned earlier. Have students apply Ptolemy's theorem to a rectangle (which is clearly a cyclic quadrilateral). For rectangle ABCD, Ptolemy's theorem appears as (AC)(BD) = (AD)(BC) + (AB)(DC).

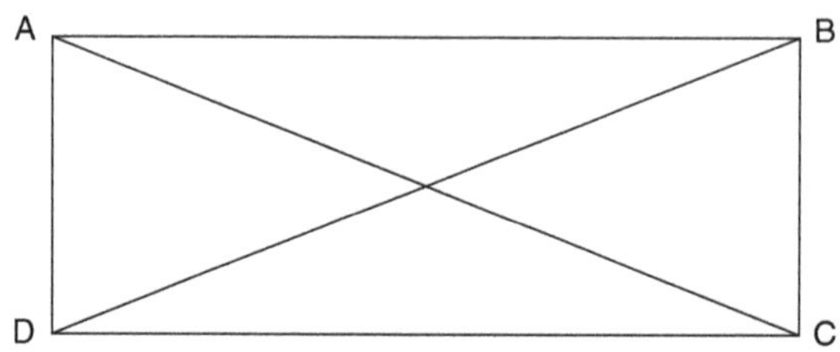

However, in the rectangle AB = DC, AD = BC, and AC = BD. Therefore by substitution, $(\text{AC})^2 = (\text{AD})^2 + (\text{DC})^2$, which is the Pythagorean theorem.

Now have students consider a rather simple application of this celebrated theorem.

Problem: If point P is on arc AB of the circumscribed circle of equilateral ΔABC, and AP = 3 while BP = 4, find the length of $\overline{\text{CP}}$.

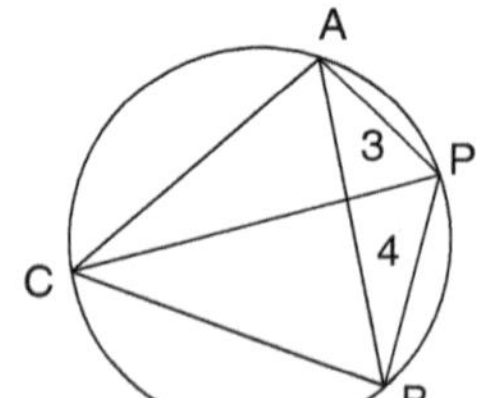

Solution: Let t represent the length of a side of equilateral ΔABC. Since quadrilateral APBC is cyclic, we may apply Ptolemy's theorem, which yields

$$(\text{CP})(t) = (\text{AP})(t) + (\text{BP})(t).$$

Therefore, CP = AP + BP = 3 + 4 = 7.

Students should be encouraged to investigate similar problems where the equilateral triangle is replaced with other regular polygons.

Often problems appear to be easier than they actually are. The next problem seems to be easily solvable by simply using the Pythagorean theorem. However, in solution, it becomes useful to employ Ptolemy's theorem.

Problem: On side $\overline{AB}$ of square ABCD, a right ΔABF with hypotenuse $\overline{AB}$, is drawn externally to the square. If AF $= 6$ and BF $= 8$, find EF, where E is the point of intersection of the diagonals of the square.

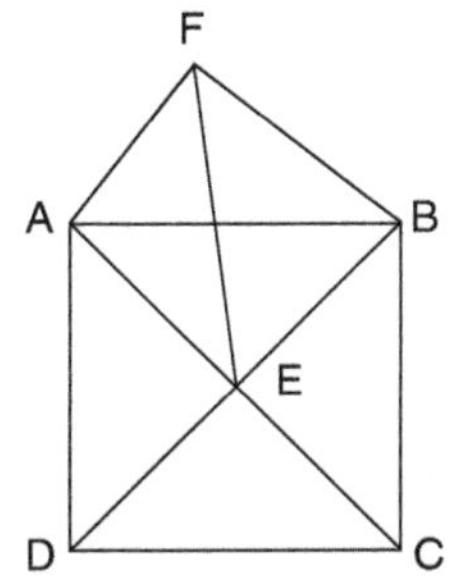

Solution: Applying the Pythagorean theorem to right ΔAFB, we get AB $=$ 10, and to right ΔAEB, we get AE $=$ BE $= 5\sqrt{2}$. Since m$\angle$AFB $=$ $m\angle$AEB $= 90$, quadrilateral AFBE is cyclic. Now we may apply Ptolemy's theorem to quadrilateral AFBE, to get (AB)(EF) $=$ (AF)(BE) $+$ (AE)(BF). Substituting the appropriate values gives us $(10)(\text{EF}) = (6)(5\sqrt{2}) + (5\sqrt{2})(8)$, or EF $= 7\sqrt{2}$.

Students should be encouraged to reconsider this problem with right ΔABF drawn internally to the square. In that case, EF $= \sqrt{2}$.

Postassessment

Have students solve each of the following problems:

1. E is a point on side $\overline{AD}$ of rectangle ABCD, so that DE $= 6$, while DA $= 8$, and DC $= 6$. If $\overline{CE}$ extended meets the circumcircle of the rectangle at F, find the measure of chord $\overline{DF}$.

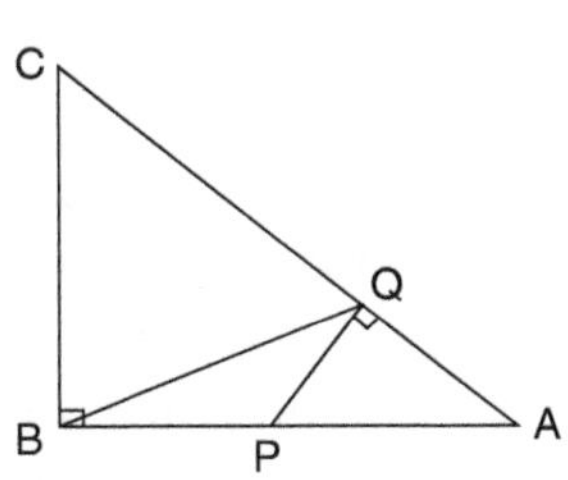

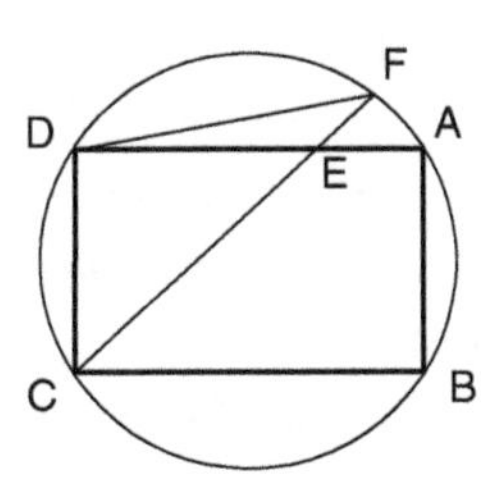

2. Point P on side $\overline{AB}$ of right ΔABC is placed so that $BP = PA = 2$. Point Q is on hypotenuse $\overline{AC}$ so that $\overline{PQ}$ is perpendicular to $\overline{AC}$. If $CB = 3$, find the measure of $\overline{BQ}$, using Ptolemy's theorem.

References

Posamentier, A. S. *Advanced Euclidean Geometry: Excursions for Secondary Teachers and Students*. New York, NY: John Wiley Publishing, 2002.
Posamentier, A. S. and C. T. Salkind. *Challenging Problems in Geometry*. New York: Dover, 1996.

Unit 59

Constructing π

Performance Objectives

1. *Students will demonstrate a clear knowledge of the π, ratio and its relationship to the circle.*
2. *Students will construct π in more than one way.*

Preassessment

Before beginning a discussion of π, review with students the meaning of diameter and circumference. Have students measure the diameter and circumference of a 25-cent piece. Also ask them to obtain similar measurements of other circular objects. Stress the importance of accurate measurement.

Teaching Strategies

Begin the lesson by writing the following chart on the chalkboard:

Object	C	D	C + D	C − D	C · D	$\frac{C}{D}$

Record some of the measurements that the students obtained. They should all have found that the diameter of the quarter is *about* 2.4 mm long and that its circumference is *approximately* 7.8 mm. Have students then fill in the rest of the chart for the objects that they have measured. Ask them if any column seems to result in approximately the same value for each object measured and have them take the average of the numbers in that column.

Their averages should be close to 3.14 (i.e., $\frac{C}{D} \approx 3.14$). Reemphasize that all the other columns produced varying results, whereas in the last column $\frac{C}{D}$ was the same regardless of the size of the object.

In 1737, this ratio was given the special name of "π" by Leonhard Euler, a famous Swiss mathematician. The exact value of π can never be determined; only approximations can be established. Here is the value of π correct to 50 decimal places:

$$\pi = 3.14159265358979323846264338327950288419716939937 51\ldots$$

Throughout the years, many attempts have been made to compute π, both algebraically and geometrically. This unit presents some of the geometric constructions involving π.

One of the first serious attempts to compute π to a certain degree of accuracy goes back to Archimedes, who tried to exactly determine π. His method was based on the fact that the perimeter of a regular polygon of n sides is smaller than the circumference of the circle circumscribed about it, while the perimeter of a similar polygon circumscribed about the circle is greater than the circle's circumference. By successively repeating this situation for larger values of n, the two perimeters will approach the circumference from both sides. Archimedes started with a regular hexagon and each time doubled the number of sides until he obtained a polygon of 96 sides. He was then able to determine that the ratio of the circumference of a circle to its diameter, or π, is less than $3\frac{10}{70}$ but greater than $3\frac{10}{71}$. We can write this in decimal notation as $3.14085 < \pi < 3.142857$. To aid in the students' understanding of this method, you might illustrate with a few diagrams. The following chart might also aid in explaining this concept, as the students will see that as the number of sides of the regular polygon increases, π is more accurately approximated.

Number of sides	Perimeter of circumscribed polygon	Perimeter of inscribed polygon
4	4.0000000	2.8284271
8	3.3137085	3.0614675
16	3.1825979	3.1214452
32	3.1517249	3.1365485
64	3.1441184	3.1403312
128	3.1422236	3.1412773
256	3.1417504	3.1415138
512	3.1416321	3.1415729
1024	3.1416025	3.1415877
2048	3.1415951	3.1415914

Students will now see how they can actually construct a line segment whose length closely approximates π. This construction was developed in the mid-1800s and involves the ratio $\frac{355}{113}$ (which had been previously discovered by a Chinese astronomer in the fifth century). $\frac{355}{113} = 3 + \frac{16}{113} = 3.1415929\cdots$ which is a correct approximation of π to six decimal places. The construction begins with a quadrant of unit radius. AO is $\frac{7}{8}$ of the radius, $\overline{AB}$ is drawn and a point C is marked off so that CB $= \frac{1}{2}$ of the radius. $\overline{CD}$ is drawn parallel to $\overline{AO}$ and $\overline{CE}$ is drawn parallel to $\overline{AD}$.

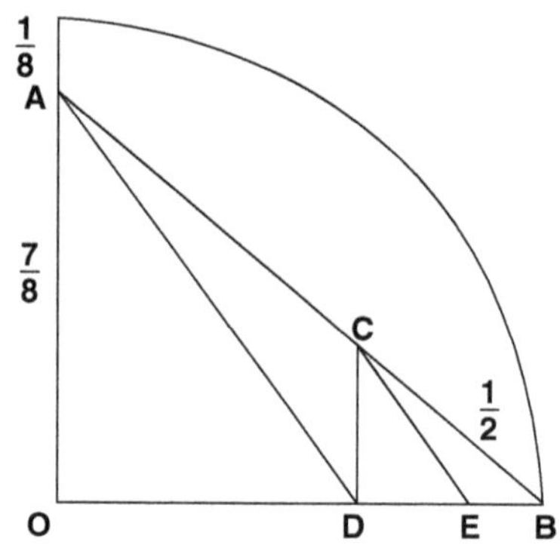

Have students find AB : $\left(\frac{7}{8}\right)^2 + 1^2 = (\text{AB})^2$. Therefore, AB $= \frac{\sqrt{113}}{8}$.

Using similar triangles, the following relationships can easily be seen. (Have students explain why ΔCDB $\sim \Delta$AOB and ΔCEB $\sim \Delta$ADB.)

$$\frac{\text{DB}}{\text{OB}} - \frac{\text{CB}}{\text{AB}} \quad \text{and} \quad \frac{\text{EB}}{\text{DB}} = \frac{\text{CB}}{\text{AB}}$$

Multiplying these expressions, we obtain

$$\frac{\text{EB}}{\text{OB}} = \frac{\text{CB}^2}{\text{AB}^2} = \frac{\frac{1}{4}}{\frac{113}{64}} = \frac{16}{113}$$

but since OB $= 1$, we get

$$\frac{\text{EB}}{1} = \frac{16}{113} \text{ or EB } = \frac{16}{113} \text{ or } \approx .1415929204\cdots$$

Since $\frac{355}{113} = 3 + \frac{16}{113}$, a line segment can now be drawn that is 3 times the radius extended by the distance EB. This will give us a line segment that differs from π by less than a millionth of a unit.

A slightly more difficult geometric approximation of π was developed in 1685 by Father Adam Kochansky, a librarian to King John III of Poland. A circle of unit radius is drawn. Then draw a tangent segment $\overline{QR}$, equal in length to three times the radius. Draw a diameter perpendicular to $\overline{QR}$ at Q, the point of tangency. Now draw a line, d, tangent at the other end of the diameter such that the measure of central angle = 30°. Connect points and extend line segments to form the figure pictured in the following figure. The students are now ready to calculate the value of π. (It will be shown that if the length of the radius is 1, line c approximates π.)

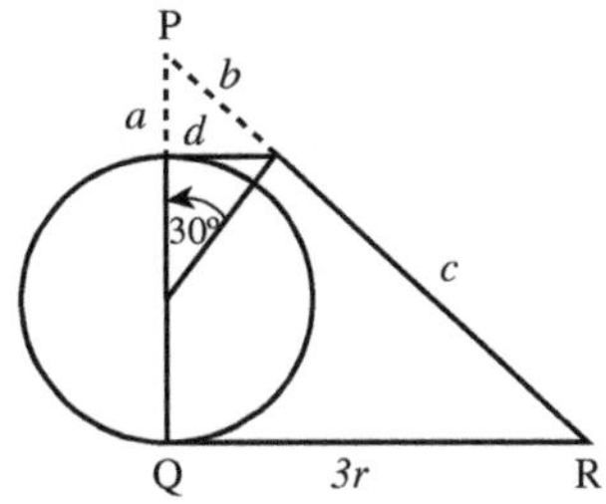

If $r = 1$, in ΔPQR, $(a+2)^2 + (3)^2 = (b+c)^2$. (1)

Also, using similar triangles, we have $\frac{a}{a+2} = \frac{d}{3}$ (2)

and $\frac{b}{b+c} = \frac{d}{3}$. (3)

From Equation (2), we obtain $3a = ad + 2d$ or $a = \frac{2d}{3-d}$. But $\tan 30° = \frac{d}{1} = d = \frac{\sqrt{3}}{3}$.

Therefore, $a = \dfrac{2\frac{\sqrt{3}}{3}}{3 - \frac{\sqrt{3}}{3}}$ or $a = \dfrac{2\sqrt{3}}{9-\sqrt{3}}$. (4)

Similarly, from Equation (3), we can obtain

$$b = \frac{cd}{3-d} = \frac{c\sqrt{3}}{9-\sqrt{3}}. \qquad (5)$$

Substituting Equations (4) and (5) into Equation (1), we now have

$$\left(\frac{2\sqrt{3}}{9-\sqrt{3}} + 2\right)^2 + 9 = \left(\frac{c\sqrt{3}}{9-\sqrt{3}} + c\right)^2.$$

Students should be able to solve this equation for c and obtain $c = \sqrt{\frac{40}{3} - 2\sqrt{3}}$.

Have students simplify this radical to obtain 3.141533 as an approximate value for c.

Throughout the lesson, it should be emphasized that these are all *approximations* of the value π, since it is impossible to construct π with straightedge and compasses.

Postassessment

1. Find the diameter of a circle whose circumference is 471 feet.
2. Construct a geometric approximation of π in more than one way.

References

Posamentier, A. S. and L. Iehmann. π: *A Biography of the World's Most Mysterious Number*. Amherst, NY:Prometheus Books, 2004.

Posamentier, A. S. and N. Gordon. An astounding revelation on the history of π. *The Mathematics Teacher* **77**(1), 1984, p. 52.

Unit 60

The Arbelos

The region bounded by three semicircles in a manner resembling a shoemaker's knife has some rather interesting properties. This region, often called an arbelos, is the topic of this unit. Here the student will be introduced to this geometric figure with the intention of pursuing its properties further.

Performance Objectives

1. *Students will identify the arbelos.*
2. *Students will solve problems involving the arbelos.*

Preassessment

This unit should be presented to students who have studied geometry (or arc currently enrolled in the last term of a geometry course). They should be able to compute lengths of arcs, areas of triangles, and areas of circles.

Teaching Strategies

Have students draw a semicircle with center O and diameter $\overline{AB}$. Let AB = 2R. Have them then mark off a point C, between A and B. Then have them let $\overline{AC}$ and $\overline{CB}$ be diameters of semicircles D and E, respectively (see Figure 1). Let AC = $2r_1$ and BC = $2r_2$. The shaded portion of the figure is known as the *Arbelos or Shoemaker's Knife*. It has some very interesting properties that were considered by Archimedes, the famous Greek mathematician.

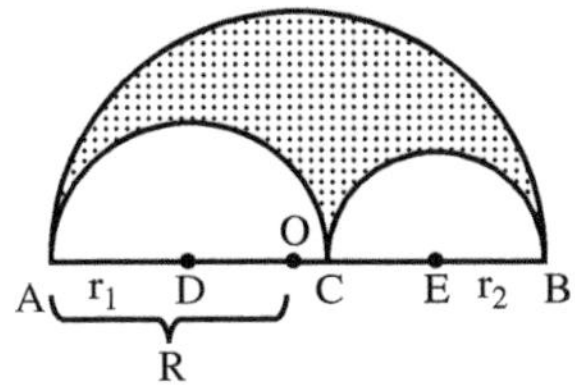

Figure 1.

You should now direct student attention to the diagram. Try to elicit from your students the following property of the arbelos: that $m\widehat{AB} = m\widehat{AC} + m\widehat{CB}$. Once students understand this property, a proof should be established. In a circle, the length of an arc $= \frac{n}{360} \times 2\pi r$ (where n is the number of degrees of the arc and r is the length of the radius), we have

$$m\widehat{AB} = \frac{180}{360} \cdot 2\pi R = \pi R$$

$$m\widehat{AC} = \frac{1}{2} \cdot 2\pi r_1 = \pi r_1$$

$$m\widehat{CB} = \frac{1}{2} \cdot 2\pi r_2 = \pi r_2$$

Also $R = r_1 + r_2$, therefore multiplying by π we get $\pi R = \pi r_1 + \pi r_2$ or $m\widehat{AB} = m\widehat{AC} + m\widehat{CB}$. Have students consider the case where three semicircles (instead of two) are taken on $\overline{AB}$. Would a similar relationship hold
true?

Students should now draw the perpendicular to $\overline{AB}$ at point C, which meets the circle at H. Also draw the common tangent to circles D and E and call the points of tangency F and G, respectively. Denote the point where these two segments intersect as S (see Figure 2). Since a line segment drawn perpendicular to a diameter is the geometric mean between the segments of the diameter, we have that $(HC)^2 = 2r_1 \cdot 2r_2 = 4r_1r_2$. Also FG = JE (have students explain why from the diagram). Since $JD = r_1 - r_2$ and $DE = r_1 + r_2$, then $(JE)^2 = (r_1+r_2)^2 - (r_1-r_2)^2 = r_1^2 + 2r_1r_2 + r_2^2 - r_1^2 + 2r_1r_2 - r_2^2 = 4r_1r_2$. Therefore, $(FG)^2 = 4r_1r_2$ or $(HC)^2 = (FG)^2 = 4r_1r_2$.

Ask your students if they can suggest another relationship that exists between $\overline{HC}$ and $\overline{FG}$. Once someone gives the response that $\overline{HC}$ and $\overline{FG}$ bisect each other at S, have the students try to prove it by themselves. $\overline{SC}$ is

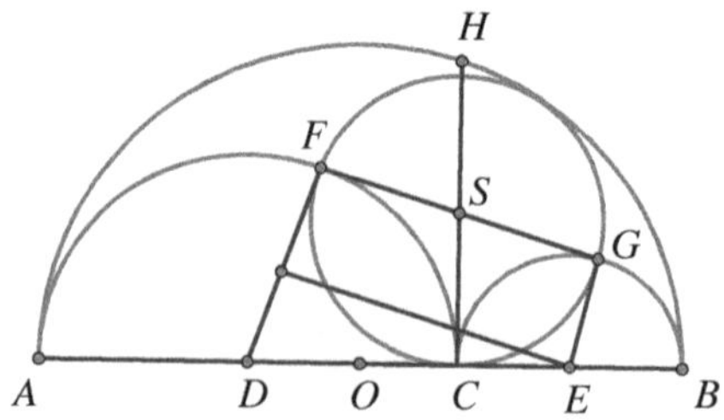

Figure 2.

a common internal tangent to both circles, therefore, FS = SC and SC = SG, which gives us FS = SG. But since HC = FG (have students explain why), we also know that HS = SC. Also since FS = SG = HS = SC, the points F, H, G, C determine a circle with center S.

A very interesting property of the arbelos is one that involves this circle, which has $\overline{HC}$ and $\overline{FG}$ as diameters. Have students try to express the area of the arbelos in terms of r_1 and r_2. Area of the arbelos = Area of semicircle ABH − (Area of semicircle AFC + Area of semicircle CGB).

Since area of a semicircle $= \frac{\pi r^2}{2}$, we have

$$\text{Area of the arbelos} = \frac{\pi \mathrm{R}^2}{2} - \left(\frac{\pi r_1^2}{2} + \frac{\pi r_2^2}{2}\right)$$
$$= \frac{\pi}{2}(\mathrm{R}^2 - r_1^2 - r_2^2).$$

We know that $R = r_1 + r_2$ and substituting we get area of the arbelos

$$= \frac{\pi}{2}((r_1 + r_2)^2 - r_1^2 - r_2^2)$$
$$= \frac{\pi}{2}(r_1^2 + 2r_1r_2 + r_2^2 - r_1^2 - r_2^2)$$
$$= \frac{\pi}{2}(2r_1r_2) = \pi r_1 r_2.$$

Have the students now find the area of circle S. The diameter HC $= 2\sqrt{r_1r_2}$, therefore the radius $= \sqrt{r_1r_2}$. The area of the circle then $= \pi(\sqrt{r_1r_2})^2 = \pi r_1 r_2$. It is now apparent that they must have proved that the area of the arbelos is equal to the area of circle S.

You may wish to introduce another interesting arbelos.

Let P and R be the midpoints of arcs $\widehat{AC}$ and $\widehat{CB}$, respectively. Let Q be the midpoint of the semicircle below $\overline{AB}$. Connect points P and R to C and to Q. A concave quadrilateral PQRC is formed (see Figure 3).

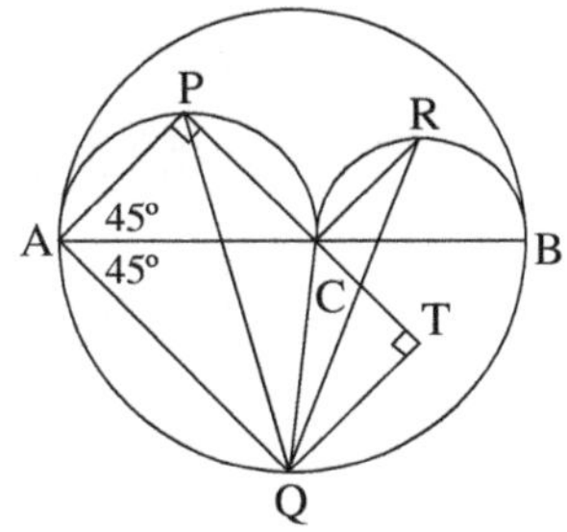

Figure 3.

The area of this quadrilateral is equal to the sum of the squares of the radii, r_1 and r_2, of the two smaller semicircles.

A proof follows: The quadrilateral can be divided into two triangles by drawing $\overline{\text{CQ}}$. The area of ΔQCP can be shown to be equal to the area of right ΔAPC. The two triangles have a common base $\overline{\text{CP}}$, therefore their heights must be proved to be equal. To do this, draw $\overline{\text{AP}}, \overline{\text{AQ}}$ and draw $\overline{\text{QT}}$ perpendicular to $\overline{\text{PC}}$ extended (see Figure 3). Since Q is the midpoint of semicircle AB, $m\widehat{\text{QB}} = 90°$. Therefore, $m\angle\text{QAB} = 45°$. Also since ΔAPC is an isosceles right triangle, $m\angle\text{PAB} = 45°$, which gives us that $m\angle\text{PAQ} = 90°$. But since $m\angle\text{APC} = 90°$ and $m\angle\text{PTQ}$ is also quadrilateral, APTQ is a rectangle and AP = QT.

Therefore, $\Delta\text{QCP} = \frac{\text{CP}\cdot\text{PA}}{2}$.

Since in isosceles right triangle APC, $(\text{CP})^2+(\text{PA})^2 = (2\text{r}_1)^2$ or $2(\text{CP})^2 = (2r_1)^2$, therefore $(\text{CP})^2 = 2r_1^2$ or $\frac{\text{CP}\cdot\text{PA}}{2} = r_1^2$.

We therefore have that area of $\Delta\text{QCP} = r_1^2$.

Similarly, it can be shown that area of $\Delta\text{QCR} = \frac{\text{CR}\cdot\text{RB}}{2} = r_2^2$.

Therefore, area of the quadrilateral $= r_1^2 + r_2^2$.

Postassessment

1. If $r_1 = 16$ and $r_2 = 4$, show that $\widehat{\text{AB}} = \widehat{\text{AC}} + \widehat{\text{CB}}$; find the radius of circle S; find the area of the arbelos.
2. Describe semicircle D below $\overline{\text{AB}}$ (Figure 4). Let $\overline{\text{AN}}$ be tangent to circle E. Show that the area of the shaded region is equal to the area of the circle, which has $\overline{\text{AN}}$ as its diameter.
3. Find the area of quadrilateral PQRC (Figure 3), if $r_1 = 8$ and $r_2 = 5$.
4. What is the relationship between the arbelos in Figure 3 and Fibonacci numbers.

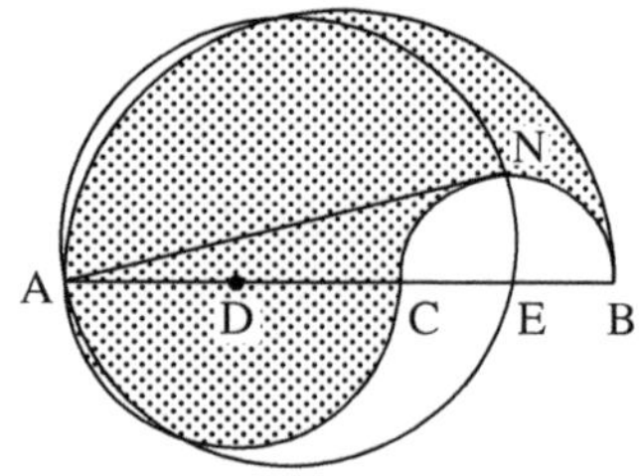

Figure 4.

Reference

Mathematical Games: The diverse pleasures of circles that are tangent to one another. *Scientific 3 American* **240**(1), 1979, pp. 18–28.

Unit 61

The Nine-Point Circle

An often neglected concept in the high school geometry curriculum is that of establishing points concyclic (on the same circle). This unit presents one of the more famous sets of concyclic points.

Performance Objectives

1. *Students will define and construct the nine-point circle.*
2. *Students will locate the center of the nine-point circle.*

Preassessment

Students should be aware of elementary methods of proving four points concyclic. For example, they should be aware of at least the following two theorems:

1. If one side of a quadrilateral subtends congruent angles at the two non-adjacent vertices, then the quadrilateral is cyclic (may be inscribed in a circle).
2. If a pair of opposite angles of a quadrilateral are supplementary, then the quadrilateral is cyclic.

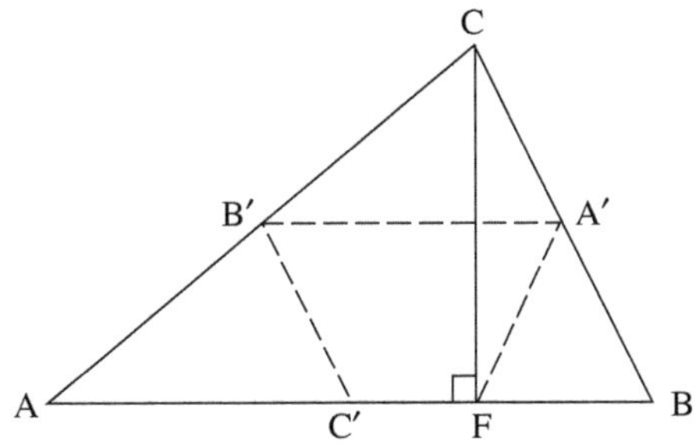

Figure 1.

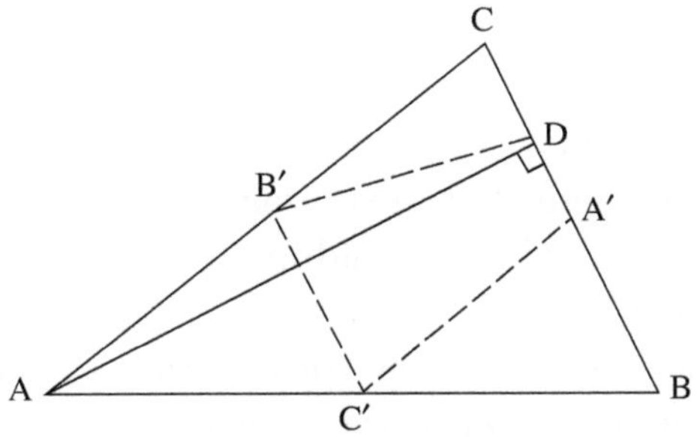

Figure 2.

Teaching Strategies

Present students with a ΔABC, with midpoints of its sides A', B', C' (see Figure 1). Draw altitude $\overline{CF}$. Ask students to prove that quadrilateral $A'B'C'D'$ is an isosceles trapezoid. To do this, they should realize that since $\overline{A'B'}$ is a segment joining the midpoints of two sides of a triangle, it is parallel to the third side of the triangle. Since $\overline{B'C'}$ joins the midpoints of $\overline{AC}$ and $\overline{AB}$, $B'C' = \frac{1}{2}(BC)$. Since the median to the hypotenuse of a right triangle is half the length of the hypotenuse, $A'F = \frac{1}{2}(BC)$. Therefore, $B'C' = A'F$, and trapezoid $A'B'C'F$ is isosceles.

Now have students prove that an isosceles trapezoid is always cyclic (using Figure 2, above).

To avoid confusion, redraw ΔABC with altitude $\overline{AD}$ as shown in the following.

In the same way as for altitude $\overline{CF}$, have students independently prove that the points B', C', A', and D are concyclic. This should be done with the above proof as a guide.

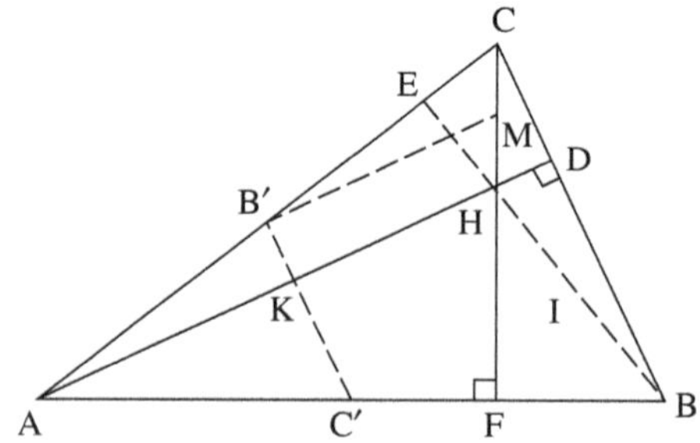

Figure 3.

Students should now be prepared to generalize a statement about the points B′, C′, A′, and E, for altitude $\overline{BE}$. This will lead to the conclusion that the points D, F, and E each lie on the unique circle determined by points A′, B′, C′. Thus, students can summarize that the feet of the altitudes of a triangle are concyclic with the midpoints of the sides. So far they have established a "six-point circle."

By this time, students should have proved that the altitudes of a triangle are concurrent. This point is called the *orthocenter*. Have them consider the orthocenter, H, of ΔABC, and the midpoint M of $\overline{CH}$.

$\overline{B'M}$ is a segment joining the midpoints of two sides of $\angle$ACH. Therefore, $\overline{B'M}//\overline{AH}$. Similarly in $\angle$ABC, we have B′C′//BC. Since altitude $\overline{AD} \perp \overline{BC}$, $\overline{B'M} \perp \overline{B'C'}$, or $m\angle MB'C' = 90°$ (Figure 3). Remember that $m\angle AFC = 90°$. Therefore, quadrilateral MB′C′F is cyclic, since its opposite angles are supplementary. This is the same circle established above, since three vertices (B′, C′, and F) are common with the six concyclic points, and three points determine a unique circle. Thus, a "seven-point circle" has been established.

To reinforce this proof, students should now prove that K and L (the midpoints of $\overline{AH}$ and $\overline{BH}$, respectively) also lie on this circle. To do this, they merely need to repeat the above procedure for points K, C′, A′, D and for points L, C′, B′, E. A brief review of the entire proof thus far will reveal a *nine-point circle*.

Have students consider $\overline{MC'}$ in Figure 4. Since it subtends right angles at points B′ and F, it must be the diameter of the circle through B′, C′, F, and M. To locate the center N of this circle, simply tell students to find the midpoint of $\overline{MC'}$. This is the center of the nine-point circle.

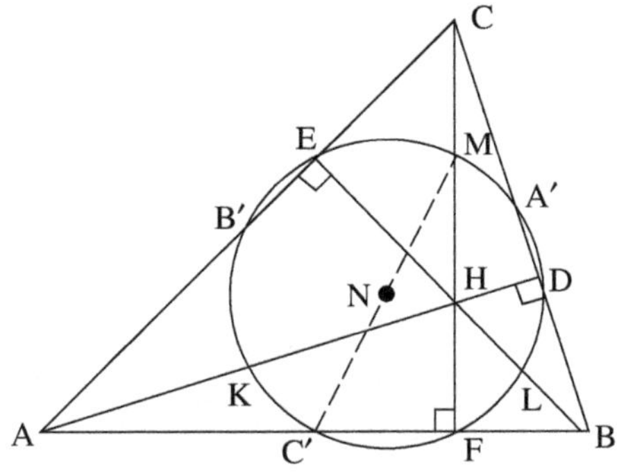

Figure 4.

Postassessment

To conclude the lesson, ask students to do the following:

1. Define the nine-point circle.
2. Construct the nine-point circle using straightedge and compasses.
3. Locate the center of the nine-point circle.

Interesting relationships involving the nine-point circle can be found in the accompanying unit, the Euler Line.

Many other interesting relationships involving the nine-point circle can be found in the book by Posamentier (2002).

Reference

Posamentier, A. S. *Advanced Euclidean Geometry: Excursions for Secondary Teachers and Students.* New York, NY: John Wiley Publishing, 2002.

Unit 62

The Euler Line

This unit should be presented to students *after* they have studied the unit (61) on the nine-point circle. This unit uses some of the material developed there and relates it to other points of a triangle.

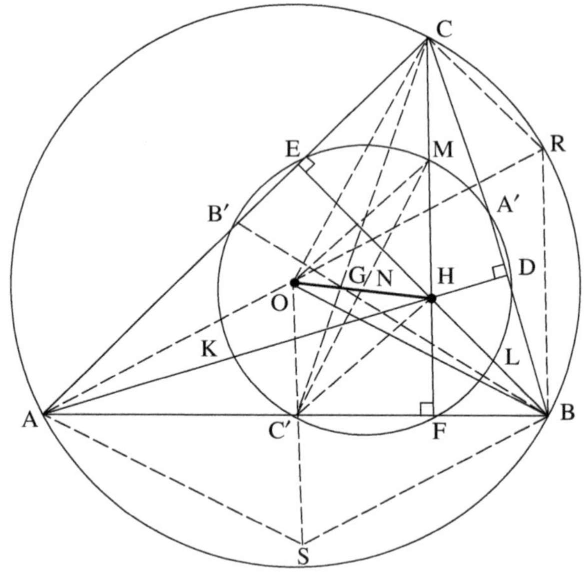

Figure 1.

Performance Objectives

1. *Students will locate the Euler line of a triangle.*
2. *Students will establish a relationship among the circumcenter, orthocenter, centroid, and the center of the nine-point circle of a triangle.*

Preassessment

Have students draw a scalene triangle and construct its nine-point circle as well as the circumcircle of the triangle.

Teaching Strategies

To facilitate the discussion, students should label their construction as in Figure 1.

Students should now draw $\overline{\text{OH}}$, the segment joining the orthocenter (the point of intersection of the altitudes) and the circumcenter (the point of intersection of the perpendicular bisectors of the sides of the triangle). This is the *Euler line.*

Have students locate the center of the nine-point circle by finding the midpoint of $\overline{\text{MC}'}$ (this was proved in *The Nine-Point Circle*). An accurate construction should place this point on the midpoint of the Euler line $\overline{OH}$. Student curiosity should now request a proof of this astonishing occurrence.

1. Draw $\overleftrightarrow{OA}$ to intersect circle O at R.
2. $\overline{OC'} \perp \overline{AB}$ (since O is on the perpendicular bisector of $\overline{AB}$ and C′ is the midpoint of $\overline{AB}$).
3. $m\angle ABR = 90°$ (an angle inscribed in a semicircle).
4. Therefore, $\overline{OC'}//\overline{RB}$ (both are perpendicular to $\overline{AB}$).
5. Similarly, $\overline{RB}//\overline{CF}$ and $\overline{RC}//\overline{BE}$.
6. $\Delta AOC' \sim \Delta ARB$ (with a ratio of similitude of $\frac{1}{2}$).
7. Therefore, $OC' = \frac{1}{2}(RB)$.
8. Quadrilateral RBHC is a parallelogram (both pairs of opposite sides are parallel).
9. Therefore, $RB = HC$, and $OC' = \frac{1}{2}(HC) = HM$.
10. Quadrilateral OC′HM is a parallelogram (one pair of sides is both congruent and parallel).
11. Therefore, since the diagonals of a parallelogram bisect each other, N (the midpoint of $\overline{MC'}$), is the midpoint of $\overline{OH}$.

So far we proved that the center of the nine-point circle bisects the Euler line. At this point, we can easily prove that the radius of the nine-point circle is half the length of the radius of the circumcircle. Since $\overline{MN}$ is a line segment joining the midpoints of two sides of ΔCOH, it is half the length of the third side $\overline{OC}$. Thus the radius, $\overline{MN}$, of the nine-point circle is half the length of the radius, $\overline{OC}$, of the circumcircle.

In 1765, Leonhard Euler proved that the centroid of a triangle (the point of intersection of the medians) trisects the line segment joining the orthocenter and the circumcenter (the Euler line).

Since $\overline{OC'}//\overline{CH}, \Delta OGC' \sim \Delta HGC$.

Earlier we proved that $OC' = \frac{1}{2}(HC)$.

Therefore, $OG = \frac{1}{2}(GH)$,

$$\text{or} \quad OG = \left(\frac{1}{3}\right)(OH).$$

The only thing remaining is to show that G is the centroid of the triangle.

Since $\overline{CC'}$ is a median and $\overline{GC'} = \frac{1}{2}(GC)$,

G must be the centroid since it appropriately trisects the median. Thus, G trisects $\overline{OH}$.

Ask students why the median $\overline{BB'}$ also trisects $\overline{OH}$ (because it contains G, the centroid).

To this point, we have bisected and trisected the Euler line with significant triangle points. Before ending the discussion of the Euler line, an interesting vector application should be considered. Review the concept of a vector and a parallelogram of forces. We shall show that $\overrightarrow{\mathrm{OH}}$ is the resultant of $\overrightarrow{\mathrm{OA}}, \overrightarrow{\mathrm{OB}}$, and $\overrightarrow{\mathrm{OC}}$. This was first published by James Joseph Sylvester (1814–1897).

Consider the point S on $\overline{\mathrm{OC'}}$, where $\mathrm{OC'} = \mathrm{SC'}$.

Since $\overline{\mathrm{OC'S}}$ is the perpendicular bisector of $\overline{\mathrm{AB}}$, quadrilateral AOBS is a parallelogram (rhombus).

Therefore, vectors $\overrightarrow{\mathrm{OS}} = \overrightarrow{\mathrm{OA}} + \overrightarrow{\mathrm{OB}}$, or $\overrightarrow{\mathrm{OC'}} = \frac{1}{2}(\overrightarrow{\mathrm{OA}} + \overrightarrow{\mathrm{OB}})$.

Since $\Delta\mathrm{OGC'} \sim \Delta\mathrm{HGC}, \mathrm{CH} = 2(\mathrm{OC'})$.

Thus, $\overrightarrow{\mathrm{CH}} = \overrightarrow{\mathrm{OA}} + \overrightarrow{\mathrm{OB}}$.

Since $\overrightarrow{\mathrm{HO}}$ is the resultant of $\overrightarrow{\mathrm{OC}}$ and $\overrightarrow{\mathrm{CH}}, \overrightarrow{\mathrm{HO}} = \overrightarrow{\mathrm{OC}} + \overrightarrow{\mathrm{CH}}$.

Therefore, $\overrightarrow{\mathrm{HO}} = \overrightarrow{\mathrm{OC}} + \overrightarrow{\mathrm{OA}} + \overrightarrow{\mathrm{OB}}$. (Substitution)

Postassessment

At the conclusion of this lesson, ask students:

1. To construct the Euler line of a given scalene triangle, and
2. To state a relationship that exists among the circumcenter, orthocenter, centroid, and the center of the nine-point circle of a given scalene triangle.

Reference

Posamentier, A. S. *Advanced Euclidean Geometry: Excursions for Secondary Teachers and Students.* New York, NY: John Wiley Publishing, 2002.

Unit 63

The Simson Line

One of the more famous sets of collinear points is known as the *Simson line.* Although this line was discovered by William Wallace in 1797, careless misquotes have, in time, attributed it to Robert Simson (1687–1768). This unit will present, prove, and apply the Simson theorem.

Performance Objectives

1. *Students will construct the Simson line.*
2. *Students will prove that the three points that determine the Simson line are, in fact, collinear.*
3. *Students will apply the properties of the Simson line to given problems.*

Preassessment

When students are presented with this unit, they should be well into the high school geometry course, having already studied angle measurement with a circle. Students should also review cyclic quadrilaterals (quadrilaterals that may be inscribed in a circle) before beginning this unit.

Teaching Strategies

Have each student construct a triangle inscribed in a circle. Then, from any convenient point on the circle (but not at a vertex of the triangle), have students construct perpendicular segments to each of the three sides of the triangle. Now, ask the class what relationship seems to be true about the three feet of the perpendiculars. If the constructions were done accurately, everyone should notice that these three points determine the *Simson line.*

The obvious question should be quickly forthcoming: "Why are these points collinear?" This is where you begin your proof.

> *Simson's theorem: The feet of the perpendiculars drawn from any point on the circumcircle of a given triangle to the sides of the triangle arc collinear.*

Given: ΔABC is inscribed in circle O.
P is on circle O.
$\overleftrightarrow{PY} \perp \overleftrightarrow{AC}$ at Y, $\overleftrightarrow{PZ} \perp \overleftrightarrow{AB}$ at Z, and
$\overleftrightarrow{PX} \perp \overleftrightarrow{BC}$ at X.
Prove: Points X, Y, and Z are collinear.

Proof:

1. $\angle$ PYA is supplementary to $\angle$PZA (both are right angles).
2. Quadrilateral PZAY is cyclic (opposite angles are supplementary).
3. Draw $\overline{PA}$, $\overline{PB}$, and $\overline{PC}$.
4. m$\angle$PYZ = m$\angle$PAZ (both are inscribed in the same arc).
5. $\angle$PYC is supplementary to $\angle$PXC (both are right angles).

6. Quadrilateral PXCY is cyclic (opposite angles are supplementary).
7. m∠PYX = m∠PCB (both are inscribed in the same arc).
8. m∠PAZ(m∠PAB) = m∠PCB (both are inscribed in the same arc of circle O).
9. m∠PYZ = m∠PYX (transitivity with steps 4, 7, and 8).
10. Since both angles, ∠PYZ and ∠PYX, share the same ray $\overleftrightarrow{YP}$, and have the same measure, their other rays $\overleftrightarrow{YZ}$ and $\overleftrightarrow{YX}$ must coincide. Therefore, points X, Y, and Z are collinear.

Present carefully to students this technique for proving collinearity. Although it is a somewhat unusual approach, it should prove quite useful to them in later work.

To strengthen the impact of the Simson line, show students a proof of the converse of the above theorem.

Given: ΔABC is inscribed in circle O.
Points X, Y, and Z are collinear.
$\overleftrightarrow{PY} \perp \overleftrightarrow{AC}$ at Y, $\overleftrightarrow{PZ} \perp \overleftrightarrow{AB}$ at Z, and $\overleftrightarrow{PX} \perp \overleftrightarrow{BC}$ at X.
Prove: P is on the circumcircle of ΔABC.

Proof:

1. Draw PA, PB, and PC (see Figure 1).

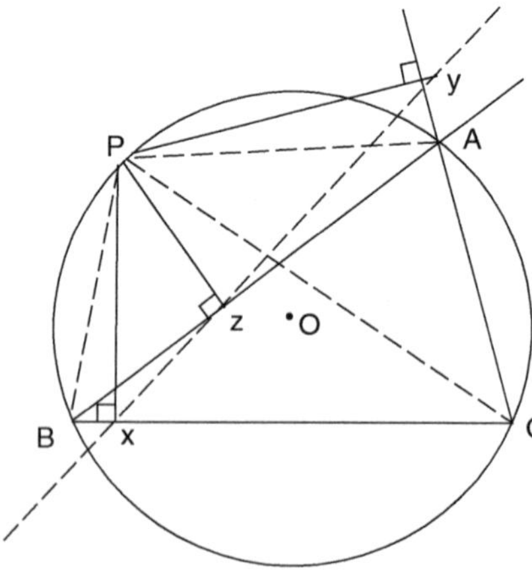

Figure 1.

2. m∠PZB = 90° = m∠PXB.

3. Quadrilateral PZXB is cyclic ($\overline{PB}$ subtends two congruent angles in the same half-plane).
4. $\angle$PBX is supplementary to $\angle$PZX (opposite angles of a cyclic quadrilateral).
5. $\angle$PZX is supplementary to $\angle$PZY (points X, Y, and Z are collinear).
6. Therefore, m$\angle$PBX = m$\angle$PZY (both are supplementary to $\angle$PZX).
7. Quadrilateral PZAY is cyclic (opposite angles, $\angle$PYA and $\angle$PZA, are supplementary).
8. m$\angle$PZY = m$\angle$PAY (both are inscribed in the same arc of the circumcircle of quadrilateral PZAY).
9. Therefore, m$\angle$PBX = m$\angle$PAY (transitivity of steps 6 and 8).
10. Thus, $\angle$PBC is supplementary to $\angle$PAC (since $\overleftrightarrow{YAC}$ is a line).
11. Quadrilateral PACB is cyclic (opposite angles are supplementary), and, therefore, P is on the circumcircle of ΔABC.

Students should now be ready to apply the Simson line to a geometric problem.

> *Sides $\overleftrightarrow{AB}$, $\overleftrightarrow{BC}$, and $\overleftrightarrow{CA}$ of ΔABC are cut by a transversal at points Q, R, and S, respectively. The circumcircles of ΔABC and ΔSCR intersect at P. Prove that quadrilateral APSQ is cyclic.*

Draw perpendiculars $\overline{PX}, \overline{PY}, \overline{PZ}$, and $\overline{PW}$ to $\overleftrightarrow{AB}$, $\overleftrightarrow{AC}$, $\overleftrightarrow{QR}$, and $\overleftrightarrow{BC}$, respectively, as in Figure 2. Since point P is on the circumcircle of ΔABC, points X, Y, and W are collinear (Simson's theorem). Similarly, since point P is on the circumcircle of ΔSCR, points Y, Z, and W are collinear. It then follows that points X, Y, and Z are collinear. Thus, P must lie on the circumcircle of ΔAQS (converse of Simson's theorem), or quadrilateral APSQ is cyclic.

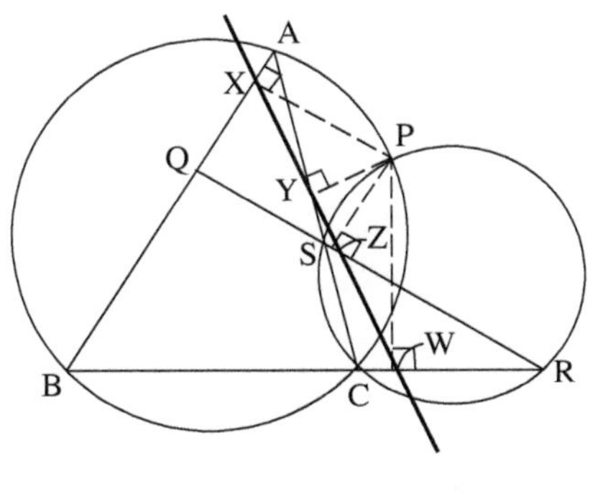

Figure 2.

Postassessment

Have students complete the following exercises.

1. Construct a Simson line of a given triangle.
2. How many Simson lines does a triangle have?
3. Prove Simson's theorem.
4. From a point P on the circumference of circle O, three chords are drawn meeting the circle in points A, B, and C. Prove that the three points of intersection of the three circles with $\overline{PA}, \overline{PB}$, and, $\overline{PC}$ as diameters, are collinear.

References

Posamentier, A. S. *Advanced Euclidean Geometry.* New York, NY: John Wiley Publishing, 2002.

Posamentier, A. S. and C. T. Salkind. *Challenging Problems in Geometry.* New York: Dover, 1996.

Unit 64

The Butterfly Problem

One of the most intriguing geometric relationships involves a figure that resembles a butterfly. Most students will easily understand the problem and think it just as simple to prove. But this is where the problem begins to generate further interest since the proof is somewhat elusive. This unit will suggest ways of presenting the problem to your class and provide a number of different proofs of this celebrated theorem.

Performance Objectives

1. *Students will state the Butterfly Problem.*
2. *Students will prove the Butterfly Problem valid.*

Preassessment

Students should have mastered most of the high school geometry course (especially the study of circles and similarity).

Teaching Strategies

Use a duplicating machine to prepare a sheet of paper for each student with a large circle containing a chord, $\overline{AB}$, (not the diameter), and its midpoint, M, clearly marked. Tell students to draw *any* two chords, $\overline{EF}$, and $\overline{CD}$, containing M. Now have them draw the chords $\overline{CE}$ and $\overline{FD}$ which intersect $\overline{AB}$ at points Q and P, respectively. Their diagrams should resemble Figure 1.

Ask your class to measure any segments that appear congruent in their diagrams, and to list the pairs. You should find that most students will have included on their lists the segments $\overline{AP} \cong \overline{BQ}$ and $\overline{MP} \cong \overline{MQ}$. Remind students that they all started their diagrams with *different* segments $\overline{CE}$ and $\overline{FD}$, and, although their diagrams resemble a butterfly in a circle, their art may differ substantially from their classmates'. This should dramatize the most astonishing result of this situation, that *everyone's* $\overline{MP} \cong \overline{MQ}$!

Students will now want to prove this remarkable result. Toward this end, a number of proofs of this celebrated theorem are presented here.

Proof I: With M the midpoint of $\overline{AB}$ and chords $\overline{FME}$ and $\overline{CMD}$ drawn, we now draw $\overline{DH}//\overline{AB}, \overline{MN} \perp \overline{DH}$, and line segments $\overline{MH}, \overline{QH}$, and $\overline{EH}$. Since $\overline{MN} \perp \overline{DH}$ and $\overline{DH}//\overline{AB}$, MN $\perp$ AB (Figure 2).

$\overline{MN}$, the perpendicular bisector of $\overline{AB}$, must pass through the center of the circle. Therefore, $\overline{MN}$ is the perpendicular bisector of $\overline{DH}$, since a line through the center of the circle and perpendicular to a chord, bisects it.

Thus MD = MH, and ΔMND $\cong$ ΔMNH (H.L). m$\angle$DMN = m$\angle$HMN, so m$\angle$x = m$\angle$y (they are the complements of congruent angles). Since $\overline{AB}//\overline{DH}$, m$\widehat{AD}$ = m$\widehat{BH}$, m$\angle$x = $\frac{1}{2}$(m$\widehat{AD}$ + m$\widehat{CB}$) (angle formed by two chords) m$\angle$x = $\frac{1}{2}$(m$\widehat{BH}$+m$\widehat{CB}$) (substitution). Therefore, m$\angle$y = $\frac{1}{2}$(m$\widehat{BH}$+ m$\widehat{CB}$). But m$\angle$CEH = $\frac{1}{2}$(m$\widehat{CAH}$) (inscribed angle). Thus, by addition,

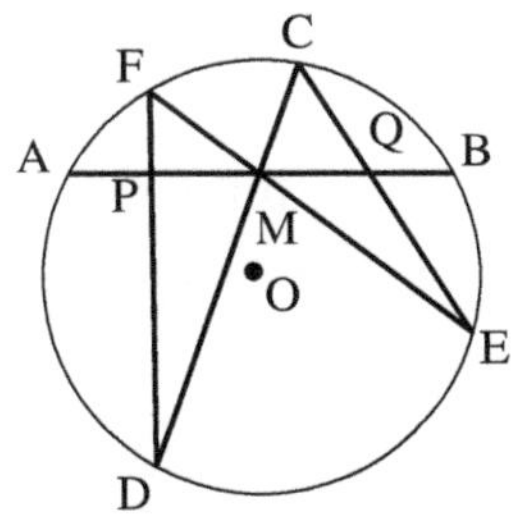

Figure 1.

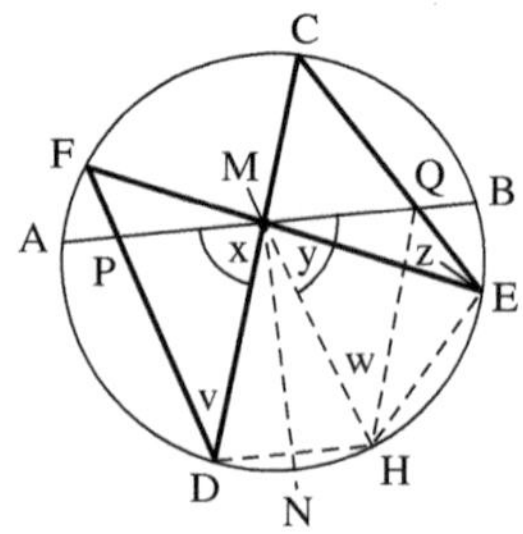

Figure 2.

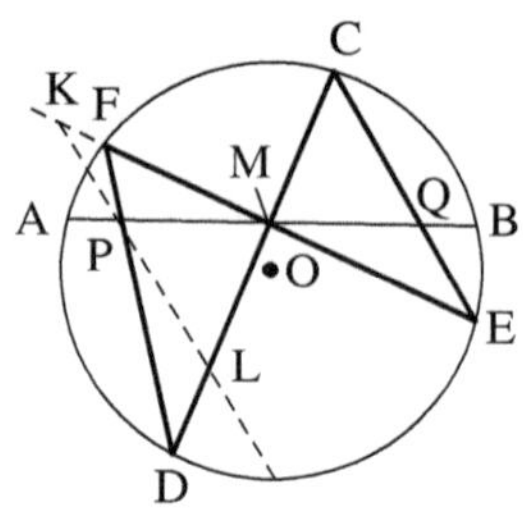

Figure 3.

$m\angle y + m\angle CEH = \angle\frac{1}{2}(m\widehat{BH} + m\widehat{CB} + m\widehat{CAH})$. Since $m\widehat{BH} + m\widehat{CB} + m\widehat{CAH} = 360°$, $m\angle y + m\widehat{CEH} = 180°$. It then follows that quadrilateral MQEH is inscriptable, that is, a circle may be circumscribed about it. Imagine a drawing of this circle. $\angle w$ and $\angle z$ are measured by the same arc, $\widehat{MQ}$ (inscribed angle), and thus $m\angle w = m\angle z$.

Now considering our original circle $m\angle v = m\angle z$, since they are measured by the same arc, $\widehat{FC}$ (inscribed angle). Therefore, by transitivity, $m\angle v = m\angle w$, and $\Delta MPD \cong \Delta MQH$(A.S.A). Thus, $MP = MQ$.

Proof II: Extend $\overline{EF}$ through F.

Draw $\overline{KPL} // \overline{CE}$ (Figure 3).

$m\angle PLC = m\angle ECL$ (alternate interior angles),

therefore, $\Delta PML \sim \Delta QMC$(A.A), and $\frac{PL}{CQ} = \frac{MP}{MQ}$

$m\angle K = m\angle E$ (alternate interior angles),

therefore, $\Delta KMP \sim \Delta EMQ$(A.A.), and $\frac{KP}{QE} = \frac{MP}{MQ}$. By multiplication,

$$\frac{(PL)(KP)}{(CQ)(QE)} = \frac{(MP)^2}{(MQ)^2}. \tag{I}$$

Since $m\angle D = m\angle E$ (inscribed angle), and $m\angle K = m\angle E$ (alternate interior angles), $m\angle D = m\angle K$. Also, $m\angle KPF = m\angle DPL$ (vertical angles).

Therefore, ΔKFP $\sim$ ΔDLP (A. A.), and $\frac{PL}{DP} = \frac{FP}{KP}$; and so

$$(PL)(KP) = (DP)(FP). \tag{II}$$

In Equation (I), $\frac{(MP)^2}{(MQ)^2} = \frac{(PL)(KP)}{(CQ)(QE)}$
we substitute from Equation (II) to get

$$\frac{(MP)^2}{(MQ)^2} = \frac{(DP)(FP)}{(CQ)(QE)}.$$

Since (DP) (FP) = (AP)(PB), and (CQ)(QE) = (BQ)(QA) (product of segments lengths of intersecting chords),

$$\frac{(MP)^2}{(MQ)^2} = \frac{(AP)(PB)}{(BQ)(QA)} = \frac{(MA - MP)(MA + MP)}{(MB - MQ)(MB + MQ)} = \frac{(MA)^2 - (MP)^2}{(MB)^2 - (MQ)^2}.$$

Then $(MP)^2(MB)^2 = (MQ)^2(MA)^2$.
But MB = MA. Therefore, $(MP)^2 = (MQ)^2$, or MP = MQ.

Proof III: Draw a line through E parallel to $\overline{AB}$ meeting the circle at G, and draw $\overline{MN} \perp \overline{GE}$. Then draw $\overline{PG}, \overline{MG}$, and $\overline{DG}$ (Figure 4).
m∠GDP(∠GDF) = m∠GEF (inscribed angles). (I)
m∠PMG = m∠MGE (alternate interior angles). (II)
Since the perpendicular bisector of $\overline{AB}$ is also the perpendicular bisector of $\overline{GE}$,
then GM = ME, and m∠GEF = m∠MGE (base angles). (III)
From (I), (II), and (III), m∠GDP = m∠PMG. (IV)

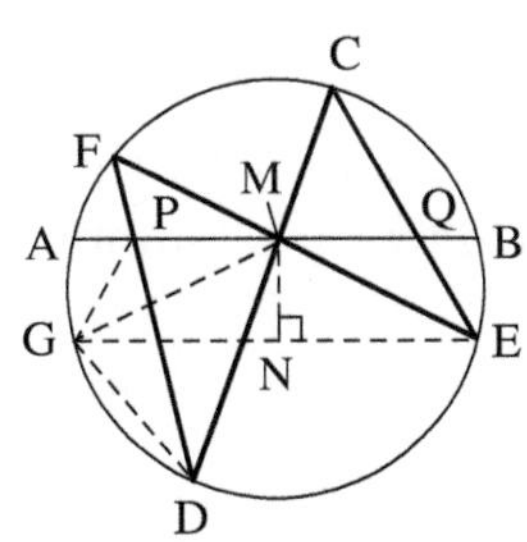

Figure 4.

Therefore, points P, M, D, and G are concyclic. (A quadrilateral is cyclic if one side subtends congruent angles at the two opposite vertices.) Hence, m∠PGM = m∠PDM (inscribed angles, in a new circle). (V)

However, m∠CEF = m∠PDM(∠FDM) (inscribed angles). (VI)

From (V) and (VI), M∠PGM = m∠QEM(∠CEF).

From (II), we know that m∠PMG = m∠MGE.

Thus, m∠QME = m∠MEG (alternate interior angles), and m∠MGE = m∠MEG (base angles).

Therefore, m∠PMG = m∠QME and ΔPMG ≅ ΔQME(A.S.A). It follows that PM = QM.

Although these proofs of the Butterfly Problem are not of the sort the average student is likely to discover independently, they do provide a very rich learning experience in a well-motivated setting.

Postassessment

Ask students to:

1. State the Butterfly Problem.
2. Explain why the Butterfly Problem is true. (Students should either present one of the above proofs or one of their own.)

Reference

Posamentier, A. S. and C. T. Salkind. *Challenging Problems in Geometry.* New York: Dover, 1996.

Unit 65

Equicircles

Equicircles is a term used to refer to both the inscribed and escribed circles of a triangle. This unit will develop a number of fascinating relationships between these circles.

Performance Objectives

1. *Students will define equicircles.*
2. *Students will state at least four properties involving equicircles.*

3. *Students will state and prove one property of equicircles.*

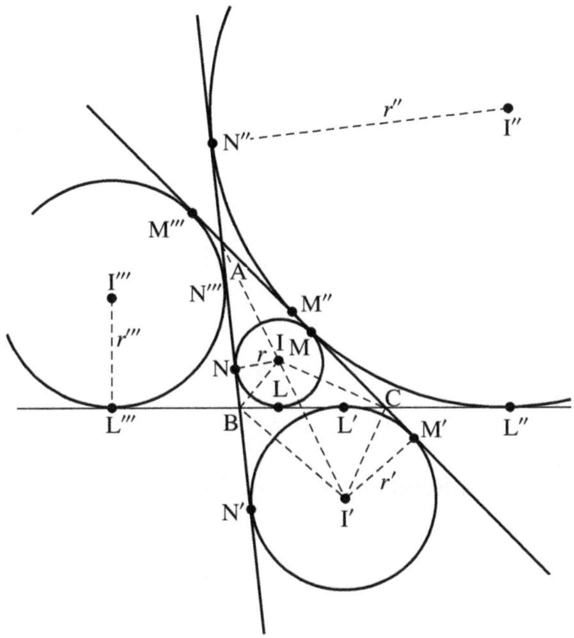

Preassessment

Students should have mastered the topic of circles in their high school geometry course.

Present the following figure to your students and ask them to find the length of $\overline{AN'}$, if the perimeter of $\Delta ABC = 16$. (Points M', N', and L' are points of tangency.)

Teaching Strategies

Although the problem posed above is quite simple, its approach is rather unusual and therefore could cause your students some difficulty. The only theorem they need to recall is that two tangent segments from an external point to a circle are congruent. Applying this theorem to the above problem, we get

$$BN' = BL' \text{ and } CM' = CL'.$$

$$\begin{aligned}\text{The perimeter of } \Delta ABC &= AB + BC + AC \\ &= AB + (BL' + CL') + AC\end{aligned}$$

which by substitution yields

$$\mathrm{AB} + \mathrm{BN}' + \mathrm{CM}' + \mathrm{AC}$$

or

$$\mathrm{AN}' + \mathrm{AM}'.$$

However, $\mathrm{AN}' = \mathrm{AM}'$ (they too are tangent segments from the same external point to the same circle).

Therefore, $\mathrm{AN}' = \frac{1}{2}($ perimeter of $\Delta\mathrm{ABC}) = 8$. By summarizing this rather fascinating fact, students will be motivated toward pursuing further relationships in this figure.

Next, let $s =$ semiperimeter of $\Delta\mathrm{ABC}$

$$a = \mathrm{BC}; b = \mathrm{AC}; c = \mathrm{AB}.$$

With your guidance, students should now be able to establish the following relationship:

$$\mathrm{BN}' = \mathrm{BL}' = \mathrm{AN}' - \mathrm{AB} = s - c$$

$$\mathrm{CM}' = \mathrm{CL}' = \mathrm{AM}' - \mathrm{AC} = s - b.$$

At this point, you ought to indicate to students that these are just a few of the segments that will be expressed in terms of the lengths of the sides of $\Delta\mathrm{ABC}$. Here the relationship of the two circles to the triangle should be defined. Students will recognize circle I as the *inscribed* circle of $\Delta\mathrm{ABC}$. Most likely they are not familiar with circle $1'$. This circle, which is also tangent to the lines of the three sides of $\Delta\mathrm{ABC}$, yet contains no interior points of the triangle, is called an *escribed* circle. A triangle has four equicircles, one inscribed and three escribed. The center of an escribed circle, called an *excenter*, is the point of intersection of two exterior angle bisectors and one interior angle bisector.

Students should again achieve further familiarity with these circles by expressing other segments in terms of the lengths of the sides of $\Delta\mathrm{ABC}$. Once again, provide guidance where necessary.

$$\begin{aligned}\mathrm{AN} + \mathrm{AM} &= (\mathrm{AB} - \mathrm{NB}) + (\mathrm{AC} - \mathrm{MC})\\ &= (\mathrm{AB} - \mathrm{LB}) + (\mathrm{AC} - \mathrm{LC})\\ &= (\mathrm{AB} + \mathrm{AC}) - (\mathrm{LB} + \mathrm{LC})\\ &= c + b - a.\end{aligned}$$

Challenge your students to show that

$$c + b - a = 2(s - a).$$

Therefore, $\mathrm{AN} + \mathrm{AM} = 2(s - a)$.

However, $\mathrm{AN} = \mathrm{AM}$, thus $\mathrm{AN} = s - a$.

Have your students conjecture how BN and CL can be expressed in terms of the lengths of the sides of $\Delta\mathrm{ABC}$.

$$\mathrm{BN} = s - b$$

$$\mathrm{CL} = s - c.$$

We are now ready to apply some of these expressions to establish two interesting relationships. These are $\mathrm{BL} = \mathrm{CL}'$ and $\mathrm{LL}' = b - c$, the difference between the lengths of the other two sides of $\Delta\mathrm{ABC}$.

Since both BL and CL′ were shown to be equal to $s - b, \mathrm{BL} = \mathrm{CL}'$.

Consider LL′, which equals $\mathrm{BC} - \mathrm{BL} - \mathrm{CL}'$.

By substitution, $\mathrm{LL}' = a - 2(s - b) = b - c$.

We can now prove rather easily that the length of the common external tangent segment of an inscribed and escribed circle of a triangle equals the length of the side contained in the line that intersects the tangent segment.

The proof proceeds as follows: $\mathrm{NN}' = \mathrm{AN}' - \mathrm{AN}$.

Earlier we showed that $\mathrm{AN}' = s$, and $\mathrm{AN} = s - a$.

By substitution, $\mathrm{NN}' = s - (s - a) = a$.

The same argument holds true for MM′.

Another interesting theorem states that the length of the common external tangent segment of two escribed circles of a triangle equals the sum of the lengths of the two sides that intersect it.

To prove this theorem, have students recall that $\mathrm{BL}'' = s$ and $\mathrm{CL}''' = s$. This was proved when the *Preassessment* problem was solved. Therefore,

$$\begin{aligned} \mathrm{L}'''\mathrm{L}'' &= \mathrm{BL}'' + \mathrm{CL}''' - \mathrm{BC} \\ &= s + s - a \\ &= b + c. \end{aligned}$$

We can also show that the length of each of the common internal tangent segments of two escribed circles of a triangle equals the length of the side opposite the vertex they determine. The proof is rather simple:

$$\mathrm{L}'\mathrm{L}'' = \mathrm{BL}'' - \mathrm{BL}' = \mathrm{BL}'' - \mathrm{BN}' = s - (s - c) = c.$$

Encourage students to investigate the above figure and discover other relationships. A consideration of the radii of the equicircles will produce some interesting results. These radii are called *equiradii.*

A theorem states that the radius of the inscribed circle of a triangle equals the ratio of the area to the semiperimeter. That is

$$\mathscr{A}\Delta\text{ABC} = \mathscr{A}\Delta\text{BCI} + \mathscr{A}\Delta\text{CAI} + \mathscr{A}\Delta\text{ABI}$$

(*Note*: $\mathscr{A}$ reads "area of")

$$\begin{aligned}\mathscr{A}\Delta\text{ABC} &= \frac{1}{2}ra + \frac{1}{2}rb + \frac{1}{2}rc\\ &= \frac{1}{2}r(a+b+c) = sr.\end{aligned}$$

$$\text{Therefore, } r = \frac{\mathscr{A}\Delta\text{ABC}}{s}.$$

A natural extension of this theorem states that the radius of an escribed circle of a triangle equals the ratio of the area of the triangle to the difference between the semiperimeter and the length of the side to which the escribed circle is tangent.

To prove this, have students consider

$$\begin{aligned}\mathscr{A}\Delta\text{ABC} &= \mathscr{A}\Delta\text{ABI}' + \mathscr{A}\Delta\text{ACI}' - \mathscr{A}\Delta\text{BCI}'\\ &= \frac{1}{2}r'c + \frac{1}{2}r'b - \frac{1}{2}r'a\\ &= \frac{1}{2}r'(c+b-a)\\ &= r'(s-a)\end{aligned}$$

$$\text{Therefore, } r' = \frac{\mathscr{A}\Delta\text{ABC}}{s-a}.$$

In a similar manner, students should show that

$$r'' = \frac{\mathscr{A}\Delta\text{ABC}}{s-b}$$

and

$$r''' = \frac{\mathscr{A}\Delta\text{ABC}}{s-c}.$$

To conclude this discussion, have students find the product of all the equiradii of a circle. All they need to do is multiply the last few expressions:

$$rr'r''r''' = \frac{(\mathscr{A}\Delta\text{ABC})^4}{s(s-a)(s-b)(s-c)}.$$

However, by Heron's formula

$$\mathscr{A}\Delta\text{ABC} = \sqrt{s(s-a)(s-b)(s-c)}.$$

Therefore, $rr'r''r''' = (\mathscr{A}\Delta\text{ABC})^2$.

At this point, ask students to summarize the various theorems and relationships developed in this unit.

Postassessment

To conclude the lesson, have students complete the following exercises:

1. Define equicircles and equiradii.
2. State four properties of equicircles.
3. State and prove one property of equicircles.

Reference

Posamentier, A. S. *Advanced Euclidean Geometry: Excursions for Secondary Teachers and Students.* New York, NY: John Wiley Publishing, 2002.

Unit 66

The Inscribed Circle and the Right Triangle

After having completed a unit on circles and a separate unit dealing with right triangles, students may enjoy seeing some relationships that integrate these units. This unit will deal with some interesting properties of the radius of an inscribed circle of a right triangle.

Performance Objectives

1. *Given a right triangle with integral length sides, students will be able to show that the inradius is an integer.*

2. *Students will be able to explain how the altitude drawn to the hypotenuse of a right triangle is related to the inradii of the triangles formed.*
3. *Students will know and be able to derive a formula relating the inradius to the area and perimeter of a right triangle.*
4. *Given a particular integral inradius, students will be able to determine the number of right triangles with integral relatively prime sides having this given inradius.*
5. *Students will be able to give one possible triple of the lengths of sides of sides of a right triangle when given a positive integral value of the inradius.*

Preassessment

Have students try the following problems:

1. Find the radius of a circle inscribed in a right triangle whose sides have lengths $3, 4, 5$.
2. Repeat this problem for a triangle whose sides have lengths $5, 12, 13$.

Teaching Strategies

After having completed the above problems either individually or as a class, students will want to consider the following question: "Given a right triangle of integral sides, will this guarantee that the radius of the inscribed circle is also an integer?" To prove that the answer is affirmative, consider the following diagram. Here, r is the inradius (i.e., the radius of the inscribed circle), and ΔABC has a right angle at C and sides of lengths a, b, c. The proof involves finding a relationship between r, a, b, and c. If the center of the circle is joined to each of the three vertices, three triangles are formed. The area of

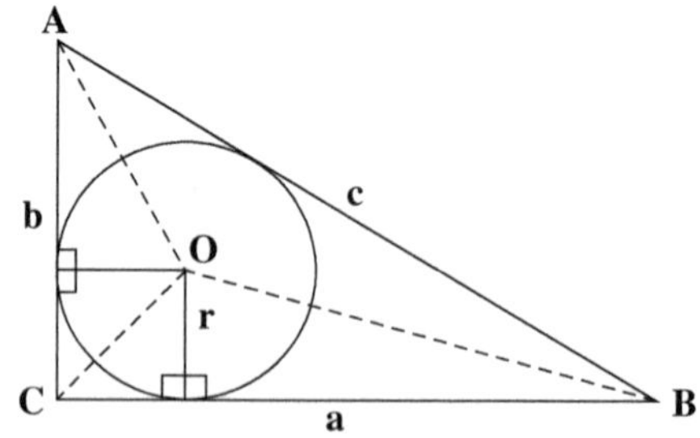

one triangle is $\frac{1}{2}ra$, the second triangle's area is $\frac{1}{2}rb$, and the third's is $\frac{1}{2}rc$. The area of ΔABC is $\frac{1}{2}ab$. Challenge the students to set up a relationship between r and a, b, c. By adding areas one gets $\frac{1}{2}ra + \frac{1}{2}rb + \frac{1}{2}rc = \frac{1}{2}ab$, which is the area of ΔABC. Thus, $r = \frac{ab}{a+b+c}$. But this only seems to make r rational values of a, b, and c. At this point, remind students (or show them for the first time) how integral values of a, b, and c are obtained from a formula. That is, show them this generating formula for sides of right triangles.

$$a = (m^2 - n^2)$$
$$b = 2mn$$
$$c = (m^2 + n^2)$$

where $m > n$ and m and n are relatively prime positive integers of different parity.

Using $r(a + b + c) = ab$, substitute a, b, and c. Thus, $2r(m^2 + mn) = 2mn(m^2 - n^2)$ or $r = n(m - n)$.

Since m and n are integers, $m > n$, then r is also an integer. Therefore, *whenever a right triangle has integral sides it also has an integral inradius.*

As a result of the above, a concise formula can be established relating the inradius to the area and perimeter of a right triangle. Since $r = \frac{ab}{a+b+c}$, substitute p (perimeter) for $a + b + c$. Also note that the area of ΔABC $= \frac{ab}{2}$ or $2\mathscr{A}\Delta\text{ABC} = ab$ (where $\mathscr{A}$ represents "area of"). Thus, $r = \frac{2\mathscr{A}\Delta}{p}$. For practice, have students find the inradius, given various values of $\mathscr{A}\Delta$ and p.

Students have probably worked for some time with right triangles whose altitude to the hypotenuse is drawn (see following figure). Now they can relate the inradius to this familiar diagram.

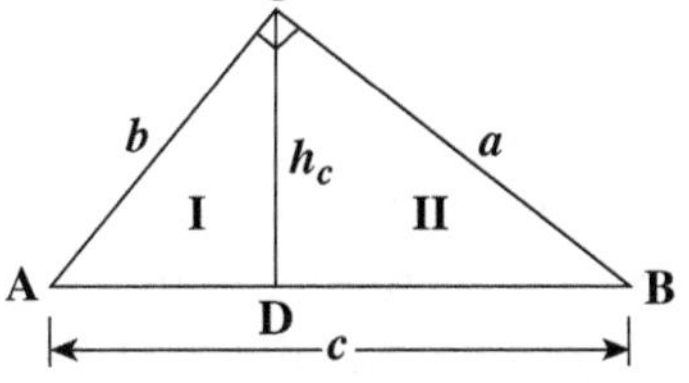

Let ΔADC be called ΔI with inradius r_{I}. Similarly, ΔDCB (ΔII) has inradius r_{II} and ΔABC (ΔIII) has inradius r_{III}. It can be shown that the sum of the inradii of ΔI, ΔII, and ΔIII equals the length of the altitude from C, which will be called h_c. Note that $\Delta\text{ADC} \sim \Delta\text{DCB} \sim \Delta\text{ABC}$. Since the corresponding inradii of the similar triangles are in the same ratio as any pair of corresponding sides, $\frac{r_{\text{I}}}{r_{\text{III}}} = \frac{b}{c}$ or $r_{\text{I}} = \frac{b}{c} r_{\text{III}}$.

In the same manner, $r_{\text{II}} = \frac{a}{c}r_{\text{III}}$. Therefore, $r_1 + r_{\text{II}} + r_{\text{III}} = \frac{a+b+c}{c}r_{\text{III}}$. Recalling that $r = \frac{2\mathscr{A}\Delta\text{III}}{p}$, $\frac{a+b+c}{c}r_{\text{III}} = \left(\frac{a+b+c}{c}\right)\left(\frac{2\mathscr{A}\Delta\text{III}}{p}\right) = \frac{2\mathscr{A}\Delta\text{III}}{\frac{p}{c}}$. But $\mathscr{A}\Delta\text{III} = \frac{1}{2}h_c c$. Thus $\frac{2\mathscr{A}\Delta\text{III}}{C} = h_c$ making $r_1 + r_{\text{II}} + r_{\text{III}} = h_c$, which is what was to be proved.

One can also use the above to prove that the area of the inscribed circle in ΔI plus the area of the inscribed circle in ΔII equals the area of the inscribed circle in ΔIII. This can be seen by recalling that it has been shown that $r_{\text{I}} = \frac{b}{c}r_{\text{III}}$ and $r_{\text{II}} = \frac{a}{c}r_{\text{III}}$. Thus, $r_{\text{I}}^2 + r_{\text{II}}^2 = \frac{b^2}{c^2}r_{\text{III}}^2 + \frac{a^2}{c^2}r_{\text{III}}^2 = \frac{a^2+b^2}{c^2}r_{\text{III}}^2 = r_{\text{III}}^2$ (since $a^2 + b^2 = c^2$). Multiplying by π, one gets $\pi r_{\text{I}}^2 + \pi r_{\text{II}}^2 = \pi r_{\text{III}}^2$, which is the theorem to be proved.

Another interesting relationship concerning the inradius is: "The number of primitive Pythagorean Triples is 2^ℓ where ℓ is the number of odd prime divisions of $r(\ell \geq 0)$, and r is the length of the corresponding inradius." The full meaning of this theorem should be clear to students before embarking on the proof. Show students that for every natural number r there exists at least one right triangle of sides $2r + 1$, $2r^2 + 2r$, and $2r^2 + 2r + 1$ where r is the inradius. Students should be able to check that this satisfies the Pythagorean theorem. As an example, have them try various values for r. For $r = 1$, one gets a triangle of side lengths 3, 4, and 5.

Getting back to proving the above theorem, let a, b, and c be sides of a right triangle with sides of integral length, where b is even, a, b, and c are relatively prime. The inradius of this triangle is the positive integral r. Recall that $r = \frac{ab}{a+b+c}r$ can also be written as $\frac{1}{2}(a + b - c)$ by noting that $\frac{ab}{a+b+c} = \frac{a+b-c}{2}$ is an identity. Students should be urged to verify this identity remembering that $a^2 + b^2 = c^2$. From the original generating formula, substitute for a, b, and c. Students should obtain $r = (m-n)n$. Since m and n are relatively prime, then $(m - n)$ and n are also relatively prime. (*Note*: $(m - n)$ is odd, m and n are relatively prime and of opposite parity.) Thus, the inradius can be decomposed into a product of two positive integers that are relatively prime and where the factor $(m - n)$ is odd.

Now consider r as any positive integer where $r = xy$ is any decomposition of r into a product of two relatively prime positive integers where one is odd. Let $m = x + y$, $n = y$. Then m and n are also relatively prime. Also, since x is odd, if $n = y$ is odd, then $m = x + y$ is even. Similarly, if m is odd, n must be even. Thus, one of the numbers m and n is even.

Recall $m > n$. Letting $a = m^2 - n^2$, $b = 2mn$, $c = m^2 + c^2$ one obtains the type of triangle desired with inradius $r = (m - n)'nab$. Therefore, every decomposition of the number r into a product of two relatively prime numbers where one is odd will determine the type of triangle desired of inradius r. It can be shown that if $\ell \geq 0$ where

$$r = 2p_1^{x1}p_2^{x2}p_3^{x3} \cdot p_t^{x\ell} \quad \text{with } p_t$$

being an odd prime integer (i a positive integer), then the number of decompositions of r is 2^{ℓ}. Thus, 2_{ℓ} must be the number of decompositions of r into two relatively prime factors where one is odd.

Thus, for every positive integer r, there exists as many distinct right triangles whose sides have lengths that are relatively prime integers with inradius r as there are distinct decompositions of r into a product of two relatively prime factors of which one is odd. The numbers of such triangles are 2_{ℓ}. This completes the proof.

Students desiring to look into the matter might try to prove that if r is a positive even integer, then the total number of right triangles with integral length sides that are not necessarily relatively prime, having r as an inradius, is given by $(x+1)(2x_1+1)(2x_2+1)\cdots(2x_{\ell}+1)$ where x and ℓ are the numbers found by decomposing r into $2^1p_1^{x1}p_2^{x2} \cdot p_t^{x\ell}$, $x \geq 0$, $\ell \geq 0$ and p_{ℓ} = odd prime, $x_1 \geq 1$ and $2 < p_1 < p_2 < \cdots < p_{\ell}$. Any positive integer can be so decomposed.

Other interesting relationships concerning the inradius might be researched by the students. For example, they might try to prove the formula (for any triangle) that the inradius r of ΔXYZ (sides x, y, z and $s = \frac{x+y+z}{2}$) is $r = \sqrt{\frac{(s-x)(s-y)(s-z)}{5}}$. Other investigations should prove to be challenging to the class.

Postassessment

1. If a right triangle has sides of 5, 12, and 13, does this guarantee r will also be an integer? If so, which integer is it? If not, explain why.
2. If an altitude drawn to the hypotenuse of a right triangle creates three similar triangles of inradii 2, 3, and 4, find the length of this altitude.
3. Find the number of distinct right triangles whose sides have lengths that are relatively prime integers having 70 as its inradii.
4. If the inradius equals 3, find the lengths of the sides of one right triangle with this inradius.
5. ΔXYZ has an area of 6 and a perimeter of 12. Find the length of its inradius.

Unit 67

The Golden Rectangle

In this unit, the concept of the golden ratio will be introduced together with some of its elementary algebraic and geometric ramifications.

Performance Objectives

1. *Students will construct a golden rectangle.*
2. *Students will state the golden ratio.*
3. *Students will demonstrate certain properties of the golden rectangle and the golden ratio.*

Preassessment

Some knowledge of geometry and intermediate algebra is necessary.

Teaching Strategies

Have your students draw a golden rectangle using the following construction. Given square ABCD, with each side one unit long, locate the midpoint, M, of $\overline{AD}$. Draw $\overline{MC}$. By the Pythagorean theorem, $MC = \frac{\sqrt{5}}{2}$. With center of compasses at M and radius $\overline{MC}$, have students describe an arc cutting $\overrightarrow{AD}$ at E (Figure 1). Then,

$$\begin{aligned} DE = ME - MD &= \frac{\sqrt{5}}{2} - \frac{1}{2}, \\ &= \frac{\sqrt{5}-1}{2}. \end{aligned} \tag{1}$$

From this result, it follows that

$$AE = AD + DE = 1 + \frac{\sqrt{5}-1}{2}, \quad \text{or} = \frac{\sqrt{5}+1}{2} = 1.61803\cdots$$

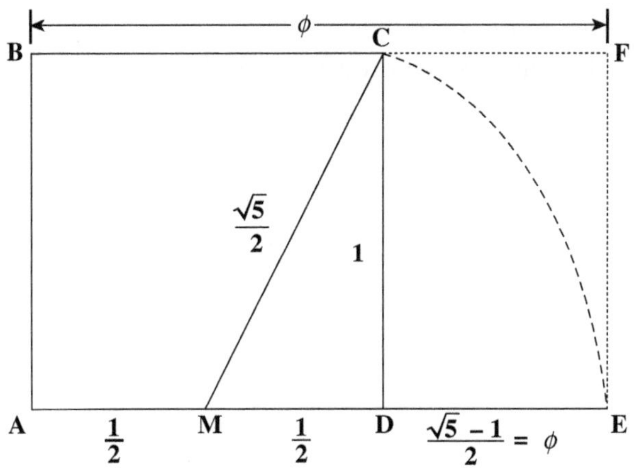

Figure 1.

By erecting a perpendicular at E to meet $\overrightarrow{\mathrm{BC}}$ at F, rectangle ABFE is constructed, where the ratio of length to width is

$$\frac{\mathrm{AE}}{\mathrm{AB}} = \frac{\frac{\sqrt{5}+1}{2}}{1} = \frac{\sqrt{5}+1}{2}. \tag{2}$$

The ratio (2) is called *the golden ratio or golden section*, denoted by the Greek letter phi (ϕ), and a rectangle having such a ratio of length to width is called a *golden rectangle*. Note that the value of (2), $\phi = 1.61803\cdots$, is an irrational number approximately equal to $\frac{8}{5}$. A rectangle with such a ratio of length to width was thought by the ancient Greeks, and corroborated experimentally by the psychologist Fechner in 1876, to be the most pleasing and harmoniously balanced rectangle to the eye.

Have your students solve the equation $x^2 - x - 1 = 0$, solutions for which are

$$r_1 = \frac{\sqrt{5}+1}{2} \quad \text{and} \quad r_2 = \frac{-\sqrt{5}+1}{2}. \tag{3}$$

From (2), $r_1 = \phi$, and r_2, when evaluated, is equal to $-1.61803\cdots$.

A relation between ϕ and r_2 will become apparent if we first evaluate the reciprocal of ϕ, i.e. determine $\frac{1}{\phi}$. From (2),

$$\frac{1}{\phi} = \frac{1}{\frac{\sqrt{5}+1}{2}} = \frac{\sqrt{5}-1}{2} = 0.61803\cdots.$$

The ratio $\frac{\sqrt{5}-1}{2}$ is denoted by ϕ'. Thus, from (3), $r_2 = \frac{-\sqrt{5}+1}{2}$ is the additive inverse of ϕ' and is denoted by $-\phi'$. In summary, then,

$$\phi = \frac{\sqrt{5}+1}{2} = 1.61803\cdots, \tag{4}$$

$$-\phi = \frac{\sqrt{5}-1}{2} = -1.61803\cdots, \tag{5}$$

$$\frac{1}{\phi} = \phi' = \frac{\sqrt{5}+1}{2} = 0.61803\cdots, \tag{6}$$

$$-\phi' = \frac{\sqrt{5}+1}{2} = -0.61803\cdots. \tag{7}$$

Keep in mind that the ratio of width to length for a golden rectangle is ϕ', whereas the ratio of length to width is ϕ. Thus, in Figure 1, $\frac{\text{DE}}{\text{DC}} = \phi'$, so that CDEF is a golden rectangle.

Some rather unique relationships can be derived from (4)–(7). For example, using (4) and (6),

$$\phi \cdot \phi' = 1, \tag{8}$$

and

$$\phi - \phi' = 1. \tag{9}$$

ϕ and ϕ' are the only two numbers in mathematics that bear the distinction of having both their products and differences equal to one!

$$\phi^2 = \left(\frac{\sqrt{5}+1}{2}\right)^2 = \frac{5+2\sqrt{5}+1}{4} = \frac{3+\sqrt{5}}{2}. \tag{10}$$

But,

$$\phi + 1 = \frac{\sqrt{5}+1}{2} + 1 = \frac{3+\sqrt{5}}{2}. \tag{11}$$

Thus, from (10) and (11),

$$\phi^2 = \phi + 1. \tag{12}$$

Furthermore, by using (6) and (12),

$$\begin{aligned}(\phi')^2 + \phi &= \frac{1}{\phi^2} + \phi = \frac{1}{\phi+1} + \phi = \frac{1+\phi^2+\phi}{\phi+1} \\ &= \frac{\phi^2+\phi^2}{\phi^2} = \frac{2\phi^2}{\phi^2} = 2. \end{aligned} \tag{13}$$

Again, using (6) and (12),

$$\phi^2 - \phi^1 = \phi + 1 - \frac{1}{\phi} = \frac{\phi^2 + \phi - 1}{\phi}$$

$$= \frac{\phi + 1 + \phi - 1}{\phi} = 2. \tag{14}$$

Hence, from (13) and (14),

$$(\phi')^2 + \phi = \phi^2 - \phi'. \tag{15}$$

Powers of ϕ: A fascinating occurrence of the Fibonacci series can be obtained if we derive powers of ϕ in terms of ϕ and take note of the coefficients and constants that arise.

For example, using (12),

$$\phi^3 = \phi^2 \cdot \phi = (\phi + 1)\phi = \phi^2 + \phi$$

$$= \phi + 1 + \phi = 2\phi + 1, \tag{16}$$

$$\phi^4 = \phi^3 \cdot \phi = (2\phi + 1)\phi = 2\phi^2 + \phi$$

$$= 2(\phi + 1) + \phi = 2\phi + 2\phi$$

$$= 3\phi + 2, \tag{17}$$

$$\text{and} \quad \phi^5 = \phi^4 \cdot \phi = (3\phi + 2)\phi = 3\phi^2 + 2\phi$$

$$= 3(\phi + 1) + 2\phi = 3\phi + 3 + 2\phi$$

$$= 5\phi + 3. \tag{18}$$

Have students generate further powers of ϕ

$$\begin{aligned}
\phi^1 &= 1\phi + 0\\
\phi^2 &= 1\phi + 1\\
\phi^3 &= 2\phi + 1\\
\phi^4 &= 3\phi + 2\\
\phi^5 &= 5\phi + 3\\
\phi^6 &= 8\phi + 5\\
\phi^7 &= 13\phi + 8\\
\phi^8 &= 21\phi + 13\\
\vdots \quad & \quad \vdots \qquad \vdots
\end{aligned} \tag{19}$$

Let us return to Figure 1. If, along $\overline{\text{CD}}$, the length DE $= \phi'$ is marked off, we obtain square DEGH, each side equal in length to ϕ'.

Thus, CH $= 1 - \phi'$ (remember that originally CD $=$ one unit). But $1 - \phi' = 1 - \frac{1}{\phi} = \frac{\phi - 1}{\phi} = \frac{\frac{1}{\phi}}{\phi} = \frac{1}{\phi^2} = (\phi')^2 = \frac{1}{\phi^2}$. With CF (or GH) $= \phi' = \frac{1}{\phi}$, $\frac{\text{CH}''}{\text{CF}} = \frac{(\phi')^2}{\phi'} = \phi'$.

Thus, CFGH is also a golden rectangle.

In like fashion, a square, each of whose sides is $(\phi')^2$ units in length, can be partitioned alongside CF of CFGH, whereby we obtain another golden rectangle CJIH. CJIH can similarly be partitioned to obtain square GJKL, leaving golden rectangle IKLH. This process of partitioning squares from golden rectangles to obtain another golden rectangle can be indefinitely continued in the manner suggested in Figure 2.

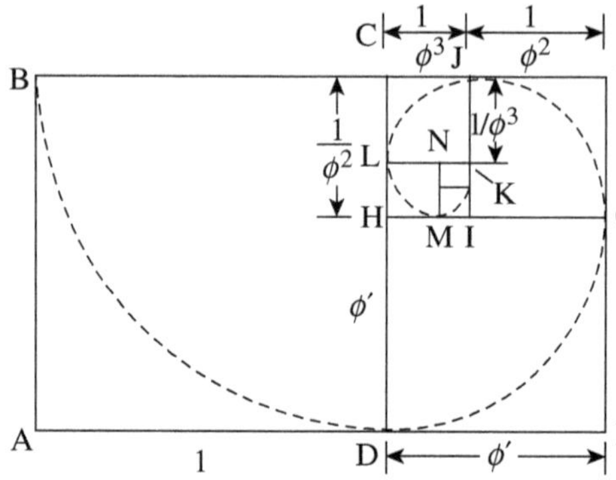

Figure 2.

If points B, D, G, J, L, M are connected by a smooth curve (see Figure 2), a spiral-shaped curve will result. This is part of an equiangular spiral, a detailed discussion of which is not possible at this time.

Postassessment

1. A line segment $\overline{\text{AE}}$ said to be divided into *extreme and mean ratio* if a point D can be located on $\overline{\text{AE}}$ such that

$$\frac{\text{AE}}{\text{AD}} = \frac{\text{AD}}{\text{DE}}. \tag{20}$$

In Figure 1, let AE $= x$ and AD $= 1$. Then, from (20), derive the quadratic equation that was used to determine the value of ϕ in (3).

References

Dunlap, R. A. *The Golden Ration and Fibonacci Numbers.* River Edge, NJ: World Scientific Publishing Co., 1997.
Livio, M. *The Golden Ratio.* New York: Broadway Books, 2002.
Posamentier, A. S. *Advanced Euclidean Geometry: Excursions for Secondary Teachers and Students.* New York, NY: John Wiley Publishing, 2002.
Posamentier, A. S. and I. Lehmann. *The Fabulous Fibonacci. Numbers.* Amherst, NY: Prometheus Books, 2007.
Posamentier, A. S. and I. Lehmann. *The Glorious Golden Ratio.* Amherst, NY: Prometheus Books, 2012.
Walser, H. *The Golden Section.* Washington, DC: Mathematical Association of America, 2001.

Unit 68

The Golden Triangle

This unit will help to develop student understanding in areas of mathematics not usually dealt with.

Performance Objectives

1. *Students will demonstrate understanding of various relationships among the pentagon, the pentagram, and the golden ratio.*
2. *Students will construct a golden triangle.*
3. *Students will demonstrate certain properties of the golden triangle with trigonometric functions.*

Preassessment

Some knowledge of geometry and intermediate algebra is necessary.

Teaching Strategies

Have your students construct a regular pentagon ABCDE by any method, after which they should draw pentagram ACEBD (see Figure 1). Let each side of the pentagon be ϕ units in length. Review with your students the various angle measures and isosceles triangles formed by the pentagram and pentagon. Particular note should be made of similar isosceles triangles BED

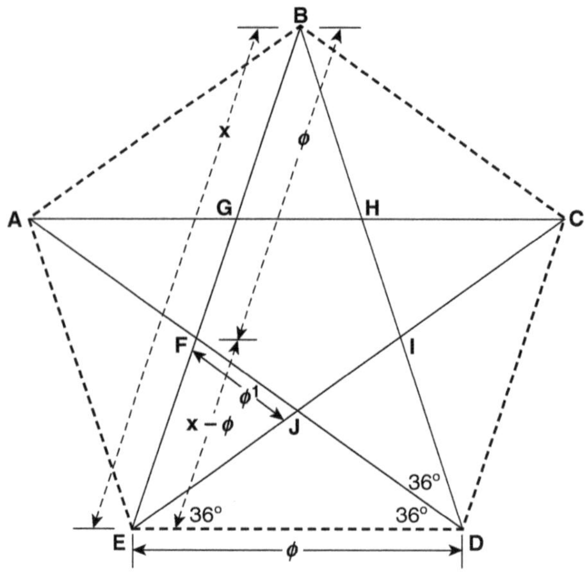

Figure 1.

and DEF, since they will be chosen, quite arbitrarily, from the many similar triangles in Figure 1, for the discussion that follows.

With $\overline{\text{DF}}$ the bisector of $\angle$BDE, triangles DEF and BDF are isosceles, so that

$$\text{ED} = \text{DF} = \text{FB} = \phi. \tag{1}$$

Let BE = BD = x. Then FE = $x - \phi$, and

$$\frac{\text{BF}}{\text{FE}} = \frac{\text{BD}}{\text{ED}}, \frac{\phi}{x-\phi} = \frac{x}{\phi}, \tag{2}$$

so that

$$x^2 - \phi x - \phi^2 = 0. \tag{3}$$

The positive root of (3), from the quadratic formula, is

$$x = \phi\left(\frac{1+\sqrt{5}}{2}\right). \tag{4}$$

But by definition, $\frac{1+\sqrt{5}}{2} = \phi$.

Hence, from (4),

$$x = \phi \cdot \phi = \phi^2 = \text{BE}, \tag{5}$$

and

$$EF = x - \phi = \phi^2 - \phi = \phi + 1 - \phi = 1. \tag{6}$$

Thus, in ΔBED, the ratio of leg to base, using (5), is $\frac{BE}{ED} = \frac{\phi^2}{\phi} = \phi$,
and in ΔDEF, the ratio of leg to base is again ϕ,
since $\frac{DE}{EF} = \frac{\phi}{1} = \phi$,
so that in any $72° - 72° - 36°$ isosceles triangle (hereafter referred to as the golden triangle), the ratio of

$$\frac{\text{leg}}{\text{base}} = \phi. \tag{7}$$

This is the same ratio of length to width defined for the golden rectangle.

In isosceles ΔEFJ, FJ $= \phi'$, since, using (6) and (7),

$$\frac{EF}{FJ} = \frac{1}{FJ} = \phi, \quad \text{implying that } FJ = \frac{1}{\phi} = \phi'. \tag{8}$$

Thus, regular pentagon FGHIJ has side length of ϕ'.

Returning to isosceles ΔDEF, it is apparent that $\overline{EJ}$ is the bisector of $\angle$DEF. In Figure 2, let $\overline{FK}$ be the bisector of $\angle$EFJ. Then FJ $=$ FK $= \frac{1}{\phi}$ and base JK $= \frac{1}{\phi^2}$. Moreover, $\overline{FK} \| \overline{BD}$, since $m\angle KFJ = m\angle JDB$.

In like fashion, the bisector of $\angle$FJK is parallel to $\overline{ED}$ and meets $\overline{FK}$ at L, forming another golden triangle, ΔJKL. This process of bisecting a base angle of a golden triangle can be continued indefinitely to produce a series of smaller and smaller golden triangles, which converge to a limiting point, 0. This point, comparable to that obtained in the golden rectangle, is the pole of an equiangular spiral which passes through the vertices $B, D, E, F, J, K, L, \ldots$ of each of the golden triangles.

A number of additional properties of the golden triangle are worth mentioning:

1. In Figure 2, let ML be of unit length. Then, from (7),

$$\begin{aligned} LK &= \phi = 1\phi + 0 \\ KJ &= \phi^2 = 1\phi + 1 \\ JF &= \phi^3 = 2\phi + 1 \\ FE &= \phi^4 = 3\phi + 2 \\ ED &= \phi^5 = 5\phi + 3 \\ DB &= \phi^6 = 8\phi + 5 \end{aligned} \tag{9}$$

forming a Fibonacci series.

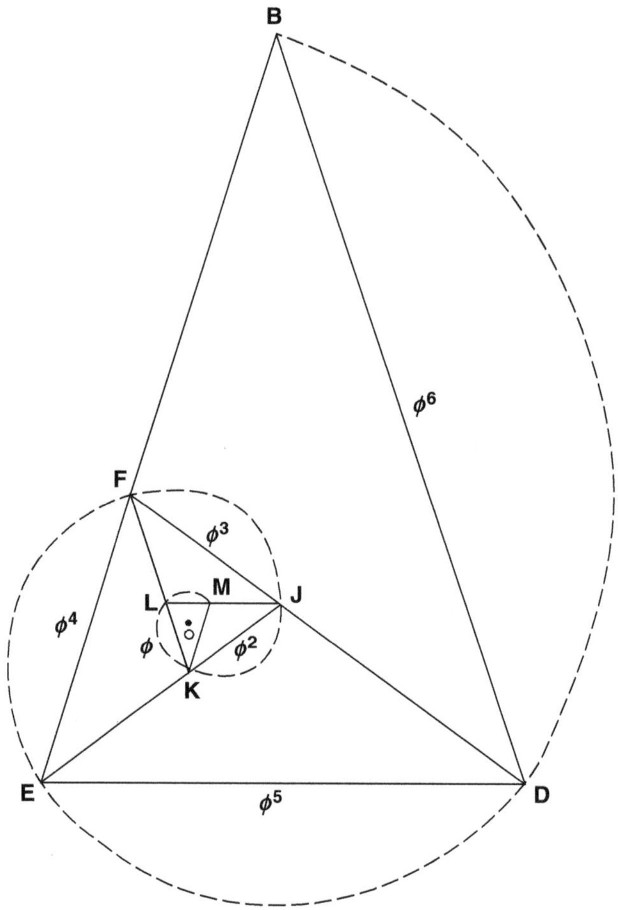

Figure 2.

2. The bisector of the vertex angle of a golden triangle divides the bisectors of the base angles in the golden ratio (see Figure 1). Since the angle bisectors of a triangle are concurrent, the bisector of $\angle$EBD must pass through J. However, from (6), EF $= 1 =$ EJ $=$ JD, and from (8) FJ $= \phi'$. Hence, $\frac{\text{JD}}{\text{FJ}} = \frac{1}{\phi'} = \phi$.
3. The golden triangle can be used to represent certain trigonometric functions in terms of ϕ (see Figure 3). Let ΔABC be $36° - 36° - 108°$ isosceles triangle, with AC $=$ CB $= 1$. Let one of the trisectors of angle C meet $\overline{\text{AB}}$ at D. Then ΔACD is a golden triangle with $m\angle\text{CDA} = m\angle\text{DCA} = 72°$. Since AC $= 1$, then AD $= 1$, and from (7), CD $= \frac{1}{\phi}$. Furthermore, ΔBCD is isosceles with $m\angle\text{BCD} = m\angle\text{DBC} = 36°$. Thus, CD $=$ DB $= \frac{1}{\phi}$, and AB $=$ AD $+$ DB $= 1 + \frac{1}{\phi} = \phi$.

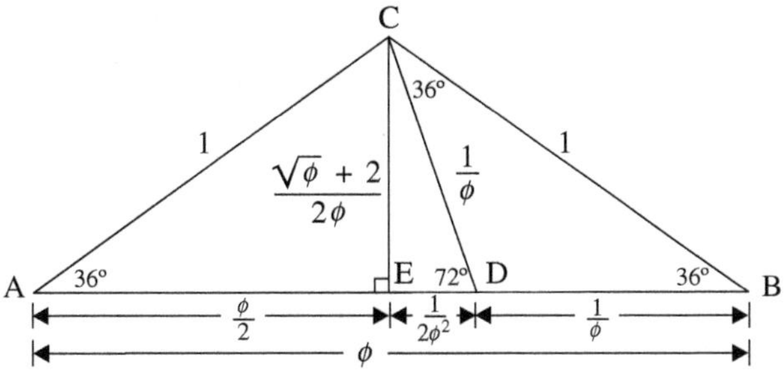

Figure 3.

From C, drop a perpendicular to meet $\overline{\text{AB}}$ at E. This makes AE = EB = $\frac{\phi}{2}$. Immediately, in right ΔACE $\cos 36° = \frac{\phi}{2}$, implying that (10)

$$\sin 54° = \frac{\phi}{2}. \tag{11}$$

Furthermore,

$$\text{ED} = \text{AD} - \text{AE} = 1 - \frac{\phi}{2} = \frac{2-\phi}{2} = \frac{1}{2\phi^2}. \tag{12}$$

Now in right ΔCED,

$$\cos 72° = \frac{\text{ED}}{\text{CD}} = \frac{\frac{1}{2\phi^2}}{\frac{1}{\phi}} = \frac{1}{2\phi} = \sin 18°. \tag{13}$$

Postassessment

1. Using the reciprocal trigonometric identities, determine values in terms of ϕ for tan, cot, sec, and csc for the angle measures indicated in (10), (11), and (13) above.
2. Using the half-angle formulas, determine trigonometric function values for 18° and 27° of ϕ.

This unit should be used in conjunction with "The Golden Rectangle."

References

Dunlap, R. A. *The Golden Ration and Fibonacci Numbers.* River Edge, NJ: World Scientific Publishing Co., 1997.

Huntley, H. E. *The Divine Proportion.* New York: Dover, 1970.

Posamentier, A. S. and I. Lehmann. *The Fabulous Fibonacci Numbers.* Amherst, NY: Prometheus Books, 2007.

Posamentier, A. S. and I. Lehmann. *The Glorious Golden Ratio.* Amherst, NY: Prometheus Books, 2012.
Walser, H. *The Golden Section.* Washington, DC: Mathematical Association of America, 2001.

Unit 69

Geometric Fallacies

Geometry students studying proofs using auxiliary sets often question the need for a rigorous reason for that set's existence. Often they don't appreciate the need for proving the existence and uniqueness of these sets. Students also develop a dependence on a diagram without analyzing its correctness. This unit introduces fallacious proofs to students in the hope that they can better grasp the need for such rigor.

Performance Objective

Given a geometric fallacy, students will determine where the fallacy occurs.

Preassessment

Students should be well acquainted with geometric proofs of both congruent and similar triangles.

Present your students with the following proof. They will recognize that it contains a fallacy. Ask them to try to determine where the error occurs.

Given: ABCD is a rectangle
$\overline{FA} \cong \overline{BA}$
R is the midpoint of $\overline{BC}$
N is the midpoint of $\overline{CF}$

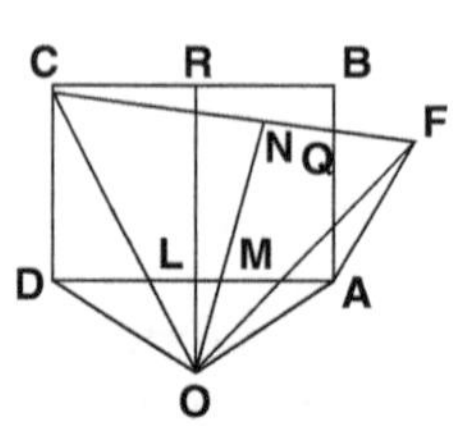

To Prove: A right angle is equal in measure to an obtuse angle ($\angle$CDA $\cong$ $\angle$FAD).

Draw $\overrightarrow{RL}$ perpendicular to $\overline{CB}$.

Draw $\overrightarrow{NM}$ perpendicular to $\overline{CF}$.

$\overrightarrow{RL}$ and $\overrightarrow{NM}$ intersect at point O. If they didn't intersect, $\overline{RL}$ and $\overline{NM}$ would be parallel and this would mean $\overline{CB}$ is parallel to $\overline{CF}$, which is impossible.

Draw $\overline{DO}$, $\overline{CO}$, $\overline{FO}$, and $\overline{AO}$.

Since $\overline{RO}$ is the perpendicular bisector of $\overline{CB}$ and $\overline{DA}, \overline{DO} \cong \overline{AO}$.

Since $\overline{NO}$ is the perpendicular bisector of $\overline{CF}$, $\overline{CO} \cong \overline{FO}$.

And, since $\overline{FA} \cong \overline{BA}$ and $\overline{BA} \cong \overline{CD}$, we have $\overline{FA} \cong \overline{CD}$.

Therefore, ΔCDO $\cong$ ΔFAO(SSS $\cong$ SSS), so $\angle$ODC $\cong$ $\angle$OAF.

Since OD $\cong$ $\overrightarrow{OA}$, we have $\angle$ODA $\cong$ $\angle$OAD.

Now, $m\angle ODC - m\angle ODA = m\angle OAF - m\angle OAD$ or $m\angle CDA = m\angle FAD$.

Teaching Strategies

When students have inspected the proof and have found nothing wrong with it, ask them to use rulers and compasses to reconstruct the diagram. The correct diagram looks like this:

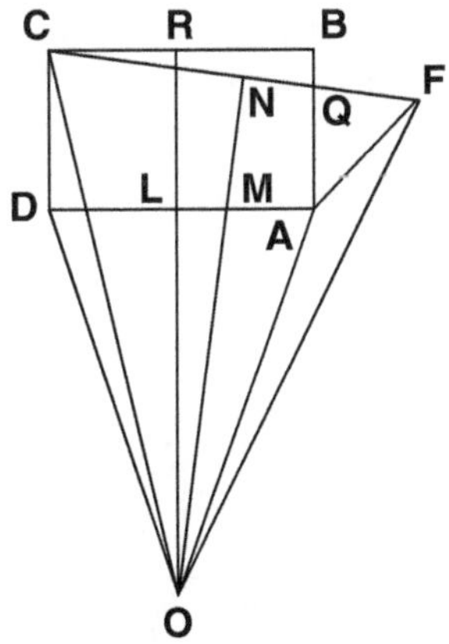

Although the triangles are congruent, our ability to subtract the specific angles no longer exists. Thus, the difficulty with this proof lies in its dependence on an incorrectly drawn diagram.

To show that ∠OAF cannot be obtuse, we must show that when $\overline{OF}$ intersects $\overline{AB}$ and $\overline{AD}$, and point O is on the perpendicular bisector of $\overline{CF}$, then point O cannot be on the perpendicular bisector of $\overline{AD}$.

Suppose point O is the intersection of the two perpendicular bisectors (as in the original diagram). Since $\overline{CD}//\overline{AQ}$, ∠DCF ≅ ∠AQF. In isosceles ΔABF, ∠ABF ≅ ∠AFB. But $m\angle AFB > m\angle AFC$; by substitution m∠ABF > $m\angle AFC$. Since ∠AQF is an exterior angle of ΔBQF, $m\angle AQF > m\angle ABF$. Therefore, $m\angle AQF > m\angle ABF > m\angle AFC$, or $m\angle AQF > m\angle AFC$. By substitution, $m\angle DCF > m\angle AFC$. Since ∠OCF ≅ ∠OFC, by subtraction, we have $m\angle DCO > m\angle OFA$. Thus, DO > OA. This is a contradiction, since DO = AO, if O is bisector of $\overline{DA}$ on the perpendicular. Therefore, point O cannot be on the perpendicular bisector of $\overline{DA}$ for any point O such that ∠OAF is obtuse.

The above proof also holds true for the following diagram, which shows that ∠OAF cannot be acute.

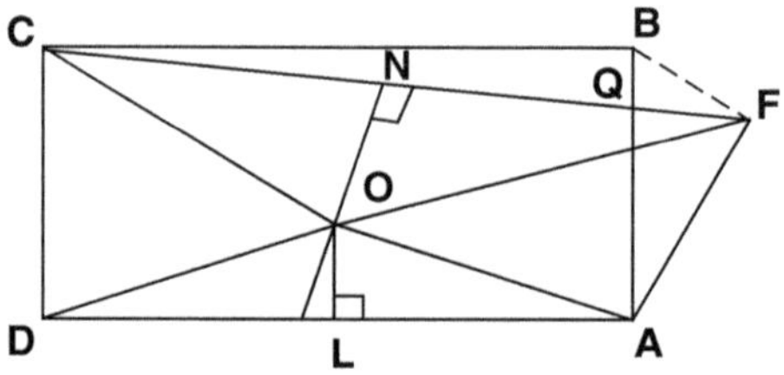

Now present your students with the following proof that any point in the interior of a circle is also on the circle.

Given: Circle O, with radius r Let A be any point in the interior of the circle distinct from O.

Prove: A is on the circle.

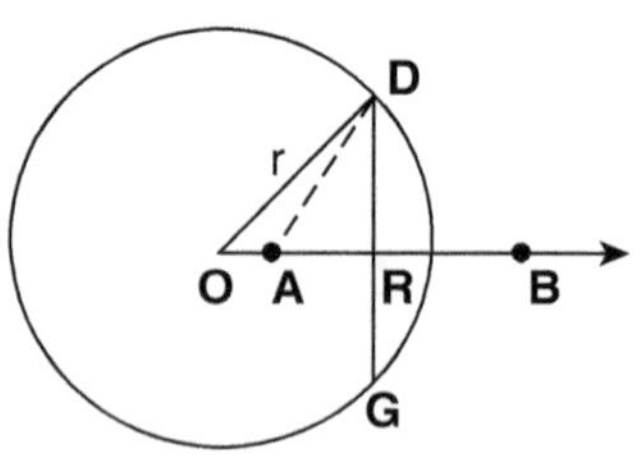

Let B be on the extension of $\overline{OA}$ through A such that $OA \cdot OB = r^2$. (Clearly OB is greater than r since OA is less than r.) Let the perpendicular bisector of $\overline{AB}$ meet the circle in points D and G, where R is the midpoint of $\overline{AB}$.

We now have $OA = OR - RA$ and $OB = OR + RB = OR + RA$.

Therefore,

$$r^2 = (OR - RA)(OR + RA)$$
$$r^2 = OR^2 - RA^2$$
$$r^2 = (r^2 - DR^2) - (AD^2 - DR^2) \quad \text{by the Pythagorean theorem}$$
$$r^2 = r^2 - AD^2$$

Therefore, $AD^2 = 0$.

Therefore, A coincides with D, and lies on the circle.

The fallacy in this proof lies in the fact that we drew an auxiliary line ($\overleftrightarrow{DRG}$) with *two* conditions — that $\overline{DRG}$ is the perpendicular bisector of $\overline{AB}$ and that it intersects the circle. Actually, all points on the perpendicular bisector of $\overline{AB}$ lie in the exterior of the circle and therefore cannot intersect the circle.

$$r^2 = OA\ (OB)$$
$$r^2 = OA\ (OA + AB)$$
$$r^2 = OA^2 + (OA)(AB)$$

Now, the proof assumes $OA + \frac{AB}{2} < r$

$$2(OA) + AB < 2r$$
$$4(OA)^2 + 4(OA)(AB) + AB^2 < 4r^2$$
$$\text{Since } r^2 = OA^2 + (OA)(AB)$$
$$\text{we have } 4r^2 + AB^2 < 4r^2$$
$$AB^2 < 0,$$
which is impossible.

This proof points to the care we must take when drawing auxiliary sets in using *one* condition only.

Here is a "triangle" consisting of four right triangles, four rectangles, and a "hole."

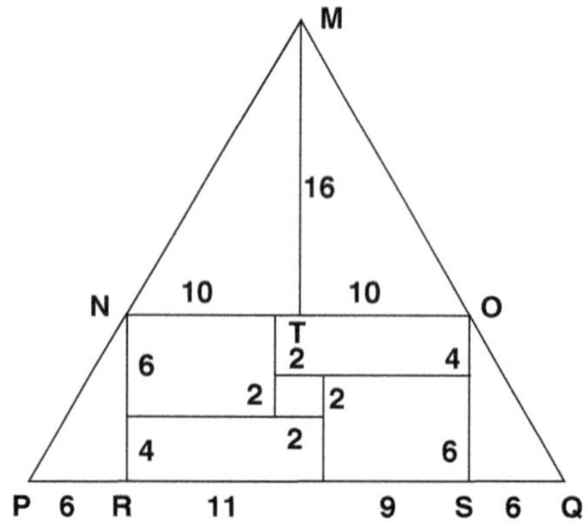

1. Have your students calculate the area of the eight regions (not the hole) [416].
2. Now have them calculate the area of the entire figure. [Since PQ = 32 and height = 26, $\frac{1}{2}\text{PQ} \cdot h = 416$.] We are now faced with this problem: How did we arrive at the same area with and without the hole?

The fallacy occurs because of an error in 2. The figure is *not* a triangle, since points M, N, and P are not collinear.

If points M, N, and P were collinear,

since ∠RNO is a right angle, ∠PNR is the complement of ∠MNT.
since ∠NRP is a right angle, ∠PNR is the complement of ∠RPN.

$$\therefore \angle\text{MNT} \cong \angle\text{RPN}$$
$$\therefore \Delta\text{MNT} \sim \Delta\text{NPR}$$

But, this is not the case.

The same argument holds for points M, O, and Q. Therefore, the figure is a pentagon; thus the formula we used in 2 is incorrect.

Postassessment

Have students select a geometric fallacy from any of the following books and explain the "error" in the proof.

References

Maxwell, E. A. *Fallacies in Mathematics.* Cambridge University Press, 1963.
Northrop, E. P. *Riddles in Mathematics.* D. Van Nostrand Co. 1944.
Posamentier, A. S. *Advanced Euclidean Geometry: Excursions for Secondary Teachers and Students.* New York, NY: John Wiley Publishing, 2002.

Posamentier, A. S., J. H. Banks, and R. L. Bannister. *Geometry, Its Elements and Structure*. 2nd edn. McGraw-Hill, 1977, pp. 240–244, 270–271.
Posamentier, A. S. and I. Lehmann. *Magnificent Mistakes in Mathematics*. Amherst, NY: Prometheus Books, 2013.

Unit 70

Regular Polyhedra

This unit will present a method that can be used to prove that there are not more than five regular polyhedra.

Performance Objective

Students will define a regular polyhedron, identify all regular polyhedra, and explain why no more than five regular polyhedra exist.

Preassessment

Display physical models of various polyhedra and have students count the number of vertices (V), the number of edges (E), and the number of faces (F) of each polyhedron. After tabulating their results, they should notice the relationship:

$$\mathrm{V}+\mathrm{F}=\mathrm{E}+2.$$

Teaching Strategies

Having empirically established Euler's theorem, ($\mathrm{V}+\mathrm{F}=\mathrm{E}+2$), students may wish to apply it to reach other conclusions about polyhedra. Depending on class interest, proof of this theorem may be in order. One source for the proof is *Geometry, Its Elements and Structure* by A. S. Posamentier and R. L. Bannister, pp. 574–576 (Dover, 2014).

One interesting application of this theorem is the proof that more than five *regular* polyhedra cannot exist. You should begin by defining a regular polyhedron as *a solid figure bounded by portions of planes called faces, each of which is a regular polygon* (congruent sides and angles). The cube is a common example of a regular polyhedron.

To begin the proof that *there are only five regular polyhedra*, let s represent the number of sides of each face and let t represent the number of faces at each vertex.

Since there are t faces at each vertex, students should realize that there are also t edges at each vertex. Suppose in counting the number of edges (E) of a given polyhedron, the number of edges at each vertex were counted and then multiplied by the number of vertices (V). This would produce *twice* the number of edges (2E) of the polyhedron, as each edge was counted twice, once at each of the two vertices it joins. Hence,

$$t\text{V} = 2\text{E}, \quad \text{or} \quad \frac{\text{V}}{\frac{1}{t}} = \frac{\text{E}}{\frac{1}{2}}.$$

Similarly, in counting the number of edges of the polyhedron, the number of sides (s) of each face were counted and then multiplied by the number of faces (F) of the polyhedron. This would also produce *twice* the number of edges of the polyhedron, as each side (edge) counted belongs to two faces. Hence, $s\text{F} = 2\text{E}$, or $\frac{\text{F}}{\frac{1}{s}} = \frac{\text{E}}{\frac{1}{2}}$.

Therefore,

$$\frac{\text{V}}{\frac{1}{t}} = \frac{\text{E}}{\frac{1}{2}} = \frac{\text{F}}{\frac{1}{s}}.$$

Students should recall the following theorem on proportions: $\frac{a}{b} = \frac{c}{d} = \frac{e}{f} = \frac{a+c+e}{b+d+f}$. Then have them apply it to the following:

$$\frac{\text{V}}{\frac{1}{t}} = \frac{-\text{E}}{-\frac{1}{2}} = \frac{\text{F}}{\frac{1}{s}} = \frac{\text{V} - \text{E} + \text{F}}{\frac{1}{t} - \frac{1}{2} + \frac{1}{s}}.$$

However, by Euler's theorem ($\text{V} - \text{E} + \text{F} = 2$),

$$\frac{\text{V}}{\frac{1}{t}} = \frac{\text{E}}{\frac{1}{2}} = \frac{\text{F}}{\frac{1}{s}} = \frac{2}{\frac{1}{t} - \frac{1}{2} + \frac{1}{s}}.$$

Students may now solve for V, E, and F:

$$\text{V} = \frac{4s}{2s + 2t - st}$$

$$\text{E} = \frac{2st}{2s + 2t - st}$$

$$F = \frac{4t}{2s + 2t - st}$$

Students should be asked to inspect the nature of V, E, and F. Realizing that these numbers must be positive, elicit from students that the denominators must be positive (since s and t are positive, as well as the numerators). Thus,

$$2s + 2t - st > 0.$$

To enable factoring, add -4 to both members of the inequality to get

$$2s + 2t - st - 4 > -4.$$

Then multiply both sides by -1:

$$-2s - 2t + st + 4 < 4, \quad \text{or} \quad (s-2)(t-2) < 4.$$

At this point, have students place restrictions on s and t. They should be quick to state that no polygon may have less than three sides; hence $s \geq 3$. Also, they should realize that at each vertex of the polyhedron there must at least be three faces; hence $t \geq 3$.

These facts will indicate that $(s-2)$ and $(t-2)$ must be positive. Since their product must be less than four, students should be able to generate the following table:

$(s-2)(t-2)$	$(s-2)$	$(t-2)$	s	t	V	E	F	Name of Polyhedron
1	1	1	3	3	4	6	4	Terrahedron
2	2	1	4	3	8	12	6	Hexahedron (cube)
2	1	2	3	4	6	12	8	Octahedron
3	3	1	5	3	20	30	12	Dodecahedron
3	1	3	3	5	12	30	20	Icosahedron

Since there are no other possible values for s and t, the above table is complete, and hence there are only five regular polyhedra. More able students should be encouraged to investigate the existence of these five regular polyhedra. One source is Euclid's *Elements*, Book XIII.

Further inspection of the above table reveals an interesting symmetry between the hexahedron and the octahedron as well as the dodecahedron and the icosahedron. That is, if s and t are interchanged, these symmetries will be highlighted. Furthermore, this table indicates that the faces of these regular polyhedra are either equilateral triangles, squares, or regular pentagons (see column s). Students should also be encouraged to further verify Euler's theorem ($V + F = E + 2$) with the data in the above table.

The figure shows the five regular polyhedra as well as "patterns" that can be used to construct (by cutting them out and appropriately folding them) these polyhedra.

Although often referred to as the five Platonic Solids, it is believed that three (tetrahedron, hexahedron, and dodecahedron) of the five solids were due to Pythagoreans, and the remaining two solids (octahedron and icosahedron) were due to the efforts of Theaetetus (414–369 B.C.). There is enough history about these solids to merit a brief report by one of the students.

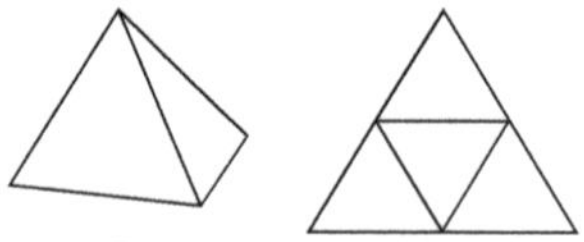

Tetrahedron {3, 3}

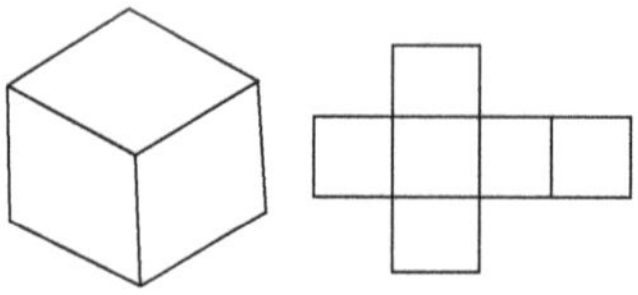

Cube {4, 3}

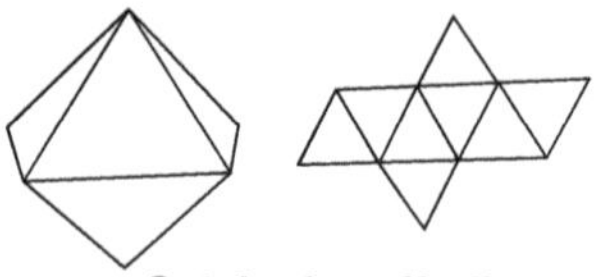

Octahedron {3, 4}

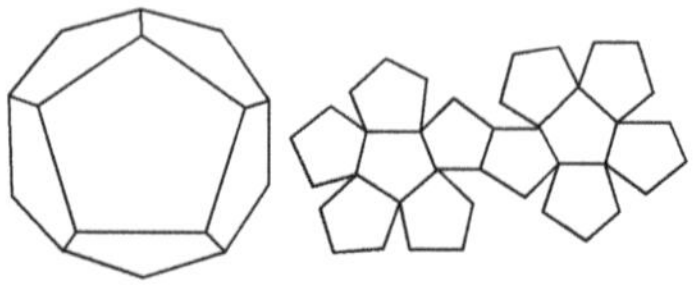

Dodecahedron {5, 3}

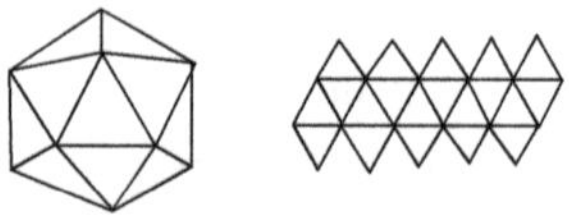

Icosahedron {3, 5}

Postassessment

1. Have students define and identify regular polyhedra.
2. Have students explain why more than five regular polyhedra cannot exist.

Unit 71

An Introduction to Topology

A lesson on topology can be taught as an enrichment of geometry. This unit will present some basic concepts of topology and their applications.

Performance Objectives

1. *Given two geometric drawings, students will determine whether they are topologically equivalent.*
2. *Given a polyhedron or a plane figure, students can show that* $V+F-E=2$ *(space), and* $V+F-E=1$ *(plane).*

Preassessment

A basic knowledge of seventh-grade geometry is desirable preceding this unit.

Teaching Strategies

Have students draw several closed curves. Then have them distinguish between those that are simple closed curves and those that are not. Some possible student responses may be the following. Ask students to redraw each of the figures without lifting their pencils off their paper. Students should realize that if Figure 1 were drawn on a rubber sheet they could twist and bend it into Figures 2, 3, 4, 5, and 6.

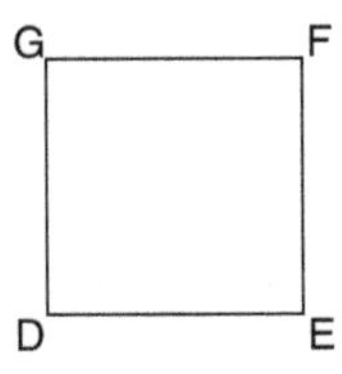

Figure 1.

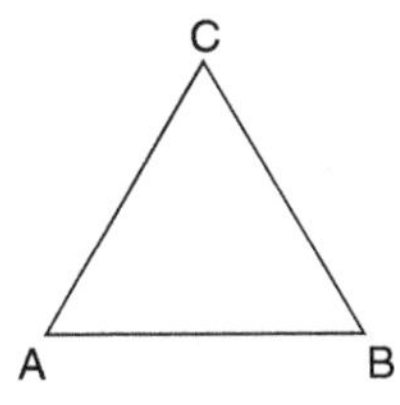

Figure 2.

Figure 3.

Suppose they now consider some geometric figures in space. Have them draw a cube (Figure 7).

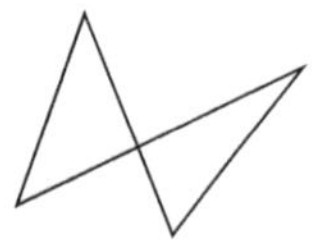

Figure 4.

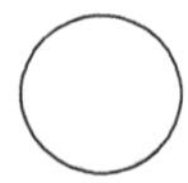

Figure 5.

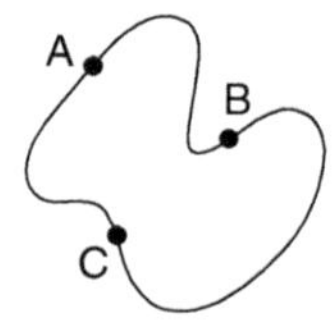

Figure 6.

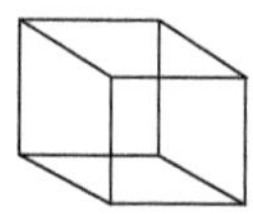

Figure 7.

Ask students if this cube can be transformed into any of Figures 8, 9, or 10 by twisting or bending it.

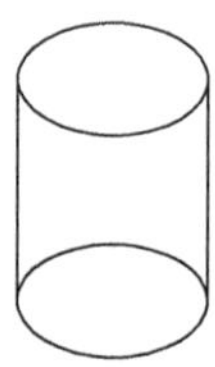

Figure 8.

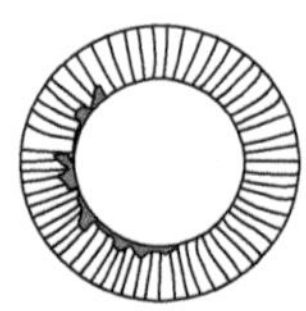

Figure 9.

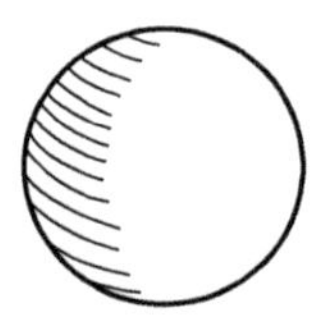

Figure 10.

They should find that the only figure of these that the cube can be transformed into is a sphere.

Therefore, a cube is topologically equivalent to a sphere. Tell students that studying figures in this way leads to a branch of mathematics called *topology* or "rubber sheet geometry."

One of the more fascinating relationships in geometry is directly taken from topology. This relationship involves the vertices (V), edges (E), and faces (F) of a polyhedron or polygon (see Unit 70 of Chapter 8). It reads: $V+F-E = 2$ (in three-dimensional space) or $V+F-E = 1$ (in a plane). Have students consider a pentagon. A pentagon has five vertices, five edges, and one face; hence $V + F - E = 5 + 1 - 5 = 1$. Students may now wish to consider a three-space figure. The cube (Figure 7) has 8 vertices, 6 faces, and 12 edges. Therefore, $V + F - E = 8 - 12 + 6 = 2$. These relationships can be demonstrated to the class by using overhead projector transparencies or with physical models.

Suppose a plane were to cut all edges of one of the trihedral angles of a cube (a piece of clay in the form of a cube would be useful here). This plane would then separate one of the vertices from the cube. However, in the process, there would be added to the original polyhedron: 1 face, 3 vertices, and 3 edges. Thus, for this new polyhedron, V is increased by 2. F is increased by 1, and E is increased by 3; yet, $V + F - E$ remains unchanged. More such experiments should be encouraged.

Figures are topologically equivalent if one can be made to coincide with the other by distortion, shrinking, stretching, or bending. If one face of a polyhedron is removed, the remaining figure is topologically equivalent to a region of a plane. This new figure (see Figure 11) will not have the same shape or size, but its *boundaries* are preserved. The edges will become sides of polygonal regions, and there will be the same number of edges and vertices in the plane figure as in the polyhedron. Each polygon that is not a triangle can be cut into triangles, or triangular regions, by drawing diagonals. Each time a diagonal is drawn, we increase the number of edges by one, but we also increase the number of faces by one. The value of $V - E + F$ is preserved. Triangles on the outer edge of the region will have either one edge on the boundary of the region, as ΔABC in Figure 11, or have two edges on the boundary, as ΔDEF. Triangles such as ΔABC can be removed by removing the one boundary side (i.e., $\overline{AC}$). By doing this, we decrease the number of faces by one and the number of edges by one. Still $V - E + F$ is unchanged. Triangles such as DEF can be removed by removing two edges (i.e., $\overline{DF}$ and $\overline{EF}$). By doing this, we decrease the number of edges by two, the number of faces by one, and the number of vertices by one. $V - E + F$ is still preserved.

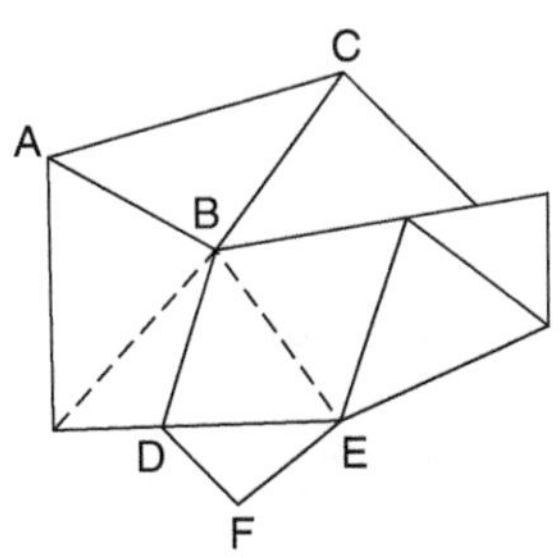

Figure 11.

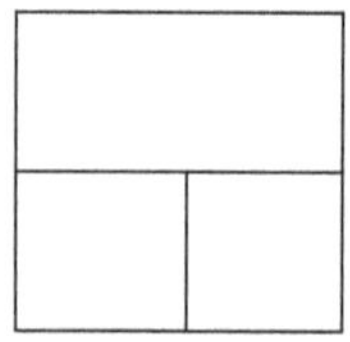

Figure 12a.

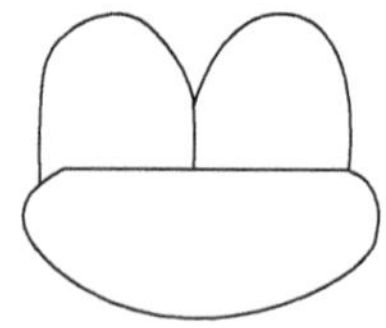

Figure 12b.

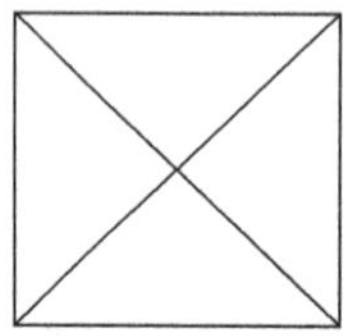

Figure 13a.

Figure 13b.

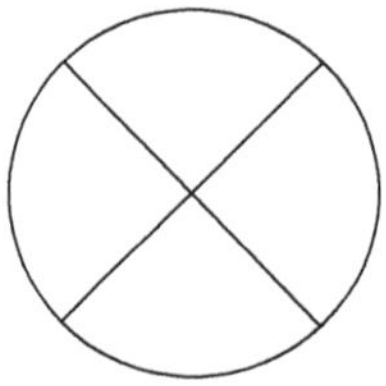

Figure 14a.

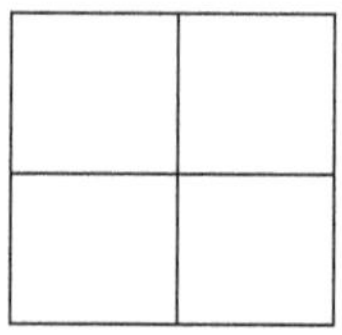

Figure 14b.

We continue in this manner until we are left with one triangle. This triangle has three vertices, three edges, and one face. Hence, $V - E + F = 1$ in the plane. We conclude that when we replace the face we removed we have $V - E + F = 2$ for a polyhedron in space.

After students have had a chance to familiarize themselves with this theorem, they should be encouraged to test it empirically with student constructed polyhedra. Clay is a good medium for this activity. Students may want to record their results on a chart.

Postassessment

1. Have students decide if any of Figures 12a, 12b, 13a, 13b, or 14a, 14b can be bent into the other figures.
2. Show that $V + F - E = 2$ holds for a tetrahedron and an octahedron.
3. Show that $V + F - E = 1$ holds for a hexagon and a dodecagon.

References

Posamentier, A. S. and C. Spreitzer, *The Mathematics Everyday Life*, Amherst, New York: Prometheus Books, 2018.

Unit 72

Angles on a Clock

This unit can be used in earlier junior high school grades as a recreational activity where some interesting relationships can be discovered, or it can be used as an enrichment application for beginning algebra students studying the topic of uniform motion.

Performance Objectives

1. *Students will determine the precise time that the hands of a clock form a given angle.*
2. *Students will solve problems related to the positions of the hands on a clock.*

Preassessment

Ask students at what time (exactly) will the hands of a clock overlap after 4 o'clock.

Teaching Strategies

Your students' first reaction to the solution to this problem will be that the answer is simply 4:20. When you remind them that the hour hand moves uniformly, they will begin to estimate the answer to be between 4:21 and 4:22. They will realize that the hour hand moves through an interval between minute markers every 12 minutes. Therefore, it will leave the interval 4:21–4:22 at 4:24. This however doesn't answer the original question about the exact time of this overlap.

In a beginning algebra class studying uniform motion problems, have students consider this problem in that light. The best way to have student begin to understand the movement of the hands of a clock is by having them consider the hands traveling independently around the clock at uniform speeds. The minute markings on the clock (from now on referred to as "markers") will serve to denote distance as well as time. An analogy should be drawn here to the uniform motion of automobiles (a popular and overused topic for verbal problems in an elementary algebra course). A problem involving a fast automobile overtaking a slower one would be appropriate.

Experience has shown that the analogy should be drawn between specific cases rather than mere generalizations. It might be helpful to have the class find the distance necessary for a car traveling at 60 m.p.h. to overtake a car with a head start of 20 miles and traveling at 5 m.p.h.

Now have the class consider 4 o'clock as the initial time on the clock. Our problem will be to determine exactly when the minute hand will overtake the hour hand after 4 o'clock. Consider the speed of the hour hand to be r, then the speed of the minute hand must be $12r$. We seek the distance, measured by the number of markers traveled, that the minute hand must travel to overtake the hour hand.

Let us refer to this distance as d markers. Hence, the distance that the hour hand travels is $d - 20$ markers, since it has a 20-marker head start over the minute hand. For this to take place, the times required for the minute hand, $\frac{d}{12r}$, and for the hour hand, $\frac{d-20}{r}$, are the same. Therefore, $\frac{d}{12r} = \frac{d-20}{r}$, and $d = \frac{12}{11} \cdot 20 = 21\frac{9}{11}$. Thus, the minute hand will overtake the hour hand at exactly $4{:}21\frac{9}{11}$.

Consider the expression $d = \frac{12}{11} \cdot 20$. The quantity 20 is the number of markers that the minute hand had to travel to get to the desired position, assuming the hour hand remained stationary. However, quite obviously, the hour hand does not remain stationary. Hence, we must multiply this quantity by $\frac{12}{11}$ as the minute hand must travel $\frac{12}{11}$ as far. Let us refer to this fraction $\left(\frac{12}{11}\right)$ as the correction factor. Have the class verify this correction factor both logically and algebraically.

To begin to familiarize the students with use of the correction factor, choose some short and simple problems. For example, you may ask them to find the exact time when the hands of a clock overlap between 7 and 8 o'clock. Here the students would first determine how far the minute hand would have to travel from the "12" position to the position of the hour hand, assuming again that the hour hand remains stationary. Then by multiplying the number of markers, 35, by the correction factor, $\frac{12}{11}$, they will obtain the exact time $\left(7{:}38\frac{2}{11}\right)$ that the hands will overlap.

To enhance students' understanding of this new procedure, ask them to consider a person checking a wristwatch against an electric clock and noticing that the hands on the wristwatch overlap every 65 minutes (as measured by the electric clock). Ask the class if the wristwatch is fast, slow, or accurate.

You may wish to have them consider the problem in the following way. At 12 o'clock, the hands of a clock overlap exactly. Using the previously described method, we find that the hands will again overlap at exactly

$1{:}05\frac{5}{11}$, and then again at exactly $2{:}10\frac{10}{11}$, and again at exactly $3{:}16\frac{4}{11}$, and so on.

Each time there is an interval of $65\frac{5}{11}$ minutes between overlapping positions. Hence, the person's watch is inaccurate by $\frac{5}{11}$ of a minute. Have students now determine if the wristwatch is fast or slow.

There are many other interesting, and sometimes rather difficult, problems made simple by this correction factor. You may very easily pose your own problems. For example, you may ask your students to find the exact times when the hands of a clock will be perpendicular (or form a straight angle) between, say, 8 and 9 o'clock.

Again, you would have the students determine the number of markers that the minute hand would have to travel from the "12" position until it forms the desired angle with the stationary hour hand. Then have them multiply this number by the correction factor $\left(\frac{12}{11}\right)$ to obtain the exact actual time. That is, to find the exact time that the hands of a clock are *first* perpendicular between 8 and 9 o'clock, determine the desired position of the minute hand when the hour hand remains stationary (here, on the 25-minute marker).

Then, multiply 25 by $\frac{12}{11}$ to get $8{:}27\frac{3}{11}$, the exact time when the hands are *first* perpendicular after 8 o'clock.

For students who have not yet studied algebra, you might justify the $\frac{12}{11}$ correction factor for the interval between overlaps in the following way:

> Think of the hands of a clock at noon. During the next 12 hours (i.e., until the hands reach the same position at midnight) the hour hand makes one revolution, the minute hand makes 12 revolutions, and the minute hand coincides with the hour hand 11 times (including mid-night, but not noon, starting just after the hands separate at noon). Because each hand rotates at a uniform rate, the hands overlap each $\frac{12}{11}$ of an hour, or $65\frac{5}{11}$ minutes.

This can be extended to other situations.

Your students should derive a great sense of achievement and enjoyment as a result of employing this simple procedure to solve what usually appears to be a very difficult clock problem.

Postassessment

1. At what time will the hands of a clock overlap after 2 o'clock?
2. At what time will the hands of a clock be perpendicular after 3 o'clock?
3. How would the "correction factor" change if our clock were a 24-hour cycle clock?

4. What would the "correction factor" be if we sought the exact time when the second hand and the minute hand were perpendicular after (*fill in a specified time*)?
5. What angle is determined by the hands of a clock at (*fill in a specified time*)?
6. What is the first time (exactly) when the second hand bisects the angle formed by the minute and hour hands of a clock after (*fill in a specified time*)?

References

Posamentier, A. S. and C. Spreitzer, *The Mathematics Everyday Life*, Amherst, New York: Prometheus Books, 2018.

Unit 73

Averaging Rates — The Harmonic Mean

This unit will present a shortcut method for determining the average of two or more rates (rates of speed, cost, production, etc.).

Performance Objectives

1. *Given various rates for a common base, students will find the average of these rates.*
2. *Given a problem calling for the average of given rates, students will correctly apply the concept of a harmonic mean when applicable.*

Preassessment

Have students solve the following problem:

> Noreen drives from her home to work at the rate of 30 mph. Later she returns home from work over the same route at the rate of 60 mph. What is her average rate of speed for both trips?

Teaching Strategies

The preceding problem should serve as an excellent motivation for this unit. Most students will probably incorrectly offer 45 mph as their answer to this problem. Their explanation will be that 45 is the average of 30 and 60. True! You must convince them that since the numbers 30 and 60 represent rates, they cannot be treated as simple quantities. Students will wonder what difference this should make.

The first task is to convince your students that their original answer, 45 mph, is incorrect. Have them realize that when Noreen drove from home to work, she had to drive twice as much time than she did on her return trip. Hence, it would be incorrect to give both rates the same "weight." If this still does not convince your students, ask them that if their test scores throughout the semester were 90, 90, 90, 90, and 40, which of the following methods would they use to find their average:

$$\begin{array}{l} 90 \text{ (average of first four tests)} \\ \underline{+40} \text{ (their last test score)} \\ 130 \div 2 = 65 \end{array}$$

Or: $90 + 90 + 90 + 90 + 40 = 400$; $400 \div 5 = 80$.

It would be expected that students would now suggest that the answer to the original problem could be obtained by: $\frac{30+30+60}{3} = 40$. This is perfectly correct; however, a simple solution as this would hardly be expected if one rate were not a multiple of the other. Most students would now welcome a more general method of solution. One such solution is based on the relationship: *Rate* × *Time* = *Distance.* Consider the following:

$$T_1 \text{ (time going to work)} = \frac{D}{30}$$

$$T_2 \text{ (time returning home)} = \frac{D}{60}$$

$$T \text{ (total time for both trips)} = T_1 + T_2 = \frac{D}{20}$$

$$R \text{ (rate for the entire trip)} = \frac{2D}{T} = \frac{2D}{\frac{D}{20}} = 40$$

R is actually the *average rate* for the entire trip, since problems of this nature deal with uniform motion.

Of particular interest are those problems where the rates to be averaged are for a common base (e.g., the same distance for various rates of speed). Have students consider the original problem in general terms, where the given rates of speed are R_1 and R_2 (instead of 30 and 60), each for a distance D.

Therefore, $T_1 = \frac{D}{R_1}$ and $T_2 = \frac{D}{R_2}$, so that

$$T = T_1 + T_2 = D\left(\frac{1}{R_1} + \frac{1}{R_2}\right) = \frac{D(R_1 + R_2)}{R_1 R_2}.$$

However, have students consider:

$$R = \frac{2D}{T} = \frac{2D}{D\left(\frac{1}{R_1} + \frac{1}{R_2}\right)} = \frac{2}{\frac{1}{R_1} + \frac{1}{R_2}} = \frac{2R_1R_2}{R_1 + R_2}. \tag{1}$$

They should notice that $\frac{2R_1R_2}{R_1+R_2}$ is actually the reciprocal of the average of the reciprocals of R_1 and R_2. Such an average is called the *harmonic mean.*

Perhaps a word about the harmonic mean would be in order. A progression of numbers is said to be harmonic if any three consecutive members of the progression (a, b, and c) have the property that

$$\frac{a}{c} = \frac{a-b}{b-c}. \tag{2}$$

This relationship may also be written as

$$a(b-c) = c(a-b). \tag{3}$$

Dividing by abc, we get

$$\frac{1}{c} - \frac{1}{b} = \frac{1}{b} - \frac{1}{a}. \tag{4}$$

This relationship shows that the reciprocals of a harmonic progression are in an arithmetic progression, as with $\frac{1}{a}$, $\frac{1}{b}$, and $\frac{1}{c}$. When three terms are in an arithmetic progression, the middle term is their mean. Thus, $\frac{1}{b}$ is the arithmetic mean between $\frac{1}{a}$ and $\frac{1}{c}$; b is the harmonic mean between a and c.

Expressing b in terms of a and c in Equation (4):

$$\frac{2}{b} = \frac{1}{a} + \frac{1}{c}, \quad \text{and} \quad b = \frac{2ac}{a+c}. \tag{5}$$

Have students compare (1) with (5)!

In a similar manner, you may wish to have the class consider the harmonic mean of three numbers, r, s, and t:

$$\frac{3}{\frac{1}{r} + \frac{1}{s} + \frac{1}{t}} = \frac{3rst}{st + rt + rs}.$$

Students may even wish to extend this to determine a "formula" for the harmonic mean of four numbers, k, m, n, and p:

$$\frac{4}{\frac{1}{k}+\frac{1}{m}+\frac{1}{n}+\frac{1}{p}}=\frac{4kmnp}{mnp+knp+kmp+kmn}.$$

Have students consider the following problem:

> Lisa bought 2 dollars worth of each of three different kinds of pencils, priced at 2¢, 4¢, and 5¢ each, respectively. What is the average price paid per pencil?

The answer to this question is $3\frac{3}{19}$, the harmonic mean of 2, 4, and 5. Stress the point that this was possible since each rate acted on the same base, 2 dollars. Similar problems (see the Postassessment) may be posed and solved by the students at this time.

You may wish to consider a geometric illustration of the concept. Although the harmonic mean enjoys the most prominence geometrically in projective geometry, it might be more appropriate to give an illustration of the harmonic mean in synthetic geometry.

Have the students consider the length of the segment containing the point of intersection of the diagonals of a trapezoid and parallel to the bases, with its endpoints in the legs (see $\overline{EGF}$ in the following figure). The length of this segment, $\overline{EGF}$, is the harmonic mean between the lengths of the bases, $\overline{AD}$ and $\overline{BC}$. In the following figure, ABCD is a trapezoid, with $\overline{AD}//\overline{BC}$ and diagonals intersecting at G. Also $\overline{EGF}//\overline{BC}$ and $\overline{DEC}$ and $\overline{AFB}$.

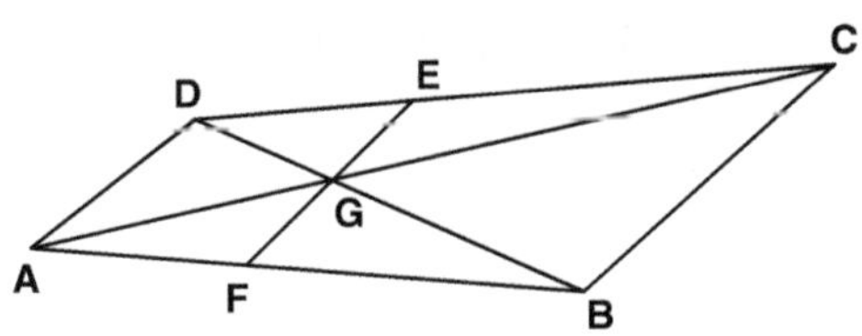

Since $\overline{GF}//\overline{BC}$, $\Delta AFG \sim \Delta ABC$, and $\frac{AF}{FG}=\frac{AB}{BC}$. Similarly, because $\overline{GF}//\overline{AD}$, $\Delta GBF \sim \Delta DBA$, and $\frac{BF}{FG}=\frac{AB}{AD}$. Therefore, $\frac{AF}{FG}+\frac{BF}{FG}=\frac{AB}{BC}+\frac{AB}{AD}$. Because $AF+BF=AB$, $\frac{AB}{FG}=\frac{AB}{BC}+\frac{AB}{AD}$, or $FG=\frac{(BC)(AD)}{BC+AD}$. In a similar manner, it can be shown that $EG=\frac{(BC)(AD)}{BC+AD}$. Therefore, $EF=FG+EG=\frac{2(BC)(AD)}{BC+AD}$; hence, EF is the harmonic mean between BC and AD.

Postassessment

1. If a jet flies from New York to Rome at 600 mph and back along the same route at 500 mph, what is the average rate of speed for the entire trip?

2. Alice buys 2 dollars worth of each of three kinds of nuts, priced at 40¢, 50¢, and 60¢ per pound, respectively. What is the average price Alice paid per pound of nuts?
3. In June, Willie got 30 hits for a batting average of .300; however, in May he got 30 hits for a batting average of .400. What is Willie's batting average for May and June?
4. Find the harmonic mean of 2, 3, 5, 6, 2, and 9.

References

Posamentier, A. S. and S. Krulik. *Teachers! Prepare Your Students for the Mathematics for SAT 1: Methods and Problem Solving Strategies.* Thousand Oaks, CA: Corwin, 1996.

Posamentier, A. S. and C. T. Salkind. *Challenging Problems in Algebra.* New York: Dover, 1996.

Posamentier, A. S. and C. T. Salkind. *Challenging Problems in Geometry.* New York: Dover, 1996.

Posamentier, A. S. The harmonic mean and its place among means in *Readings for Enrichment in Secondary School Mathematics.* Ed. by Max A. Sobel. Reston, VA: NCTM, 1988.

Unit 74

Howlers

In *Fallacies in Mathematics*, E. A. Maxwell refers to the following cancellations as *howlers*:

$$\frac{1\not{6}}{\not{6}4} = \frac{1}{4}$$

$$\frac{2\not{6}}{\not{6}5} = \frac{2}{5}$$

This unit will offer a method of presenting these howlers to elementary algebra students to help them better understand number concepts.

Performance Objectives

1. *Students will develop a howler not already presented in class.*
2. *Students will explain why there are only four howlers composed of two-digit fractions.*

Preassessment

Students should be able to reduce fractions to lowest terms. They should also be familiar with such concepts as factor and prime number, and be able to perform all operations on fractions.

Teaching Strategies

Begin your presentation by asking students to reduce to lowest terms the following fractions: $\frac{16}{64}$, $\frac{19}{95}$, $\frac{26}{65}$, $\frac{49}{98}$. After they have reduced to lowest terms each of the fractions in the usual manner, tell them that they did a lot of unnecessary work. Show them the following cancellations:

$$\frac{1\not{6}}{\not{6}4} = \frac{1}{4}$$
$$\frac{1\not{9}}{\not{9}5} = \frac{1}{5}$$
$$\frac{2\not{6}}{\not{6}5} = \frac{2}{5}$$
$$\frac{4\not{9}}{\not{9}8} = \frac{4}{8} = \frac{1}{2}.$$

At this point, your students will be somewhat amazed. Their first reaction is to ask if this can be done to any fraction composed of two-digit numbers. Challenge your students to find another fraction (comprised of two-digit numbers) where this type of cancellation will work. Students might cite $\frac{5\not{5}}{\not{5}5} = \frac{5}{5} = 1$ as an illustration of this type of cancellation.

Indicate to them that although this will hold true for all multiples of eleven, it is trivial, and our concern will be only with proper fractions (i.e., whose value is less than one).

After they are thoroughly frustrated, you may begin a discussion on why the four fractions above are the only ones (composed of two-digit numbers) where this type of cancellation will hold true.

Have students consider the fraction $\frac{10x+a}{10a+y}$.

The nature of the above four cancellations was such that when canceling the a's, the fraction was equal to $\frac{x}{y}$. Therefore, $\frac{10x+a}{10a+y} = \frac{x}{y}$.

This yields:

$$
\begin{aligned}
y(10x + a) &= x(10a + y) \\
\text{or} \quad 10xy + ay &= 10ax + xy \\
9xy + ay &= 10ax \\
\text{and} \quad y &= \frac{10ax}{9x + a}.
\end{aligned}
$$

At this point, have students inspect this relationship. They should realize that it is necessary that x, y, and a be integers, since they were digits in the numerator and denominator of a fraction. It is now their task to find the values of a and x for which y will also be integral.

To avoid a lot of algebraic manipulation, you might have students set up a chart that will generate values of y from $y = \frac{10ax}{9x+a}$. Remind them that x, y, and a must be *single digit* integers. Below is a portion of the table they will construct. Notice that the cases where $x = a$ are excluded since $\frac{x}{a} = 1$.

The portion of the chart pictured above already generated two of the four integral values of y; that is, when $x = 1$, $a = 6$, then $y = 4$, and when $x = 2$, $a = 6$ and $y = 5$. These values yield the fractions $\frac{16}{64}$, and $\frac{26}{65}$, respectively. The remaining two integral values of y will be obtained when $x = 1$, $a = 9$, then $y = 5$; and when $x = 4$, $a = 9$, then $y = 8$. These yield the fractions $\frac{19}{95}$ and $\frac{49}{98}$, respectively. This should convince students that there are only four such fractions composed of two-digit numbers.

		x \ a	1	2	3	4	5	6	...	9
		1		$\frac{20}{11}$	$\frac{30}{12}$	$\frac{40}{13}$	$\frac{50}{14}$	$\frac{60}{15} = 4$		
		2	$\frac{20}{19}$		$\frac{60}{21}$	$\frac{80}{22}$	$\frac{100}{23}$	$\frac{120}{24} = 5$		
		3	$\frac{30}{28}$	$\frac{60}{29}$		$\frac{120}{31}$	$\frac{150}{32}$	$\frac{180}{33}$		
		:								
		.								
		9								

Students may now wonder if there are fractions composed of numerators and denominators of more than two digits where this strange type of cancellation holds true. Have students try this type of cancellation with $\frac{4\cancel{9}\cancel{9}}{\cancel{9}\cancel{9}8}$. They should find that, in fact,

$$\frac{499}{998} = \frac{4}{8} = \frac{1}{2}.$$

Soon they will realize that

$$\frac{49}{98} = \frac{499}{998} = \frac{4999}{9998} = \frac{49999}{99998} = \cdots .$$

$$\frac{16}{64} = \frac{166}{664} = \frac{1666}{6664} = \frac{16666}{66664} = \frac{166666}{666664} = \cdots .$$

$$\frac{19}{95} = \frac{199}{995} = \frac{1999}{9995} = \frac{19999}{99995} = \frac{199999}{999995} = \cdots .$$

$$\frac{26}{65} = \frac{266}{665} = \frac{2666}{6665} = \frac{26666}{66665} = \frac{266666}{666665} = \cdots .$$

Students with higher ability may wish to justify these extensions of the original howlers.

Students who at this point have a further desire to seek out additional fractions that permit this strange cancellation should be shown the following fractions. They should verify the legitimacy of this strange cancellation and then set out to discover more such fractions.

$$\frac{3\cancel{3}2}{8\cancel{3}0} = \frac{32}{80} = \frac{2}{5}$$

$$\frac{3\cancel{8}5}{8\cancel{8}0} = \frac{35}{80} = \frac{7}{16}$$

$$\frac{1\cancel{3}8}{\cancel{3}45} = \frac{18}{45} = \frac{2}{5}$$

$$\frac{2\cancel{7}5}{7\cancel{7}0} = \frac{25}{70} = \frac{5}{14}$$

$$\frac{1\cancel{6}\cancel{3}}{\cancel{3}2\cancel{6}} = \frac{1}{2}$$

$$\frac{2\cancel{0}3}{6\cancel{0}9} = \frac{1}{3}.$$

Postassessment

Have students:

1. Generate a "howler" not already presented in this discussion.
2. Explain why there are only four howlers, each composed of two-digit numbers.

Reference

Posamentier, A. S. and I. Lehmann. *Magnificent Mistakes in Mathematics.* Amherst, NY: Prometheus Books, 2013.

Unit 75

Digit Problems Revisited

Problems involving the digits of a number as presented in the elementary algebra course are usually very straightforward and somewhat dull. Often they serve merely as a source of drill for a previously taught skill. This unit shows how digit problems (perhaps somewhat "off the beaten path" in nature) can be used to improve a student's concept of numbers.

Performance Objectives

1. *Students will solve problems involving the digits of a number.*
2. *Students will analyze a mathematical fact about the nature of certain numbers.*

Preassessment

Students should be able to solve simple linear equations as well as simple simultaneous equations.

Teaching Strategies

Begin your presentation by asking your students to select any three-digit number in which the hundreds digit and units digit are unequal. Then have them write the number whose digits are in the reverse order from the selected number. Now tell them to subtract these two numbers (the smaller number from the greater one). Once again tell them to take this difference, reverse its digits, and add the "new" number to the *original difference.* They all should end up with 1,089.

For example, suppose a student selected the number 934. The number with the digits reversed is 439. The computation would appear as follows:

$$\begin{array}{rl} 934 & \\ 439 & \\ \hline 495 & \text{(difference)} \\ 594 & \text{(reversed digits)} \\ \hline 1089 & \text{(sum)} \end{array}$$

When students compare results, they will be amazed to discover uniformity in their answers. At this point, they should be quite eager to find out why they all came up with the same result.

Begin by letting them represent the original number by $100h + 10t + u$, where h, t, and u represent the hundreds, tens, and units digits, respectively. Let $h > u$, which would have to be true in one of the original numbers. In subtracting, $u - h < 0$; therefore, take 1 from the tens place to make the units place $10 + u$ (of the minuend).

Since the tens digits of the two numbers to be subtracted are equal, and 1 was taken from the tens digit of the minuend, then the value of this digit is $10(t - 1)$. The hundreds digit of the minuend is $h - 1$, since 1 was taken away to enable subtraction in the tens place, making the value of the tens digit $10(t - 1) + 100 = 10(t + 9)$. Pictorially this appears as

$$\begin{array}{lll} 100\,(h-1) & +10\,(t+9) & +\,(u+10) \\ 100\,u & +10t & +\,h \\ \hline 100\,(h-u-1) & +10\,(9) & +\,u-h+10. \end{array}$$

Reversing the digits of this difference yields

$$100(u - h + 10) + 10(9) + h - u - 1.$$

Adding the last two lines yields

$$100(9) + 10(18) + (10 - 1) = 1089.$$

Another problem involving the digits of a number and presenting a somewhat unusual twist follows:

> Seven times a certain two-digit number equals a three-digit number. When the digit 6 is written after the last digit of the three-digit number, the three-digit number is increased by 1,833. Find the two-digit number.

The major obstacle students encounter in the solution of this problem is how to indicate placing a 6 after a number. Let students represent the two-digit number by a. Therefore, the three-digit number is $7a$. Now to place a 6 after a number is to multiply the number by 10 and add the 6. The required equation is then $70a + 6 = 7a + 1833$, and $a = 29$.

To further exhibit the usefulness of working algebraically with a digital expression of a number, you may find it exciting to show students why a number is divisible by 9 (or 3) if the sum of its digits is divisible by 9 (or 3). Have them consider any five-digit number, say, $ab{,}cde$, that is $10{,}000a + 1{,}000b + 100c + 10d + e$. Since this number may be rewritten as $(9{,}999+1)a + (999+1)b + (99+1)c + (9+1)d + e$, or $9{,}999a + 999b + 99c + 9d + a + b + c + d + e$, and the sum of the first four terms is divisible by 9 (or 3), the sum of the remaining terms must also be divisible by 9 (or 3). That is, for the number to be divisible by 9 (or 3), $a + b + c + d + e$ must be divisible by 9 (or 3).

Another digit problem with a rather nonroutine solution should now be presented. Students will find the following analysis a bit unusual.

> Find the two-digit number N such that when it is divided by 4 the remainder is zero, and such that all of its positive integral powers end in the same two digits as the original number, N.

Students will naturally want to begin solving this problem by letting $\mathrm{N} = 10t + u$. Since $10t + u = 4m$ (i.e., a multiple of 4), u is even. Ask students which even digits have squares terminating with the same digit as the original digit. Once students establish that only 0 and 6 satisfy this property, therefore $u = 0$ or 6.

The case $u = 0$ implies that $t = 0$ so that $\mathrm{N} = 00$, a trivial case, for if $t = 0$, N will terminate in 0 while its square will terminate in 00.

Now have students consider $u = 6$. Then $\mathrm{N} = 10t + 6 = 4m$, or $5t + 3 = 2m$. This indicates that $t = 1, 3, 5, 7$, or 9. But $\mathrm{N}^2 = (10t + 6)^2 = 100t^2 + 120t + 36 = 100t^2 + 100d + 10e + 36$, where $120t = 100d + 10e$. Since the last two digits of N^2 are the same as those of N, $10e + 36 = 10t + 6$, and $t = e + 3$, so that $t \geq 3$. Also, $120t = 100d + 10(t - 3)$, whereby $11t = 10d - 3$, and $11t \leq 87$, or $t \leq 7$.

Have students try $t = 3, 36^2 = 1296$ (reject) then try $t = 5, 56^2 = 3136$ (reject). Finally try $t = 7$, then $76^2 = 5776$ (accept since $N = 76$).

There are many other problems involving the digits of numbers that you may wish to present to your class to advance the number theory introduction which this model provided.

Postassessment

1. Show, using a digital representation of a number, that a given number is divisible by 8, if the last three digits (considered as a new number) are divisible by 8.
2. There are two numbers formed of the same two digits — one in reverse of the other. The difference between the squares of the two numbers is 7,128 and the sum of the number is 22 times the difference between the two digits. What are the two numbers?
3. By shifting the initial digit 6 of the positive integer N to the end, we obtain a number equal to $\frac{1}{4}N$. Find the smallest possible value of N that satisfies the conditions.

Unit 76

Algebraic Identities

This unit will present a geometric process for carrying out algebraic identities. With only the representation of a number by a length, and lacking sufficient algebraic notation, early Greeks devised the method of application of areas to prove these identities.

Performance Objective

Students will geometrically establish algebraic identities using the method of application of areas.

Preassessment

1. Have students expand $(a + b^2)$.
2. Have students expand $a(b + c)$.
3. Have students expand $(a - b)^2$.
4. Elicit for what values of a and b each of the above generated equalities are true.

Teaching Strategies

After students have considered the questions above, they should be reacquainted with the properties of an identity. Once students understand the concept of an identity, introduce the method of application of areas by illustrating geometrically the identity $(a+b)^2 = a^2 + 2ab + b^2$. To begin, have students draw a square of side length $(a+b)$. The square should then be partitioned into various squares and rectangles (see Figure 1). The lengths of the various sides are appropriately labeled.

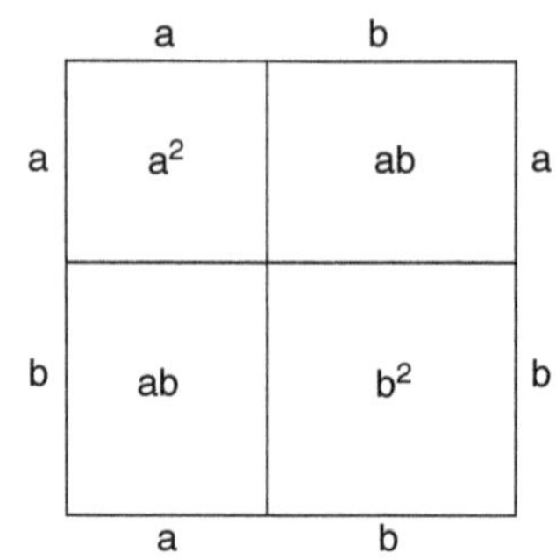

Figure 1.

Students can easily determine the area of each region. Since the area of the large square equals the areas of the four quadrilaterals into which it was partitioned, students should get

$$(a+b)^2 = a^2 + ab + ab + b^2 = a^2 + 2ab + b^2.$$

A more rigorous proof can be found in Euclid's *Elements*, Proposition 4, Book II.

Next illustrate geometrically the identity $a(b+c) = ab + ac$. To begin, have students draw a rectangle whose adjacent sides are of lengths a and $(b+c)$. The rectangle should then be partitioned into various rectangles (see Figure 2). The lengths of these sides are also labeled.

Students can easily determine the area of each partition. Elicit from students that since the area of the large rectangle equals the areas of the two quadrilaterals into which it was partitioned, the diagram illustrates $a(b+c) = ab + ac$.

Have students consider the following identity $(a+b) \times (c+d) = ac + ad + bc + bd$. Guide students to draw the appropriate rectangle with side lengths $(a+b)$ and $(c+d)$. The rectangle should be partitioned into various rectangles (see Figure 3). The lengths of sides and areas of regions have been labeled. As in the other cases, the area of the large rectangle equals the areas of the four quadrilaterals into which it was partitioned.

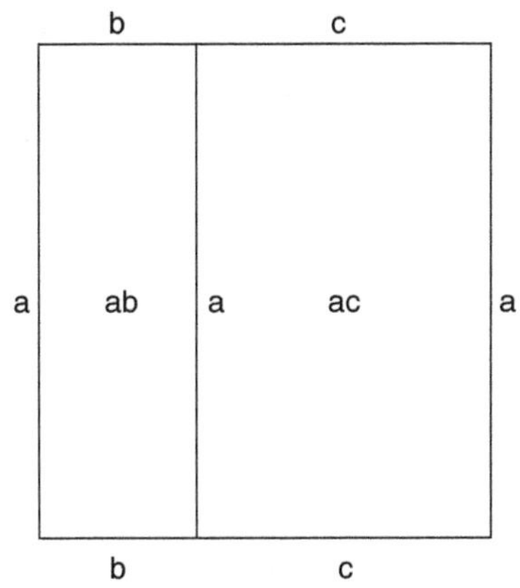

Figure 2.

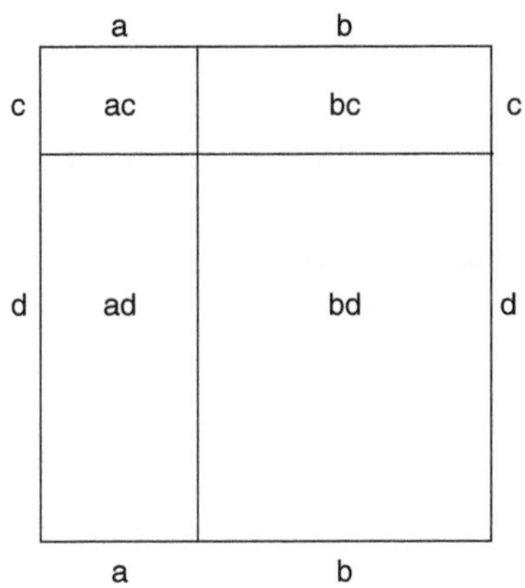

Figure 3.

Figure 3 illustrates the identity $(a + b) \times (c + d) = ac + ad + bc + bd$. Explain to students that the method of application of areas can be used to prove most algebraic identities. The difficulty will lie in their choice of dimensions for the quadrilateral and the partitions made.

After students feel comfortable using areas to represent algebraic identities, have them consider the Pythagorean relationship, $a^2+b^2 = c^2$. Although this is not an identity, the application of areas is still appropriate. Have students draw a square of side length $(a + b)$. Show students how to partition

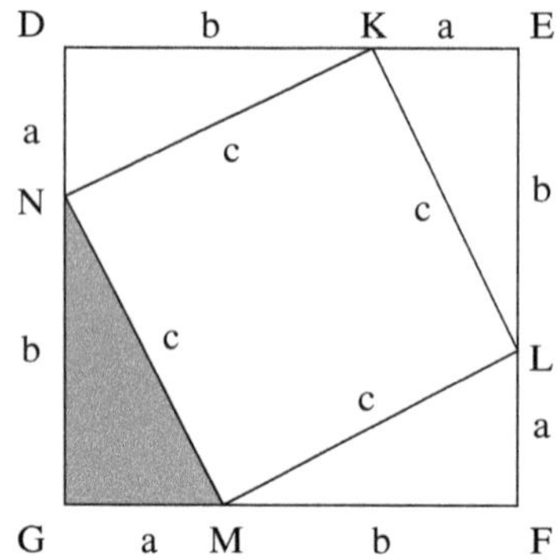

Figure 4.

this square into four congruent triangles and a square (see Figure 4). The lengths of the sides have been labeled.

Figure 4 illustrates:

1. Area DEFG = 4 × (Area of △GNM) + Area KLMN.
2. Therefore, $(a + b)^2 = 4\left(\frac{1}{2}ab\right) + c^2$.
3. If we now substitute the identity for $(a + b)^2$, which was proven before, we obtain, $a^2 + 2ab + b^2 = 2ab + c^2$.

Elicit from students the remainder of the proof to conclude that $a^2+b^2 = c^2$. Students should now be able to pose and solve geometrically their own identities.

Postassessment

1. Have students indicate how to establish the following algebraic identities geometrically.

   a. $(a - b)^2 = a^2 - 2ab + b^2$
   b. $a^2 - b^2 = (a + b) \times (a - b)$.

2. Have students determine other identities that can be proved using the method of application of areas.

Reference

Posamentier, A. S. and H. A. Hauptman. 101+ *Great Ideas for Introducing Key Concepts in Mathematics*. Thousand Oaks, CA: Corwin, 2006.

Unit 77

A Method for Factoring Trinomials of the Form $ax^2 + bx + c$

This unit presents a rather unusual method for factoring, when possible, trinomials of the form $ax^2 + bx + c$, where a, b, and c are integers. This technique is especially helpful when the coefficient a of $ax^2 + bx + c$ is different from 1, because in this case the usual method based on trial and error is rather tedious for most of the trinomials.

Performance Objectives

1. *Given various trinomials of the form $ax^2 + bx + c$, students will analyze and factor them.*
2. *Students will be able to apply this technique to the solution of quadratic equations.*

Preassessment

Students should be familiar with multiplications and factorizations of binomials and with factorizations of perfect square trinomials.

Teaching Strategies

Begin this lesson by giving several examples of multiplications of binomials: $(x + 5)(x + 2)$, $(2x - 3)(x + 1), (5x - 2)(3x - 7)$, etc. Have students notice the following properties of these multiplications:

(a) They always yield trinomials of the form $ax^2 + bx + c$, where a, b, and c are integers.
(b) The product of the first terms of the binomials is the first term of the trinomial.
(c) It is impossible for a to obtain the value of zero from the product of any two binomials. Thus, a is always different from zero in the trinomial of the form $ax^2 + bx + c$.

Once students have practiced these multiplications, have them consider the inverse operation. That is, given trinomials of the form $ax^2 + bx + c$, have students factor them as the product of two binomials. Ask for suggestions on how to factor different trinomials; for example, $x^2+5x+6, 2x^2-7x-4$, and so on. Then have students consider the factorization of the general trinomial ax^2+bx+c in the following fashion:

$$ax^2+bx+c=\frac{a(ax^2+bx+c)}{a}=\frac{a^2x^2+abx+ac}{a}$$

this being possible because a is always different from zero.

If $a^2x^2+abx+ac$ can be factored, one factorization could be $(ax+y)(ax+z)$, where y and z are to be determined. Thus, we have

$$\begin{aligned}ax^2+bx+c&=\frac{a^2x^2+abx+ac}{a}=\frac{(ax+y)(ax+z)}{a}\\&=\frac{a^2x^2+a(y+z)x+yz}{a}.\end{aligned}$$

If the second and fourth equalities are now compared, we notice that $y+z=b$ and $yz=ac$. Thus, to factor a trinomial of the form ax^2+bx+c, it is only necessary to express it as the product $\frac{(ax+y)(ax+z)}{a}$, where y and z can be determined by noticing that their sum must be b and their product must be ac. Also have students notice that because $\frac{(ax+y)(ax+z)}{a}=\frac{a^2x^2+abx+ac}{a}$, it follows that the numerator is a multiple of a, and therefore, it will always be possible to cancel the constant a.

Example 1

Factor $5x^2+8x+3$

We have $5x^2+8x+3=\frac{(5x+y)(5x+z)}{5}$, where $y+z=8$ and $yz=(5)(3)=15$. An analysis of the constant term 15 reveals that the possible pairs of numbers y and z whose product is 15 are 15 and $1, -15$ and $-1, 5$ and 3, and -5 and -3; but because their sum must be 8, the only possible combination of y and z is 5 and 3. Therefore,

$$\begin{aligned}5x^2+8x+3&=\frac{(5x+5)(5x+3)}{5}\\&=\frac{\cancel{5}(x+1)(5x+3)}{\cancel{5}}\\&=(x+1)(5x+3).\end{aligned}$$

Example 2

Factor $6x^2+5x-6$

We have $(6x^2+5x-6)=\frac{(6x+y)(6x+z)}{6}$, where $y+z=5$ and $yz=(6)(-6)=-36$.

An analysis of the product yz, that is, of -36 reveals that the possible pairs of numbers whose product is -36 are 36 and $-1, -36$ and $1, 18$ and $-2, -18$ and $2, 12$ and $-3, -12$ and $3, 9$ and $-4, -9$ and 4, and 6 and -6. But because the algebraic sum of y and z must be $+5$, we have that the only possible combination is 9 and -4. Therefore,

$$\begin{aligned} 6x^2 + 5x - 6 &= \frac{(6x+9)(6x-4)}{6} \\ &= \frac{3(2x+3)2(3x-2)}{6} \\ &= (2x+3)(3x-2). \end{aligned}$$

If $a = 1$, we have the simpler form $x^2 + bx + c$. Thus, $x^2 + bx + c = \frac{(1x+y)(1x+z)}{1} = (x+y)(x+z)$, where $y + z = b$ and $yz = c$.

Example 3

Factor $x^2 - 4x - 5$

We have $x^2 - 4x - 5 = (x+y)(x+z)$, where $y+z$ is -4 and yz is -5. Thus, the possible pairs of numbers are 5 and -1, and -5 and 1; but because the algebraic sum is -4, the only possible combination is -5 and $+1$. Therefore, $x^2 - 4x - 5 = (x-5)(x+1)$. This technique is also applicable to the solution of quadratic equations, that is, equations of the form $ax^2 + bx + c = 0$.

Example 4

Solve $2x^2 - 7x - 4 = 0$

We first factor $2x^2 - 7x - 4$. Thus, $2x^2 - 7x - 4 = \frac{(2x+y)(2x+z)}{2}$, where $y + z$ is -7 and yz is -8. Because the product is -8, we find that the possible pairs are 8 and $-1, -8$ and 1, 4 and -2, and 4 and 2. But because the algebraic sum is -7, the only possible combination is -8 and 1. Thus, $2x^2 - 7x - 4 = \frac{(2x-8)(2x+1)}{2} = (x-4)(2x+1)$. Therefore, we have $2x^2 - 7x - 4 = (x-4)(2x+1) = 0$ and the roots of this quadratic equation will be 4 and $-\frac{1}{2}$. It is important for students to understand that there is no guarantee that "any" given trinomials can be factored; $x^2 - 5x - 7$ and $x^3 - 5x - 6$ *cannot* be factored.

Postassessment

Have students complete the following exercises.

1. Factor the following trinomials:

a. $x^2 - 8x + 12$	b. $4x^2 + 4x - 3$
c. $x^2 + 10x + 25$	d. $3x^2 - 5x$
e. $2r^2 + 13r - 7$	f. $9m^2 - 1$

2. Solve the following quadratic equations:

a. $x^2 - 3x - 4 = 0$ b. $6x^2 + x = 2$

Reference

Posamentier, A. S. and H. A. Hauptman. *101+ Great Ideas for Introducing Key Concepts in Mathematics.* Thousand Oaks, CA: Corwin, 2006.

Unit 78

Solving Quadratic Equations

This unit presents four new methods for solving quadratic equations.

Performance Objective

Students will solve a given quadratic equation in at least four different ways.

Preassessment

Students should be able to solve the equation:

$$x^2 - 7x + 12 = 0.$$

Teaching Strategies

In all likelihood, most of your students solved the above equation by the *Factoring* method. That is, to solve they performed the following operations:

$$x^2 - 7x + 12 = 0$$

$$(x - 3)(x - 4) = 0$$

$$\begin{array}{l|l} x - 3 = 0 & x - 4 = 0 \\ x = 3 & x = 4 \end{array}$$

This method cannot be used to solve all types of quadratic equations. If the trinomial $ax^2 + bx + c$ from the equation $ax^2 + bx + c = 0$ is not factorable, then this method cannot be used to solve the equation.

The rest of this lesson develops four new methods for solving quadratic equations.

Completing the Square

Consider the equation $ax^2 + bx + c = 0$, where a, b, and c are integers and $a \neq 0$.

$$ax^2 + bx + c = x^2 + \frac{b}{a}x + \frac{c}{a} = 0.$$

Add the square of one-half the coefficient of x to both sides:

$$x^2 + \frac{b}{a}x + \left(\frac{b}{2a}\right)^2 = -\frac{c}{a} + \left(\frac{b}{2a}\right)^2$$

$$\left(x + \frac{b}{2a}\right)^2 = -\frac{c}{a} + \frac{b^2}{4a^2}.$$

Take the square root of both sides:

$$x + \frac{b}{2a} = \pm\sqrt{\frac{b^2 - 4ac}{4a^2}}$$

$$x = \frac{-b \pm \sqrt{b^2 - 4ac}}{2a}.$$

This is the quadratic formula.

Example: Solve $x^2 - 7x + 12 = 0$

$$x^2 - 7x + \left(\frac{-7}{2}\right)^2 = -12 + \left(\frac{-7}{2}\right)^2$$

$$\left(x - \frac{7}{2}\right)^2 = -12 + \frac{49}{4} = \frac{-48 + 49}{4} = \frac{1}{4}$$

$$x - \frac{7}{2} = \pm\sqrt{\frac{1}{4}} = \pm\frac{1}{2}$$

$$x = \frac{7}{2} \pm \frac{1}{2} \quad x = 3, 4.$$

Splitting the Difference

Let x_1 and x_2 be the roots of the given equation $ax^2 + bx + c = 0$. Then $x^2 + \frac{b}{a}x + \frac{c}{a} = 0$ or

$$(x - x_1)(x - x_2) = 0.$$

We know that the sum of the roots $x_1 + x_2 = \frac{-b}{a}$ and that the product of the roots $x_1 x_2 = \frac{c}{a}$.

Let $x_1 = \frac{-b}{2a} + \mathrm{N}$, where N is some rational number and $x_2 = \frac{-b}{2a} - \mathrm{N}$.

Then the product of the roots, $\frac{c}{a} = x_1 x_2 = \left(-\frac{b}{2a} + \mathrm{N}\right)\left(-\frac{b}{2a} - \mathrm{N}\right)$.

In solving for N, we get

$$\mathrm{N} = \pm\frac{\sqrt{b^2 - 4ac}}{2a}.$$

Therefore, the roots are $x = \frac{-b}{2a} \pm \frac{\sqrt{b^2-4ac}}{2a}$.

Example: Solve $x^2 - 7x + 12 = 0$

Students will establish that the sum of the roots $x_1 + x_2 = 7$.

Therefore, one root must be $\frac{7}{2} + \mathrm{N}$ and the other must be $\frac{7}{2} - \mathrm{N}$, where N is some rational number.

Because the product of the roots is $12, x_1 x_2 = \left(\frac{7}{2} + \mathrm{N}\right)\left(\frac{7}{2} - \mathrm{N}\right) = \frac{49}{4} - \mathrm{N}^2 = 12$.

Therefore, $\mathrm{N}^2 = \frac{1}{4}$, and $\mathrm{N} = \pm\frac{1}{2}$.

$$\text{Thus, the roots are } x_1 = \frac{7}{2} + \mathrm{N} = \frac{7}{2} + \frac{1}{2} = 4$$

$$x_2 = \frac{7}{2} - \mathrm{N} = \frac{7}{2} - \frac{1}{2} = 3.$$

Method of Simultaneous Equations

Rather than developing the general case first, we shall first solve the given equation $x^2 - 7x + 12 = 0$. This order should be easier to follow for this method.

Example: Solve $x^2 - 7x + 12 = 0$

Consider the sum and product of the roots $x_1 + x_2 = 7$ and $x_1 x_2 = 12$.

Square the sum $(x_1 + x_2)^2 = 49$.

Multiply the product by -4: $-4x_1 x_2 = -48$.

By addition $(x_1 + x_2)^2 - 4x_1 x_2 = 49 - 48 = 1$.

However, the left side simplifies to $(x_1 - x_2)^2$.
Therefore, $x_1 - x_2 = \pm\sqrt{1} = \pm 1$.
Remember that $x_1 + x_2 = 7$.
Now solving these equations simultaneously:

$$2x_1 = 8 \quad x_1 = 4 \quad x_2 = 3.$$

The general case for $ax^2 + bx + c = 0$ follows: The square of the sum of the roots

$$(x_1 + x_2)^2 = x_1^2 + 2x_1x_2 + x_2^2 = \frac{b^2}{a^2}.$$

The product of the roots and -4 is

$$-4x_1x_2 = \frac{-4c}{a}.$$

As above, we add these last two equations:

$$x_1^2 - 2x_1x_2 + x_2^2 = \frac{b^2}{a^2} - \frac{4c}{a}$$

$$(x_1 - x_2)^2 = \frac{b^2 - 4ac}{a^2}$$

Therefore,

$$x_1 - x_2 = \pm\frac{\sqrt{b^2 - 4ac}}{a}.$$

Since $x_1 + x_2 = \frac{-b}{a}$,

$$x_1 \text{ or } x_2 = \frac{1}{2}\left(\frac{-b}{a} \perp \frac{\sqrt{b^2 - 4ac}}{a}\right)$$

$$= \frac{-b \pm \sqrt{b^2 - 4ac}}{2a}.$$

Method of Root Reduction

Again we start the discussion with the solution for a specific equation before considering the general form.

Example: Solve $x^2 - 7x + 12 = 0$

Let $r = x - n$, then $x = r + n$, and $x^2 = (r + n)^2 = r^2 + 2rn + n^2$.
We now substitute the appropriate values in the original equation.

$$(r^2 + 2rn + n^2) - 7(r + n) + 12 = 0$$

$$r^2 + r(2n - 7) + (n^2 - 7n + 12) = 0.$$

If $2n - 7 = 0$, then the r term is annihilated.

This will happen when $n = \frac{7}{2}$.
We then have $r^2 + (n^2 - 7n + 12) = 0$ or by substituting

$$n = \frac{7}{2}: \quad r^2 + \left(\frac{49}{4} - 7\left(\frac{7}{2}\right) + 12\right) = 0$$

$$r^2 = \frac{49}{4} - 12 = \frac{1}{4}; \quad \text{and} \quad r = \pm\frac{1}{2}.$$

Thus, the roots $(x = r + n)$ are

$$x_1 = +\frac{1}{2} + \frac{7}{2} = 4; \quad x_2 = -\frac{1}{2} + \frac{7}{2} = 3.$$

The general case proceeds in a similar manner. Consider the equation $ax^2 + bx + c = 0$. Let $r = x - n$, then $x = r + n$ and

$$x^2 = (r + n)^2 = r^2 + 2rn + n^2.$$

Now substitute these values into the original equation:

$$x^2 + \frac{bx}{a} + \frac{c}{a} = 0$$

$$(r^2 + 2rn + n^2) + \frac{b}{a}(r + n) + \frac{c}{a} = 0$$

$$\text{or} \quad r^2 + r\left(2n + \frac{b}{a}\right) + \left(n^2 + \frac{bn}{a} + \frac{c}{a}\right) = 0.$$

To annihilate the r term, we let

$$2n + \frac{b}{a} = 0, \quad \text{or} \quad n = \frac{-b}{2a}.$$

This then gives us

$$r^2 + \left(n^2 + \frac{b}{a}n + \frac{c}{a}\right) = 0$$

$$\text{or} \quad r^2 = -\left(n^2 + \frac{b}{a}n + \frac{c}{a}\right).$$

However, because $n = \frac{-b}{2a}$

$$r^2 = -\left(\frac{b^2}{4a^2} - \frac{b^2}{2a^2} + \frac{c}{a}\right)$$

$$r^2 = \frac{b^2 - 4ac}{4a^2} \quad \text{and} \quad r = \pm\frac{\sqrt{b^2 - 4ac}}{2a}.$$

Therefore, since $x = r + n$, $x = \frac{-b}{2a} \pm \frac{\sqrt{b^2-4ac}}{2a}$ or $x = \frac{-b\pm\sqrt{b^2-4ac}}{2a}$.

Although some of these methods for solving quadratic equations are not too practical, they do offer students a better understanding of many of the underlying concepts.

Postassessment

Ask students to use at least four of the methods presented in this lesson to solve the following equations.

1. $x^2 - 11x + 30 = 0$
2. $x^2 + 3x - 28 = 0$
3. $6x^2 - x - 2 = 0$

Unit 79

The Euclidean Algorithm

This unit presents a method of introducing students to the Euclidean Algorithm for finding the greatest common divisor of two given integers.

Performance Objectives

1. *Given any two integers, students will determine the greatest common divisor of the two integers, regardless of the magnitude of the two integers.*
2. *Having determined the greatest common divisor, the students will then be able to express the greatest common divisor in terms of the two integers.*

Preassessmcnt

Ask students how they would weigh 12 ounces, 2 ounces, 3 ounces, 4 ounces, 1 ounce, and 11 ounces using only a set of two pan balance scales and some 5 and 7 ounce weights.

Teaching Strategies

Students should be able to suggest weighing the weights in the following manner:

1. 12 ounces: Place one 5 oz. and one 7 oz. weight on the same pan, and the 12 ounces can be weighed on the other pan.
2. 2 ounces: Place one 7 oz. weight on one pan and a 5 oz. weight on the other pan. Then the desired 2 oz. weight is that which must be placed on the pan containing the 5 oz. weight in order to balance the scales.

3. 3 ounces: Place two 5 oz. weights on one pan and a 7 oz. weight on the other pan. The desired 3 oz. weight is that which must be added to the 7 oz. weight in order to balance the scales.
4. 4 ounces: Place two 5 oz. weights on one pan and two 7 oz. weights on the other. The desired weight is that which must be added to the two 5 oz. weights in order to balance the scales.
5. 1 ounce: Place three 5 oz. weights on one pan and two 7 oz. weights on the other. The desired weight is that which must be added to the two 7 oz. weights in order to balance the scales.
6. 11 ounces: Place five 5 oz. weights on one pan and two 7 oz. weights on the other pan. The desired weight is that which must be added to the two 7 oz. weights in order to balance the scales.

Students should then be asked to weigh 1 ounce, 2 ounces, 3 ounces, and 4 ounces using other combinations of given weights. They should soon be able to discover that the smallest weight that can be weighed using any combination of given weights is equal to the *greatest common divisor* of the two weights:

Given weights	G.C.D.	Minimum weighable
2 and 3	1	1
2 and 4	2	2
3 and 9	3	3
8 and 20	4	4
15 and 25	5	5

The greatest common divisor of A and B will be referred to as either G.C.D. of A and B or (A, B).

To find $(945, 219)$, we can use the *Euclidean Algorithm.* The Euclidean Algorithm is based on a lemma, which states: A and B are integers where A does not equal zero. If B is divided by A, quotient Q and remainder R is obtained ($\mathrm{B} = \mathrm{QA}+\mathrm{R}$), then $(\mathrm{B}, \mathrm{A}) = (\mathrm{A}, \mathrm{R})$. Using the following procedure, the G.C.D. of 945 and 219 can be found:

$$\text{Divide 945 by 219: } 945 = (4)(219) + 69 \tag{1}$$

$$\text{Divide 219 by 69: } 219 = (3)(69) + 12 \tag{2}$$

now continue this process

$$69 = (5)(12) + 9 \tag{3}$$

$$12 = (1)(9) + 3 \tag{4}$$

$$9 = (3)(3) + 0 \ldots \text{until R}$$

$$\text{equals } 0.$$

Therefore, the G.C.D. of 945 and 219 is 3, which was the last nonzero remainder in the successive divisions. This method may be used to find (A, B) where A and B are any two integers. Have students practice this algorithm with some exercise before continuing the lesson.

For stronger students in the class (or just for your interest), a proof of this algorithm is provided. The following is a statement and proof of the Euclidean Algorithm.

For given nonzero integers a and b, divide a by b to get remainder r_1: divide b by r_1 to get remainder r_2. This is continued so that when remainder r_k is divided by r_{k+1} the remainder r_{k+2} is obtained. Eventually there will be an r_n such that $r_{n+1} = 0$. It follows that $|r_n|$ is the greatest common divisor of a and b.

Proof: The division algorithm will determine integers $q_1, r_1, q_2, r_2, q_3, r_3, \cdots,$ where

$$\begin{aligned} a &= q_1 b + r_1 \\ b &= q_2 r_1 + r_2 \\ r_1 &= q_3 r_2 + r_3 \\ &\vdots \end{aligned}$$

and where $0 \leq \cdots < r_3 < r_2 < r_1 < |b|$. There are only $|b|$ nonnegative integers less than $|b|$.

Therefore, there must be an $r_{n+1} = 0$ for $n + 1 \leq |b|$. If $r_1 = 0$, then $(a, b) = b$. If $r_1 \neq 0$, then

$$\begin{aligned} a &= q_1 b + r_1 \\ b &= q_2 r_2 + r_2 \\ r_1 &= q_3 r_2 + r_3 \\ &\vdots \\ r_{n-2} &= q_n r_{n-1} + r_n \\ r_{n-1} &= q_{n+1} r_n. \end{aligned}$$

Let $d = (a, b)$. Since $d|a$, and $d|b$, then $d|r_1$. Similarly, since $d|b$ and $d|r_1$, then $d|r_2$. Again, since $d|r_1$ and $d|r_2$, then $d|r_3$. Continuing this reasoning eventually yields $d|r_{n-2}$, and $d|r_{n-1}$, then $d|r_n$.

Since $r_n \neq 0, r_n|r_{n-1}$. Also $r_n|r_n$; therefore $r_n|r_{n-2}$. Similarly, $r_n|r_{n-3}$, $r_n|r_{n-4}, \cdots r_n|r_2, r_n|r_1, r_n|b$, and $r_n|a$. Since $r_n|a$ and $r_n|b$, therefore $r_n|d$. Thus, $r_n|d$ and $d|r_n$, it follows that $r_n = d$, or $r_n = (a, b)$.

At this juncture, it would be nice to be able to express the G.C.D. of two integers in terms of the two integers, that is, $\mathrm{MA} + \mathrm{NB} = (\mathrm{A}, \mathrm{B})$, where M and N are integers. In the earlier case of $(945, 219), 3 = \mathrm{M}(219) + \mathrm{N}(945)$. By working backwards ("up" the Euclidean Algorithm), we can accomplish the following:

From line (4) above: $3 = 12 - 9$.

Substituting for 9 from line (3) above:

$$3 = 12 - (69 - 5 \cdot 12), \quad 3 = 6 \cdot 12 - 69.$$

Substituting for 12 from line (2):

$$3 = 6(219 - 3 \cdot 69) - 69.$$

Substituting for 69 from line (1):

$$3 = 6 \cdot 219 - 19(945 - 4 \cdot 219), \text{ or}$$

$$3 = 82(219) - 19(945).$$

Earlier students have determined the minimum that can be weighed by a 945 oz. and a 219 oz. weight by finding $(945, 219)$. Now they can also determine how many 945 oz. weights to place on one pan and how many 219 oz. to place on the other pan, by expressing $(945, 219)$ in terms of 945 and 219. That is, they must place 82 quantities of the 219 oz. weights on one pan and 19 quantities of the 945 oz. weights on the other pan. The desired weight is that which must be added to the 19 quantities of the 945 oz. weights in order to balance the weights. This scheme may be used to develop an understanding of Diophantine Equations.

Postassessment

Students should be able to compute the G.C.D. of the following pairs of integers and express the G.C.D. in terms of the two integers:

1. 12 and 18
2. 52 and 86
3. 865 and 312
4. 120 and 380

Reference

Posamentier, A. S. and I. Lehmann. *Mathematical Curiosities: A Treasure Trove of Unexpected Entertainments.* Amherst, NY: Prometheus Books, 2014.

Unit 80

Prime Numbers

This unit will introduce students to fascinating facts concerning prime numbers.

Performance Objectives

1. *Given a number, students will use Euler's ϕ function to find the number of positive integers less than the given number that are relatively prime to it.*
2. *Students will explain why it is not possible for a polynomial with integral coefficients to exist that will generate only primes.*

Preassessment

Ask students to identify which of the following are prime numbers:

(a) 11 (b) 27 (c) 51
(d) 47 (e) 91 (f) 1

Teaching Strategies

Mathematicians have spent years trying to find a general formula that would generate primes. There have been many attempts, but none have succeeded.

Have students test the expression $n^2 - n + 41$ by substituting various positive values for n. Make a chart on the chalkboard recording their findings. As they proceed, they should begin to notice that as n ranges in value from 1 through 40, only prime numbers are being produced. (If they have not substituted $n = 40$, have them do so.) Then ask them to try $n = 41$. The value of $n^2 - n + 41$ is $(41)^2 - 41 + 41 = (41)^2$, which is not prime. A similar expression, $n^2 - 79n + 1601$, produces primes for all values of n up to 80. But for $n = 81$, we have $(81)^2 - 79 \cdot 81 + 1601 = 1763 = 41 \cdot 43$, which is not a prime. Students might now wonder if it is possible to have a polynomial in n with integral coefficients whose values would be primes for every positive integer n. Advise them not to try to find such an expression; Leonhard Euler (1707–1783) proved that none can exist. Euler showed that any proposed expression will produce at least one nonprime.

Euler's proof follows. First, assume that such an expression exists, being in the general form: $a + bx + cx^2 + dx^3 + \cdots$ (understanding that some of the coefficients may be zero). Let the value of this expression be s when $x = m$. Therefore, $s = a + bm + cm^2 + dm^3 + \cdots$. Similarly, let t be the value of the expression when $x = m+ns$: $t = a+b(m+ns)+c(m+ns)^2+d(m+ns)^3 \cdots$. This may be transformed to

$$t = (a + bm + cm^2 + dm^3 + \cdots) + \mathrm{A},$$

where A represents the remaining terms all of which are multiples of s. But the expression within the parentheses is, by hypothesis, equal to s. This makes the whole expression a multiple of s, and the number produced is not a prime. Every such expression will produce at least one prime, but not necessarily more than one. Consequently, no expression can generate primes exclusively.

Although this last statement was recognized early in mathematical history, mathematicians continued to conjecture about forms of numbers that generated only primes.

Pierre de Fermat (1601–1665), who made many significant contributions to the study of number theory, conjectured that all numbers of the form $F_n = 2^{2^n} + 1$, where $n = 0, 1, 2, 3, 4, \ldots$ were prime numbers. Have students find F_n for $n = 0, 1, 2$. They will see that the first three numbers derived from this expression are $3, 5$, and 17. For $n = 3$, students will find that $F_n = 257$; by telling them $\mathrm{F}_4 = 65{,}537$, they should notice these numbers are increasing at a very rapid rate. For $n = 5, \mathrm{F}_n =$ 4,294,967,297, and Fermat could not find any factor of this number. Encouraged by his results, he expressed the opinion that all numbers of this form are probably also prime. Unfortunately he stopped too soon, for in 1732, Euler showed that $\mathrm{F}_5 = 4{,}294{,}967{,}297 = 641 \times 6{,}700{,}417$ (not a prime!). It was not until 150 years later that the factors of F_6 were found: $18{,}446{,}744{,}073{,}709{,}551{,}617 = 247{,}177 \times 67{,}280{,}421{,}310{,}721$. As far as is now known, many more numbers of this form have been found but *none* of them have been prime. It seems that Fermat's conjecture has been completely turned around, and one now wonders if any primes beyond F_4 exist.

Euler also continued further into his study of primes. He began examining those integers that are *relatively prime* (two integers are relatively prime if they have no common positive factor except 1). Have students write down the number 12 and the positive integers less than 12. Tell them to cross out 12 itself and then all the integers that have a factor greater than 1 in common with 12.

$$1 \quad \not{2} \quad \not{3} \quad \not{4} \quad 5 \quad \not{6} \quad 7 \quad \not{8} \quad \not{9} \quad \not{1}0 \quad 11 \quad \not{1}2$$

They will see that $1, 5, 7, 11$ are the only integers remaining. Therefore, there are four positive integers less than 12 that are relatively prime to it. The number of such integers is denoted by $\phi(n)$ and is known as Euler's ϕ function. For $n = 1$, we have $\phi(n) = 1$. For $n > 1, \phi(n)$ = the number of positive integers less than n and relatively prime to it. Thus, as we just saw, $\phi(12) = 4$.

Let students now find values of $\phi(n)$ for $n = 1, 2, 3, 4, 5$. A chart, such as Figure 1 is convenient for this.

n	integers relatively prime to and less than *n*	$\phi(n)$
1		1
2	1 2	1
3	1 2 3	2
4	1 2 3 4	2
5	1 2 3 4 5	4
6	1 2 3 4 5 6	2
7	prime	6
8	1 2 3 4 5 6 7 8	4
9	1 2 3 4 5 6 7 8 9	6
10	1 2 3 4 5 6 7 8 9 10	4
11	prime	10
12	1 2 3 4 5 6 7 8 9 10 11 12	4

Figure 1.

Students should notice that when n is prime, it is not necessary to list all the numbers. Since a prime is relatively prime to all positive integers less than it, we therefore have $\phi(n) = n - 1$, for n a prime.

Have students continue to find $\phi(n)$ for $n = 6$ through 12. Looking down the $\phi(n)$ column, it does not seem that any particular pattern is emerging. We would like to, though, obtain an expression for the general term, so that $\phi(n)$ can be calculated for any number. We have already stated that if n is prime, then $\phi(n) = n - 1$. To discover an expression for $\phi(n)$ if n is not prime, we will look at a particular case. Let $n = 15$. Decomposing 15 into primes, we obtain $15 = 3 \cdot 5$. We can write this as $n = p \cdot q$, where $n = 15, p = 3$, and $q = 5$. Next, have students write down 15 and all positive integers less than 15. Have them cross out all integers having 3 (which is p) as a factor:

$$1 \quad 2 \quad \not{3} \quad 4 \quad 5 \quad \not{6} \quad 7 \quad 8 \quad \not{9} \quad 10 \quad 11 \quad 1\not{2} \quad 13 \quad 14 \quad \not{1}5$$

They will see that there are 5 or $5 = \frac{15}{3} = \frac{n}{p}$ of these.

There are 10 numbers remaining or $10 = 15 - \frac{15}{3} = n - \frac{n}{p} = n\left(1 - \frac{1}{p}\right)$.

From these 10 integers, have students cross out those having 5 (which is q) as a factor:

$$1 \quad 2 \quad 4 \quad \not{5} \quad 7 \quad 8 \quad 1\not{0} \quad 11 \quad 13 \quad 14.$$

There are only 2 of these or $2 = \frac{1}{5}(10) = \frac{1}{q}\left[n\left(1 - \frac{1}{p}\right)\right]$.

There are now 8 numbers left:
$8 = 10 - \frac{1}{5}(10) = n\left(1 - \frac{1}{p}\right) - \frac{1}{q}\left[n\left(1 - \quad \frac{1}{p}\right)\right] \cdot n\left(1 - \frac{1}{p}\right)$ is a factor of both terms of the expression. We have therefore established a formula for the number of positive integers less than n and relatively prime to it:

$$\phi(n) = n\left(1 - \frac{1}{p}\right)\left(1 - \frac{1}{q}\right).$$

The number n, though, might have more than 2 factors in its prime decomposition, so let us now state a more general formula (given without proof). Let the number n be decomposed into its prime factors $p, q, r, \ldots, w$. Then $n = p^a \cdot q^b \cdot r^c \cdots w^h$, where $a, b, c, \ldots, h$ are positive integers (which may or may not be all 1s). Then $\phi(n) = n\left(1 - \frac{1}{p}\right)\left(1 - \frac{1}{q}\right)\left(1 - \frac{1}{r}\right)\cdots\left(1 - \frac{1}{w}\right)$. The teacher should also show the students that if n is a prime, the formula still holds since $\phi(n) = n - 1 = n\left(\frac{n-1}{n}\right) = n\left(1 - \frac{1}{n}\right)$. To see how the formula works, work together with students in finding the following: $\phi(21), \phi(43), \phi(78)$.

Solutions: $\phi(21) = \phi(7 \cdot 3) = 21\left(1 - \frac{1}{7}\right)\left(1 - \frac{1}{3}\right) = 21.$

$$\left(\frac{6}{7}\right)\left(\frac{2}{3}\right) = 12$$

$$\phi(43) = 43 - 1 \text{ (since 43 is a prime)} = 42$$

$$\phi(78) = \phi(2 \cdot 3 \cdot 13) = 78\left(1 - \frac{1}{2}\right)\left(1 - \frac{1}{3}\right)$$

$$\left(1 - \frac{1}{13}\right) = 78\left(\frac{1}{2}\right)\left(\frac{2}{3}\right)\left(\frac{12}{13}\right) = 24.$$

At this point, some students might have noticed that every value of $\phi(n)$ is even. Justification of this may serve as a springboard for further investigation.

Postassessment

1. Find each of the following: a. $\phi(13)$ b. $\phi(14)$ c. $\phi(48)$ d. $\phi(73)$ e. $\phi(100)$.
2. Have students explain why there is no polynomial with integral coefficients that will generate only primes.

Reference

Posamentier, A. S. and I. Lehmann. *Mathematical Curiosities: A Treasure Trove of Unexpected Entertainments.* Amherst, NY: Prometheus Books, 2014.

Unit 81

Algebraic Fallacies

All too often students make errors in their mathematics work that are subtle than an error in computation or some other careless act. To prevent errors that are the results of violations of mathematical definitions of concepts, it would be wise to exhibit such flaws beforehand. This is the main mission of this unit.

Performance Objective

Given an algebraic fallacy, students will analyze and determine where the fallacy occurs in the algebraic "proof."

Preassessment

Students should be familiar with the basic algebraic operations normally taught in the high school elementary algebra course.

Teaching Strategy

When the theory behind mathematical operations is poorly understood, there exists the possibility that the operations will be applied in a formal and perhaps illogical way. Students, not aware of certain limitations on these operations, are likely to use them where they do not necessarily apply. Such improper reasoning leads to an absurd result called a fallacy. The following paradoxes will illustrate how such fallacies can arise in algebra when certain algebraic operations are performed without realizing the limitations on those operations.

Almost everyone who has been exposed to elementary algebra will come across, at one time or another, a proof that 2 = 1 or 1 = 3, etc. Such a "proof" is an example of a fallacy.

"Proof":

1.	Let	$a = b$
2.	Multiply both sides by a:	$a^2 = ab$
3.	Subtract b^2 from both sides:	$a^2 - b^2 = ab - b^2$
4.	Factoring:	$(a+b)(a-b) = b(a-b)$
5.	Dividing both sides by $(a-b)$:	$(a+b) = b$
6.	Since $a = b$, then	$2b = b$
7.	Dividing both sides by b:	$2 = 1$

Ask the students to analyze the "proof" and find out where the reasoning breaks down. Of course, the trouble is in the fifth step. Since $a = b$, then $a - b = 0$. Therefore, division by zero was performed, which is *not permissible*. It would be appropriate at this time to discuss what division means in terms of multiplication. To divide a by b implies that there exists a number y such that $b \cdot y = a$, or $y = \frac{a}{b}$. If $b = 0$, there are two possibilities, either $a \neq 0$ or $a = 0$. If $a \neq 0$, then $y = \frac{a}{0}$ or $0 \cdot y = a$. Ask your students if they can find a number that when multiplied by zero will equal a. Your students should conclude that there is no such number y. In the second case, where $a = 0, y = \frac{0}{0}$ or $0 \cdot y = 0$.

Here any number for y will satisfy the equation, hence any number multiplied by zero is zero. Therefore, we have the "rule" that division by zero is not permissible. There are other fallacies based on division by zero. Have your students discover for themselves where and how the difficulty occurs in each of the following examples.

1. To "prove" that any two unequal numbers are equal. Assume that $x = y + z$, and x, y, z are positive numbers. This implies $x > y$. Multiply both sides by $x - y$. Then $x^2 - xy = xy + xz - y^2 - yz$. Subtract xz from both sides:

$$x^2 - xy - xz = xy - y^2 - yz.$$

Factoring, we get $x(x-y-z) = y(x-y-z)$.
Dividing both sides by $(x-y-z)$ yields $x = y$. Thus x, which was assumed to be greater than y, has been shown to be equal y. The fallacy occurs in the division by $(x-y-z)$, which is equal to zero.

2. To "prove" that all positive whole numbers are equal. By doing long division, we have, for any value of x

$$\frac{x-1}{x-1} = 1$$
$$\frac{x^2-1}{x-1} = x+1$$
$$\frac{x^3-1}{x-1} = x^2+x+1$$
$$\frac{x^4-1}{x-1} = x^3+x^2+x+1$$
$$\vdots$$
$$\frac{x^n-1}{x-1} = x^{n-1}+x^{n-2}+\cdots+x^2+x+1.$$

Letting $x = 1$ in all of these identities, the right side then assumes the values $1, 2, 3, 4, \ldots, n$. The left side members are all the same. Consequently, $1 = 2 = 3 = 4 = \cdots = n$. In this example, the left-hand side of each of the identities assumes the value $\frac{0}{0}$ when $x = 1$. This problem serves as evidence that $\frac{0}{0}$ can be any number.

Consider the following, and ask your students if they would agree with the following statement, "If two fractions are equal and have equal numerators, then they also have equal denominators." Let the students give illustrations using any fractions they choose. Then have them solve the following equation:

$$6 + \frac{8x-40}{4-x} = \frac{2x-16}{12-x}. \tag{1}$$

Add terms on the left-hand side, to get

$$\frac{6(4-x)+8x-40}{4-x} = \frac{2x-16}{12-x}. \tag{2}$$

Simplifying: $\dfrac{2x-16}{4-x} = \dfrac{2x-16}{12-x}$

Since the numerators are equal, this implies $4 - x = 12 - x$. Adding x to both sides, $4 = 12$. Again, as in some of the previous examples, the division by zero is disguised. Have students find the error. Point out that the axioms cannot be blindly applied to equations without considering the values of the variables for which the equations are true. Thus, Equation (1) is not an identity true for all values of x, but it is satisfied only by $x = 8$. Have students solve $(12 - x)(2x - 16) = (4 - x) \times (2x - 16)$ to verify this. Thus $x = 8$ implies that the numerators are zero. You may also have the students prove the general case for $\frac{a}{b} = \frac{a}{c}$, to show that a cannot be zero.

Another class of fallacies includes those that neglect to consider that a quantity has two square roots of equal absolute value; however, one is positive and the other is negative. As an example, take the equation $16-48 = 64-96$. Adding 36 to both sides $16 - 48 + 36 = 64 - 96 + 36$. Each member of the equation is now a perfect square, so that $(4 - 6)^2 = (8 - 6)^2$. Taking the square root of both sides, we get $4 - 6 = 8 - 6$, which implies $4 = 8$. Ask the students where the fallacy occurs. The fallacy in this example lies in taking the improper square root. The correct answer should be $(4 - 6) = -(8 - 6)$.

The following fallacies are based on the failure to consider all the roots of a given example.

Have students solve the equation $x + 2\sqrt{x} = 3$ in the usual manner. The solutions are $x = 1$ and $x = 9$. The first solution satisfies the equation, while the second solution does not. Have students explain where the difficulty lies.

A similar equation is $x - a = \sqrt{x^2 + a^2}$. By squaring both sides and simplifying, we get $-2ax = 0$, or $x = 0$. Substituting $x = 0$ in the original equation, we find that this value of x does not satisfy the equation. Have the students find the correct root of the given equation.

So far we have dealt with square roots of positive numbers. Ask the students what happens when we apply our usual rules to radicals containing imaginary numbers, in light of the following problem. The students have learned that $\sqrt{a} \cdot \sqrt{b} = \sqrt{ab}$, for example, $\sqrt{2} \cdot \sqrt{5} = \sqrt{2 \cdot 5} = \sqrt{10}$. But this gives then, $\sqrt{-1} \cdot \sqrt{-1} = \sqrt{(-1)(-1)} = \sqrt{1} = 1$. However, $\sqrt{-1} \cdot \sqrt{-1} = (\sqrt{-1})^2 = -1$. It therefore may be concluded that $1 = -1$, since both equal $\sqrt{-1} \times \sqrt{-1}$. Students should try to explain the error. They should realize that we cannot apply the ordinary rules for multiplication of radicals to imaginary numbers.

Another proof that can be used to show $-1 = +1$ is the following:

$$\sqrt{-1} = \sqrt{-1}$$
$$\sqrt{\frac{1}{-1}} = \sqrt{\frac{-1}{1}}$$
$$\frac{\sqrt{1}}{\sqrt{-1}} = \frac{\sqrt{-1}}{\sqrt{1}}$$
$$\sqrt{1} \cdot \sqrt{1} = \sqrt{-1} \cdot \sqrt{-1}$$
$$1 = -1.$$

Have students replace i for $\sqrt{-1}$, and -1 for i^2 to see where the flaw occurs.

Before concluding the topic on algebraic fallacies, it would be appropriate to consider a fallacy involving simultaneous equations. The students, by now, should realize that in doing the preceding proofs a certain law or operation was violated. Consider an example where hidden flaws in equations can bring about ludicrous results. Have students solve the following pairs of equations by substituting for x in the first equation:

$$2x + y = 8 \text{ and } x = 2 - \frac{y}{2}. \text{ The result will be } 4 = 8.$$

Have students find the error. When students graph these two equations, they will find the two lines to be parallel and therefore have *no* points in common.

Further exhibition of such fallacies will prove a worthwhile activity due to the intrinsically dramatic message they carry.

Postassessment

Have students determine where and how the fallacy occurs in the following examples.

1. a. $x = 4$
   b. $x^2 = 16$
   c. $x^2 - 4x = 16 - 4x$
   d. $x(x - 4) = 4(4 - x)$
   e. $x(x - 4) = -4(x - 4)$
   f. $x = -4$
2. a. $(y + 1)^2 = y^2 + 2y + 1$
   b. $(y + 1)^2 - (2y + 1) = y^2$
   c. $(y + 1)^2 - (2y + 1) - y(2y + 1) = y^2 - y(2y + 1)$

d. $(y+1)^2 - (y+1)(2y+1) + \frac{1}{4}(2y+1)^2$
$= y^2 - y(2y+1) + \frac{1}{4}(2y+1)^2$

e. $\left[(y+1) - \frac{1}{2}(2y+1)\right]^2 = \left[y - \frac{1}{2}(2y+1)\right]^2$

f. $y+1-\frac{1}{2}(2y+1) = y - \frac{1}{2}(2y+1)$

g. $y+1=y$

Other fallacies/paradoxes can be found in: *Math Wonders to Inspire Teachers and Students* by A. S. Posamentier (Association for Supervision and Curriculum Development, 2007).

References

A. S. Posamentier and I. Lehmann. *Magnificent Mistakes in Mathematics*, Amherst, New York: Prometheus Books, 2013.

Unit 82

Sum Derivations with Arrays

Performance Objectives

1. *Students will derive the formula for the sum of the first* n *natural numbers, triangular numbers, square numbers, or pentagonal numbers.*
2. *Given any integral value of* n, *students will apply the proper formula to find the sum of the first* n *figurate numbers.*

Preassessment

Before beginning this lesson, be sure students are familiar with the meanings of *figurate numbers* and *formation of sequences of figurate numbers.* They should also have some knowledge of elementary algebra.

Teaching Strategies

To begin to familiarize students with this topic, have them construct dot arrays on graph paper to illustrate the first few terms in the sequences of various figurate numbers.

Discuss with the class the visual relationships. Most students will clearly see that we can represent the sum of the first n natural numbers as follows:

$$
\begin{aligned}
\mathrm{N}_n &= \text{Sum of the first } n \text{ natural numbers} \\
&= 1 + 2 + 3 + \cdots + n \\
&= 1 + (1+1) + (1+1+1) + \cdots \\
&\quad + \underbrace{(1 + 1 + \cdots + 1)}_{n}
\end{aligned}
$$

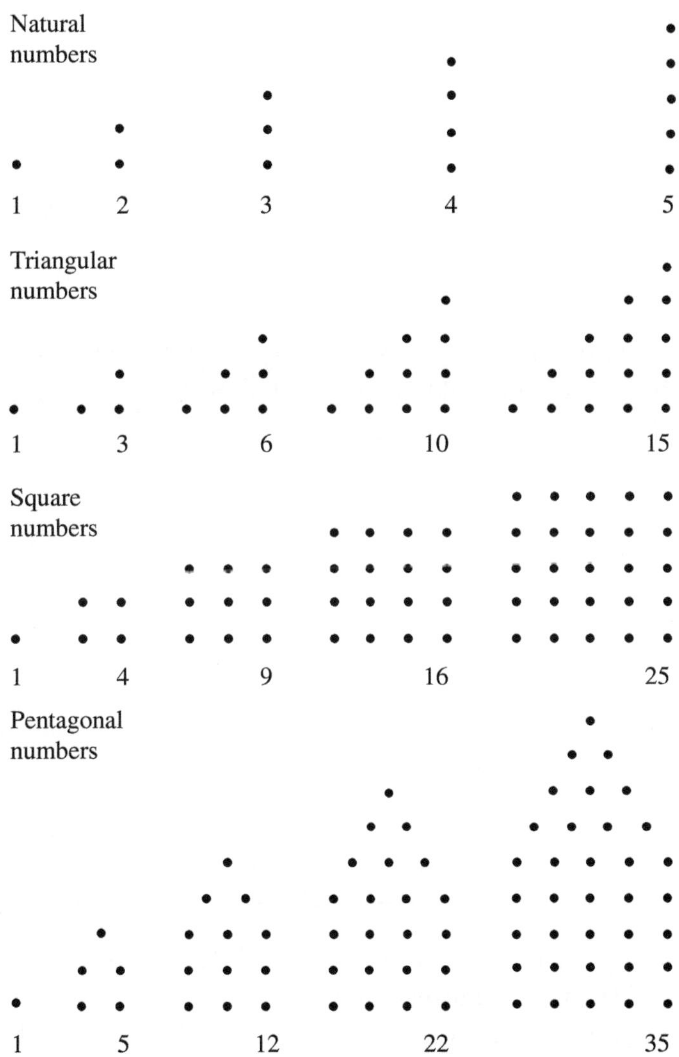

N_n can also be represented as the sum of the numbers in an array:

```
                    n
N_n=   1 1 1 1 • • • 1
         1 1 1 • • • 1
           1 1 • • • 1   n
             1 • • • 1
               •     •
                 •   •
                   • •
                     1
```

By switching the rows with the columns, N_n can be made to look slightly different:

```
N_n=   1
       1 1
       1 1 1
       1 1 1 1
    n  • • • • •
       • • • •   •
       • • • •     •
       1 1 1 1 • • • 1
              n
```

These two representations of N_n in array form can now be combined to produce an array for $2N_n$ as shown in the following:

```
2N_n=  1 1 1 1 • • • 1
       1 1 1 1 • • • 1
       1 1 1 1 • • • 1
       1 1 1 1 • • • 1
       1 1 1 1 •     •   (n+1)
       • • • • • •   •
       • • • •   • • •
       • • • •     • 1
       1 1 1 1 • • • 1
             n
```

Students can distinctly see that $2N_n = n(n+1)$ from an inspection of the array for $2N_n$. We therefore have

$$\boxed{N_n = \frac{n(n+1)}{2}}$$

This resulting formula for N_n can be applied whenever it is required to find the sum of the first n natural numbers.

Have students consider trying to derive a formula for the first n triangular numbers. Clearly, from the dot arrays presented earlier, the following can be established:

$$\begin{aligned}
\mathrm{T}_n &= \text{Sum of the first } n \text{ triangular numbers}\\
&= 1 + 3 + 6 + \cdots + \mathrm{N}_n\\
&= 1 + (1+2) + (1+2+3) + \cdots + (1+2+3+\cdots+n)
\end{aligned}$$

T_n can now be represented as the sum of the numbers in an array:

$$\mathrm{T}_n = \begin{array}{cccccccccc}
 & 1 & & & & & & & \\
 & 1 & 2 & & & & & & \\
 & 1 & 2 & 3 & & & & & \\
n & 1 & 2 & 3 & 4 & & & & \\
 & \cdot & \cdot & \cdot & \cdot & \cdot & & & \\
 & \cdot & \cdot & \cdot & \cdot & & \cdot & & \\
 & \cdot & \cdot & \cdot & \cdot & & & \cdot & \\
 & 1 & 2 & 3 & 4 & \cdot & \cdot & \cdot & n \\
 & & & & n & & & &
\end{array}$$

By applying the previously determined formula for N to each row of this array, we obtain

$$\begin{aligned}
\mathrm{T}_n = \frac{1(1+1)}{2}
&+ \frac{2(1+2)}{2}\\
&+ \frac{3(1+3)}{2}\\
&+ \frac{4(1+4)}{2}\\
&+ \cdots\\
&+ \frac{n(1+n)}{2}
\end{aligned}$$

Students should now see that

$$2\mathrm{T}_n = 1(2) + 2(3) + 3(4) + 4(5) + \cdots + n(n+1),$$

which can be represented in a very convenient form as the sum of the numbers in an array:

$$2\mathrm{T}_n = \begin{array}{ccccccccc}
2 & 3 & 4 & 5 & \cdot & \cdot & \cdot & (n+1) & \\
 & 3 & 4 & 5 & \cdot & \cdot & \cdot & (n+1) & \\
 & & 4 & 5 & \cdot & \cdot & \cdot & (n+1) & \\
 & & & 5 & \cdot & \cdot & \cdot & (n+1) & n \\
 & & & & \cdot & & & \cdot & \\
 & & & & & \cdot & & \cdot & \\
 & & & & & & \cdot & \cdot & \\
 & & & & & & & (n+1) &
\end{array}$$

The combination of the array for T_n and the array for $2T_n$ produces an array for $3T_n$, which is easy to sum up:

$$
\begin{array}{ll}
 & \qquad\qquad (n+1) \\
3T_n = & \\
 & \begin{array}{ccccccccc}
1 & 2 & 3 & 4 & 5 & \cdot & \cdot & \cdot & (n+1) \\
1 & 2 & 3 & 4 & 5 & \cdot & \cdot & \cdot & (n+1) \\
1 & 2 & 3 & 4 & 5 & \cdot & \cdot & \cdot & (n+1) \\
1 & 2 & 3 & 4 & 5 & \cdot & \cdot & \cdot & (n+1) \quad n \\
\cdot & \cdot & \cdot & \cdot & \cdot & \cdot & & & \cdot \\
\cdot & \cdot & \cdot & \cdot & & \cdot & \cdot & & \cdot \\
\cdot & \cdot & \cdot & \cdot & & & \cdot & \cdot & \cdot \\
1 & 2 & 3 & 4 & \cdot & \cdot & \cdot & n & (n+1)
\end{array}
\end{array}
$$

Our formula for N_n directly yields

$$3T_n = n\frac{(n+1)(1+[n+1])}{2}$$

$$\boxed{T_n = \frac{n(n+1)(n+2)}{6}}$$

Students are now ready to consider the sum of the first n square numbers.

$$
\begin{aligned}
S_n &= \text{Sum of the first } n \text{ square numbers} \\
&= 1^2 + 2^2 + 3^2 + 4^2 + \cdots + n.
\end{aligned}
$$

In array form, this appears as

$$
\begin{array}{ll}
 & \qquad\quad n \\
S_n = & \begin{array}{cccccccc}
1 & 2 & 3 & 4 & \cdot & \cdot & \cdot & n \\
 & 2 & 3 & 4 & \cdot & \cdot & \cdot & n \\
 & & 3 & 4 & \cdot & \cdot & \cdot & n \\
 & & & 4 & \cdot & \cdot & \cdot & n \quad n \\
 & & & & & \cdot & & \cdot \\
 & & & & & & \cdot & \cdot \\
 & & & & & & \cdot & \cdot \\
 & & & & & & & n
\end{array}
\end{array}
$$

Combining the array for T_n with this array for S_n, we obtain an array for $S_n + T_n$:

$$
\begin{array}{ll}
 & \qquad\quad n \\
 & \begin{array}{cccccccc}
1 & 2 & 3 & 4 & \cdot & \cdot & \cdot & n
\end{array} \\
S_n+T_n = & \begin{array}{cccccccc}
1 & 2 & 3 & 4 & \cdot & \cdot & \cdot & n \\
1 & 2 & 3 & 4 & \cdot & \cdot & \cdot & n \\
1 & 2 & 3 & 4 & \cdot & \cdot & \cdot & n \quad (n+1) \\
1 & 2 & 3 & 4 & \cdot & & & \cdot \\
\cdot & \cdot & \cdot & \cdot & \cdot & \cdot & & \cdot \\
\cdot & \cdot & \cdot & \cdot & & \cdot & \cdot & \cdot \\
1 & 2 & 3 & 4 & & & \cdot & n \\
1 & 2 & 3 & 4 & \cdot & \cdot & \cdot & n
\end{array}
\end{array}
$$

Students should observe that each row of the array for $\mathrm{S}_n + \mathrm{T}_n$ is the sum of the first n natural numbers. In the present notation, this is N_n. Because the array has $(n+1)$ rows, we clearly obtain

$$\mathrm{S}_n + \mathrm{T}_n = (n+1)\mathrm{N}_n.$$

Substituting the formulas previously derived for T_n and N_n is an exercise in elementary algebra which quickly yields

$$\boxed{S_n = \frac{n(n+1)(2n+1)}{6}}.$$

Postassessment

Have each student:

1. Derive a formula for the sum of the first n pentagonal numbers using arrays.
2. Apply the various formulas the class has derived to find the sum of the first n figurate numbers for several integral values of n.

Unit 83

Pythagorean Triples

While teaching the Pythagorean theorem at the secondary school level, teachers often suggest that students recognize (and memorize) certain common ordered sets of three numbers that can represent the lengths of the sides of a right triangle. Some of these ordered sets of three numbers, known as Pythagorean triples, are: $(3,4,5), (5,12,13), (8,15,17)$, and $(7,24,25)$. The student is asked to discover these Pythagorean triples as they come up in selected exercises. How can one generate more triples without a trial and error method? This question, often asked by students, will be answered in this unit.

Performance Objectives

1. *Students will generate six primitive Pythagorean triples using the formulas developed in this unit.*

2. *Students will state properties of various members of a primitive Pythagorean triple.*

Preassessment

Students should be familiar with the Pythagorean theorem. They should be able to recognize Pythagorean triples and distinguish between primitive Pythagorean triples and others.

Teaching Strategies

Ask your students to find the missing member of the following Pythagorean triples:

1. $(3, 4, \_\_\_\_)$
2. $(7, \_\_\_\_, 25)$
3. $(11, \_\_\_\_, \_\_\_\_)$

The first two triples can be easily determined using the Pythagorean theorem. However, this method will not work with the third triple. At this point, you can offer your students a method for solving this problem.

Before beginning the development of the desired formulas, we must consider a few simple lemmas.

Lemma 1: When 8 divides the square of an odd number, the remainder is 1.

Proof: We can represent an odd number by $2k+1$, where k is an integer, $(2k+1)^2 = 4k^2 + 4k + 1 = 4k(k+1) + 1$.
Since k and $k+1$ are consecutive, one of them must be even.
Therefore, $4k(k+1)$ must be divisible by 8.
Thus $(2k+1)^2$, when divided by 8, leaves a remainder of 1.

The following lemmas follow directly.

Lemma 2: When 8 divides the sum of two odd square numbers, the remainder is 2.

Lemma 3: The sum of two odd square numbers cannot be a square number.

Proof: Since the sum of two odd square numbers, when divided by 8, leaves a remainder of 2, the sum is even but not divisible by 4. It therefore cannot be a square number.

We are now ready to begin our development of formulas for Pythagorean triples. Let us assume that (a, b, c) is a primitive Pythagorean triple. This implies that a and b are relatively prime. Therefore, they cannot both be even. Can they both be odd?

If a and b are both odd, then by Lemma 3: $a^2+b^2 \neq c^2$. This contradicts our assumption that (a, b, c) is a Pythagorean triple; therefore a and b cannot both be odd. Therefore, one must be odd and one even.

Let us suppose that a is odd and b is even. This implies that c is also odd.

We can rewrite $a^2 + b^2 = c^2$ as

$$b^2 = c^2 - a^2$$
$$b^2 = (c+a) \cdot (c-a).$$

Since the sum and difference of two odd numbers is even,

$$c + a = 2p \text{ and } c - a = 2q \, (p \text{ and } q \text{ are natural numbers}).$$

By solving for a and c, we get

$$c = p + q \quad \text{and} \quad a = p - q.$$

We can now show that p and q must be relatively prime. Suppose p and q were not relatively prime; say $g > 1$ was a common factor. Then g would also be a common factor of a and c. Similarly g would also be a common factor of $c + a$ and $c - a$. This would make g^2 a factor of b^2, since $b^2 = (c+a) \cdot (c-a)$. It follows that g would then have to be a factor of b. Now if g is a factor of b and also a common factor of a and c, then a, b, and c are not relatively prime. This contradicts our assumption that (a, b, c) is a *primitive* Pythagorean triple. Thus, p and q must be relatively prime.

Since b is even, we may represent b as

$$b = 2r.$$

But $b^2 = (c+a)(c-a)$.

Therefore, $b^2 = (2p) \cdot (2q) = 4r^2$, or $pq = r^2$.

If the product of two relatively prime natural numbers (p and q) is the square of a natural number (r), then each of them must be the square of a natural number. Therefore, we let

$$p = m^2 \quad \text{and} \quad q = n^2,$$

where m and n are natural numbers. Since they are factors of relatively prime numbers (p and q), they (m and n) are also relatively prime.

$$\text{Since } a = p - q \quad \text{and} \quad c = p + q$$

$$a = m^2 - n^2 \quad \text{and} \quad c = m^2 + n^2$$

$$\text{Also, since } b = 2r \quad \text{and} \quad b^2 = 4r^2 = 4pq = 4m^2n^2$$

$$b = 2mn.$$

To summarize, we now have formulas for generating Pythagorean triples:

$$a = m^2 - n^2 \quad b = 2mn \quad c = m^2 + n^2.$$

The numbers m and n cannot both be even, since they are relatively prime. They cannot both be odd, for this would make $c = m^2 + n^2$ an even number, which we established earlier as impossible. Since this indicates that one must be even and the other odd, $b = 2mn$ must be divisible by 4. Therefore, no Pythagorean triple can be composed of three prime numbers. This does *not* mean that the other members of the Pythagorean triple may not be prime.

Let us reverse the process for a moment. Consider relatively prime numbers m and n (where $m > n$) and where one is even and the other odd. We will now show that (a, b, c) is a primitive Pythagorean triple where $a = m^2 - n^2, b = 2mn$, and $c = m^2 + n^2$.

It is simple to verify algebraically that

$$\left(m^2 - n^2\right)^2 + (2mn)^2 = \left(m^2 + n^2\right)^2$$

thereby making it a Pythagorean triple. What remains is to prove that (a, b, c) is a *primitive* Pythagorean triple.

Suppose a and b have a common factor $h > 1$. Since a is odd, h must also be odd. Because $a^2 + b^2 = c^2$, h would also be a factor of c. We also have h a factor of $m^2 - n^2$ and $m^2 + n^2$ as well as of their sum, $2m^2$, and their difference $2n^2$.

Since h is odd, it is a common factor of m^2 and n^2. However, m and n (and as a result m^2 and n^2) are relatively prime. Therefore, h cannot be a common factor of m and n. This contradiction establishes that a and b are relatively prime.

Having finally established a method for generating primitive Pythagorean triples, students should be eager to put it to use. The following table gives some of the smaller primitive Pythagorean triples.

Pythagorean Triples

m	n	a	b	c
2	1	3	4	5
3	2	5	12	13
4	1	15	8	17
4	3	7	24	25
5	2	21	20	29
5	4	9	40	41
6	1	35	12	37
6	5	11	60	61
7	2	45	28	53
7	4	33	56	65
7	6	13	84	85

A fast inspection of the above table indicates that certain primitive Pythagorean triples (a, b, c) have $c = b + 1$. Have students discover the relationship between m and n for these triples.

They should notice that for these triples $m = n + 1$. To prove this will be true for other primitive Pythagorean triples (not in the table), let $m = n + 1$ and generate the Pythagorean triples.

$$a = m^2 - n^2 = (n+1)^2 - n^2 = 2n + 1$$
$$b = 2mn = 2n(n+1) = 2n^2 + 2n$$
$$c = m^2 + n^2 = (n+1)^2 + n^2 = 2n^2 + 2n + 1.$$

Clearly $c = b + 1$, which was to be shown!

A natural question to ask your students is to find all primitive Pythagorean triples that are consecutive natural numbers. In a method similar to that used above, they ought to find that the only triple satisfying that condition is $(3, 4, 5)$.

Other investigations can be proposed for student consideration. In any case, students should have a far better appreciation for Pythagorean triples and elementary number theory after completing this unit.

Postassessment

1. Find six primitive Pythagorean triples that are not included in the above table.

2. Find a way to generate primitive Pythagorean triples of the form (a, b, c) where $b = a + 1$.
3. Prove that every primitive Pythagorean triple has one member that is divisible by 3.
4. Prove that every primitive Pythagorean triple has one member that is divisible by 5.
5. Prove that for every primitive Pythagorean triple the product of its members is a multiple of 60.
6. Find a Pythagorean triple (a, b, c) where $a^2 = b + 2$.

References

Posamentier, A. S. *The Pythagorean Theorem: The Story of Its Power and Beauty*, Amherst, New York: Prometheus Books, 2010.

Unit 84

Divisibility

The unit will present methods for finding divisors without doing division.

Performance Objectives

1. *Given any integer, students will determine its prime factors, without doing any division.*
2. *Students will produce rules for testing divisibility by all natural numbers less than* 49, *and some greater than* 49.

Preassessment

Have students indicate without doing any division which of the following numbers are divisible by 2, by 3, and by 5.

a. 792 b. 835 c. 356 d. 3890 e. 693 f. 743

Teaching Strategies

Students are probably aware that any even number is divisible by 2; hence of the above numbers, (a), (c), and (d) are divisible by 2. Many will also recognize that a number whose terminal digit (units digit) is either 5 or 0, is divisible by 5; hence, (b) and (d) are divisible by 5. At this point, students will be eager to extend this rule to hold true for testing divisibility by 3. Of the above numbers, (c), (e), and (f) are the only numbers whose terminal digit is a multiple of 3; yet only one of these numbers, 693, is in fact divisible by 3. This should stir up sufficient curiosity so as to create a desire among the students to develop rules to test divisibility by numbers other than 2 and 5.

There are various ways to develop rules for testing divisibility by various numbers. They may be developed in order of magnitude of the numbers. This method may be appealing to some; however, it detracts from the various patterns that students so often appreciate in the development of mathematics. In this unit, we shall consider the rules in groups of related methods.

Divisibility by powers of 2: A given number is divisible by 2^1 *(or* $2^2, 2^3, \ldots 2^n$*, respectively) if the last 1 (or* $2, 3, \ldots n$*, respectively) digit(s) is (are) divisible by* 2^1 *(or* $2^2, 2^3, \ldots 2^n$*, respectively).*

Proof: Consider the following n-digit number:

$$a_{n-1}a_{n-2}a_{n-3}\cdots a_2a_1a_0,$$

which can be written as

$$10^{n-1}a_{n-1} + 10^{n-2}a_{n-2} + \cdots + 10^2a_2 + 10^1a_1 + 10^0a_0.$$

Since all terms except the last are always divisible by 2, we must be assured of the divisibility of the last term when testing divisibility by 2. Similarly, since all the terms except the last two are always divisible by 2^2, we must merely determine if the last two digits (considered as a number) is divisible by 2^2. This scheme may easily be extended to the nth case.

Divisibility by powers of 5: A given number is divisible by 5^1 *(or* $5^2, 5^3, \ldots 5^n$*, respectively) if the last 1 (or* $2, 3, \ldots n$*, respectively) digit(s) is (are) divisible by* 5^1 *(or* $5^2, 5^3, \ldots 5^n$*, respectively).*

Proof: The proof of these rules follows the same scheme as the proof for testing divisibility by powers of two, except the 2 is replaced by a 5.

Divisibility by 3 and 9: A given number is divisible by 3 (or 9) if the sum of the digits is divisible by 3 (or 9).

Proof: Consider the number $a_8a_7a_6a_5a_4a_3a_2a_1a_0$ (the general case $a_na_{n-1}\ldots a_3a_2a_1a_0$ is similar). This expression may be written as

$$a_8(9+1)^8 + a_7(9+1)^7 + \cdots + a_1(9+1) + a_0.$$

Using the expression $M_i(9)$ to mean a multiple of 9, for $i = 1, 2, 3, \ldots 7, 8$, we can rewrite the number as

$$a_8[\mathrm{M}_8(9) + 1] + a_7[\mathrm{M}_7(9) + 1] + \cdots + a_1[\mathrm{M}_1(9) + 1] + a_0.$$

(A mention of the binomial theorem may be helpful here.) The number equals $\mathrm{M}(9) + a_8 + a_7 + a_6 + a_5 + a_4 + a_3 + a_2 + a_1 + a_0$, where $\mathrm{M}(9)$ is a multiple of 9. Thus, the number is divisible by 9 (or 3) if the sum of the digits is divisible by 9 (or 3).

A rule for testing divisibility by 11 is proved in a manner similar to the proof for divisibility by 3 and 9.

Divisibility by 11: A given number is divisible by 11 if the difference of the two sums of alternate digits is divisible by 11.

Proof: Consider the number $a_8a_7a_6a_5a_4a_3a_2a_1a_0$ (using the general case is similar). This expression may be written as

$$\begin{aligned} &a_8(11-1)^8 + a_7(11-1)^7 + \cdots + a_1(11-1) + a_0 \\ &\quad = a_8[\mathrm{M}_8(11) + 1] + a_7[\mathrm{M}_7(11) - 1] + \cdots + a_1[\mathrm{M}_1(11) - 1] + a_0. \end{aligned}$$

This expression then equals $\mathrm{M}(11) + a_8 - a_7 + a_6 - a_5 + a_4 - a_3 + a_2 - a_1 + a_0$. Thus, the number is divisible by 11 if $a_8 + a_6 + a_4 + a_2 + a_0 - (a_7 + a_5 + a_3 + a_1)$ is divisible by 11.

You would be wise to indicate the extensions in bases other than ten of each of the previously mentioned divisibility rules. Often students are able to make these generalizations on their own (especially with appropriate coaxing). The remainder of this unit will deal with rules for testing divisibility of primes ≥ 7 and composites.

Divisibility by 7: Delete the last digit from the given number, then subtract twice this deleted digit from the remaining number. If the result is divisible by 7, the original number is divisible by 7. This process may be repeated if the result is too large for simple inspection of divisibility by 7.

Proof: To justify the technique, consider the various possible terminal digits and the corresponding subtraction:

Terminal digit	Number subtracted from original	Terminal digit	Number subtracted from original
1	$20 + 1 = 21 = 3 \cdot 7$	5	$100 + 5 = 105 = 15 \cdot 7$
2	$40 + 2 = 42 = 6 \cdot 7$	6	$120 + 6 = 126 = 18 \cdot 7$
3	$60 + 3 = 63 = 9 \cdot 7$	7	$140 + 7 = 147 = 21 \cdot 7$
4	$80 + 4 = 84 = 12 \cdot 7$	8	$160 + 8 = 168 = 24 \cdot 7$
		9	$180 + 9 = 189 = 27 \cdot 7$

In each case, a multiple of 7 is being subtracted one or more times from the original number. Hence, if the remaining number is divisible by 7, then so is the original number.

> *Divisibility by* 13*: This is the same as the rule for testing divisibility by* 7*, except that the* 7 *is replaced by* 13 *and instead of subtracting* twice *the deleted digit, we subtract* nine times *the deleted digit each time.*

Proof: Once again consider the various possible terminal digits and the corresponding subtraction:

Terminal digit	Number subtracted from original	Terminal digit	Number subtracted from original
1	$90 + 1 = 91 = 7 \cdot 13$	5	$450 + 5 = 455 = 35 \cdot 13$
2	$180 + 2 = 182 = 14 \cdot 13$	6	$540 + 6 = 546 = 42 \cdot 13$
3	$270 + 3 = 273 = 21 \cdot 13$	7	$630 + 7 = 637 = 49 \cdot 13$
4	$360 + 4 = 364 = 28 \cdot 13$	8	$720 + 8 = 728 = 56 \cdot 13$
		9	$810 + 9 = 819 = 63 \cdot 13$

In each case, a multiple of 13 is being subtracted one or more times from the original number. Hence, if the remaining number is divisible by 13, then the original number is divisible by 13.

> *Divisibility by* 17*: This is the same as the rule for testing divisibility by* 7 *except that the* 7 *is replaced by* 17 *and instead of subtracting* twice *the deleted digit, we subtract* five times *the deleted digit each time.*

Proof: The proof for the rule for divisibility by 17 follows a similar pattern to those for 7 and 13.

The patterns developed in the preceding three divisibility rules (for 7, 13, and 17) should lead students to develop similar rules for testing divisibility by larger primes. The following chart presents the "multipliers" of the deleted digits for various primes.

To test divisibility by	7	11	13	17	19	23	29	31	37	41	43	47
Multiplier	2	1	9	5	17	16	26	3	11	4	30	14

To fill in the gaps in the set of integers, a consideration of divisibility of composites is necessary.

> *Divisibility by composites: A given number is divisible by a composite number if it is divisible by each of its relatively prime factors.* The following chart offers illustrations of this rule. You or your students should complete the chart to 48.

To be divisible by	6	10	12	15	18	21	24	26	28
The number must be divisible by	2,3	2,5	3,4	3,5	2,9	3,7	3,8	2,13	4,7

At this juncture, the student has not only a rather comprehensive list of rules for testing divisibility, but also an interesting insight into elementary number theory. Have students practice using these rules (to instill greater familiarity) and try to develop rules to test divisibility by other numbers in base ten and to generalize these rules to other bases. Unfortunately lack of space prevents a more detailed development here.

Postassessment

Have students do these exercises.

1. State a rule for testing divisibility by
   a. 8 b. 18 c. 13 d. 23 e. 24 f. 42
2. Determine the prime factors of
   a. 280 b. 1001 c. 495 d. 315 e. 924

References

Posamentier, A. S. *Math Wonders to Inspire Students and Teachers.* Association for Supervision and Curriculum Development, 2003.

Posamentier, A. S. and S. Krulik. *Teachers! Prepare Your Students for the Mathematics for SAT 1: Methods and Problem Solving Strategies.* Thousand Oaks, CA: Corwin, 1996.

Posamentier, A. S. and C. T. Salkind. *Challenging Problems in Algebra.* New York: Dover, 1996.

Unit 85

Fibonacci Sequence

Performance Objectives

Students will:

1. *Define the Fibonacci Sequence.*
2. *Find sums of various Fibonacci numbers.*
3. *Find the sum of squares of the first Fibonacci numbers.*
4. *Discover properties of Fibonacci numbers.*

Preassessment

Have students try to solve the following problem:

How many pairs of rabbits will be produced in a year, beginning with a single pair, if in one month each pair bears a new pair which becomes productive from the second month on?

Teaching Strategies

Italian mathematician Leonardo of Pisa (he was the son, figlio, of Bonaccio, hence the name Fibonacci) presented the above problem in his book *LIBER ABACI* published in 1202. Consider its solution with students. Begin by drawing a chart as shown in Figure 1.

Start with first month and proceed to the next months explaining the procedure as you go along. Remind students that a baby pair must mature one month before becoming productive.

Continue the chart until the twelfth month where it will be discovered that 377 pairs of rabbits are produced in a year. Now focus students' attention on the third column (Number of As), the Fibonacci Sequence. Have them try to discover a rule for continuing this sequence. Tell students to notice that each term is the sum of the two preceding terms. This can be written as a general expression: $f_n = f_{n-1} + f_{n-2}$, where f_n stands for the nth Fibonacci number. For example, $f_3 = f_1 + f_2$; $f_4 = f_2 + f_3$; $f_7 = f_5 + f_6$. Also $f_1 = f_2 = 1$.

The Fibonacci Sequence has many interesting properties that students can observe by studying the relationships between the terms. It can be proved

Let A = Adult pairs, B = Baby pairs

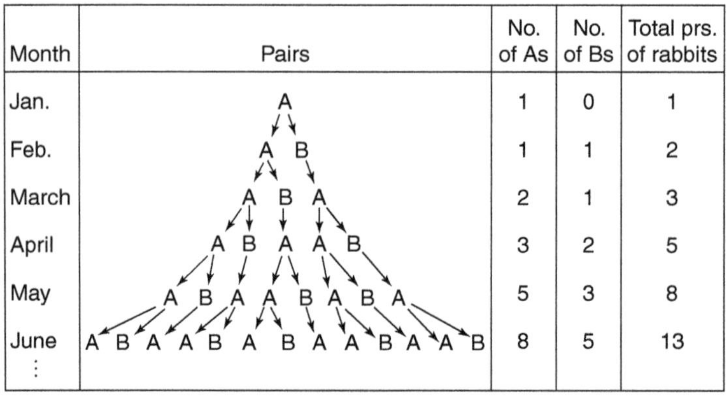

Figure 1.

that the sum of the first n Fibonacci numbers,

$$f_1 + f_2 + \cdots + f_n = f_{n+2} - 1. \tag{A}$$

It has already been noted that the following relations hold: $f_1 = f_3 - f_2$ (since $f_3 = f_1 + f_2$)

$$\begin{aligned} f_2 &= f_4 - f_3 \\ f_3 &= f_5 - f_4 \\ &\vdots \\ f_{n-1} &= f_{n+1} - f_n \\ f_n &= f_{n+2} - f_{n+1}. \end{aligned}$$

By *termwise* addition of all these equations, it follows that $f_1 + f_2 + f_3 + \cdots + f_n = f_{n+2} - f_2$, but we know that $f_2 = 1$. Therefore, $f_1 + f_2 + f_3 + \cdots + f_n = f_{n+2} - 1$.

In a similar manner, we can find an expression for the sum of the first n Fibonacci numbers with odd indices:

$$f_1 + f_3 + f_5 + \cdots + f_{2n-1} = f_{2n}. \tag{B}$$

To do this, we write

$$\begin{aligned}
f_1 &= f_2 \\
f_3 &= f_4 - f_2 (\text{because } f_4 = f_2 + f_3) \\
f_5 &= f_6 - f_4 \\
f_7 &= f_8 - f_6 \\
f_{2n-3} &= f_{2n-2} - f_{2n-4} \\
f_{2n-1} &= f_{2n} - f_{2n-2}.
\end{aligned}$$

Again by termwise addition, we obtain

$$f_1 + f_3 + f_5 + \cdots + f_{2n-1} = f_{2n}.$$

The sum of the first n Fibonacci numbers with even indices is

$$f_2 + f_4 + f_6 + \cdots + f_{2n} = f_{2n+1} - 1. \quad \text{(C)}$$

To prove this, we subtract equation (B) from twice equation (A), that is, $f_1 + f_2 + f_3 + \cdots + f_{2n} = f_{2n+2} - 1$, to obtain $f_2 + f_4 + \cdots + f_{2n} = f_{2n+2} - 1 - f_{2n} = f_{2n+2} - f_{2n} - 1 = f_{2n+1} - 1$ (because $f_{2n+2} = f_{2n} + f_{2n+1}$ and $f_{2n+1} = f_{2n+2} - f_{2n}$), which is what we wanted to prove.

By yet another application of the process of termwise addition of equations, we can derive a formula for the sum of the squares of the first n Fibonacci numbers. We must first note that for $k > 1$

$$f_k f_{k+1} - f_{k-1} f_k = f_k (f_{k+1} - f_{k-1}) = f_k \cdot f_k = f_k^2.$$

This gives us the following relations:

$$\begin{aligned}
f_{1^2} &= f_f f_2 - f_0 f_1 (\text{where } f_0 = 0) \\
f_{2^2} &= f_2 f_3 - f_1 f_2 \\
f_{3^2} &= f_3 f_4 - f_2 f_3 \\
&\vdots \\
f_{n-1^2} &= f_{n-1} f_n - f_{n-2} f_{n-1} \\
f_{n^2} &= f_n f_{n+1} - f_{n-1} f_n.
\end{aligned}$$

By adding termwise, we obtain

$$f_{1^2} + f_{2^2} + f_{3^2} + \cdots + f_{n^2} = f_n \cdot f_{n+1}.$$

The Fibonacci Sequence is also connected to a famous, ancient topic in mathematics. Examining the ratios of the first successive pairs of numbers in the sequence, we obtain the following:

$$\begin{array}{ll}
\frac{1}{1} = 1.0000 & \frac{2}{1} = 2.0000 \\
\frac{3}{2} = 1.5000 & \frac{5}{3} = 1.6667 \\
\frac{8}{5} = 1.6000 & \frac{13}{8} = 1.6250 \\
\frac{21}{13} = 1.6154 & \frac{34}{21} = 1.6190 \\
\frac{55}{34} = 1.6176 & \frac{89}{55} = 1.6182 \\
\frac{144}{89} = 1.6180 & \frac{233}{144} = 1.6181.
\end{array}$$

The ratios $\frac{f_n}{f_{n-1}} (n > 0)$ form a decreasing sequence for the odd values of n and an increasing sequence for the even values of n. Each ratio on the right-hand side is larger than each corresponding ratio on the left-hand side. The ratio approaches a limiting value between 1.6180 and 1.6181. It can be shown that this limit is $\frac{1+\sqrt{5}}{2}$ or approximately 1.61803 to five decimal places.

The ratio was so important to the Greeks that they gave it a special name, the "golden ratio" or the "golden section." They did not express the relationship in decimal form but with a geometric construction in which two line segments are in the exact golden ratio, 1.61803... to 1.

The golden ratio yields the basic connection between the Fibonacci Sequence and geometry. Consider again the ratios of consecutive Fibonacci numbers. As we said earlier, the table of fractions above seems to be approaching the golden ratio. Let us investigate this notion further by considering the line segment $\overline{\text{APB}}$, with P partitioning $\overline{\text{AB}}$ so that $\frac{\text{AB}}{\text{AP}} = \frac{\text{AP}}{\text{PB}}$.

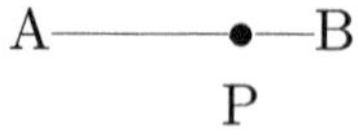

Let $x = \frac{\text{AB}}{\text{AP}}$. Therefore, $x = \frac{\text{AB}}{\text{AP}} = \frac{\text{AP+PB}}{\text{AP}} = 1 + \frac{\text{PB}}{\text{AP}} = 1 + \frac{\text{AP}}{\text{AB}} = 1 + \frac{1}{x}$. Thus, $x = 1 + \frac{1}{x}$ or $x^2 - x - 1 = 0$.

The roots of this equation are

$$a = \frac{1+\sqrt{5}}{2} \approx 1.6180339887, \text{ and}$$

$$b = \frac{1-\sqrt{5}}{2} \approx -.6180339887.$$

Since we are concerned with lengths of line segments, we shall use only the positive root, a. As a and b are roots of the equation $x^2 - x - 1 = 0$, $a^2 = a + 1(1)$ and $b^2 = b + 1(2)$.

Multiplying (1) by a^n (where n is an integer), $a^{n+2} = a^{n+1} + a^n$. Multiplying (2) by b^n (where n is an integer), $b^{n+2} = b^{n+1} + b^n$. Subtracting equation (2) from equation (1):

$$a^{n+2} - b^{n+2} = \left(a^{n+1} - b^{n+1}\right) + (a^n - b^n).$$

Now dividing by $a - b = \sqrt{5}$ (nonzero!):

$$\frac{a^{n+2} - b^{n+2}}{a - b} = \frac{a^{n+1} - b^{n+1}}{a - b} + \frac{a^n - b^n}{a - b}.$$

If we now let $t_n = \dfrac{a^n - b^n}{a - b}$, then $t_{n+2} = t_{n+1} + t_n$ (same as the Fibonacci Sequence definition). All that remains to be shown to establish t_n as the nth Fibonacci number, f_n, is that $t_1 = 1$ and $t_2 = 1$:

$$t_1 = \frac{a^1 - b^1}{a - b} = 1$$

$$t_2 = \frac{a^2 - b^2}{a - b} = \frac{(a - b)(a + b)}{a - b} = \frac{(\sqrt{5})(1)}{(\sqrt{5})} = 1.$$

Therefore, $f_n = \dfrac{a^n - b^n}{a - b}$, where $a = \dfrac{1 + \sqrt{5}}{2}, b = \dfrac{1 - \sqrt{5}}{2}$, and $n = 1, 2, 3, \ldots$

Postassessment

1. Find the sum of the first 9 Fibonacci numbers.
2. Find the sum of the first 5 Fibonacci numbers with odd indices.

References

Bicknell, M. and E. H. Verner, Jr. *A Primer for the Fibonacci Numbers.* San Jose, California: The Fibonacci Association, 1972.

Brother, U. A. *An Introduction to Fibonacci Discovery.* San Jose, Calif.: The Fibonacci Association, 1965.

Garland, T. H. *Fascinating Fibonaccis.* Palo Alto, CA: Dale Seymour Public, 1987.

Hoggatt, V. E., Jr. *Fibonacci and Lucas Numbers.* Boston: Houghton Mifflin, 1969.

Posamentier, A. S. *Advanced Euclidean Geometry: Excursions for Secondary Teachers and Students.* New York, NY: John Wiley Publishing, 2002.

Posamentier, A. S. and I. Lehmann. *The Fabulous Fibonacci Numbers.* Amherst, NY: Prometheus Books, 2007.

Vorob'ev, N. N. *Fibonacci Numbers.* New York: Blaisdell Publishing, 1961.

Unit 86

Diophantine Equations

This unit may be presented to any class having mastered the fundamentals of elementary algebra.

Performance Objectives

1. *Given an equation with two variables, students will find integral solutions (if they exist).*
2. *Given a verbal problem that calls for a solution of a Diophantine equation, students will determine (where applicable) the number of possible solutions.*

Preassessment

Have students solve the following problem: Suppose you are asked by your employer to go to the post office and buy 6-cent and 8-cent stamps. He gives you 5 dollars to spend. How many combinations of 6-cent and 8-cent stamps could you select from to make your purchase?

Teaching Strategies

Most students will promptly realize that there are two variables that must be determined, say x and y. Letting x represent the number of 8-cent stamps and y represent the number of 6-cent stamps, the equation: $8x+6y=500$ should follow. This should then be converted to $4x+3y=250$. At this juncture, the student should realize that although this equation has an infinite number of solutions, it may or may not have an infinite number of *integral* solutions; moreover, it may or may not have an infinite number of *positive integral* solutions (as called for by the original problem). The first problem to consider is whether integral solutions in fact, exist.

For this, a useful theorem may be employed. It states that if the greatest common factor of a and b is also a factor of k, where a, b, and k are integers, then there exist an infinite number of integral solutions for x and y in $ax+by=k$. Equations of this type solutions must be integers are known as *Diophantine equations* in honor of the Greek mathematician Diophantus, who wrote about them.

Since the greatest common factor of 3 and 4 is 1, which is a factor of 250, there exist an infinite number of integral solutions to the equation $4x + 3y = 250$. The question now facing your students is how many (if any) *positive* integral solutions exist?

One possible method of solution is often referred to as Euler's method (Leonhard Euler, 1707–1783). To begin, students should solve for the variable with the coefficient of least absolute value; in this case, y.

Thus, $y = \frac{250-4x}{3}$. This is to be rewritten to separate the integral parts as

$$y = 83 + \frac{1}{3} - x - \frac{x}{3} = 83 - x + \frac{1-x}{3}.$$

Now introduce another variable, say t; and let $t = \frac{1-x}{3}$.

Solving for x yields $x = 1 - 3t$. Since there is no fractional coefficient in this equation, the process does *not* have to be repeated as it otherwise would have to be (i.e., each time introducing new variables, as with t above). Now substituting for x in the above equation yields $y = \frac{250-4(1-3t)}{3} = 82 + 4t$. For various integral values of t, corresponding values for x and y will be generated. A table of values such as that below might prove useful.

t	...	-2	-1	0	1	2	...
x	...	7	4	1	-2	-5	...
y	...	74	78	82	86	90	...

Perhaps by generating a more extensive table, students will notice for what values of t positive integral values for x and y may be obtained. However, this procedure for determining the number of positive integral values of x and y is not very elegant. The students should be guided to the following inequalities to be solved simultaneously:

$$1 - 3t > 0 \quad \text{and} \quad 82 + 4t > 0.$$

$$\text{Thus, t} < \frac{1}{3} \quad \text{and} \quad t > -20\frac{1}{2},$$

or $-20\frac{1}{2} < t < \frac{1}{3}$. This indicates that there are 21 possible combinations of 6-cent and 8-cent stamps that can be purchased for 5 dollars.

Students might find it helpful to observe the solution to a more difficult Diophantine equation. The following is such an example:

Solve the Diophantine equation $5x - 8y = 39$.

1. Solve for x, since its coefficient has the lower absolute value of the two coefficients:

$$x = \frac{8y + 39}{5} = y + 7 + \frac{3y + 4}{5}.$$

2. Let $t = \frac{3y+4}{5}$, then solve for y:

$$y = \frac{5t - 4}{3} = t - 1 + \frac{2t - 1}{3}.$$

3. Let $u = \frac{2t-1}{3}$, then solve for t:

$$t = \frac{3u + 1}{2} = u + \frac{u + 1}{2}.$$

4. Let $v = \frac{u+1}{2}$, then solve for u:

$$u = 2v - 1.$$

We may now reverse the process because the coefficient of v is an integer.

5. Now substituting in the reverse order:

$$t = \frac{3u + 1}{2}.$$

Therefore: $t = \frac{3(2-1)+1}{2} = 3\nu - 1$.

Also: $y = \frac{5t-4}{3}$.

Therefore: $y = \frac{5(3v-1)-4}{3} = \boxed{5\nu - 3 = y}$.

Similarly: $x = \frac{8y+39}{5}$

Therefore: $x = \frac{8(5\nu-3)+39}{5} = \boxed{8\nu + 3 = x}$.

v	...	−2	−1	0	1	2	...
x	...	−13	−5	3	11	19	...
y	...	−13	−8	−3	2	7	...

This table indicates how the various solutions of this Diophantine equation may be generated. Students should be urged to inspect the nature of the members of the solution set.

Another method of solving Diophantine equations is presented in Unit 87.

Postassessment

Have students solve each of the following Diophantine equations and then determine the number of positive integral solutions (if any).

1. $2x + 11y = 35$
2. $3x - 18y = 40$
3. $7x - 3y = 23$
4. $4x - 17y = 53$

References

Posamentier, A. S. and C. T. Salkind. *Challenging Problems in Algebra*. New York: Dover, 1996.

Unit 87

Continued Fractions and Diophantine Equations

This lesson should be considered after the accompanying unit, "Diophantine Equations" is presented. This unit describes another method of solving Diophantine equations.

Performance Objectives

1. *Given an equation with two variables, students will find integral solutions (if they exist).*
2. *Given a verbal problem that calls for a solution of a Diophantine equation, students will determine (where applicable) the number of possible solutions.*
3. *Given an improper fraction, students will write an equivalent continued fraction.*

Preassessment

Students should have successfully mastered the concepts of the unit "Diophantine Equations."

Teaching Strategies

Before discussing this method of solution of Diophantine equations, an excursion into continued fractions would be appropriate. Every improper fraction (reduced to lowest terms) has an equivalent continued fraction. For example,

$$\frac{11}{7} = 1 + \frac{4}{7} = 1 + \frac{1}{\frac{7}{4}} = 1 + \cfrac{1}{1 + \frac{3}{4}}$$

$$= 1 + \cfrac{1}{1 + \cfrac{1}{\frac{4}{3}}} = 1 + \cfrac{1}{1 + \cfrac{1}{1 + \cfrac{1}{3}}}.$$

The last expression is called a *simple continued fraction*, since all the numerators after the first term are 1. These are the only types of continued fractions we shall consider here.

Consider a general improper fraction (reduced to lowest terms) and its equivalent simple continued fraction:

$$\frac{r}{s} = a_1 + \cfrac{1}{a_2 + \cfrac{1}{a_3 + \cfrac{1}{a_4 + \cfrac{1}{a_5}}}}$$

We shall call $c_1 = a_1$ the first convergent;

$$c_2 = a_1 + \frac{1}{a_2}, \text{ the second convergent;}$$

$$c_3 = a_1 + \cfrac{1}{a_2 + \cfrac{1}{a_3}}, \text{ the third convergent;}$$

$$c_4 = a_1 \cfrac{1}{a_2 + \cfrac{1}{a_3 + \cfrac{1}{a_4}}}, \text{ the fourth convergent;}$$

$$c_5 = a_1 + \cfrac{1}{a_2 + \cfrac{1}{a_3 + \cfrac{1}{a_4 + \cfrac{1}{a_5}}}}, \quad \text{the last convergent.}$$

For example, for the above continued fraction equivalent to $\frac{11}{7}$,

$$c_1 = 1; c_2 = 2; c_3 = \frac{3}{2}; c_4 = \frac{11}{7}.$$

It would be appropriate at this juncture to derive a method for finding the nth convergent of a general continued fraction.

Let $c_n = \frac{r_n}{s_n}$ (the nth convergent)

$$c_1 = a_1, \text{ therefore } r_1 = a_1 \text{ and } s_1 = 1.$$

$$c_2 = a_1 + \frac{1}{a_2} = \frac{a_1a_2 + 1}{a_2}.$$

Therefore, $r_2 = a_1a_2 + 1$ and $s_2 = a_2$.

$$c_3 = a_1 + \frac{1}{a_2 + \frac{1}{a_3}} = a_1 + \frac{1}{\frac{a_2a_3+1}{a_3}}$$

$$= a_1 + \frac{a_3}{a_2a_3 + 1} = \frac{a_1a_2a_3 + a_1 + a_3}{a_2a_3 + 1}$$

$$= \frac{a_3(a_1a_2 + 1) + a_1}{a_3a_2 + 1}, \text{ because}$$

$a_1a_2 + 1 = r_2; a_1 = r_1; a_2 = s_2; 1 = s_1;$ we get

$$c_3 = \frac{a_3r_2 + r_1}{a_3s_2 + s_1}.$$

Therefore, $r_3 = a_3r_2 + r_1$ and $s_3 = a_3s_2 + s_1$.

Similarly $c_4 = \frac{a_4r_3 + r_2}{a_4s_3 + s_2}$. Following this pattern,

$$\boxed{c_n = \frac{a_nr_{n-1} + r_{n-2}}{a_ns_{n-1} + s_{n-2}} = \frac{r_n}{s_n}}.$$

(This can be proved by mathematical induction. Now consider the general case for $n = 2$:

$$c_2 = \frac{a_2 r_1 + r_0}{a_2 s_1 + s_0}. \text{ Earlier } c_2 \text{ was found to}$$

$$\text{equal } \frac{a_1 a_2 + 1}{a_2}.$$

Equating corresponding parts yields

$$a_2 r_1 + r_0 = a_1 a_2 + 1.$$

Therefore, $r_1 = a_1$ and $r_0 = 1$

$$\text{also, } a_2 s_1 + s_0 = a_2.$$

Therefore, $s_1 = 1$ and $s_0 = 0$.

In a similar way, consider the general case for $n = 1$:

$$c_1 = \frac{a_1 r_0 + r_{-1}}{a_1 s_0 + s_{-1}}. \text{ Earlier this was found equal to } \frac{a_1}{1}.$$

Equating corresponding parts yields:

$$a_1 r_0 + r_{-1} = a_1.$$

Therefore, $r_0 = 1$ and $r_{-1} = 0$

$$\text{also, } a_1 s_0 + s_{-1} = 1.$$

Therefore, $s_0 = 0$ and $s_{-1} = 1$.

Have students convert $\frac{117}{41}$ to the equivalent continued fraction, $2 + \frac{1}{1+\frac{1}{5+\frac{1}{1+\frac{1}{5}}}}$.

Now set up a table:

Convergents

n	-1	0	1	2	3	4	5
a_n			2	1	5	1	5
$c_n = \frac{r_n}{s_n}$	0	1	2	3	17	20	117
	1	0	1	1	6	7	41

The first two columns for r_n and s_s are constant. However, the other values vary with the particular fraction. The values of a_n are taken directly from the continued fractions. Each value of r_n and s_n is obtained from the general formula derived earlier. To check if this chart was constructed properly, students should notice that the last convergent is in fact the original improper fraction.

An inspection of the various cross-products suggests $r_n \cdot s_{n-1} - r_{n-1} \cdot s_n = (-1)^n$.

With this background material learned, the students are now ready to apply their knowledge of continued fractions to solving Diophantine equations of the form $ax + by = k$, where the greatest common factor of a and b is a factor of k. First they should form an *improper* fraction using the two coefficients, say $\frac{a}{b}$. Then convert this fraction to a continued fraction: $\frac{a}{b} = \frac{r_n}{s_n}$.

Using the previously discovered formula,

$$r_n \cdot s_{n-1} - r_{n-1} \cdot s_n = (-1)^n$$

and substituting $a \cdot s_{n-1} - b \cdot r_{n-1} = 1$ (or multiply by -1).

Now multiplying by k: $a(k \cdot s_{n-1}) - b(k \cdot r_{n-1}) = k$.

Thus, $x = k \cdot s_{n-1}$, and $y = -k \cdot r_{n-1}$ is a solution of the Diophantine equation.

For example, consider the Diophantine equation

$$41x - 117y = 3.$$

After setting up the table mentioned earlier, the $n - 1$ convergent is used. That is, $r_{n-1} = 20$ and $s_{n-1} = 7$. The above relationship

$$a(k \cdot s_{n-1}) - b(k \cdot r_{n-1}) = k$$

yields with appropriate substitution:

$$41(3 \cdot 20) - 117(3 \cdot 7) = 3.$$

Thus, one solution of $41x - 117y = 3$ is $x = 60$ and $y = 21$.

To find the remaining solutions, the following scheme is used.

Subtract $41(60) - 117(21) = 3$ from $41x - 117y = 3$ to obtain $41(x - 60) - 117(y - 21) = 0$.

Therefore, $41(x - 60) = 117(y - 21)$, or $\dfrac{x - 60}{117} = \dfrac{y - 21}{41} = t$.

Thus, $t = \dfrac{x - 60}{117}$ and $\boxed{x = 117t + 60}$.

Also, $t = \dfrac{y - 21}{41}$ and $\boxed{y = 41t + 21}$.

A table of solutions may then be constructed.

t	$\cdots$	-2	-1	0	1	2	$\cdots$
x	$\cdots$	-174	-57	60	177	294	$\cdots$
y	$\cdots$	-61	-20	21	62	103	$\cdots$

Postassessment

Have students change each of the following improper fractions to equivalent continued fractions.

1. $\frac{37}{13}$ 2. $\frac{47}{23}$ 3. $\frac{173}{61}$

Have students solve each of the following Diophantine equations and then determine how many (if any) positive solutions exist.

4. $7x - 31y = 2$
5. $18x - 53y = 3$
6. $5x - 2y = 4$
7. $123x - 71y = 2$

Unit 88

Simplifying Expressions Involving Infinity

This unit presents simple algebraic methods (appropriate for elementary algebra students) to solve seemingly difficult problems involving infinity.

Performance Objective

Given an algebraic problem involving infinity, students will use a simple algebraic method to solve the problem.

Preassessment

Students should be able to work with radical equations and quadratic equations.

Teaching Strategies

Offer the following problem to your students for solution:

Find the value of x if

$$x^{x^{x^{x}}} = 2.$$

Most students' first reaction will be one of bewilderment. Since they have probably never worked with an infinite expression, they are somewhat overwhelmed. Students may try to substitute into the expression values for x_2 in order to estimate an answer to the problem. Before they are entirely frustrated, begin by explaining the infinite nature of the expression. Explain also that

$$3^{3^{3}} \neq 27^{3} = 19,683 \quad \text{but rather } 3^{3^{3}} = 3^{27} = 7{,}625{,}597{,}484{,}989.$$

Now have your students inspect the original expression in the following way: if $x^{x^2} = 2$, then, since there are an infinite number of xs, one x less would not affect the expression. Therefore, the exponent of the first x (lowest base) is 2.

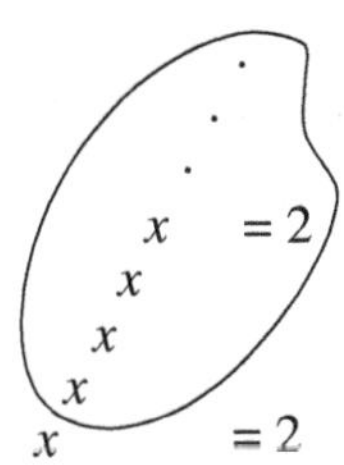

Thus, this expression simplifies to $x^2 = 2$, and $x = \sqrt{2}$. Students should be asked to consider the possibility of $x < 0$.

Students will naturally wonder if they could compose a similar problem by replacing 2 with, say, 5 or 7. Without elaborating, indicate to them that values to replace 2 may not be chosen at random, and that in fact these replacement values may not exceed e (i.e., the base of the natural system of logarithms, approximately $2.7182818284\cdots$).

To reinforce the scheme used in the solution of the above problem, have students consider the value of the nest of radicals

$$\sqrt{5+\sqrt{5+\sqrt{5+\sqrt{5+\sqrt{5+}}}}}\cdots.$$

To find x, where

$x = \sqrt{5+\sqrt{5+\sqrt{5+\sqrt{5+\sqrt{5+\cdots}}}}}$ have students realize that nothing is lost by deleting the first 5 of this nest of radicals, since there are an infinite number of them. Thus,

$$x = \sqrt{5+\underbrace{\sqrt{5+\sqrt{5+\sqrt{5+\sqrt{5+\cdots}}}}}_{=x}}$$

or $x = \sqrt{5+x}$, which is a simple radical equation. Students merely square both sides of the equation and solve the resulting quadratic equation:

$$\begin{aligned} x^2 &= 5 + x \\ x^2 &= x - 5 = 0 \\ x &= \frac{1 \pm \sqrt{21}}{2}. \end{aligned}$$

Since x is positive, $x = \frac{1+\sqrt{21}}{2} \approx 2.79$.

An alternative approach to evaluating a nest of radicals is to first square both sides of the original equation to get $x^2 = 5+\sqrt{5+\sqrt{5+\sqrt{5+\sqrt{5+\cdots}}}}$ and then substitute x, so that $x^2 = 5 + x$. The rest is as in the previous method.

It is important that you stress inspecting the reasonability of the value of the nest of radicals. That is, should the value be positive or negative, real or imaginary, etc.

Another application of this method of evaluating expressions involving infinity is with continued fractions. Before introducing infinite continued fractions, you ought to refresh your students' memories about continued fractions. You may wish to have them write $\frac{13}{5}$ as a continued fraction:

$$\begin{aligned} \frac{13}{5} &= 2 + \frac{3}{5} = 2 + \frac{1}{\frac{5}{3}} = 2 + \frac{1}{1+\frac{2}{3}} \\ &= 2 + \frac{1}{1+\frac{1}{\frac{3}{2}}} = 2 + \frac{1}{1+\frac{1}{1+\frac{1}{2}}}. \end{aligned}$$

Further, you may also want to have them simplify the continued fraction

$$1 + \cfrac{1}{2 + \cfrac{1}{3 + \cfrac{1}{4}}}$$

$$1+\cfrac{1}{2+\cfrac{1}{3+\cfrac{1}{4}}}=1+\cfrac{1}{2+\cfrac{1}{\frac{13}{4}}}=1+\cfrac{1}{2+\cfrac{4}{13}}$$

$$=1+\cfrac{1}{\frac{30}{13}}=1+\frac{13}{30}=\frac{43}{13}$$

Now have students consider the infinite continued fraction

$$1+\cfrac{1}{1+\cfrac{1}{1+\cfrac{1}{1+\cfrac{1}{1+\cdots}}}}$$

They will soon realize that the previous method of simplification will no longer work. At this point you would show them the following method:

$$\text{Let } x=1+\cfrac{1}{1+\cfrac{1}{1+\cfrac{1}{1+\cdots}}}.$$

Once again deleting the first "part" of the infinite continued fraction will not affect its value (because of the nature of infinity).

$$\text{Therefore } x=1+\cfrac{1}{1+\cfrac{1}{1+\cfrac{1}{1+\cdots}}}\quad(=x)$$

or $x=1+\frac{1}{x}$, which yields $x^2=x+1$

$$x^2-x-1=0$$

and $x=\frac{1\pm\sqrt{5}}{2}$; however, since $x>0, x=\frac{1+\sqrt{5}}{2}$. Some of your students may recognize this value as that of the golden ratio.

More advanced students might wonder how a nonrepeating infinite expression is evaluated. For these students, you may wish to present the following:

Evaluate $\sqrt{1+2\sqrt{1+3\sqrt{1+4\sqrt{1+5\sqrt{1+}}}}}\cdots$.

To evaluate this expression, some preliminary work must first be done. Since

$(n+2)^2 = n^2+4n+4 = 1+(n+1)(n+3)$,

$n+2 = \sqrt{1+(n+1)(n+3)}$. $n(n+2) = n\sqrt{1+(n+1)(n+3)}$.

Let $f(n) = n(n+2)$

then $f(n+1) = (n+1)(n+3)$.

Thus, $f(n) = n\sqrt{1+(n+1)(n+3)}$

$$f(n) = n\sqrt{1+f(n+1)}$$

$$f(n) = n\sqrt{1+(n+1)\sqrt{1+f(n+2)}}$$

$$f(n) = n\sqrt{1+(n+1)\sqrt{1+(n+2)\sqrt{1+f(n+3)}}}, \text{ and so on.}$$

Now if $n=1$, then $f(n) = 1(1+2) = 3$ and $3=$

$$1\sqrt{1+(1+1)\sqrt{1+(1+2)\sqrt{1+(1+3)\sqrt{1+}}}}\cdots$$

$$=1\sqrt{1+2\sqrt{1+3\sqrt{1+4\sqrt{1+}}}}\cdots$$

As a result of presenting the methods considered in this unit, your students should have a more solid concept of infinite expressions.

Postassessment

1. Simplify: $\sqrt{7+\sqrt{7+\sqrt{7+\sqrt{7+}}}}\cdots$

2. Simplify: $2+\cfrac{1}{3+\cfrac{1}{3+\cfrac{1}{3+\cfrac{1}{3+\cdots}}}}$

3. Simplify: $1+\cfrac{1}{2+\cfrac{1}{1+\cfrac{1}{2+\cfrac{1}{1+\cfrac{1}{2+\cdots}}}}}$

Unit 89

Continued Fraction Expansion of Irrational Numbers

Performance Objectives

1. *Given an irrational number, students will write an equivalent continued fraction.*
2. *Given an infinite expansion, students will get back to the irrational number.*

Preassessment

Students should be familiar with continued fractions.

Teaching Strategies

The procedure for expanding an irrational number is essentially the same as that used for rational numbers. Let x be the given irrational number. Find a_1, the greatest integer less than x, and express x in the form

$$x = a_1 + \frac{1}{x_2}, \quad 0 < \frac{1}{x_2} < 1,$$

where the number $x_2 = \frac{1}{x-a_1} > 1$ is irrational: for, if an integer is subtracted from an irrational number, the difference and the reciprocal of the difference are irrational.

Find a_2, the largest integer less than x_2 and express x_2 in the form

$$x_2 = a_2 + \frac{1}{x_3}, 0 < \frac{1}{x_3} < 1, \quad a_2 \geq 1,$$

where again the number

$$x_3 = \frac{1}{x_2 - a_2} > 1.$$

This calculation may be repeated indefinitely, producing in succession the equations

$$x = a_1 + \frac{1}{x_2}, x_2 > 1$$

$$x_2 = a_2 + \frac{1}{x_3}, x_3 > 1, a_2 \geq 1$$

$$x_3 = a_3 + \frac{1}{x_4}, x_4 > 1, a_3 \geq 1$$

$$\vdots \qquad \vdots \qquad \vdots$$

$$x_n = a_n + \frac{1}{x_{n+1}}, x_{n+1} > 1, a_n \geq 1$$

$$\vdots \qquad \vdots \qquad \vdots$$

where $a_1, a_2, a_3, \ldots, a_n, \ldots$ are all integers and the numbers $x, x_2, x_3, \ldots$ are all irrational. This process cannot end because the only way this could happen would be for some integer a_n to be equal to x_n. This is impossible since each successive x_i is irrational.

Substituting x_2 from the second equation above into the first equation, then x_3 from the third into this result, and so on, produces the required infinite simple continued fraction

$$x = a_1 + \frac{1}{x_2} = a_1 + \frac{1}{a_2 + \frac{1}{x_3}} = a_1 + \frac{1}{a_2 + \frac{1}{a_3 + \frac{1}{x_4}}}$$

or sometimes written as $x = [a_1, a_2, a_3, a_4, \ldots]$, where the three dots indicate that the process is continued indefinitely.

Example 1

Expand $\sqrt{3}$ into an infinite simple continued fraction.

Solution: The largest integer less than $\sqrt{3}$ is 1. Therefore, $a_1 = 1$ and

$$\sqrt{3} = 1 + \frac{1}{x_2}.$$

Solving this equation for x_2, we get

$$x_2 = \frac{1}{\sqrt{3} - 1} \cdot \frac{\sqrt{3} + 1}{\sqrt{3} + 1} = \frac{\sqrt{3} + 1}{2}.$$

Therefore, $\sqrt{3} = a_1 + \frac{1}{x_2} = 1 + \frac{1}{\frac{\sqrt{3}+1}{2}}$.

$x_2 = a_2 + \frac{1}{x_3}$ where $a_2 = 1$, since it is the largest integer less than $\frac{\sqrt{3}+1}{2}$. Therefore,

$$x_3 = \frac{1}{\frac{\sqrt{3}+1}{2} - 1} = \frac{2}{\sqrt{3}-1} \cdot \frac{\sqrt{3}+1}{\sqrt{3}+1} = \sqrt{3}+1$$

$$\sqrt{3} = 1 + \cfrac{1}{1 + \cfrac{1}{\sqrt{3}+1}}.$$

Continuing this process,

$x_3 = 2 + \frac{1}{x_4}$, $a_3 = 2$ since 2 is the largest integer less than $\sqrt{3}+1$.

$$x_4 = \frac{1}{\sqrt{3}-1} \cdot \frac{\sqrt{3}+1}{\sqrt{3}+1} = \frac{\sqrt{3}+1}{2}$$

$$\sqrt{3} = 1 + \cfrac{1}{1 + \cfrac{1}{2 + \cfrac{1}{\cfrac{\sqrt{3}+1}{2}}}}.$$

Since $x_4 = \frac{\sqrt{3}+1}{2}$ is the same as $x_2 = \frac{\sqrt{3}+1}{2}$, x_3 will produce the same result as x_3, namely $\sqrt{3}+1$. All the following partial quotients will be $1, 2, 1, 2$ and the infinite expansion of $\sqrt{3}$ will be

$$\sqrt{3} = 1 + \cfrac{1}{1 + \cfrac{1}{2 + \cfrac{1}{1 + \cfrac{1}{2 + \ddots}}}} = [1, 1, 2, 1, 2 \ldots] = [1, \overline{1, 2}]$$

The bar over the 1 and 2 indicates that the numbers 1 and 2 are repeated indefinitely.

Example 2

Find the infinite continued fraction expansion for

$$x = \frac{\sqrt{30}-2}{13}$$

Solution: Since $\sqrt{30}$ is between 5 and 6, the largest integer less than x is $a_1 = 0$. Then

$$x = \frac{\sqrt{30}-2}{13} = 0 + \frac{1}{x_2},$$

where $x_2 = \frac{1}{x} = \frac{13}{\sqrt{30}-2} \cdot \frac{\sqrt{30}+2}{\sqrt{30}+2} = \frac{\sqrt{30}+2}{2} > 1$.

The largest integer less than x_2 is $a_2 = 3$, therefore,

$$x_2 = a_2 + \frac{1}{x_3} = 3 + \frac{1}{x_3}.$$

$$\text{Therefore, } x_3 = \frac{1}{x_2 - 3} = \frac{1}{\frac{\sqrt{30}+2}{2} - 3}$$

$$= \frac{2}{\sqrt{30}-4} \cdot \frac{\sqrt{30}+4}{\sqrt{30}+4}$$

$$= \frac{2(\sqrt{30}+4)}{14} = \frac{\sqrt{30}+4}{7}.$$

The largest integer less than x_3 is $a_3 = 1$.

$$\text{Therefore, } x_4 = \frac{1}{x_3 - 1} = \frac{1}{\frac{\sqrt{30}+4}{7} - 1}$$

$$= \frac{7}{\sqrt{30}-3} \cdot \frac{\sqrt{30}+3}{\sqrt{30}+3} = \frac{\sqrt{30}+3}{3}.$$

In a similar way, we get $x_5 = \frac{\sqrt{30}+3}{7}$, $x_6 = \frac{\sqrt{30}+4}{2}$, and $x_7 = \frac{\sqrt{30}+4}{7} = x_3$.

Further investigation will show that the sequence $1, 2, 1, 4$ repeats. The required expansion is

$$x = 0 + \frac{1}{x_2} = 0 + \cfrac{1}{3 + \cfrac{1}{x_3}} = 0 + \cfrac{1}{3 + \cfrac{1}{1 + \cfrac{1}{x_4}}}$$

$$= 0 + \cfrac{1}{3 + \cfrac{1}{1 + \cfrac{1}{2 + \cfrac{1}{x_5}}}} \qquad = 0 + \cfrac{1}{3 + \cfrac{1}{1 + \cfrac{1}{2 + \cfrac{1}{1 + \cfrac{1}{x_6}}}}}$$

$$= 0 + \cfrac{1}{3 + \cfrac{1}{1 + \cfrac{1}{2 + \cfrac{1}{1 + \cfrac{1}{4 + \cfrac{1}{x_7}}}}}}$$

so finally we obtain

$$x = \frac{\sqrt{30} - 2}{13} = [0, 3, \overline{1, 2, 1, 4}].$$

Students may prove that a given infinite continued fraction actually represents an irrational number. Consider showing that $[2, \overline{2, 4}]$ represents $\sqrt{6}$. Begin by writing:

Let $x = 2 + \cfrac{1}{2 + \cfrac{1}{4 + \cfrac{1}{2 + \cfrac{1}{4 + \ddots}}}}$

where $y = 2 + \cfrac{1}{4 + \cfrac{1}{2 + \cfrac{1}{4 + \ddots}}}$

Therefore, $y = 2 + \cfrac{1}{4 + \cfrac{1}{y}} = 2 + \dfrac{y}{4y + 1}$.

Solving for y yields: $\dfrac{2 + \sqrt{6}}{2}$.

However, $x = 2 + \dfrac{1}{y} = 2 + \dfrac{2}{2 + \sqrt{6}}$.

Hence, $x = 2 + \sqrt{6} - 2 = \sqrt{6}$.

Postassessment

Have students change each of the following into an infinite simple continued fraction.

1. $\sqrt{2}$ 2. $\sqrt{43}$ 3. $\dfrac{25+\sqrt{53}}{22}$

Have students show that the infinite continued fraction $[\overline{3,6}] = \sqrt{10}$.

Unit 90

The Farey Sequence

This unit presents a discussion of a rather unusual sequence of numbers. This topic can be presented to students at various levels. However, the emphasis will change with the various ability and maturity levels of the students.

Performance Objectives

1. *Students will show that the fraction immediately before* $\frac{1}{2}$ *and its immediate successor of the Farey Sequence are complementary.*
2. *Students will establish the relationship between* π *and the number of terms in the Farey Sequence.*

Preassessment

Using the following sequence of fractions, have students find the sum of the two fractions:

a. fifth term to the left and third term to the right of $\frac{1}{2}$,
b. third term to the left and third term to the right of $\frac{1}{2}$,
c. second term to the left and second term to the right of $\frac{1}{2}$,

$$\frac{1}{7}, \frac{1}{6}, \frac{1}{5}, \frac{1}{4}, \frac{2}{7}, \frac{1}{3}, \frac{2}{5}, \frac{3}{7}, \frac{1}{2}, \frac{4}{7}, \frac{3}{5}, \frac{2}{3}, \frac{5}{7}, \frac{3}{4}$$

Ask students to generalize their results.

Teaching Strategies

A review of the preassessment activity will indicate that the three sums students were asked to find all resulted in 1. That is,

$$\frac{1}{4}+\frac{3}{4}=1, \quad \frac{1}{3}+\frac{2}{3}=1, \quad \text{and} \quad \frac{2}{5}+\frac{3}{5}=1.$$

We shall refer to a pair of fractions whose sum is 1 as *complementary.* Let us now inspect the given sequence.

If we list all proper common fractions in their lowest terms in order of magnitude up to some arbitrarily assigned limit — such as with denominators not exceeding 7 — we have the 17 fractions

$$\frac{1}{7}, \frac{1}{6}, \frac{1}{5}, \frac{1}{4}, \frac{2}{7}, \frac{1}{3}, \frac{2}{5}, \frac{3}{7}, \frac{1}{2}, \frac{4}{7}, \frac{3}{5}, \frac{2}{3}, \frac{5}{7}, \frac{5}{4}, \frac{3}{4}, \frac{4}{5}, \frac{5}{6}, \frac{6}{7}.$$

This is called the Farey Sequence. The Farey Sequence F_n, of order n, is defined as the *ordered set consisting of* 0, *the irreducible proper fractions with denominators from* $\frac{1}{2}$ to n, *arranged in order of increasing magnitude, and* $\frac{1}{1}$. There are many characteristic properties of the Farey Sequence. One is the relationship students discovered earlier; fractions equidistant from $\frac{1}{2}$ are complementary; that is, their sum equals one. Another interesting relationship involves the number of terms in the Farey Sequence of order n and π.

Before beginning a development of this sequence, students should be given some background on the Farey Sequence. They should be told that Farey, in 1816, discovered the sequence while perusing lengthy tables of decimal quotients. Apparently the numerator of any fraction in the Farey Sequence is obtained by adding the numerators of the fractions on each side of it, and similarly for the denominators. Since the result must be in lowest terms, this holds true for triplet:

$$\frac{1}{3}, \frac{2}{5}, \frac{3}{7} \quad \text{where} \quad \frac{3+1}{3+7}=\frac{4}{10}=\frac{2}{5}.$$

Students will see that the sum of fractions equidistant from $\frac{1}{2}$ equals 1. This can be proved many ways.

Suppose that $\frac{\ell}{n}$ is a number of the series that is less than $\frac{1}{2}$ and such that ℓ and n are relatively prime. Comparing the corresponding number of the other side of $\frac{1}{2}$, we find $\frac{(n-\ell)}{n}$. Since this belongs to the Farey Sequence, it is necessary that g.c.d. $(n-\ell, n)=1$. Supposing that $(n-\ell)$ and n are not relatively prime, then $n-\ell=qd$ and $n=qd+\ell$. Also $n=rd$ and thus $rd=qd+\ell$. Therefore, d divides ℓ and consequently d divides $(n-\ell)$ for

d divides n. This however contradicts the fact that $\frac{(n-\ell)}{n}$ was in its lowest terms (which is the definition of terms in the Farey Sequence) and therefore g.c.d. $(n-\ell, n) = 1$.

Now to prove that $\frac{\ell}{n} + \frac{a}{b} = 1$. Let $\frac{\ell}{n}$ be the immediate predecessor of $\frac{1}{2}$. If there was another term immediately succeeding $\frac{1}{2}$ and belonging to F_n, then the fractions are arranged as follows:

$$\frac{\ell}{n}, \frac{1}{2}, \frac{a}{b} \quad \text{where} \quad \frac{1}{2} = \frac{\ell+a}{n+b}$$

(one of the properties of the sequence). To prove that $\frac{\ell}{n}+\frac{a}{n} = 1$, $\frac{a}{b}$ must be in lowest terms if it belongs to F_n. If two fractions whose sum is 1 are in lowest terms, then their denominators are equal. Thus, if $\frac{\ell}{n} + \frac{a}{b} = 1, b$ must equal n. Therefore, $\frac{\ell+a}{n} = 1$ and $l + a = n$, or $a = n - \ell$. But $\frac{\ell}{n}$ was the immediate predecessor of $\frac{1}{2}$ and $\frac{a}{b} = \frac{n-\ell}{n}$. Thus, the immediate predecessor of $\frac{1}{2}$ and its immediate successor are complementary because their sum equals 1.

Another very interesting property results between π and the sum of the terms in the Farey Sequence. The number of fractions of order n is obtained as follows. Since the fractions are all in lowest terms, it follows that for a given denominator b, the number of numerators is the number of integers less than, and prime to, b. Students should then see the number of fractions N, in the Farey Sequence is equal to $\phi(2)+\phi(3)+\phi(4)+\cdots+\phi(n)$, where $\phi(n)$ is the number of positive integers less than or equal to n that are relatively prime to n. If $n = 7$, we have $\text{N} = \phi(2) + \phi(2) + \phi(3) + \phi(4) + \phi(5) + \phi(6) + \phi(7) = 1+2+2+4+2+6$. The value of N increases rapidly as n increases and when $n = 100$, $\text{N} = 3043$. Thus, there are many irreducible common fractions with numerators and denominators not exceeding 100.

There is a remarkable formula involving the ϕ function and π (the ratio of the circumference to the diameter of the circle).

The ϕ function refers to Euler's function. The sum of the Farey Sequence can be written by using a formula in terms of Euler's function ϕ. If $\frac{h}{k}$ is a term in the Farey Sequence, then g.c.d. $(h, k) = 1$. For any fixed number $k > 1$, the number of terms of the form $\frac{h}{k}$ is $\phi(k)$. It can be demonstrated that the sum $\phi(1) + \phi(2) + \cdots + \phi(n)$ is approximated by the expression $\frac{3n^2}{\pi^2}$, the approximation becoming more and more accurate as n increases. Except for the first term this sum represents the number of terms, N in a Farey Sequence of order n. Since we know the value of π to any desired degree of accuracy, this means we can find approximately the number of terms in a Farey Sequence without evaluating separately $\phi(1), \phi(2), \phi(3) \ldots \phi(n)$. Thus, for $n = 100$, we would have $\text{N} = \frac{3 \cdot 100^2}{\pi^2} = 3039.6355\ldots$ whereas the true value is 3043.

Thus, the number of terms of the Farey series approaches $\frac{3n^2}{\pi^2}$.

Postassessment

1. Given $n = 200$, $n = 8$, find the number of terms in the Farey Sequence using the expression $\frac{3n^2}{\pi^2}$.
2. Have students find other properties of the Farey Sequence.

Unit 91

The Parabolic Envelope

This unit describes briefly the mechanical construction of the parabolic envelope and shows how students can use the envelope to derive a host of related curves.

Performance Objective

Using the envelope as a foundation, students will draw a variety of curves by different techniques without point by point plotting from an equation. In the process, they will be introduced to the visual concepts of an envelope, evolute and pedal to a given curve.

Preassessment

Students should have completed a basic geometry course and be familiar with the conic sections.

Teaching Strategies

Have your students construct tangents to a parabola in the following manner:

Draw an angle of any measure, and divide each side of the angle into the same number of equally-spaced intervals. In Figure 1, we have an angle, A, of measure 50°, divided on each side into 17 equally-spaced, numbered intervals. Starting, as we did, at the lower left of the angle, lines were drawn connecting points 1 to $17, 2$ to $16, 3$ to 15, and so on, terminating with

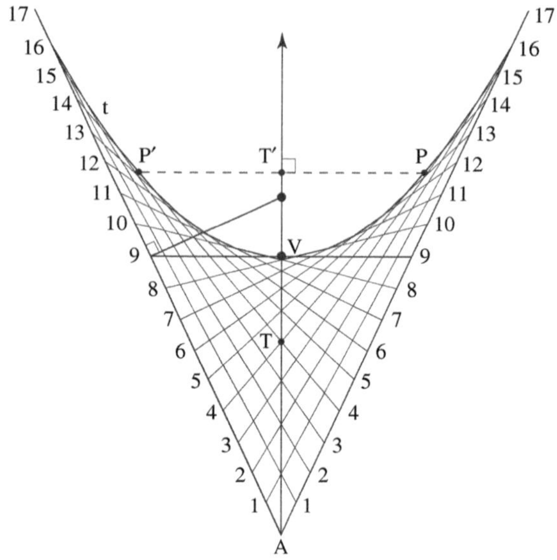

Figure 1.

17-1, (where the notation "17-1" means the segment connecting point 17 to point 1, or conversely). The resulting array of lines are tangents, which *envelope* a parabola.

The midpoint, V, of the line 9-9, is the *vertex* of the parabola, and 9-9 is the tangent to the parabola at V. A line from A to V, extended beyond V, is the parabola's axis of *symmetry*, and is also included in Figure 1. Ask students why 9-9 is the tangent perpendicular to $\overline{AV}$.

Erect a perpendicular to either side of the angle at point 9. We state without proof that the intersection of this perpendicular with the axis of symmetry determines the *focus*, F, of the parabola. More ambitious students may wish to prove this. At any rate, it would be meaningful to mention the reflective and locus properties of the focus.

Specific points of tangency on the parabola can be visually approximated directly from Figure 1. They can be more exactly located from the fact that a tangent to the parabola intersects the axis at a distance from the vertex equal to the ordinate of the point of tangency. As an example in Figure 1, tangent 14-4 intersects the axis at T. Locate a point T′ on the axis above V such that TV = VT′. Draw a line through T′ parallel to 9-9 so as to intersect the envelope at P and P′, then P and P′ are the points on the parabola where 14-4 and 4-14 are tangent. Other points on the parabola can be determined in the same manner.

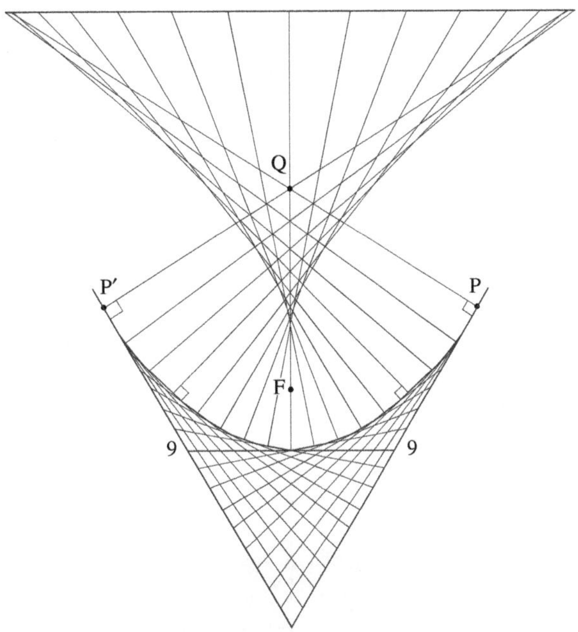

Figure 2.

Evolute to the parabola

Having located all such points of tangency P and P′ on the parabola, use a right-angled triangle or carpenter's square to erect perpendiculars at each of these points. These perpendiculars to a curve at the point of tangency are called *normals*. The envelope of all such normals defines the *cvolutc* to the curve; that is, the normals are then tangents to the evolute of the given curve. Thus, the evolute to the parabola can be shown to be a one-cusped curve called a *semi-cubic parabola*. This is shown in Figure 2.

To realize an accurately drawn evolute, utilize the symmetry of the parabola about the axis. Hence, normals to P and P intersect at Q on the axis of symmetry.

Pedal curves to the parabola

Figure 3 shows a given curve, C, and a fixed point, F, on or in the neighborhood of C. Dropping perpendiculars from F to each of the tangents to the curve, we find that the locus of the feet of the perpendiculars, P defines the *pedal* curve to the given curve with respect to F. For a given curve, different choices of F will result in different pedal curves.

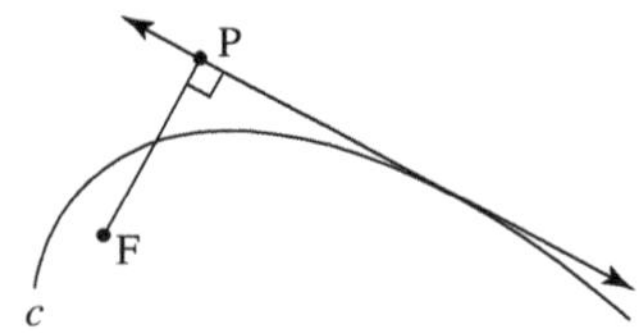

Figure 3.

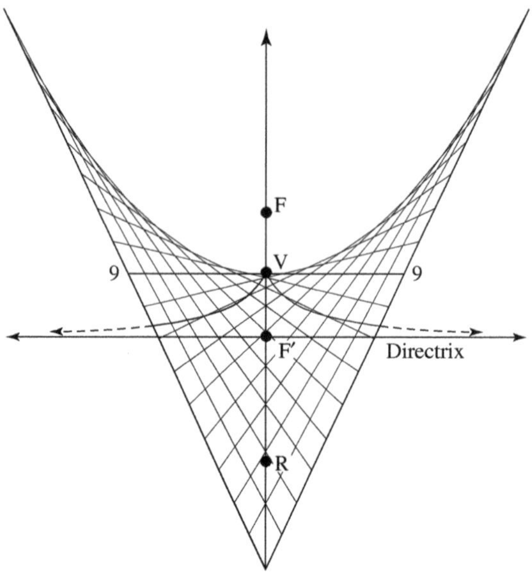

Figure 4.

Now let the focus, F of the parabola in Figure 1 be the fixed point for consideration. If perpendiculars are dropped to each of the tangents, students will note that the locus of the feet of these perpendiculars is the line 9-9. That is, the tangent to the vertex is the pedal curve to a parabola with respect to its focus. Conversely, it can be shown that a perpendicular erected to a tangent with 9-9 will pass through F. This latter fact justifies the technique used earlier to locate the parabola's focus. (Students should recall that to prove a locus, a biconditional statement must be proved.)

Next let V be the fixed point. From V, we drop perpendiculars to each of the tangents obtained in Figure 1. The locus of the feet is shown in Figure 4 as a curve having a cusp at V, and symmetric to the axis. We state without proof that the locus is the *cissoid of Diocles*. Locate F′ on the axis, below V, such that FV = VF′. Through F′, draw a line parallel to 9-9. This line is the parabola's *directrix*, and can be shown to be the asymptotic line that the cissoid approaches.

At this point, you may wish to discuss the various properties of a parabola, such as its reflective properties. You may also define the parabola in terms of a locus, now that the focus and direction have been determined. Thus, a parabola is the locus of points equidistant from a point (the focus), and a line (the directrix), not containing the point. Folding waxed paper clearly demonstrates this locus. Draw a line and an external point on a piece of wax paper. Fold the wax paper repeatedly in such a manner that the point is superimposed on the line. The creases produced form a parabolic envelope.

Postassessment

Have students draw two additional pedal curves to the parabola. Tell them to use the following as a guide:

a. Let F′ be the fixed point. The pedal curve will be seen to the *Right Strophoid.*
b. Locate R (Figure 4), the reflection of F through the directrix such that FF′ = F′R. Have R be the fixed point. The pedal curve is the *trisectrix of MacLaurin.*
c. The *contrapedal* to a given curve is the locus of the feet of the perpendiculars from a given fixed point to the normals to a given curve. In Figure 2, with F as fixed point, determine the locus of the contrapedal. From F, drop perpendiculars to each of the normals you drew to obtain the evolute. The locus of the contrapedal is a parabola whose turning point concurs with F.
d. Confirm by measurement that the contrapedal locus in (c) is identical to the locus of the midpoints of the segment of a normal from the point of tangency on the parabola to the point of intersection of the normal and the parabola's axis of symmetry.

References

Lockwood, E. H. *A Book of Curves.* Cambridge University Press, 1961.

Posamentier, A. S. and H. A. Hauptman. 101+ *Great Ideas for Introducing Key Concepts in Mathematics.* Thousand Oaks, CA: Corwin, 2006.

Zwikker, C. *The Advanced Geometry of Plane Curves and their Applications.* New York: Dover Publications, 1963.

Unit 92

Application of Congruence to Divisibility

Performance Objectives

1. *Given any integer, students will determine its prime factors without doing any division.*
2. *Students will produce rules for testing divisibility by natural numbers other than those presented in this unit.*

Preassessment

1. Have students find the prime factors of each of the following: (a) 144 (b) 840 (c) 360.
2. Have students indicate, without doing any division, which of the following numbers are divisible by 2, 3, and 5: (a) 234 (b) 315.

Teaching Strategies

Begin the lesson by introducing the concept of number congruences. Two numbers that have the same remainder when divided by 7 are said to be *congruent modulo* 7.

For example, 23 and 303 give the same remainder when divided by 7. Therefore, we say that 23 and 303 are congruent modulo 7. This statement can be represented by symbols as follows: $23 = 303 \pmod 7$.

In general, two integers a and b are congruent modulo m (written as $a \equiv b \pmod m$) if they give the same nonnegative remainder when divided by the integer $m \neq 0$.

Because of this definition, we have the following double implication:

$$a \equiv b \pmod m \Leftrightarrow \begin{cases} a = mk + r \\ b = mk' + r \end{cases} \quad 0 \leq r < |m|$$

The symbol "≡" was first used in 1801 by Carl Friedrich Gauss (1777–1855), the famous German mathematician. It is suggested by its similarity with the ordinary equality. It has nothing to do with geometric congruence. The sign $\not\equiv$ means "is not congruent to."

Example 1

$$17 \equiv -4 \pmod 7$$
$$17 = 72 + 3.$$

Because,

$$-4 = 7 \cdot (-1) + 3.$$

In this example, we must use -1 as a quotient. If we use 0, the remainder would be negative against the definition of congruence.

Example 2

$a \equiv 0 \pmod a$. This is true because they both give the same remainder 0.

Another Definition of Congruences

Two numbers are congruent modulo m, *if and only if their difference is divisible by* m. We want to prove that $a \equiv b \pmod m \Leftrightarrow a - b = \bar{m}$ ($\bar{m}$ reads "a multiple of m").

Proof:

If $a \equiv b (\text{mod}\, m) \Leftrightarrow \begin{array}{l} a = mk_1 + r \\ b = mk_2 + r \end{array}$

Subtracting: $a - b = m(k_1 - k_2)$ or $a - b = \bar{m}$

Therefore: $a \equiv b(\text{mod}\, m) \Rightarrow a - b = \bar{m}$

Conversely: if $a - b = \bar{m} \Rightarrow a = b + \bar{m}$

$$\Rightarrow a = b + km. \quad (1)$$

But $b = mk' + r(2)$. Thus, from (1) and (2) we have

$$\begin{aligned} a = b + km &= (mk' + r) + km \\ &= m(k' + k) + r \\ &= mk'' + r(3). \end{aligned}$$

From (2) and (3), we then have

$$\begin{aligned} &a = mk'' + r \\ &b = mk' + r \text{ therefore: } a \equiv b(\text{mod}\, m). \end{aligned}$$

Thus, $a - b = \bar{m} \Rightarrow a \equiv b(\text{mod}\, m)$

Therefore, $a \equiv b(\text{mod}\, m) \Rightarrow a - b = \bar{m}$, Q.E.D.

Students should now be ready to consider the following.

Some Elementary Properties of Congruences

If $a = b \pmod{m}$ and $c \equiv d \pmod{m}$, then

$$\text{(I) } a + c \equiv b + d \pmod{m}$$
$$\text{(II) } ac \equiv bd \pmod{m}$$
$$\text{(III) } ka \equiv kb \pmod{m} \text{ for every integer } k.$$

These properties follow from the definition of congruences.

We shall prove (II); the others can be proved by following the same method.

Since $a \equiv b \pmod{m} \Leftrightarrow a = b + \bar{m}$ (1)
and $c \equiv d \pmod{m} \Leftrightarrow c = d + \bar{m}$ (2)
Then, multiplying (1) and (2):

$$\begin{aligned} ac &= bd + b\bar{m} + d\bar{m} \\ &= bd + (b + d)\bar{m} \\ &= bd + \bar{m}. \end{aligned}$$

Therefore, $ac \equiv bd \pmod{m}$.

Another interesting aspect of modular systems are *power residues*. The *power residues* of a number a with respect to another number m are the *remainders obtained when the successive powers* $a^0, a^1, a^2, \ldots$ *of* a *are divided by* m.

Example 3

Find the power residues of 5 with respect to 3. Since

$$5^0 : 3 = 1 : 3 = 0 \cdot 3 + 1, \text{ therefore, } r_0 = 1$$
$$5^1 : 3 = 5 : 3 = 1 \cdot 3 + 2, \text{ therefore, } r_1 = 2$$
$$5^2 : 3 = 25 : 3 = 8 \cdot 3 + 1, \text{ therefore, } r_2 = 1$$
$$5^3 : 3 = 125 : 3 = 41 \cdot 3 + 2, \text{ therefore, } r_3 = 2$$

and so on.

Therefore, the power residues of 5 with respect to 3 will be: $1, 2, 1, 2, \ldots$. Have students consider why no number other than 1 or 2 appears in this sequence.

Example 4

Find the power residues of 10 modulo 2. Also indicate the different congruences. We have

$$10^0 : 2 = 1 : 2 = 0 \cdot 2 + 1, \text{ therefore, } r_0 = 1.$$

$$10^1 : 2 = 10 : 2 = 5 \cdot 2 + 0, \text{ therefore, } r_1 = 0.$$

$$10^2 : 2 = 100 : 2 = 50 \cdot 2 + 0, \text{ therefore, } r_2 = 0.$$

Therefore, the power residues are $1, 0, 0, \ldots$ Thus, the congruences will be

$$10^0 \equiv 1, 10^1 \equiv 0, 10^2 \equiv 0, \ldots, (\text{mod}2).$$

Students should be able to justify the appearance of this sequence.

Once students have mastered the concept of power residues, they should be ready to consider various *properties of power residues.*

(I) The power residue of a^0, when divided by m, is always 1.

Proof: We have that $a^0 : m = 1 : m = 0 \cdot 1 + 1$, i.e., a remainder of 1. Hence, $a^0 \equiv 1(\text{mod}\, m)$.

(II) If a power residue is zero, then the following power residues are also zero.

Proof: Let a^h give a zero power residue when divided by m. Then, $a^h \equiv 0$ $(\text{ mod } m)$. If both sides are multiplied by a, we have $a \cdot a^h \equiv a \cdot 0$ $(\text{ mod } m)$ or $a^{h+1} \equiv 0$ $(\text{mod}\, m)$. Therefore, $a^{h+1}, a^{h+2}, \ldots$ will give zero power residues also. This was evident in Example 4.

Criteria for divisibility

Have students consider any number $N = a_n a_{n-1} \ldots a_2 a_1 a_0$ written in base 10. Therefore, $N = a_0 10^0 + a_1 10^1 + a_2 10^2 + \cdots + a_n 10^n$. Let $r_0, r_1, \ldots, r_n$ be the power residues of $10(\text{mod}\, m)$. Therefore, $10^0 \equiv 1, 10^1 \equiv r_1, \ldots 10^n \equiv r_n(\text{mod}\, m)$. Have students multiply each congruence by $a_0, a_1, \ldots, a_n$, respectively, to get $a_0 10^0 \equiv a_0 a_1 10^1 \equiv a_1 r_1, \ldots, a_n 10^n \equiv a_n r_n(\text{mod}\, m)$. If they are added in order, we will get $a_0 10^0 + \cdots + a_n 10^n \equiv a_0 + a_1 r_1 + \cdots + a_n r_n(\text{mod}\, m)$.

Thus, $N \equiv a_0 + a_1 r_1 + a_2 r_2 + \cdots + a_n r_n$.

From this last congruence, N will be divisible by m if and only if $a_0 + a_1 r_1 + \cdots + a_n r_n$ is divisible by m. This statement can be used to find the different criteria for divisibility the following way:

Divisibility by 2 and 5

We have for any number N that

$$N = a_0 + a_1 r_1 + \cdots + a_n r_n(\text{mod} m).$$

If students consider $m = 2$ (or $m = 5$), they will have that $r_1 = 0$ because $10^1 = 0 (\bmod 2)$ (or mod 5). Therefore, $r_2 = 0, \ldots$ Hence, they will have $N \equiv a_0 (\bmod 2 \text{ or } 5)$. This means that: A number is divisible by 2 or 5, if and only if, its last digit is divisible by 2 or 5.

Divisibility by 3 and 9

We have that

$$10^0 \equiv 1, 10^1 \equiv 1, \ldots (\bmod 3 \text{ or } 9).$$

Since $N \equiv a_0 + a_1 r_1 + a_2 r_2 + \cdots + a_n r_n (\bmod m)$, thus, $N \equiv a_0 + a_1 + a_2 + \cdots + a_n (\bmod 3 \text{ or } 9)$. Therefore, a given number is divisible by 3 or 9, if and only if the sum of its digits is divisible by 3 or 9.

Divisibility by 11

Since $10^0 \equiv 1,\ 10^1 \equiv -1,\ 10^2 \equiv 1, \ldots, (\bmod 11)$.

Hence, $N \equiv a_0 - a_1 + a_2 - \cdots + (-1)^n a_n (\bmod 11)$. Therefore, a given number is divisible by 11 if and only if the difference of the two sums of the alternate digits is divisible by 11.

The previous method will lead students to develop similar rules for testing divisibility by other primes. It should be emphasized and justified that a number is divisible by a composite number if it is divisible by each of its *relative prime* factors.

Therefore, if we want to determine if a number is divisible by 6, we only have to test its divisibility by 2 and 3.

With sufficient discussion, students should be able to establish a comprehensive list of rules for testing divisibility as well as develop insight into some elementary theory of congruences.

Postassessment

Have students perform these exercises:

1. State a rule for testing divisibility by
   a. 4 and 25 b. 7 c. 13 d. 101
2. Determine the prime factors of:
   a. 1220 b. 315 c. 1001
3. Find the criteria for divisibility by 6 and 11 in base 7.

Unit 93

Problem Solving — A Reverse Strategy

Geometry teachers are frequently asked, "How did you know which approach to take in order to prove these two line segments parallel?" Generally, the teacher would like to think that experience prompted the proper conclusion. This would, of course, be of no value to the questioning student. He or she would like to learn a definite procedure to follow. The teacher would be wise to describe to the student a reverse strategy that would have the student begin with the desired conclusion and discover each preceding step in order.

Performance Objective

Given a problem situation that lends itself to a reverse strategy of solution, students will employ this strategy to successfully solve the problem.

Preassessment

Have students solve the following problem:

If the sum of two numbers is 2, and the product of the same two numbers is 3, find the sum of the reciprocals of these numbers.

Teaching Strategies

A reverse strategy is certainty not new. It was considered by Pappus of Alexandria about 320 A.D. In Book VII of Pappus' *Collection*, there is a rather complete description of the methods of "analysis" and "synthesis." T. L. Heath in his book, *A Manual of Greek Mathematics* (Oxford University Press, 1931, pp. 452–452) provides a translation of Pappus' definitions of these terms:

> *Analysis* takes that which is sought as if it were admitted and passes from it through its successive consequences to something which is admitted as the result of synthesis; for in analysis we assume that which is sought as if it were already done, and we inquire what it is from which this results, and again what is the antecedent cause of the latter, and so on, until, by so retracing our steps, we come upon something already known or belonging to the class of first

> principles, and such a method we call analysis as being solution backwards.
>
> But in *synthesis*, reversing the process, we take as already done that which was last arrived at in the analysis and, by arranging in their natural order as consequences what before were antecedents, and successively connecting them one with another, we arrive finally at the construction of that which was sought; and this we call synthesis.

Unfortunately, this method has not received its due emphasis in the mathematics classroom. This discussion will reinforce the value of reverse strategy in problem solving.

To better understand this technique for problem solving, a number of appropriate problems will be presented. Discussion of their solutions should help students attain a better grasp of this method.

Let us first consider the following simple problem from basic geometry.

Problem 1:

Given: $\overline{AB} \cong \overline{DC}$
$\overline{AB} // \overline{DC}$
$\angle BAH \cong \angle \overline{DCG}$
$\overline{BEGHFD}$
$\overline{GE} \cong \overline{HF}$
Prove: $\overline{AE} // \overline{CF}$

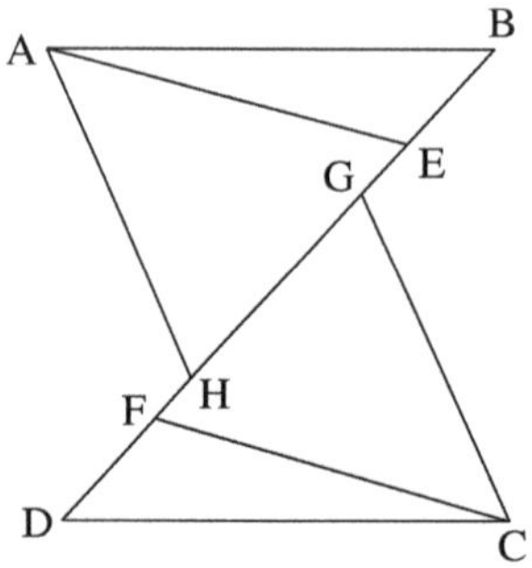

Solution: The first thoughts of a student trying to do this proof is to consider what information is given, and then what is to be proved. Having considered the given information, the poorly trained student will usually proceed blindly, proving segments, angles, and triangles congruent until (if ever) he or she reaches the desired conclusion.

On the other hand, a well-trained student, after considering the given information, will immediately look at the desired conclusion and begin working in reverse from that conclusion ("analysis"). First this student will ask what methods there are for proving lines parallel. This will for the most part lead to proving angles congruent. In this proof, clever students will realize that if they were able to prove $\angle$AED $\cong$ $\angle$CFB, they would then be able to prove $\overline{\text{AE}}$ parallel to $\overline{\text{CF}}$. But how can they prove $\angle$AED $\cong$ $\angle$CFB? Because of the type of training they receive, most students will generally react to this question by trying to find a pair of congruent triangles that have $\angle$AED and $\angle$CFB as a pair of corresponding angles. Continuing this reverse approach, students must now locate such a pair of congruent triangles. It would be helpful if students could prove $\triangle$AEH $\cong$ $\triangle$CFG, as these triangles have $\angle$AED and $\angle$CFB as a pair of corresponding angles. Can these triangles be proven congruent? Evidently not. All that students know about these triangles at this point is that $\overline{\text{HE}} \cong \overline{\text{GF}}$. Using this type of reasoning, they soon will prove that $\triangle$ABH $\cong$ $\triangle$CDG, which will help to prove $\triangle$AEH $\cong$ $\triangle$CFG. Then, by retracing steps of reverse reasoning in the opposite order ("synthesis"), students will easily attain the desired conclusion. It is clear that reverse strategy was instrumental in formulating a path to the desired conclusion.

The reverse approach to solving a problem becomes dramatically stronger, when the resulting solution becomes significantly more elegant. As an example, let us consider the following problem offered in the preassessment.

Problem 2: If the sum of two numbers is 2 and the product of these same two numbers is 3, find the sum of the reciprocals of these two numbers.

Solution: A first reaction after reading this problem would be to set up the equation $x + y = 2$, and $xy = 3$. A well-trained student of algebra would promptly set out to solve these equations simultaneously. She or he may solve the first equation for y to get $y = 2 - x$, and then substitute appropriately in the second equation so that $x(2-x) = 3$ or $x^2-2x+3 = 0$. As $x = 1\pm\sqrt{-2}$, the two numbers are $1+i\sqrt{2}$ and $1-i\sqrt{2}$. Now the sum of their reciprocals is

$$\frac{1}{1+i\sqrt{2}} + \frac{1}{1-i\sqrt{2}} = \frac{(1-i\sqrt{2})+(1+i\sqrt{2})}{(1+i\sqrt{2})\cdot(1-i\sqrt{2})} = \frac{2}{3}.$$

This solution is by no means elegant.

Had students used a reverse strategy ("analysis"), they would have first inspected the desired conclusion; that is, $\frac{1}{x} + \frac{1}{y}$.

The sum of these fractions is $\frac{x+y}{xy}$. The two original equations immediately reveal the numerator and the denominator of this fraction. This produces the answer, $\frac{2}{3}$, immediately. For this particular problem, a reverse strategy was superior to the more common, straightforward approach.

Problem 3: If the sum of two numbers is 2, and the product of the same two numbers is 3, find the sum of the squares of the reciprocals of these numbers.

Solution: To find the sum of the squares of the reciprocals (of the numbers described in the above problem) by a reverse approach, the student must first consider the conclusion, that is: $\left(\frac{1}{x}\right)^2 + \left(\frac{1}{y}\right)^2$ or $\frac{1}{x^2} + \frac{1}{y^2}$. Once again, students would be required to add the fractions to get $\frac{x^2+y^2}{x^2y^2}$. Therefore, the denominator of the answer is $(xy)^2 = 9$. However, the numerator is not as simple to evaluate as it was earlier. Students must now find the value of $x^2 + y^2$. Once again students must look backward. Can they somehow generate $x^2 + y^2$? Students will be quick to suggest that $(x + y)^2$ will yield $x^2 + y^2 + 2xy$, which in part produces $x^2 + y^2$. Besides, $(x + y)^2 = (2)^2 = 4$ and $2xy = 2 \cdot 3 = 6$. Hence, $x^2 + y^2 = -2$. The problem is therefore, solved as $\frac{1}{x^2} + \frac{1}{y^2} = \frac{x^2+y^2}{x^2y^2} = \frac{-2}{9}$.

A similar procedure can be employed to find the value of $\left(\frac{1}{x}\right)^3 + \left(\frac{1}{y}\right)^3$ from the original two equations, $x+y = 2$ and $xy = 3$. Once again, beginning with the conclusion and working in reverse, $\frac{1}{x^3} + \frac{1}{y^3} = \frac{x^3+y^3}{(xy)^3}$. Since students already know that $(xy)^3 = (3)^3 = 27$, they need only to find the value of $x^3 + y^3$. How can they generate $x^3 + y^3$?

From $(x + y)^3 = x^3 + y^3 + 3x^2y + 3xy^2$

we get $\quad x^3 + y^3 = (x + y)^3 - 3xy(x + y)$

$x^3 + y^3 = (2)^3 - 3(3)(2)$

$x^3 + y^4 = -10.$

Therefore, $\frac{1}{x^3} + \frac{1}{y^3} = \frac{x^3+y^3}{(xy)^3} = \frac{-10}{27}$.

This procedure may also be used to find the sum of higher powers of these reciprocals.

Another problem whose solution lends itself nicely to a reverse strategy ("analysis") involves geometric constructions.

Problem 4: Construct a triangle given the lengths, m_a and m_b, of two medians of a triangle and the length, c, of the side whose endpoints are each an endpoint of one of the medians.

Solution: Rather than immediately trying to perform the required construction, students would be wise to use a reverse strategy. They may assume construction and inspect the results.

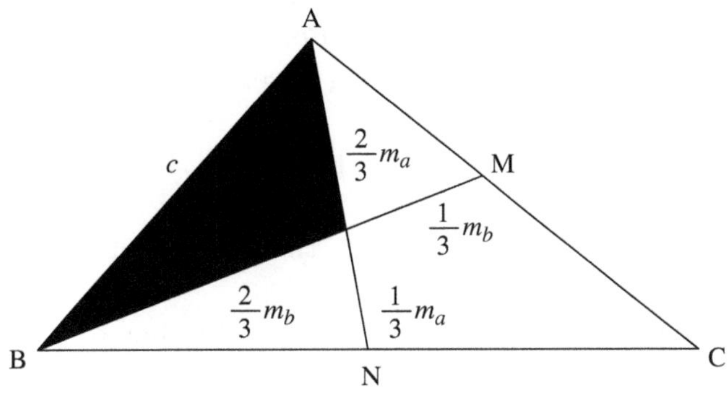

Students will soon realize that they would be able to construct the shaded triangle above as they can easily obtain the lengths of its sides $\left(c, \frac{2m_a}{3}, \frac{2m_b}{3}\right)$. Points M and N can then be located using the property of the centroid. Then point C will be determined by the intersection of $\overrightarrow{\text{AM}}$ and $\overrightarrow{\text{BN}}$. Having started from the conclusion and working in reverse, students have formulated a plan for constructing the required triangle, by merely retracing steps in the reverse direction ("synthesis").

Although there are many problems whose solutions can be significantly simplified by using a reverse strategy, there are also a great number of problems where a straightforward approach is best. It is natural for a student to approach a problem in a straightforward manner. Yet we as teachers must encourage our students to abandon the straightforward approach when a solution is not easily forthcoming, and attempt a reverse solution.

Some problems require only a partial reverse strategy. In such problems, it is useful to begin with the conclusion and work backward until a path to the conclusion is established. Let us consider the following problems.

Problem 5: Find the solution of the following: $\left(x - y^2\right)^2 + (x - y - 2)^2 = 0$, where x and y are real numbers.

Solution: Students of algebra might naturally use a straightforward approach to solve this equation. After squaring each of the polynomials as indicated, confusion would mount. Students previously exposed to a reverse strategy would then try to analyze the solution set of the equation. The values of x and y must be such that the sum of the squares of the polynomials equals zero. How can the sum of two positive numbers equal zero? Students can answer this question by saying that $x - y^2 = 0$ and $x - y - 2 = 0$. Up to this point, students used a reverse strategy ("analysis"). However, now they must proceed in a straightforward manner ("synthesis") solving the equations $x - y^2 = 0$ and $x - y - 2 = 0$ simultaneously.

In his book *How to Solve It*, George Polya discusses a backward method of problem solving that is similar to a reverse strategy discussed in this unit. Polya emphasizes the importance of the role of a teacher in presenting such methods to students when he states that "there is some sort of psychological repugnance to this reverse order which may prevent a quite able student from understanding the method if it is not presented carefully."

It is the responsibility of the mathematics teacher to make a conscious effort to stress the importance, benefits, and possible limitations of a reverse strategy in problem solving.

Postassessment

1. If $x + y = 2$ and $xy = 3$, find $\frac{1}{x^4} + \frac{1}{y^4}$.
2. Construct a triangle given the lengths of two sides and the length of an altitude to one of these sides.
3. Have students use analysis and synthesis to prove the following:

   In right $\triangle$ABC, $\overline{\text{CF}}$ the median drawn to hypotenuse $\overline{\text{AB}}$, $\overline{\text{CE}}$ is the bisector of $\angle$ACB, and $\overline{\text{CD}}$ is the altitude to $\overline{\text{AB}}$.
   Prove that $\angle$DCE $\cong$ $\angle$ECF.

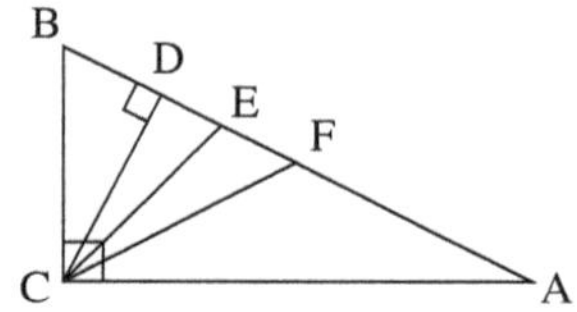

4. Evaluate $x^5 + \frac{1}{x^5}$, if $x^2 + \frac{1}{x^2} = 7$.
   (Answer: ± 123)

References

Polya, G. *How to Solve It*. Princeton, NJ: Princeton University Press, 1973.

Posamentier, A. S. and S. Krulik. *Problem-Solving Strategies of Efficient and Elegant Solutions: A Resource for the Mathematics Teacher*. Thousand Oaks, CA: Corwin, 2nd Ed., 2008.

Posamentier, A. S. and S. Krulik. *Problem-Solving Strategies in Mathematics: from Common Approaches to Exemplary Strategies*, Hackensack, New Jersey: World Scientific Publishing, 2015.

Unit 94

Decimals and Fractions in Other Bases

Performance Objective

Students will rationalize repeating decimals or repeating fractions in other bases.

Preassessment

Ask students to find the decimal number equivalent to $\frac{87}{99}\left(=\frac{87}{10^2-1}\right)$. Challenge students to represent a repeating decimal by a simple rational number.

Teaching Strategies

Decimal numbers are usually classified as repeating and nonrepeating. Repeating decimals are further partitioned into terminating and nonterminating decimals. Students are usually readily aware that a terminating decimal represents a particular rational number. But the nature of a nonterminating decimal is more intriguing. We begin this exploration by confining ourselves to nonterminating repeating decimals. Consider the repeating decimal: $.1212\overline{12}$. (The bar over the last two digits indicates the repeating digits.) What we want to do is to represent this decimal by an equivalent rational simple fraction. If we let $x = .12\overline{12}\ldots$ and then $100x = 12.12\overline{12}\ldots$, subtracting the former from the latter yields the equation: $100x - x = 12$, or $x = \frac{12}{100-1} = \frac{12}{99}$. We have now found a ratio representation for .1212.... Some further exploration is in order now. Note that $\frac{12}{99} + \frac{88}{99} = 1$. However, if we add the equivalent decimal representations $\frac{+.8787\overline{87}}{.9999\overline{99}}$ one would think that $.99999\overline{9} = 1$. Indeed, applying the above technique yields $x = .9999\overline{9}$ and $10x = 9.9999\overline{9}$; therefore, $10x - x = 9$, and $x = \frac{7}{10-1}, x = 1$.

This illustration leads us to an important theorem: any repeating decimal can be represented as a rational number (i.e., the *ratio* of two integers, the denominator not zero).

Proof: Let the repeating decimal be represented by $.a_1a_2\ldots a_n\ldots$ where a_i is a digit and n is an integer that represents the length of the repetition. As before, let

$$x = .a_1a_2\ldots a_n\ldots$$

and

$$10^n x = a_1a_2\ldots a_n a_1a_2\ldots a_n\cdots$$

Now

$$\begin{aligned} 10^n x - x &= a_1a_2\ldots a_n \\ x(10^n - 1) &= a_1a_2\ldots a_n \\ x &= \frac{a_1a_2\cdots a_n}{10^n - 1}. \end{aligned}$$

The repeating decimal is now represented by a rational number.

Students will now want to consider repeating fractions in bases other than 10 (no longer called decimals!). Suppose we have in base 3 the repeating fraction: $.12\overline{12}$. Students should be guided to ask the following:

1. Can this repeating fraction be represented by a rational number in base 3?
2. In general, can *any* repeating fraction in *any* given base be represented by a rational number?

Begin by using the approach applied earlier to repeating decimals. Let $x = .12\overline{12}$. Ask students how the ternary point may be shifted two places to the right (note the ternary point in base 3 is analogous to the decimal point in base 10). Now $3^2x = 12.12\overline{12}$. By subtracting x, we get $3^2x - x = 12$, $x(3^2 - 1) = 12, x = \frac{12}{3^2-1} = \frac{12}{22}$. Thus, the repeating fraction in base 3 can be represented by a rational number. Have students notice the analogous form in base 10.

Using these illustrations as models, we shall prove that a repeating fraction in any base can be represented by a rational number in that base.

Proof: Consider any base B and any repeating fraction in that base: $.a_1a_2\ldots a_n\ldots$ where a_1 is a digit of the number and n is an integer that represents the length of the repetition $x = .a_1a_2\ldots a_n\ldots$

$$\begin{aligned} B^n x &= a_1a_2\ldots a_n.a_1a_2\ldots a_n\ldots \\ B^n x - x &= a_1a_2\ldots a_n \\ x &= \frac{a_1a_2\ldots a_n}{B^n - 1}. \end{aligned}$$

This proves that any repeating fraction can be represented by a rational number.

Postassessment

Have students do the following exercises:

1. If $x = \frac{123}{10^3-1}$, what is its decimal representation?
2. If $x = \frac{11256}{7^4-1}$, represent x as a rational fraction.
3. Rationalize the repeating fraction $x = .23\overline{23}$ when x is in base 10, 8, and 5.

Unit 95

Polygonal Numbers

This unit can be taught to a class that has a reasonably good command of the basic algebraic skills. Since most of the unit employs intuitive thinking, a good degree of this training should result. It would be helpful if the students were familiar with arithmetic sequences and the formula for the sum of its series. However, if students are not familiar with this topic, essentials can be developed in a reasonably short time.

Performance Objectives

1. *Given the rank of any regular polygon, the student will find a number that corresponds to it.*
2. *The student will discover relations between two or more different polygonal numbers of given ranks.*

Preassessment

The ancient Babylonians discovered that some whole numbers can be broken down into patterns of units. This link between arithmetic and geometry was also of concern to the ancient Greeks. For example, the number 3 can be represented as three dots forming a triangle, as can be number 6.

Which regular polygon do you think the number 4 represents? The number 9? After the students have had some time to find such polygons, ask them to provide the answers.

Numbers that can be related to geometric figures are called *figurate* or *polygonal* numbers.

Teaching Strategies

Tell the students that it would be very easy to find the number that corresponds to a given regular polygon if we could find a formula such that given any regular polygon and its rank, we could obtain that number from it.

Begin by telling the students what the rank of a regular polygon indicates. For any regular polygon, the rank indicates, in order, the corresponding polygonal number. For example, for a triangle, rank $1 = 3$ (the first triangular number), rank $2 = 6$ (the second triangular number), rank $3 = 10$, etc.

Now draw five figures that will show how to get the first five ranks of the first five figurate numbers (triangular, square, pentagonal, hexagonal, and heptagonal). To save time, you may use an overhead projector instead, or you may distribute mimeographed sheets with the drawings. Make a corresponding table. Both the drawings and the table that follow indicate what might be shown to your students.

It should be clear to the student that making a figure to obtain every possible triangular, square, etc. number is a very tedious task. Instead we will study how consecutive polygonal numbers of a given polygon follow each other and, by looking at the sequence formed, try to obtain a formula for the rth rank of each given polygon.

If we look at the first row of figurate numbers corresponding to triangular numbers, and if we also look at their corresponding ranks (Table 1), we will notice that they can be written as

$$1 = r$$

$$3 = (r-1) + r$$

$$6 = (r-2) + (r-1) + r$$

$$10 = (r-3) + (r-2) + (r-1) + r$$
$$15 = (r-4) + (r-3) + (r-2) + (r-1) + r.$$

Table 1.

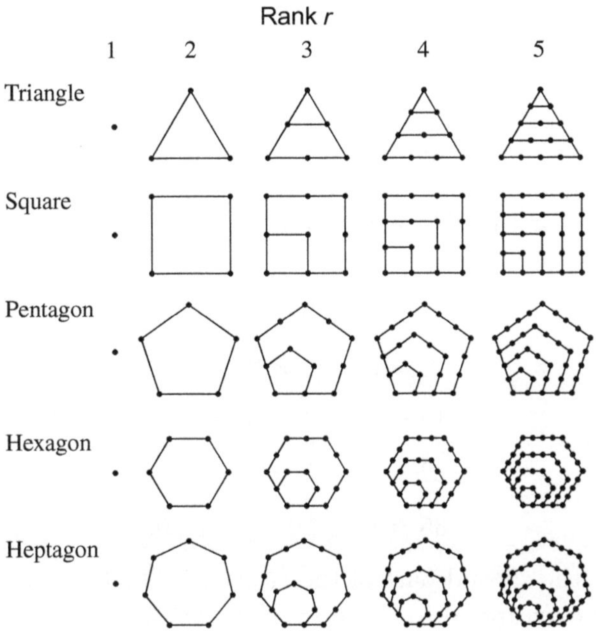

Table 2.

Figure	No. sides N	Rank r 1	2	3	4	5
Triangular	3	1	3	6	10	15
Square	4	1	4	9	16	25
Pentagonal	5	1	5	12	22	35
Hexagonal	6	1	6	15	28	45
Heptagonal	7	1	7	18	34	55

If we look at the ranks, we will also notice that their sequence forms an arithmetic sequence and that each triangular number of rank r is the sum of that arithmetic sequence $1, 2, 3, \ldots, r$ from 1 to r.

Thus, we can conclude that the rth triangular number is given by $T_r = \frac{r(r+1)}{2}$.

Next, let's look at the square numbers:

$$1 = r^2 = 1^2$$
$$4 = r^2 = 2^2$$
$$9 = r^2 = 3^2$$
$$16 = r^2 = 4^2$$
$$25 = r^2 = 5^2.$$

It is clear that each square number is the square of its corresponding rank. So the r square number is r^2.

The formula for the rth pentagonal number can be obtained if we write each number in the following way:

$$1 = r^2 + 0 = 1 + 0$$
$$5 = r^2 + 1 = 2^2 + 1$$
$$12 = r^2 + 3 = 3^2 + 3$$
$$22 = r^2 + 6 = 4^2 + 6$$
$$35 = r^2 + 10 = 5^2 + 10.$$

If we study the second part of the sums 0,1,3,6,10, we will see that each of the numbers correspond to the sum of the arithmetic sequence $0, 1, 2, \ldots,$ $(r-1)$, which is $\frac{(r-1)r}{2}$. So the rth pentagonal number is

$$r^2 + \frac{(r-1)r}{2} = \frac{2r^2 + (r-1)r}{2} = \frac{(2r^2 + r^2 - r)}{2}$$
$$= \frac{(3r^2 - r)}{2} = \frac{r(3r-1)}{2}.$$

To find a formula for the rth hexagonal number, consider the first five as follows:

$$1 = 1r$$
$$6 = 3r = 3(2)$$
$$15 = 5r = 5(3)$$
$$28 = 7r = 7(4)$$
$$45 = 9r = 9(5).$$

An inspection of the coefficients of r: $1, 3, 5, 7, 9$, would reveal that each corresponds to the sum of the corresponding rank and the rank immediately before it. That is, each coefficient is equal to $r + (r - 1)$. Therefore, the rth hexagonal number is $[r + (r - 1)]r = (2r - 1)r$.

The rth heptagonal number is found as follows. Write the first seven heptagonal numbers in the following way:

$$\begin{aligned} 1 &= 2r^2 - 1 = 2(1)^2 - 1 \\ 7 &= 2r^2 - 1 = 2(2)^2 - 1 \\ 18 &= 2r^2 + 0 = 2(3)^2 + 0 \\ 34 &= 2r^2 + 2 = 2(4)^2 + 2 \\ 55 &= 2r^2 + 5 = 2(5)^2 + 5. \end{aligned}$$

It probably will be very difficult for the students to arrive at a formula for the second part X of each number $2r^2 + X$. Therefore, after the students have looked at the numbers for a short time, the teacher should immediately point out that each X is equal to the sum of the arithmetic sequence $-1, 0, 1, 2, 3, \ldots, (r - 2)$ minus one, which is $\frac{(r-2)(r-1)}{2} - 1$.

Students should test the formula on each of the given numbers above. So the rth heptagonal number is

$$\begin{aligned} &2r^2 + \frac{(r-2)(r-1)}{2} - 1 \\ &\quad = 2r^2 + \frac{(r-2)(r-1) - 2}{2} \\ &\quad = 2r^2 + \frac{r^2 - 3r + 2 - 2}{2} = \frac{r(5r-3)}{2}. \end{aligned}$$

Call attention to the fact that we now have a formula for the rth rank of each of the first five polygonal numbers. So we are now able to find any triangular, square, pentagonal, hexagonal, and heptagonal number. But, there are regular polygons of $8, 9, \ldots 20, 100$, etc. sides, and we would also like to have a formula for the rth rank of each of them. It is our next task to find such formulas.

To do this, let's write the formulas we have already found as follows:

No. of sides	Rank=r
3	$\frac{r(r+1)}{2} = \frac{r^2+r}{2} = \frac{1r^2}{2} + \frac{r}{2}$
4	$r^2 = \frac{2r^2}{2} = \frac{2r^2}{2} + \frac{0}{2}$
5	$\frac{r(3r-1)}{2} = \frac{3r^2-r}{-2} = \frac{3r^2}{2} - \frac{r}{2}$
6	$r(2r-1) = \frac{4r^2-2r}{2} = \frac{4r^2}{2} - \frac{2r}{2}$
7	$\frac{r(5r-3)}{2} = \frac{5r^2-3r}{2} = \frac{5r^2}{2} - \frac{3r}{2}$
⋮	
⋮	
⋮	
⋮	
N	

Let's now look at the last column. We will notice that the coefficients of the $\frac{r^2}{2}$ terms can be written as $(N-2)$. Also the coefficients of the $\frac{r}{2}$ terms can be written as $-(N-4)$. Therefore, the rth rank of a N-gonal number is

$$\frac{(N-2)r^2}{2} - \frac{(N-4)r}{2} = \frac{(N-2)r^2-(N-4)r}{2} =$$
$$\left(\frac{r}{2}\right)[(N-2)r-(N-4)] = \left(\frac{r}{2}\right)[(r-1)N-2(r-2)].$$

The completed table (including the first five ranks of the N-gonal number) looks like this:

No.	Rank						
sides	1	2	3	4	5	...	r
3	1	3	6	10	15	...	$\frac{r(r+1)}{2}$
4	1	4	9	16	25	...	r^2
5	1	5	12	22	35	...	$\frac{r(3r-1)}{2}$
6	1	6	15	28	15	...	$r(2r-1)$
7	1	7	18	34	55	...	$\frac{r(5r-3)}{2}$
$\vdots$							
N	1	$\ldots\left(\frac{r}{2}\right)[(r-1)N-2(r-2)]$					

At this point, it would be instructive for the students to workout some simple examples using the formula for the $r^{\text{th}}N$-gonal number.

Examplc 1

Find the third octagonal number.

Solution: Let N = 8 and r = 3. Substitute these numbers in the formula $\frac{r}{2}$ $[(r-1)N-2(r-2)]$

$$= \frac{3}{2}[(3-1)8-2(3-2)] = \frac{3}{2}(2)8-2(1)$$

$$= \frac{3}{2}[16-2] = \frac{3\times 14}{2} = 21.$$

Example 2

To what regular polygon does the number 40 correspond if $r = 4$?

Solution: In this case, we know the rank and the number, but we must find N. We substitute into and solve the following equation:

$$\left(\frac{r}{2}\right)[(r-1)N - 2(r-2)] = 40$$

$$\left(\frac{4}{2}\right)[(4-1)N - 2(4-2)] = 40$$

$$2[3N - 2(2)] = 40, \quad \text{and} \quad N = 8.$$

So the figure is a regular octagon.

The following examples are a little more difficult for they call for applications of the formulas to find relationships between different types of polygonal numbers.

Example 3

Show that the rth pentagonal number is equal to r plus three times the $(r-1)^{\text{th}}$ triangular number.

Solution: To do this problem, we must first write the formula for the rth pentagonal number:

$$P_r = \frac{r(3r-1)}{2} = \frac{3r^2}{2} - \frac{r}{2}.$$

Rewrite

$$\frac{-r}{3} \quad \text{as} \quad \frac{-3r}{2} + r.$$

Now we have

$$\frac{3r^2}{2} - \frac{3r}{2} + r = \frac{3(r^2 - r)}{2} + r$$

$$\frac{3r(r-1)}{2} + r = 3\mathrm{T}_{r-1} + r,$$

where T_{r-1} is the $(r-1)^{\text{st}}$ triangular number.

Example 4

Show that any hexagonal number is equal to the sum of a pentagonal number of the same rank and a triangular number of the preceding rank.

Solution:
$$\begin{aligned}(\text{Hex})_r &= r(2r-1) = 2r^2 - r\\ &= \frac{3r^2-r}{2} + \frac{r^2-r}{2}\\ &= \frac{r(3r-1)}{2} + \frac{r(r-1)}{2} = P_r + T_{r-1}.\end{aligned}$$

Postassessment

1. Draw a regular octagon corresponding to the third octagonal number of Example 1 (study the drawings of the first five figurate numbers before doing this problem).
2. Find the first three decagonal (10 sides) numbers.
3. Show that any heptagonal number is equal to the sum of a hexagonal number of the same rank and a triangular number of the previous rank (i.e., show $(\text{Hep})_r = (\text{Hex})_r + T_{r-1}$).
4. Show that any N-gonal number ($N \geq 5$) is equal to the sum of the $(N-1)$-gonal number of the same rank and a triangular number of the previous rank. (*Hint*: Begin with the $(N-1)$-gonal number of rank r and T_{r-1}. Carry out the addition.)
5. Show that the sum of any number of consecutive odd integers, starting with 1, is a perfect square (i.e., a square number).

Unit 96

Networks

This unit will serve as an introductory lesson in topology.

Performance Objective

Given a closed curve, students will determine if it is traversible or non-traversible.

Preassessment

Have students try to trace with a pencil each of the following configurations without missing any part and without going over any part twice.

Ask students to determine the number of arcs or line segments that have an endpoint at each of A, B, C, D, E.

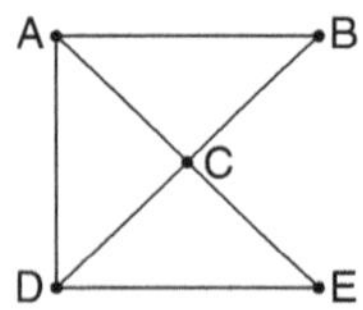

Figure 1.

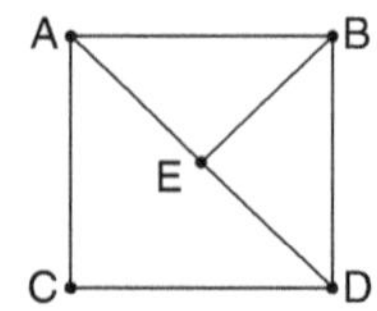

Figure 2.

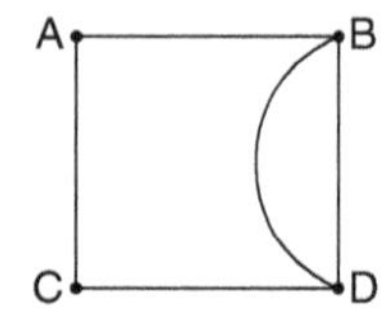

Figure 3.

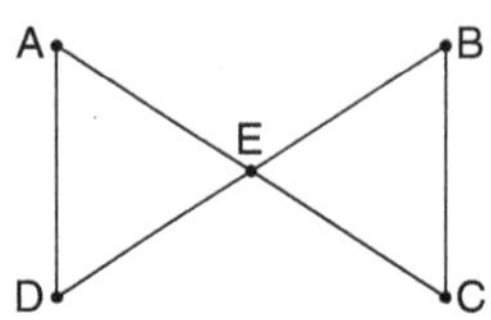

Figure 4.

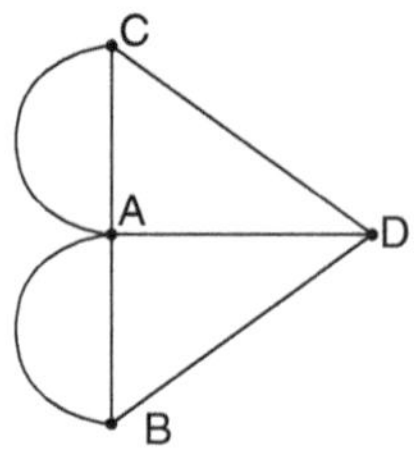

Figure 5.

Teaching Strategies

Configurations such as Figures 1–5, which are made up of line segments and/or continuous arcs are called *networks*. The number of arcs or line segments that have an endpoint at a particular vertex, is called the *degree* of the vertex.

After trying to trace these networks without taking their pencils off the paper and without going over any line more than once, students should notice two direct outcomes. The networks can be traced (or *traversed*) if they have (1) all even degree vertices or (2) exactly two odd degree vertices. The proof of these two outcomes follows.

There are an even number of odd degree vertices in a connected network.

Proof: Let V_1 be the number of vertices of degree 1, let V_3 be the number of vertices of degree 3, and V_n be the number of vertices of degree n. Also let $N = V_1 + V_3 + V_5 + \cdots + V_{2n-1}$. N is the number of odd degree vertices in a given connected network ("connected" meaning without loose ends). Since there are 3 arc endpoints at V_3, 5 at V_5, and n at V_n, the total number of arc endpoints in a connected network is $M = V_1 + 2V_2 + 3V_3 + 4V_4 + \cdots + 2nV_{2n}$.

$$\begin{aligned} M - N &= 2V_2 + 2V_3 + 4V_4 + 4V_5 + \cdots + (2n-2)V_{2n-1} + 2nV_{2n} \\ &= 2(V_2 + V_3 + 2V_4 + 2V_5 + \cdots + 2(n-1)V_{2n-1}2nV_{2n}). \end{aligned}$$

Since the difference of two even numbers is an even number, $M-(M-N) = N$ is an even number.

A connected network can be traversed only if it has at most two odd degree vertices.

Proof: On a continuous path, the inside vertices must be passed through. That is, if a line "enters" the point, another must "leave" the point. This accounts for the endpoints. The only vertices that do not conform to this rule are the beginning and endpoints in the traversing. These *two* points may be of odd order. By the previous theorem, it was established that there must be an even number of odd vertices; therefore there can only be *two* or *zero* vertices of odd order in order to traverse a network.

Have students now draw both traversible and nontraversible networks (using these two theorems). Network 1 in the Preassessment has five vertices. B, C, E are of even degree and vertices A and D are of odd degree. Since Figure 1 has exactly two odd degree vertices as well as three even degree vertices, it is traversible. If we start at A then go down to D, across to E, back up to A, across to B, and down to D we have chosen a desired route.

Network 2 has five vertices. Vertex C is the only even degree vertex. Vertices A, B, E, and D, are all of odd degree. Consequently, since the network has more than two odd vertices, it is not traversible.

Network 3 is traversible because it has two even vertices and exactly two odd degree vertices.

Network 4 has five even degree vertices and can be traversed.

Network 5 has four odd degree vertices and *cannot* be traversed.

To generate interest among your students, present them with the famous Königsberg Bridge Problem. In the eighteenth century, the small Prussian city of Königsberg, located where the Pregel River formed two branches, was faced with a recreational dilemma: Could a person walk over each of the seven bridges exactly once in a continuous walk through the city? In 1735, the famous mathematician Leonhard Euler (1707–1783) proved that this walk could not be performed. Indicate to students that the ensuing discussion will tie in their earlier work with networks to the solution of the Königsberg Bridge Problem.

Tell pupils to indicate the island by A, the left bank of the river by B, the right one by C, and the area between the two arms of the upper course by D. If we start at Holzt and walk to *Sohmede* and then through *Honig*, through *Hohe*, through *Kottel*, through *Grüne*, we will never cross Kramer. On the other hand, if we start at *Kramer* and walk to Honig, through Hohe, through Kottel, through Sohmede, through *Holzt*, we will never travel through *Grüne*.

The Königsberg Bridge Problem is the same problem as the one posed in Figure 5. Let's take a look at Figures 5 and 6 and note the similarity. There are seven bridges in Figure 6 and there are seven lines in Figure 5. In Figure 5, each vertex is of odd degree. In Figure 6, if we start at D, we have three choices, we could go to Hohe, Honig, or Holzt. If in Figure 5 we start at D, we have three line paths to choose from. In both figures if we are at C, we have either three bridges we could go on or three lines. A similar situation exists for locations A and B in Figure 6 and vertices A and B in Figure 5. Emphasize that this network *cannot* be traversed.

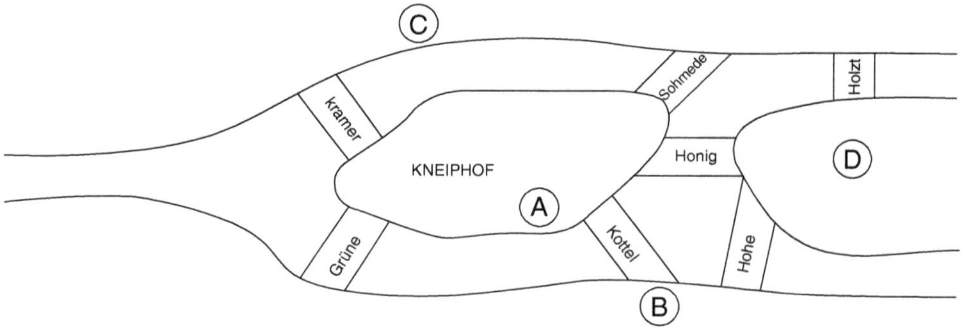

Figure 6.

Another example of a problem where the consideration of the traversibility of a network is important is the Five Room House Problem. Have students consider the diagram of a five-room house.

Each room has a doorway to each adjacent room and a doorway leading outside the house. The problem is to have a person start either inside or outside the house and walk through each doorway *exactly* once.

Students should be encouraged to try various paths. They will realize that although the number of attempts is finite, there are far too many to make a trial-and-error solution practical. They should be guided to a network diagram analogous to this problem.

Figure 7b shows various possible paths joining the five rooms A, B, C, D, and E, and the outside F. The problem now reduces to merely determining

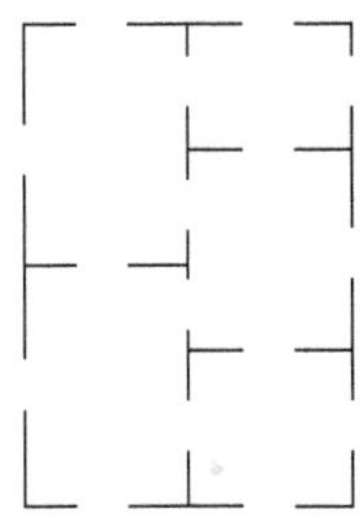

Figure 7A.

if this network is traversible. There are *four vertices* of odd degree and two vertices of even degree. Because there are not *exactly* two or zero vertices of odd order, this network *cannot* be traversed; hence the Five Room House Problem does not have a solution path.

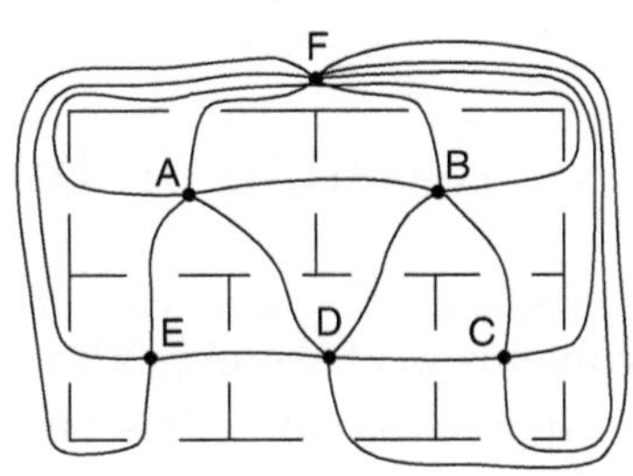

Figure 7B.

Other problems of a similar nature may now be presented to the students.

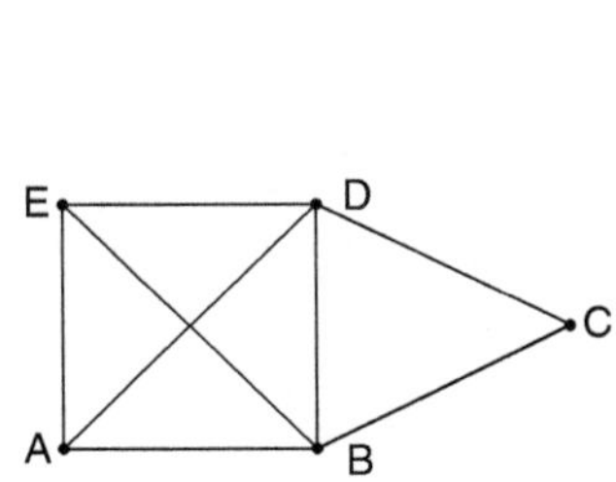

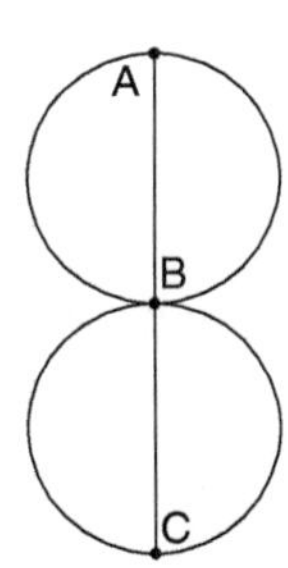

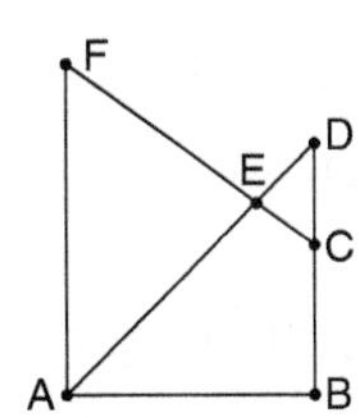

Figure 8. Figure 9. Figure 10.

Postassessment

1. Have students determine if the networks in figure 8, 9 and 10 can be traced without removing their pencils from the paper and without going over any line twice (i.e., traversed).
2. Have students draw a house floor plan and then determine if one can walk through each doorway exactly once.

Reference

Posamentier, A.S. and W. Schulz (eds.). *The Art of Problem-Solving: A Resource for the Mathematics Teacher.* Thousand Oaks, CA: Corwin, 1996.

Posamentier, A.S. and P. Poole, *Understanding Mathematics Through Problem-Solving*, Hackensack, New Jersey: World Scientific Publishing, 2020.

Unit 97

Angle Trisection — Possible or Impossible?

Of the Three Famous Problems of Antiquity, the one most instructive to a high school student is the angle trisection. This unit will present a discussion and proof that any angle cannot be trisected with only straightedge and compasses.

Performance Objective

Students will outline a proof that an angle of measure 120° *cannot be trisected.*

Preassessment

Students should be familiar with the basic geometric constructions.

Teaching Strategies

Ask students to trisect an angle of measure 90° using only straightedge and compasses. With little difficulty, they ought to be able to construct an angle of measure 60° at the vertex of the given angle. This virtually completes the trisection. However, now ask students to trisect an angle of measure 120°. This will cause difficulty because it is impossible with straightedge and compasses. At this point, begin discussion of the impossibility of angle trisection using only straightedge and compasses.

With the aid of a unit length and an angle of measure A, it is possible to construct a line segment of length cos A (see Figure 1).

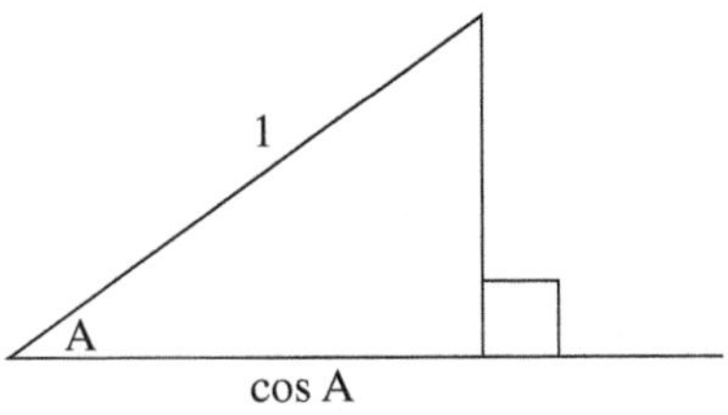

Figure 1.

If we can trisect $\angle$A, then we can also construct the $\cos\frac{A}{3}$. If we can show that $\cos\frac{A}{3}$ cannot be constructed, then we have shown that $\angle$A cannot be trisected. Here we shall let $m\angle A = 120°$, and show $\angle$A cannot be trisected.

We shall first obtain an expression for $\cos A$ in terms of $\cos\frac{A}{3}$.

$$\cos 3y = \cos(2y + y) = \cos 2y \cos y - \sin 2y \sin y.$$

But $\cos 2y = 2\cos^2 y - 1$.

Substituting

$$\begin{aligned}\cos 3y &= \cos y(2\cos^2 y - 1) - \sin 2y \sin y\\ &= [2\cos^3 y - \cos y] - \sin 2y \sin y.\end{aligned}$$

But $\sin 2y = 2\sin y \cos y$.

Therefore,

$$\cos 3y = [2\cos^3 y - \cos y] - \sin y(2\sin y \cos y)$$

$$\cos 3y = [2\cos^3 y - \cos y] - 2\sin^2 y \cos y$$

$$\cos 3y = [2\cos^3 y - \cos y] - 2\cos y(1 - \cos^2 y)$$

$$\cos 3y = [2\cos^3 y - \cos y] - 2\cos y + 2\cos^3 y$$

$$\cos 3y = 4\cos^3 y - 3\cos y.$$

Let $3y = A$ to obtain

$$\cos A = 4\cos^3\frac{A}{3} - 3\cos\frac{A}{3}.$$

Multiply by 2 and replace $2\cos\frac{A}{3}$ with x to get

$$2\cos A = x^3 - 3x.$$

Because $\cos 120^\circ = -\frac{1}{2}, x^3 - 3x + 1 = 0.$

Students should now recall that one of the criteria of constructibility indicates that constructible roots must be of the form $a + b\sqrt{c}$, where a and b are rational and c is constructible.

First, then, we must show that $x^3 - 3x + 1 = 0$ rational roots. To do this, we assume that there is a rational root, $\frac{p}{q}$, where p and q have no common factor greater than 1. Substituting for $\frac{p}{q}$, we have

$$\left(\frac{p}{q}\right)^3 - 3\left(\frac{p}{q}\right) + 1 = 0$$
$$p^3 - 3pq^2 + q^3 = 0$$
$$q^3 = 3pq^2 - p^3$$
$$q^3 = p\left(3q^2 - p^2\right).$$

This means that q^3, and hence q, has the factor p.
Therefore, p must equal ± 1. Also, solving for p^3

$$p^3 = 3pq^2 - q^3$$
$$p^3 = q^2(3p - q).$$

This means p and q must have a common factor, and hence $q = \pm 1$. We can conclude from this that the only rational root of $x^3 - 3x + 1 = 0$ is $r = \pm 1$. By substitution, we can show that neither $+1$ nor -1 is root.

Next, assume $x^3 - 3x + 1 = 0$ has a constructible root $a + b\sqrt{c}$. By substitution in the equation $x^3 - 3x + 1 = 0$, we can show that if $a + b\sqrt{c}$ is a root, then its conjugate, $a - b\sqrt{c}$, is also a root. The sum of the roots of the polynomial equation $x^n + a_1 x^{n-1} + a_2 x^{n-2} + \cdots + a_n = 0$ is

$$r_1 + r_2 + r_3 + \cdots + r_n = -a_1.$$

It follows from this that the sum of the roots $x^3 - 3x + 1 = 0$ is zero. If two roots are $a + b\sqrt{c}$ and $a - b\sqrt{c}$, with the third root r, we have

$$a + b\sqrt{c} + a - b\sqrt{c} + r = 0$$
$$r = -2a.$$

But a is rational and hence r is rational, and we have a contradiction. Hence, the angle whose measure is 120° cannot be trisected. This essentially proves that any angle cannot be trisected with only straightedge and compasses.

Postassessment

Have students write an outline of the proof presented in this unit as well as a discussion of its significance.

Unit 98

Comparing Means

This unit can be used as a major part of a lesson on statistics.

Performance Objectives

1. *Students will compare magnitudes of three means for any two or more numbers.*
2. *Students will prove comparison relationships between means.*

Preassessment

After students have reviewed the arithmetic and geometric means, have students express h in terms of a and b, where a, h, b are a harmonic sequences.

Teaching Strategies

Begin by defining the three means (arithmetic, harmonic, and geometric) in the following way.

Suppose a, m, b are an arithmetic sequence. The middle term (m) is said to be the *arithmetic mean*. Since a, m, b have a common difference $m - a =$

$b - m$, and

$$\boxed{m = \frac{a+b}{2} = \text{arithmetic mean (A.M.)}}.$$

Now suppose a, h, b are a harmonic sequence. The middle term (h) is said to be the *harmonic mean.* Since a, h, b have reciprocals with a common difference, $\frac{1}{h} - \frac{1}{a} = \frac{1}{b} - \frac{1}{h}$ and

$$\boxed{h = \frac{2ab}{a+b} = \text{harmonic mean (H.M.)}}.$$

Finally suppose a, g, b are a *geometric sequence.* Since a, g, b have a common ratio, $\frac{g}{a} = \frac{b}{g}$ and

$$\boxed{g = \sqrt{ab} = \text{geometric mean (G.M.)}}.$$

Often a pictorial model crystallizes understanding, so a geometric interpretation is appropriate here. Consider the semiclrcle with diameter $\overline{\text{AOPB}}$ with $\overline{\text{AO}} \cong \overline{\text{OB}}$ and $\overline{\text{PR}} \perp \overline{\text{APB}}$.(R is on the semicircle.) Also $\overline{\text{PS}} \perp \overline{\text{RSO}}$. Let AP $= a$ and PB $= b$.

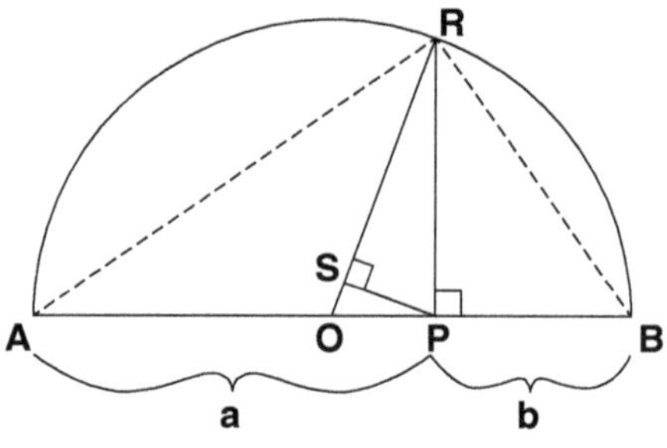

Since RO $= \frac{1}{2}$AB $= \frac{1}{2}$(AP + PB) $= \frac{1}{2}(a+b)$, RO is the *arithmetic mean* (A.M.) between a and b.

Consider right $\triangle$ARB. Since $\triangle$BPR $\sim$ $\triangle$RPA, $\frac{\text{PB}}{\text{PR}} = \frac{\text{PR}}{\text{AP}}$ or $(\text{PR})^2 =$ (AP) $\cdot$ (PB) $= ab$. Therefore, PR $= \sqrt{ab}$. Thus, PR is the *geometric mean* (G.M.) between a and b.

Because ΔRPO $\sim$ ΔRSP, $\frac{\text{RO}}{\text{PR}} = \frac{\text{PR}}{\text{RS}}$. Therefore, RS $= \frac{(\text{PR})^2}{\text{RO}}$. But $(\text{PR})^2 = ab$ and RO $= \frac{1}{2}$AB $= \frac{1}{2}(a+b)$. Thus, RS $= \frac{ab}{\frac{1}{2}(a+b)} = \frac{2ab}{a+b}$, which is the *harmonic mean* (H.M.) between a and b.

This geometric interpretation lends itself quite well to a comparison of the magnitudes of these three means. Since the hypotenuse of a right triangle is its longest side, in $\triangle$ROP, RO $>$ PR and in $\triangle$RSP, PR $>$ RS. Therefore,

RO > PR > RS. However, these triangles may degenerate, so RO $\geq$ PR $\geq$ RS, which implies that A.M. $\geq$ G.M. $\geq$ H.M.

The student is familiar with both arithmetic mean and geometric mean (sometimes called the mean proportional), but a brief introduction to the harmonic mean might be in order.

The *harmonic mean* between two numbers is the *reciprocal of the arithmetic mean between the reciprocals* of these two numbers. This is because a harmonic sequence is a sequence of reciprocals of members of an arithmetic sequence. For a and b,

$$\text{H.M.} = \frac{1}{\frac{\frac{1}{a}+\frac{1}{b}}{2}} = \frac{2ab}{a+b}.$$

Example 1: Find the harmonic mean of a, b, and c.

Solution: By the definition,

$$\text{H.M.} = \frac{1}{\frac{\frac{1}{a}+\frac{1}{b}+\frac{1}{c}}{3}} = \frac{3abc}{ab+ac+bc}.$$

Both the arithmetic mean and the geometric mean have popular applications in secondary school curriculum. The harmonic mean also has a very useful and often neglected application in elementary mathematics. The harmonic mean is the "average of rates." For example, suppose the average rate of speed for the trip to and from work is desired, when the rate of speed to work is 30 m.p.h. and returning (over the same route) is 60 m.p.h. The average speed is the harmonic mean between 30 and 60; that is,

$$\frac{(2)(30)(60)}{30+60} = 40.$$

To show that the average rate of speed of two (or more) speeds is in fact the harmonic mean of these speeds, consider rates of speed $r_1, r_2, r_3, \ldots, r_n$, each traveled for a time $t_1, t_2, t_3, \ldots, t_n$, respectively, and *each* over a distance d.

$$t_1 = \frac{d}{r_1}, t_2 = \frac{d}{r_2}, t_3 = \frac{d}{r_3}, \ldots t_n = \frac{d}{r_n}.$$

The average speed (for the entire trip) is

$$\frac{\text{total distance}}{\text{total time}} = \frac{nd}{t_1+t_2+t_3+\cdots+t_n} = \frac{nd}{\frac{d}{r_1}+\frac{d}{r_2}+\cdots+\frac{d}{r_n}} = \frac{n}{\frac{1}{r_1}+\frac{1}{r_2}+\frac{1}{r_3}+\cdots+\frac{1}{r_n}},$$

which is the harmonic mean.

Example 2

If Lisa bought \$1.00 worth of each of three kinds of candy, 15 ¢, 25¢, and 40¢ per pound, what was her average price paid per pound?

Solution: Because the harmonic mean is the average of rates (taken over the same base), the average price per pound was

$$\frac{(3)(15)(25)(40)}{(15)(25)+(15)(40)+(25)(40)} = 22\frac{62}{79}¢.$$

To complete the discussion of a comparison of the magnitude of the three means, consider with the class a more general (algebraic) discussion.

In general terms,

$$\text{A.M.} = \frac{a_1 + a_2 + a_3 + \cdots + a_n}{n}$$

$$\text{G.M.} = \sqrt[n]{a_1 \cdot a_2 \cdot a_3 \cdot \ldots \cdot a_n}$$

$$\text{H.M.} = \frac{\frac{n}{n}}{\frac{1}{a_1} + \frac{1}{a_2} + \frac{1}{a_3} + \cdots + \frac{1}{a_n}}.$$

Theorem 1: A.M. $\geq$ G.M.

Proof: Let $g = \sqrt[n]{a_1 \cdot a_2 \cdot a_3 \cdot \cdots \cdot a_n}$;

then $1 = \sqrt[n]{\frac{a_1}{g} \cdot \frac{a_2}{g} \cdot \frac{a_3}{g} \cdot \cdots \cdot \frac{a_n}{g}}$.

Therefore, $1 = \frac{a_1}{g} \cdot \frac{a_2}{g} \cdot \frac{a_3}{g} \cdot \cdots \cdot \frac{a_n}{g}$.

However, $\frac{a_1}{g} + \frac{a_2}{g} + \frac{a_3}{g} + \cdots + \frac{a_n}{g} \geq n$ since if the product of n positive numbers equals 1, their sum is *not* less than n. Therefore,

$$\frac{a_1 + a_2 + a_3 + \cdots + a_n}{n} \geq g.$$

Hence, $\frac{a_1 + a_2 + a_3 + \cdots + a_n}{n} \geq \sqrt[3]{a_1 \cdot a_2 \cdot a_3 \cdot \cdots \cdot a_n}$,
or A.M. $\geq$ G.M.

This proof for two numbers a and $b (a > b)$ is rather cute.

Since $a - b > 0$, $(a-b)^2 > 0$, or $a^2 - 2ab + b^2 > 0$. By adding $4ab$ to both sides of the inequality,

$$a^2 + 2ab + b^2 > 4ab.$$

Taking the positive square root yields

$$\frac{a + b}{2} > \sqrt{ab}.$$

Hence, A.M. > G.M. (*Note*: if $a = b$ then A.M. = G.M.)

Theorem 2: G.M. ≥ H.M.

Proof: Since A.M. ≥ G.M. for $a_1^b,\ a_2^b,\ a_3^b\ \cdots,\ a_n^b$

$$\frac{a_1^b + a_2^b + a_3^b + \cdots + a_n^b}{n} \geq \sqrt[n]{a_1^b \cdot a_2^b \cdot a_3^b \cdots a_n^b}$$

when

$$\frac{1}{b} < 0, \left[\sqrt[n]{a_1^b \cdot a_2^b \cdot a_3^b \cdot \cdots \cdot a_n^b}\right]^{\frac{1}{b}} \geq \left[\frac{a_1^b + a_2^b + a_3^b + \cdots + a_n^b}{n}\right]^{\frac{1}{b}}$$

Take $b = -1$, then

$$\sqrt[n]{a_1 \cdot a_2 \cdot a_3 \cdot \ldots a_n} \geq \left[\frac{a_1^{-1} + a_2^{-1} + a_3^{-1} + \cdots + a_n^{-1}}{n}\right]^{-1}$$

Hence,

$$\sqrt[n]{a_1 \cdot a_2 \cdot a_3 \cdot \cdots a_n} \geq \frac{n}{\frac{1}{a_1} + \frac{1}{a_2} + \frac{1}{a_3} + \cdots + \frac{1}{a_n}}, \text{ or G.M.} \geq \text{H.M.}$$

Once again for two numbers a and b $(a > b)$, the proof becomes much simpler: Because (from above) $a^2 + 2ab + b^2 > 4ab, ab(a + b)^2 > (4ab)(ab)$.

Therefore, $ab > \frac{4a^2b^2}{(a + b)^2}$, or $\sqrt{ab} > \frac{2ab}{a + b}$.

Hence, G.M. > H.M. (*Note*: if $a = b$ then G.M. = H.M.)

Postassessment

1. Find the A.M., G.M, and H.M. for each of the following:
   a. 20 and 60
   b. 25 and 45
   c. 3, 15, and 45
2. Arrange the G.M., H.M., and A.M. in ascending order of magnitude.
3. Show that for two given numbers the G.M. is the geometric mean between the A.M. and H.M.
4. Prove that the G.M. > H.M. for a, b, and c.

References

Posamentier, A. S. and H. A. Hauptman. *101+ Great Ideas for Introducing Key Concepts in Mathematics.* Thousand Oaks, CA: Corwin, 2006.

Posamentier, A. S. and I. Lehmann. *Mathematical Curiosities: A Treasure Trove of Unexpected Entertainments.* Prometheus Books, 2014.

Posamentier, A. S. and C. Spreitzer, *The Mathematics Everyday Life*, Amherst, New York: Prometheus Books, 2018.

Unit 99

Pascal's Pyramid

The ability to expand and generalize is one of the most important facilities a teacher can help a student develop. In this unit, the familiar application of Pascal's triangle to determine the coefficients of a binomial expansion $(a + b)^n$ is expanded by the use of "Pascal's pyramid" to consider the coefficients of $(a + b + c)^n$.

Performance Objectives

1. *Students will evaluate trinomial expansions* $(a+b+c)^n$ *of lower powers.*
2. *Students will discover important relationships between Pascal's triangle and Pascal's pyramid.*

Preassessment

If your students are familiar with Pascal's triangle, have them perform the following expansions:

a. $(a+b)^3$
b. $(a-b)^4$
c. $(x+2y)^5$

Ask your students to test their algebraic multiplication (and patience) by evaluating

a. $(a+b+c)^3$
b. $(a+b+c)^4$

Teaching Strategies

Begin by reviewing Pascal's triangle. You might mention that this triangle is not Pascal's alone. In fact, the triangle was well known in China before 1300 and was also known to Omar Khayyam, author of the *Rubaiyat*, almost 600 years before Pascal. Historical accuracy aside, each row of Pascal's (or Khayyam's or Ying Hui's) triangle yields the coefficients of $(a + b)^n$.

1	$(a+b)^0$
11	$(a+b)^1$
121	$(a+b)^2$
1331	$(a+b)^3$
14641	$(a+b)^4$

For example, to find $(a+b)^4$, use the coefficients in row 5 of the triangle: $a^4 + 4a^3b + 6a^2b^2 + 4ab^3 + b^4$.

Whereas a binomial expansion can be represented by a readily visible triangle, the trinomial expansion is represented by the more complex pyramid. The first expansion $(a+b+c)^0$ has the single coefficient 1. We can visualize this as the vertex of the pyramid. Each succeeding expansion is then represented by a triangular cross-section of the pyramid with the coefficient 1 at each of the vertices.

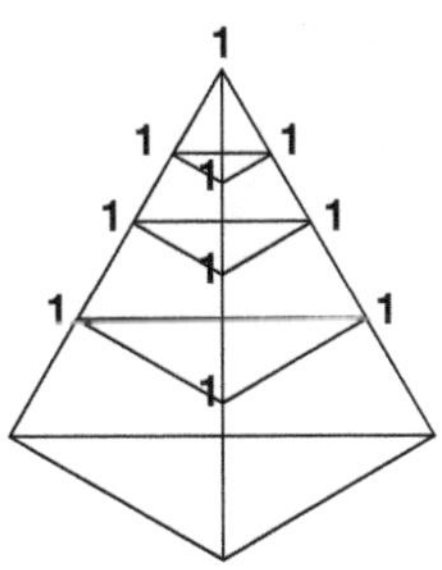

Figure 1.

Therefore, each of the lateral edges of the pyramid consists of a sequence of 1s. The second expansion $(a+b+c)^1$ has coefficients $1a + 1b + 1c$, which are represented by the first layer triangle with entries only at the vertices,

$$\begin{matrix} & 1 & \\ 1 & & 1. \end{matrix}$$

There are two methods of generating the coefficients of higher powers by means of the pyramid. In the first, again consider each expansion as a triangular cross-section of a pyramid. The numbers on the outer edge of each layer (the numbers between the vertices) are found by adding the two numbers that lie directly above. For example, $(a+b+c)^2$ has 1s at each vertex, and 2s between

$$\begin{array}{ccccc} & & 1 & & \\ & 2 & & 2 & \\ 1 & & 2 & & 1. \end{array}$$

To determine the terms in the interior of the triangle, add the three terms that lie above; for example, $(a+b+c)^3$ has the following coefficients:

$$\begin{array}{ccccccc} & & & 1 & & & \\ & & 3 & & 3 & & \\ & 3 & & 6 & & 3 & \\ 1 & & 3 & & 3 & & 1 \end{array}$$

or referring to the pyramid:

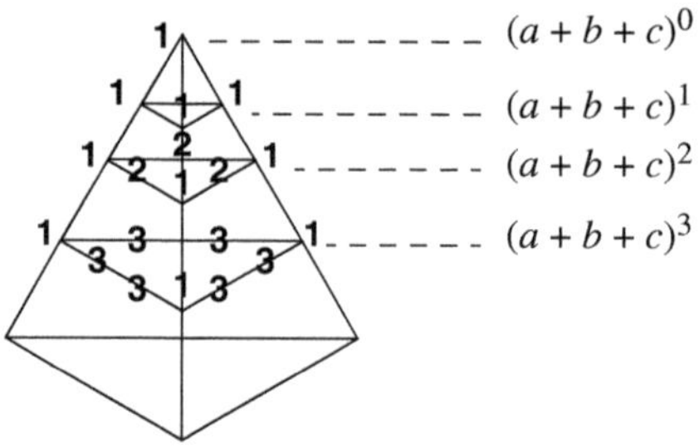

Figure 2.

To assign these coefficients to the correct variables:

1. Let the coefficient in the first row of the pyramid be "a" to the highest power of that expansion;
2. Let the elements of the second row be the coefficients of the product of "a" to the second highest power and the other variables to the first power;
3. In the third row again decrease the power of "a" and arrange the other variables such that the sum of the exponents of each term equals the power the original expansion was raised to;
4. Within each row, the powers of "a" remain the same while the power of "b" decreases left to right and the power of "c" increases.

Specifically, consider $(a+b+c)^3$, which has a coefficient configuration:

$$\begin{array}{ccccccc} & & & 1 & & & \\ & & 3 & & 3 & & \\ & 3 & & 6 & & 3 & \\ 1 & & 3 & & 3 & & 1. \end{array}$$

The full expansion would then be $a^3+3a^2b+3a^2c+3ab^2+6abc+3ac^2+b^3+3b^2c+3bc^2+c^3$.

In working with these pyramids, students may notice that the edge of each triangle corresponds exactly to a row of the Pascal triangle, i.e., the edge of $(a+b+c)^3$, 1 3 3 1, is the same as the fourth row in Pascal's triangle. This observation leads to the second method of deriving the pyramid.

Let the left edge of the trinomial expansion be represented by the corresponding row of Pascal's triangle. Then multiply each row of Pascal's triangle by the number on the left edge to generate the coefficients for the trinomial expansion. For example, the left edge of $(a+b+c)^4$ will be 1 4 6 4 1, corresponding to the fifth row of Pascal's triangle.

$$\begin{array}{cc} 1 & (1\times 1) \\ 1\ 1 & (4\times 1)(4\times 1) \\ 1\ 2\ 1 & (6\times 1)(6\times 2)(6\times 1) \\ 1\ 3\ 3\ 1 & (4\times 1)(4\times 3)(4\times 3)(4\times 1) \\ 1\ 4\ 6\ 4\ 1 & (1\times 1)(1\times 4)(1\times 6)(1\times 4)(1\times 1). \end{array}$$

Multiplying these elements along the edge by the consecutive rows of the triangle yields $(a+b+c)^4$.

$$\begin{array}{ccccccccc} & & & & 1 & & & & \\ & & & 4 & & 4 & & & \\ & & 6 & & 12 & & 6 & & \\ & 4 & & 12 & & 12 & & 4 & \\ 1 & & 4 & & 6 & & 4 & & 1. \end{array}$$

This may at first appear to be a complicated procedure, but practice in its use will clear up initial confusion and introduce an intriguing and useful technique.

Postassessment

1. Have the students compare the time required to expand $(a+b+c)^4$ algebraically versus the pyramid expansion.
2. Expand $(a+b+c)^5, (a+b+c)^6$.

3. Expand $(a + 2b + 3c)^3, (a + 4b + c)^4$.
4. Some students may be interested in constructing a working model of the pyramid, composed of detachable triangular sections with the appropriate coefficients noted on each surface.

Unit 100

The Multinomial Theorem

This unit should be used with a class that has already studied the Binomial Theorem.

Performance Objectives

1. *Students will find the coefficient of any given term of a given multinomial expansion without actually expanding it.*
2. *Students will justify the existence of the coefficients of the multinomial expansion.*
3. *Students will successfully apply the Multinomial Theorem to a given trinomial.*

Preassessment

Have students expand $(a+b)^4$ by using the Binomial Theorem. Ask students to determine the number of different arrangements that can be formed from the letters AAABBBCC.

Teaching Strategies

Begin by reviewing student responses concerning the number of arrangements of AAABBBCC. They should realize that this problem differs from asking them to determine the number of arrangements of ABCDEFGH (where each symbol to be arranged is different from the rest). In the latter case, the first place (of the eight places) can be filled in any one of eight ways, the second place can be filled in any one of seven ways, the third in six ways, the fourth in five ways, ..., the eighth in only one way. Using the counting principle, the total number of ways is $8 \cdot 7 \cdot 6 \cdot 5 \cdot 4 \cdot 3 \cdot 2 \cdot 1 = 8!$ (read "8 factorial").

Having already studied the Binomial Theorem, students should be familiar with the basic concepts of *combinations*.

That is, ${}_nC_r = \binom{n}{r} = \frac{nPr}{r!} = \frac{n!}{r!(n-r)!}$.

Students should now be ready to consider the original problem, finding the number of arrangements of AAABBBCC. They should be carefully led through the following development.

Let $\#(A)$ represent the "number of ways of selecting positions for the As." Because there are three As, from the 8 positions, 3 must be selected. This can be done ${}_8C_3$ or $\binom{8}{3}$ ways. Hence $= \binom{8}{3} = \frac{8!}{3! \cdot 5!}$.

Similarly $\#(B) = \binom{5}{3} = \frac{5!}{3! \cdot 2!}$, because 3 positions for the three Bs must be selected from the remaining 5 positions. This leaves 2 positions to be selected for the two Cs. Since only 2 positions remain, there is only one way of selecting these 2 positions, that is $\binom{2}{2} = \frac{2!}{2! \cdot 0!} = 1$. Indicate that $0! = 1$ by definition. By using the counting principle, $\#(\text{A and B and C}) = \#(\text{A}) \cdot \#(\text{B}) \cdot \#(\text{C}) =$

$$\frac{8!}{3! \cdot 5!} \cdot \frac{5!}{3! \cdot 2!} \cdot \frac{2!}{2! \cdot 0!} = \frac{8!}{3! \cdot 3! \cdot 2!}.$$

This last expression is usually symbolized as $\binom{8}{3,3,2}$, which represents the number of ways of arranging 8 items consisting of repetitions of 3 items, 3 items, and 2 items.

To reinforce this technique, ask your students to determine the number of ways in which the letters of *Mississippi* can be arranged. Taking into consideration the repetitions (i.e., 1-M, 4-Is, 4-Ss, 2-Ps), students should obtain

$$\frac{11 \quad 10 \quad 9 \quad \not{8} \quad 7 \quad \not{6} \quad 5 \quad \not{4} \quad \not{3} \quad \not{2} \quad \not{1}}{1 \quad \not{4} \quad \not{3} \quad \not{2} \quad 1 \quad \not{4} \quad \not{3} \quad \not{2} \quad 1 \quad \not{2} \quad \not{1}} = 34{,}650.$$

Students should now be guided to generalizing this scheme for counting to n items that include n_1 items of one kind, n_2 items of another, n_3 items of a third kind, $\ldots, n_r$ items of a last kind. Clearly $n_1 + n_2 + n_3 + \ldots + n_r = n$. Applying the scheme from before, $\#(N_1)$ shall represent "the number of ways in which n_1 positions may be selected from the n positions available." Hence $\#(N_1) = \frac{n!}{n_1!(n-n_1)!}$. Similarly, $\#(N_2) = \frac{(n-n_1)!}{n_2!(n-n_1-n_2)!}$, since only $n - n_1$ places remained from which to select n_2 positions. Similarly, $\#(\text{N}_r) = \frac{(n-n_1-n_2-\ldots-n_{r-1})!}{n_r!(n-n_1-n_2-\ldots-n_r)!}$. Since $n_1 + n_2 + n_3 + \cdots + n_r = n_1\#(N_r)\ldots 1$, $\frac{n_r!}{n_r! \cdot 0!} = 1$.

Using the counting principle for these r cases, the numbers of ways of arranging these n items (with r items being repeated) is obtained:

$$\frac{n!}{n_1!(n-n_1)!}\cdot\frac{(n-n_1)!}{n_2!(n-n_1-n_2)!}\cdot\frac{(n-n_1-n_2)!}{n_3!(n-n_1-n_2-n_3)!}$$

$$\ldots\cdot 1=\frac{n!}{n_1!\cdot n_2!\cdot n_3!\cdots n_r!}=\binom{n}{n_1,n_2,n_3,\ldots,n_r},$$

which is a convenient symbol to use here.

Students should apply this general formula to the case where $r = 2$. They will get $\binom{n}{n_1 n_2} = \frac{n!}{n_1!\cdot n_2!} = \frac{n!}{n_1!(n-n_1)!}$, which is the familiar ${}_nC_{n_1}$ or $\binom{n}{n_1}$.

The students should now be ready to tackle the Multinomial Theorem. In the preassessment, they were asked to expand $(a + b)^4$. They should note that certain terms appear more than once. For example, the term *aaab*, commonly written a^3b, appears $\binom{4}{3}$ times. This corresponds to the number of arrangements of *aaab*. For each such term, the same argument holds.

Students should now consider the expansion of $(a+b+c)^4$. To actually compute this expansion, students may multiply a different combination of the members of each of the factors to obtain each term. For example, some of the 81 terms will appear as *aaaa*, *aaab*, *aabb*, *abac*, *abab*, *cbcb*, ... These are commonly written as $a^4, a^3b, a^2b^2, a^2bc, a^2b^2, b^2c^2$. In the above list, a^2b^2 appeared twice; however, in the complete expansion (of 81 terms), it would appear $\binom{4}{2,2,0} = \frac{4!}{2!\ 2!\ 0!} = 6$ times. Thus, if a student were asked to find the coefficient of the term a^3bc^2 in the expansion $(a+b+c)^6$, he would merely evaluate $\binom{6}{3,1,2} = \frac{6!}{3!\cdot 1!\cdot 2!} = 60$. Hence, the entire expansion may be written as

$$(a+b+c)^4=\sum_{n_1+n_2+n_3=4}\frac{4!}{n_1!\cdot n_2!\cdot n_3!}\cdot a^{n_1}\cdot b^{n_2}\cdot c^{n_3}.$$

From here, the general Multinomial Theorem follows easily:

$$(a_1+a_2+a_3+\cdots+a_r)^n$$

$$=\sum_{n_1+n_2+\cdots+n_r=n}\frac{n!}{n_1!\cdot n_2!\cdot\ldots\cdot n_r!}\cdot a_1^{n_1}\cdot a_2^{n_2}\cdot a_3^{n_3}\cdot\ldots\cdot a_r^{n_r}.$$

Although rather cumbersome, some students might wish to prove this theorem by mathematical induction.

Following are two applications of the Multinomial Theorem.

1. Expand and simplify:

$$(2x+y-z)^3 = \binom{3}{3,0,0}(2x)^3(y)^0(-z)^0 + \binom{3}{0,3,0}(2x)^0(y)^3(-z)^0$$
$$+\binom{3}{0,0,3}(2x)^0(y)^0(-z)^3 + \binom{3}{2,1,0}(2x)^2(y)^1(-z)^0$$
$$+\binom{3}{2,0,1}(2x)^2(y)^0(-z)^1 + \binom{3}{1,1,1}(2x)^1(y)^1(-z)^1$$
$$+\binom{3}{0,2,1}(2x)^0(y)^2(-z)^1 + \binom{3}{0,1,2}(2x)^0(y)^1(-z)^2$$
$$+\binom{3}{1,2,0}(2x)^1(y)^2(-z)^0 + \binom{3}{1,0,2}(2x)^1(y)^0(-z)^2.$$

$$(2x+y-z)^3 = 8x^3 + y^3 - z^3 + 12x^2y - 12x^2z +$$
$$6xy^2 + 6xz^2 - 12xyz - 3y^2z + 3yz^2.$$

2. Find the term in the expansion of $\left(2x^2 - y^3 + \frac{1}{2}z\right)^3$ that contains x^4 and z^4. The general term of the expansion is $\binom{7}{a,\,b,\,c}(2x^2)^a(-y^3)^b\left(\frac{1}{2}z\right)^c$, where $a+b+c=7$. Thus, the terms containing x^4 and z^4 have $a=2$ and $c=4$, $\therefore\ b=1$. Substituting in the above gives $\binom{7}{2,\,1,\,4}(2x^2)^2(-y^3)^1\left(\frac{1}{2}z\right)^4 = \frac{7!}{2!1!4!}(4x^4)(-y^3)\times\left(\frac{1}{16}z^4\right) = \frac{-105}{4}x^4y^3z^4$.

Postassessment

1. Have students find the coefficient of a^2b^5d in the expansion of $(a+b-c-d)^8$.
2. Ask students to explain how the coefficients for any term of a multinomial expansion are derived.
3. Have students expand $(2x+y^2-3)^5$.

Unit 101

Algebraic Solution of Cubic Equations

People's interest in cubic equations can be traced back to the times of the early Babylonians about 1800–1600 B.C. However, the algebraic solution of third degree equations is a product of the Italian Renaissance.

The algebraic solution of cubic equations is thus associated with the names of the Italian mathematicians Scipione del Ferro, Niccolo de Brescia (called Tartaglia), Gerolamo Cardano, and Rafael Bombelli.

Performance Objectives

1. *Given some cubic equations, students will find their solutions.*
2. *Given a verbal problem that calls for a solution of a cubic equation, students will determine (where applicable) the real solutions to the problem.*

Preassessment

Students should have mastered operations with quadratic equations. They should also have a solid background in complex numbers and trigonometry.

Teaching Strategies

Review roots of complex numbers in the following fashion:

The nth root of a complex number z is obtained by taking the nth root of the absolute value r, and dividing the amplitude ϕ by n. This will give you the principal value of that root. The general formula to get all the roots of z is

$$\sqrt[n]{z} = \sqrt[n]{r}\left[\cos\frac{\phi + 2k\pi}{n} + i\sin\frac{\phi + 2k\pi}{n}\right].$$

For $k = 0$, this yields the principal value, and for $k = 1, 2, 3, \ldots, n-1$, we get the rest of the roots.

Example 1

Find the cube roots of unity. We have that $1 = \cos 0° + i\sin 0°$, therefore, $\phi = 0°$ and $r = 1$. The general formula is then, $z = \cos\frac{2k\pi}{3} + i\sin\frac{2k\pi}{3}$, where

$k = 0, 1,$ and 2.

$$\text{If } k = 0,\ z_1 = \cos 0 + i \sin 0 = 1 \text{ (Principal value).}$$

$$\text{If } k = 1,\ z_2 = \cos \frac{2\pi}{3} + i \sin \frac{2\pi}{3} = \cos 120^\circ + i \sin 120^\circ$$

$$= -\cos 60^\circ + i \sin 60^\circ = -\frac{1}{2} + \frac{\sqrt{3}}{2} i.$$

$$\text{If } k = 2,\ z_3 = \cos \frac{4\pi}{3} + i \sin \frac{4\pi}{3} = \cos 240^\circ + i \sin 240^\circ$$

$$= -\cos 60^\circ - i \sin 60^\circ = -\frac{1}{2} - \frac{\sqrt{3}}{2} i.$$

Note that each one of the complex roots of unity generates the other roots. To do so, we only have to take the second and third powers of those roots. For example,

if we take $\alpha = z_2 = -\frac{1}{2} + \frac{\sqrt{3}}{2} i$, we get

$$\alpha^2 = \left(-\frac{1}{2} + \frac{\sqrt{3}}{2} i\right)^2 = \left(-\frac{1}{2}\right)^2 + 2\left(-\frac{1}{2}\right)\left(\frac{\sqrt{3}}{2} i\right) + \left(\frac{\sqrt{3}}{2} i\right)^2$$

$$\alpha^2 = \frac{1}{4} - \frac{\sqrt{3}}{2} i + \frac{3}{4} i^2, \text{ but } i^2 = -1, \text{ thus,}$$

$$\alpha^2 = \frac{1}{4} - \frac{3}{4} - \frac{\sqrt{3}}{2} i = -\frac{1}{2} - \frac{\sqrt{3}}{2} i = z_3.$$

Similarly,

$$\alpha^3 = \alpha^2 \cdot \alpha = \left(-\frac{1}{2} - \frac{\sqrt{3}}{2} i\right)\left(-\frac{1}{2} + \frac{\sqrt{3}}{2} i\right)$$

$$\alpha^3 = \left(-\frac{1}{2}\right)^2 - \left(\frac{\sqrt{3}}{2} i\right)^2 = \frac{1}{4} - \frac{3}{4} i^2$$

$$= \frac{1}{4} + \frac{3}{4} = 1 = z_1.$$

Therefore, the three roots of unity are 1, α, and α^2 where α can be either $z_2 = -\frac{1}{2} + \frac{\sqrt{3}}{2} i$ or $z_3 = -\frac{1}{2} - \frac{\sqrt{3}}{2} i$.

Example 2

Find the cubic roots of the real number a. We have that $a = a(\cos 0^\circ + i \sin 0^\circ)$; therefore, $\sqrt[3]{a} = \sqrt[3]{r}\left(\cos \frac{2k\pi}{3} - i \sin \frac{2k\pi}{3}\right)$, where $k = 0, 1,$ or 2. But,

$\cos\frac{2k\pi}{3} - i\sin\frac{2k\pi}{3}$ where $k = 0, 1$, or 2, will give the three roots of unity (see Example 1). Thus, if the real root of a is a', the three roots of a will be $a', a'\alpha$, and $a'\alpha^2$, where α can be either $-\frac{1}{2} + \frac{\sqrt{3}}{2}i$ or $-\frac{1}{2} - \frac{\sqrt{3}}{2}i$.

Let us now consider the general cubic equation:

$$ax^3 + bx^2 + cx + d = 0,$$

where a, b, c, and d are arbitrary complex numbers. This equation can be reduced to a simpler form without the second degree term, by making the transformation $x = y - \frac{b}{3a}$. Thus, we have

$$a\left(y - \frac{b}{3a}\right)^3 + b\left(y - \frac{b}{3a}\right)^2 + c\left(y - \frac{b}{3a}\right) + d = 0$$

$$a\left(y^3 - \frac{b}{a}y^2 + \frac{b^2}{3a^2}y - \frac{b}{27a^3}\right) + b\left(y^2 - \frac{2b}{3a}y + \frac{b^2}{9a^2}\right)$$

$$c\left(y - \frac{b}{3a}\right) + d = 0, \text{ and we have}$$

$$ay^3 + \left(\frac{b^2}{3a} - \frac{2b^2}{3a} + c\right)y + \left(-\frac{b^3}{27a^2} + \frac{b^3}{9a^2} - \frac{bc}{3a} + d\right) = 0.$$

Now, if we make

$$\frac{b^2}{3a} - \frac{2b^2}{3a} + c = c' \quad \text{and} \quad \frac{-b^3}{27a^2} + \frac{b^3}{9a^2} - \frac{bc}{3a} + d = d',$$

the general equation will become

$$ay^3 + c'y + d' = 0.$$

To avoid fractions in the solution of this equation, we divide it by a, and write it in the following fashion:

$$y^3 + 3py + 2q = 0.$$

This last equation is called the reduced cubic equation and, as we have shown, any cubic equation can be reduced to that form.

To solve the reduced equation, the following identity is considered:

$$(a + b)^3 - 3ab(a + b) - (a^3 + b^3) = 0.$$

If this identity is compared with the reduced equation, we have that

$$a + b = y, \quad ab = -p, \quad \text{and} \quad a^3 + b^3 = -2q.$$

From these equations, we see that we only have to find the values of a and b to find y. This can be done by solving the following system of equations:

$$\begin{aligned} ab &= -p \\ a^3 + b^3 &= -2q \end{aligned} \quad \text{or} \quad \begin{aligned} a^3b^3 &= -p^3 \\ a^3 + b^3 &= -2q. \end{aligned}$$

From the second equation, we have that $b^3 = -2q - a^3$, and substituting this value in the first equation, we have $-a^3(2q + a^3) = -p^3$, therefore, $a^6 + 2a^3q - p^3 = 0$.

If we make $a^3 = v$, we obtain the following quadratic equation: $v^2 + 2qv - p^3 = 0$.

The roots of this quadratic equation are

$$v_1 = -q + \sqrt{q^2 + p^3} \quad \text{and}$$
$$v_2 = -q - \sqrt{q^2 + p^3}.$$

Because of the symmetry of a and b in the system, we can take v_1 or v_2 to be a^3 or b^3 randomly.

So, $a^3 = -q + \sqrt{q^2 + p^3}$ and
$b^3 = -q - \sqrt{q^2 + p^3}$.

Therefore, $a = \sqrt[3]{-q + \sqrt{q^2 + p^3}}$ and
$b = \sqrt[3]{-q - \sqrt{q^2 + p^3}}$.

But $y = a + b$, thus,

$$y = \sqrt[3]{-q + \sqrt{q^2 + p^3}} + \sqrt[3]{-q - \sqrt{q^2 + p^3}},$$

which is called Cardano's formula for the cubic.

Since a^3 and b^3 have three roots each, it seems that the equation has nine roots. This is not the case, for since $ab = -p$, the cubic roots of a^3 and b^3 are to be taken in pairs so that their product (which is ab) is a rational number $-p$.

Now, we know that the cubic roots of a^3 are a (the principal value), $a\alpha$, and $a\alpha^2$, where α is one of the complex roots of unity. Similarly, the cubic roots of b^3 are $b, b\alpha$, and $b\alpha^2$.

However, if the product of a and b must be rational, we have that the only admissible solutions are $(a, b), (a\alpha, b\alpha^2)$, and $(a\alpha^2, b\alpha)$, because

$$ab = -p$$

$$a\alpha \cdot b\alpha^2 = ab\alpha^3 = ab = -p \text{ (because } \alpha^3 = 1)$$

$$a\alpha^2 \cdot b\alpha = ab\alpha^3 = ab = -p.$$

Therefore, the values of y are $a + b$, $a\alpha + b\alpha^2$, and $a\alpha^2 + b\alpha$.

But $x = y - \frac{b}{3a}$, and so the roots of the general cubic equation will be found once we know y.

Example 3

Solve the equation $x^3 + 3x^2 + 9x - 13 = 0$. First, we must reduce this equation to eliminate the second degree term. The transformation is $x = y - \frac{b}{3a}$.

In this example, $a = 1$, $b = 3$, $c = 9$, and $d = -13$. Therefore, $x = y - \frac{3}{3(1)} = y - 1$.

Thus, substituting $y - 1$ for x in the equation, $(y - 1)^3 + 3(y - 1)^2 + 9(y - 1) - 13 = 0$, or $(y^3 - 3y^2 + 3y - 1) + 3(y^2 - 2y + 1) + 9(y - 1) - 13 = 0$. So, $y^3 - 6y - 20 = 0$ is the reduced equation.

Therefore,

$$3p = 6, \quad p = 2, \quad \text{and} \quad p^3 = 8$$

$$2q = -20, \quad q = -10, \quad \text{and} \quad q^2 = 100.$$

Thus, $\sqrt{q^2 + p^3} = \sqrt{108} = 6\sqrt{3}$, and,

$$\begin{aligned} a &= \sqrt[3]{10 + 6\sqrt{3}} = \sqrt[3]{1 + 3\sqrt{3} + 9 + 3\sqrt{3}} \\ &= \sqrt[3]{(1 + \sqrt{3})^3} = 1 + \sqrt{3} \\ b &= \sqrt[3]{10 - 6\sqrt{3}} = \sqrt[3]{1 - 3\sqrt{3} + 9 - 3\sqrt{3}} \\ &= \sqrt[3]{(1 - \sqrt{3})^3} = 1 - \sqrt{3}. \end{aligned}$$

The solutions for the reduced equation are then

$$y_1 = a + b = (1 + \sqrt{3}) + (1 - \sqrt{3}) = 2$$

$$y_2 = a\alpha + b\alpha^2 = (1 + \sqrt{3})\left(-\frac{1}{2} + \frac{\sqrt{3}}{2}i\right) + (1 - \sqrt{3})\left(-\frac{1}{2} - \frac{\sqrt{3}}{2}i\right) = -1 + 3i$$

$$y_3 = a\alpha^2 + b\alpha = (1 + \sqrt{3})\left(-\frac{1}{2} - \frac{\sqrt{3}}{2}i\right) + (1 - \sqrt{3})\left(-\frac{1}{2} + \frac{\sqrt{3}}{2}i\right) = -1 - 3i.$$

But, $x = y - 1$. Therefore,

$$x_1 = y_1 - 1 = 2 - 1 = 1$$

$$x_2 = y_2 - 1 = -1 + 3i - 1 = -2 + 3i$$

$$x_3 = y_3 - 1 = -1 - 3i - 1 = -2 - 3i.$$

Thus, in this example, we have for solutions one real and two conjugate complex roots. In this first of two units, we have studied the general solution of the cubic. In a second, we will study the different cases, reducible and irreducible, in the solution of cubic equations using Cardano's formula.

Postassessment

1. Find the roots of $x^3 + 6x^2 + 17x + 18 = 0$.
2. Solve $x^3 - 11x^2 + 35x - 25 = 0$.
3. Find the solution of $x^3 - 3x^2 + 3x - 1 = 0$.

Unit 102

Solving Cubic Equations

In the first of two units on cubic equations, we have studied the general solution of the cubic. In this second, we will study the different cases, reducible and irreducible, in the solution of cubic equations using Cardano's formula.

Performance Objectives

1. *Given some cubic equations, students will analyze them to see the kind of solutions they are going to obtain when the equation is solved.*
2. *Students will solve given cubic equations.*

Preassessment

Students should have mastered operations with complex numbers and quadratic equations. They should also have a solid background in trigonometry.

Teaching Strategies

Review the content of the previous unit on cubic equations in the following fashion:

Given a general cubic equation $\mathrm{A}x^3 + \mathrm{B}x^2 + \mathrm{C}x + \mathrm{D} = 0$, it is always possible to eliminate the second degree term by making the change of variables $x = y - \frac{\mathrm{B}}{3\mathrm{A}}$. This transformation will lead to an equation of the form $y^3 + 3py + 2q = 0$, which is called the reduced or normal cubic equation.

The solution of the reduced equation is given by the Cardano's formula $y = \sqrt[3]{-q + \sqrt{q^2 + p^3}} + \sqrt[3]{-q - \sqrt{q^2 + p^3}}$. If $a = \sqrt[3]{-q + \sqrt{q^2 + p^3}}$ and $b = \sqrt[3]{-q - \sqrt{q^2 + p^3}}$, the roots of the reduced equations are $y_1 = a + b$, $y_2 = a\alpha + b\alpha^2$, and $y_3 = y_3 = a\alpha^2 + b\alpha$, where $\alpha = -\frac{1}{2} + \frac{\sqrt{3}}{2}i$ and $\alpha^2 = \frac{1}{2} - \frac{\sqrt{3}}{2}i$ are cube roots of unity.

Once the values y_1, y_2, and y_3 are found, the solutions of the general cubic equation will be obtained by using the transformation $x = y - \frac{\mathrm{B}}{3\mathrm{A}}$.

From the Cardano's formula, it is obvious that the nature of the solutions will depend on the value of $q^2 + p^3$, which for this reason is called the discriminant of the cubic. This is so, because $q^2 + p^3$, being under a square root, will yield real or imaginary values according to the sign of the sum $q^2 + p^3$.

Before discussing the discriminant, it is useful to rewrite solutions of the reduced equation in the following fashion:

$$y_1 = a + b;$$

$$y_2 = a\alpha + b\alpha^2 = a\left(-\frac{1}{2} + \frac{\sqrt{3}}{2}i\right) + b\left(-\frac{1}{2} - \frac{\sqrt{3}}{2}i\right);$$

$$y_3 = a\alpha^2 + b\alpha = a\left(-\frac{1}{2} - \frac{\sqrt{3}}{2}i\right) + b\left(-\frac{1}{2} + \frac{\sqrt{3}}{2}i\right);$$

and simplifying:

$$y_1 = a + b; \quad y_2 = -\frac{a+b}{2} + \frac{a-b}{2}\sqrt{3}i;$$

$$y_3 = -\frac{a+b}{2} - \frac{a-b}{2}\sqrt{3}i.$$

Let us now consider the discriminant $q^2 + p^3$.

1. If $q^2 + p^3 > 0, a$ and b have each one real value, then we can suppose a and b will also be real. Consequently, $a + b$ and $a - b$ will also be real. Therefore, we have that if $a + b = m$ and $a - b = n$, the solutions of the reduced equation are

$$y_1 = a + b = m; \quad y_2 = -\frac{m}{2} + \frac{n}{2}\sqrt{3}i;$$

$$y_3 = -\frac{m}{2} - \frac{n}{2}\sqrt{3}i.$$

Thus, if $q^2 + p^3 > 0$, we have one real root and two conjugate imaginary roots.

Example 1

Solve $x^3 - 6x^2 + 10x - 8 = 0$.

First, we must eliminate the square term. The transformation for this example is

$$x = y - \frac{\mathrm{B}}{3\mathrm{A}} = y - \frac{-6}{3} = y + 2.$$

Thus, substituting $y + 2$ for x in the equation

$$(y+2)^3 - 6(y+2)^2 + 10(y+2) - 8 = 0$$

$$y^3 + 6y^2 + 12y + 8 - 6y^2 - 24y - 24 + 10y + 20 - 8 = 0$$

$$y^3 - 2y - 4 = 0 \text{ (Reduced equation).}$$

Therefore,

$$3p = -2 \quad p = -\frac{2}{3} \quad \text{and} \quad p^3 = -\frac{8}{27}$$

$$2q = -4 \quad q = -2 \quad \text{and} \quad q^2 = 4.$$

Thus, $q^2 + p^3 = 4 - \frac{8}{27} = \frac{100}{27} > 0$.

Therefore, we know that in the solution, one root must be real and two conjugate imaginary.
The values for a and b are

$$a = \sqrt[3]{-q + \sqrt{q^2 + p^3}} = \sqrt[3]{2 + \sqrt{\frac{100}{27}}}$$

$$= \sqrt[3]{2 + \frac{10}{3\sqrt{3}}}$$

$$b = \sqrt[3]{-q - \sqrt{q^2 + p^3}} = \sqrt{2 - \sqrt{\frac{100}{27}}}$$

$$= \sqrt[3]{2 - \frac{10}{3\sqrt{3}}}$$

and simplifying:

$$a = \sqrt[3]{\frac{6\sqrt{3} + 10}{3\sqrt{3}}} = \sqrt[3]{\frac{3\sqrt{3} + 9 + 3\sqrt{3} + 1}{\sqrt{27}}}$$

$$= \sqrt[3]{\frac{(3 - 1)^3}{\sqrt{27}}}$$

$$b = \sqrt[3]{\frac{6\sqrt{3} - 10}{3\sqrt{3}}} = \sqrt[3]{\frac{3\sqrt{3} - 9 + 3\sqrt{3} - 1}{\sqrt{27}}}$$

$$= \sqrt[3]{\frac{(\sqrt{3} - 1)^3}{\sqrt{27}}}$$

$$a = \frac{\sqrt{3} + 1}{\sqrt{3}} \quad \text{and} \quad b = \frac{\sqrt{3} - 1}{\sqrt{3}}.$$

The solutions for the reduced equation are then

$$y_1 = a + b = \frac{\sqrt{3} + 1}{\sqrt{3}} + \frac{\sqrt{3} - 1}{\sqrt{3}} = 2$$

$$y_2 = a\alpha + b\alpha^2 = \left(\frac{\sqrt{3} + 1}{\sqrt{3}}\right)\left(-\frac{1}{2} + \frac{\sqrt{3}}{2}i\right) + \left(\frac{\sqrt{3} - 1}{\sqrt{3}}\right)\left(-\frac{1}{2} - \frac{\sqrt{3}}{2}i\right)$$

$$y_3 = a\alpha^2 + b\alpha = \left(\frac{\sqrt{3}+1}{\sqrt{3}}\right)\left(\frac{1}{2} - \frac{\sqrt{3}}{2}i\right) + \left(\frac{\sqrt{3}-1}{\sqrt{3}}\right)\left(-\frac{1}{2} + \frac{\sqrt{3}}{2}i\right)$$

and simplifying

$$y_1 = 2; \quad y_2 = -1 + i; \quad y_3 = -1 - i.$$

Therefore, the solutions of the general equation are

$$x_1 = y_1 + 2 = 2 + 2 = 4$$
$$x_2 = y_2 + 2 = -1 + i + 2 = 1 + i$$
$$x_3 = y_3 + 2 = -1 - i + 2 = 1 - i.$$

2. If $q^2 + p^3 = 0$, a and b are equal; therefore if m represents the common real value of a and b, we have

$$y_1 = m + m = 2m$$
$$y_2 = -\frac{m+m}{2} + \frac{m-m}{2}\sqrt{3}i = -m$$
$$y_3 = -\frac{m+m}{2} - \frac{m-m}{2}\sqrt{3}i = -m.$$

Thus, in this case, we have that all the roots are real and two are equal.

Example 2

Find the roots of $x^3 - 12x + 16 = 0$. In this example, we already have the reduced equation, therefore:

$$3p = -12 \quad p = -4 \quad \text{and} \quad p^3 = -64$$
$$2q = 16 \quad q = 8 \quad \text{and} \quad q^2 = 64.$$

Thus, $q^2 + p^3 = 64 - 64 = 0$. This means that the solution will have three real roots, two of them equal. The values of a and b are

$$a = \sqrt[3]{-q + \sqrt{q^2 + p^3}} = \sqrt[3]{-8} = -2$$
$$b = \sqrt[3]{-q - \sqrt{q^2 + p^3}} = \sqrt[3]{-8} = -2.$$

Therefore, the roots are

$$y_1 = a + b = -4$$
$$y_2 = a\alpha + b\alpha^2 = -2(\alpha + \alpha^2)$$

$$= -2\left(-\frac{1}{2} + \frac{\sqrt{3}}{2}i - \frac{1}{2} - \frac{\sqrt{3}}{2}i\right) = 2$$

$$y_3 = a\alpha^2 + b\alpha = -2(\alpha^2 + \alpha)$$

$$= -2\left(-\frac{1}{2} - \frac{\sqrt{3}}{2}i - \frac{1}{2} + \frac{\sqrt{3}}{2}i\right) = 2.$$

3. If $q^2 + p^3 < 0$, a and b will be complex numbers because of the square root of the discriminant, which is negative in this case. Therefore, if the values a and b are $a = \mathrm{M} + \mathrm{N}i$ and $b = \mathrm{M} - \mathrm{N}i$, the solutions of the reduced equation will be

$$y_1 = a + b = 2\mathrm{M}$$

$$y_2 = -\frac{2\mathrm{M}}{2} + \frac{2\mathrm{N}i}{2}\sqrt{3}i = -\mathrm{M} - \sqrt{3}\mathrm{N}$$

$$y_3 = -\frac{2\mathrm{M}}{2} - \frac{2\mathrm{N}i}{2}\sqrt{3}i = -\mathrm{M} + \sqrt{3}\mathrm{N}$$

which are all real roots and unequal.

However, there is no general arithmetic or algebraic method of finding the exact value of the cubic root of complex numbers. Therefore, Cardano's formula is of little use in this case, which for this reason is called the irreducible case.

The solution of this case can be obtained with the use of trigonometry. Thus, when the Cardano's formula has the form $y = \sqrt[3]{u + vi} + \sqrt[3]{u - vi}$, we call $r = \sqrt{u^2 + v^2}$ and $\tan\Theta = \frac{v}{u}$; therefore, the cubic root of them will be

$$y = \sqrt[3]{r}\left[\cos\frac{\Theta + 2k\pi}{3} + i\sin\frac{\Theta + 2k\pi}{3}\right] + \sqrt[3]{r}\left[\cos\frac{\Theta + 2k\pi}{3} - i\sin\frac{\Theta + 2k\pi}{3}\right],$$

where $k = 0, 1$, and 2.

If we simplify this expression, we obtain $y = 2\sqrt[3]{r}\cos\frac{\Theta + 2k\pi}{3}$, where $k = 0, 1$, and 2.

Therefore, the three roots are

$$y_1 = 2\sqrt[3]{r}\cos\frac{\Theta}{3}; \quad y_2 = 2\sqrt[3]{r}\cos\frac{\Theta + 2\pi}{3};$$

$$y_3 = 2\sqrt[3]{r}\cos\frac{\Theta + 4\pi}{3}.$$

Example 3

Solve $x^3 - 6x - 4 = 0$.

From this equation, we have $3p = -6$ and $2q = -4$; therefore $p^3 = -8$, $q^2 = 4$, $p^3 + q^2 = -4$, and $\sqrt{p^3 + q^2} = 2i$.

The solution then would be $y = \sqrt{2 + 2i} + \sqrt{2 - 2i}$, thus, $r = \sqrt{4 + 4} = \sqrt{8}$, $\tan \Theta = \frac{2}{2}$ or $\tan \Theta = 1$ and $\Theta = \frac{\pi}{4}$. Therefore, the roots of the equation are

$$x_1 = 2\sqrt[3]{r}\cos\frac{\Theta}{3} = 2\sqrt[3]{\sqrt{8}}\cos\frac{\pi}{12} = 2\sqrt{2}\cos 15^\circ$$

$$x_2 = 2\sqrt[3]{r}\cos\frac{\Theta + 2\pi}{3} = 2\sqrt{2}\cos\frac{\frac{\pi}{4} + 2\pi}{3} = 2\sqrt{2}\cos 135^\circ$$

$$x_3 = 2\sqrt[3]{r}\cos\frac{\Theta + 14\pi}{3} = 2\sqrt{2}\cos\frac{\frac{\pi}{4} + 4\pi}{3} = 2\sqrt{2}\cos 255^\circ.$$

But

$$\sin\frac{y}{2} = \sqrt{\frac{1 - \cos x}{2}} \quad \text{and} \quad \cos\frac{x}{2} = \sqrt{\frac{1 + \cos x}{2}}.$$

Therefore,

$$\sin 15^\circ = \sqrt{\frac{1 - \cos 30^\circ}{2}} = \sqrt{\frac{1 - \sqrt{3}/2}{2}} = \frac{\sqrt{2 - \sqrt{3}}}{2}$$

and

$$\cos 15^\circ = \sqrt{\frac{1 + \cos 30^\circ}{2}} = \sqrt{\frac{1 + \sqrt{3}/2}{2}} = \frac{\sqrt{2 + \sqrt{3}}}{2}.$$

Thus, the roots are

$$x_1 = 2\sqrt{2}\frac{\sqrt{2 + \sqrt{3}}}{2} = \sqrt{4 + 2\sqrt{3}} = \sqrt{1 + 2\sqrt{3} + 3} = 1 + \sqrt{3}$$

$$x_2 = 2\sqrt{2}\cos 135^\circ = 2\sqrt{2}\,(-\cos 45^\circ) = 2\sqrt{2}\left(-\frac{\sqrt{2}}{3}\right) = -2$$

$$x_3 = 2\sqrt{2}\,(\sin 15^\circ) = 2\sqrt{2}\left(\frac{\sqrt{2 - \sqrt{3}}}{2}\right) = \sqrt{1 + 2\sqrt{3} + 3} = 1 - \sqrt{3}.$$

The reducible cases may also employ the aid of trigonometry.

Postassessment

Analyze and then solve the following cubics:

1. $x^3 - 6x^2 + 11x - 6 = 0$
2. $x^3 - 5x^2 + 9x - 9 = 0$
3. $x^3 - 75x + 250 = 0$
4. $x^3 - 6x^2 + 3x + 10 = 0.$

Unit 103

Calculating Sums of Finite Series

Mathematical induction has become thoroughly entrenched in secondary school curricula. Many textbooks provide a variety of applications of this technique of proof. Most popular among these applications is proving that specific series have given formulas as sums. Although most students merely work the proof as required, some may question how the sum of a particular series was actually generated.

This unit will provide you with a response to students' requests for deriving formulas for certain series summations.

Performance Objectives

1. *Given some finite series, students will find their sum.*
2. *Students will develop formulas for determining the sum of various finite series.*

Preassessment

Students should have mastered operations with algebraic expressions, functions, and concepts of finite sequence and series.

Teaching Strategies

Review concepts of sequences and series in the following fashion:

A *finite sequence* is a finite set of ordered elements or terms, each related to one or more of the preceding elements in some specifiable way.

Examples:
1. $1, 3, 5, 7, \ldots, 19$
2. $\sin x,\ \sin 2x,\ \sin 3x, \ldots, \sin 20x$
3. $2, 4, 6, 8, \ldots, 2n$.

Let us now consider any finite sequence of elements $u_1, u_2, \ldots, u_n$. We can obtain the following partial sums:

$$s_1 = u_1$$
$$s_2 = u_1 + u_2$$
$$s_3 = u_1 + u_2 + u_3$$
$$\cdots\cdots\cdots\cdots\cdots\cdots\cdots\cdots$$
$$s_n = u_1 + u_2 + \cdots + u_n.$$

We call this sum $u_1 + u_2 + \cdots + u_n$ a *finite series* of the elements of the sequence $u_1, u_2, \ldots, u_n$. s_n represents the total sum of these elements. For example, if we have the sequence $1, 2, 3, 4$, the series is $1+2+3+4$, and the sum s_4 is 10.

This example is a simple one. However, if instead of considering four terms $1, 2, 3$, and 4, we consider "n" terms $1, 2, 3, \ldots, n$, it will not be that simple to calculate their sum $s_n = 1+2+3+\cdots+n$. Sometimes, there are easy ways to calculate the sum of a specific series, but we cannot apply that particular method to all the series.

For example, the previous series $1+2+3+\cdots+n$, could be calculated by using the following artifice:

$$1 = 1 = \frac{1\cdot 2}{2}$$
$$1+2 = 3 = \frac{2\cdot 3}{2}$$
$$1+2+3 = 6 = \frac{3\cdot 4}{2}$$
$$\cdots\cdots\cdots\cdots\cdots\cdots\cdots\cdots$$
$$1+2+\cdots+n = \frac{n(n+1)}{2},$$

which is the total sum of the series.

This means that if we want to calculate the sum of the series $1+2+3+\cdots+10$, we will have $S_{10} = \frac{10(10+1)}{2} = \frac{10\cdot 11}{2} = 55$.

We cannot apply this artifice to every series, therefore, we must find a more general method that permits us to calculate the sum of several series. This method is given by the following theorem:

Theorem 1: Let us consider a finite series $u_1 + u_2 + u_3 + \cdots + u_n$. If we can find a function $F(n)$ such that $u_n = F(n+1) - F(n)$, then $u_1 + u_2 + \cdots + u_n = F(n+1) - F(1)$.

Proof: We have by hypothesis that $u_n = F(n+1) - F(n)$; therefore, if we apply it for $n-1, n-2, \ldots, 3, 2, 1$, we will get the following relations:

$$\begin{aligned} u_n &= F(n+1) - F(n) \\ u_{n-1} &= F(n) - F(n-1) \\ u_{n-2} &= F(n-1) - F(n-2) \\ &\vdots \qquad \vdots \qquad \vdots \\ u_2 &= F(3) \qquad - F(2) \\ u_1 &= F(2) \qquad - F(1). \end{aligned}$$

If we now add these relations, we will get

$$u_1 + u_2 + u_3 + \cdots + u_n = F(n+1) - F(1),$$

which proves the theorem.

Before having students embark on applications, have them consider the following examples:

1. Find the sum of the series $1+2+3+\cdots+n$. Because $u_n = n$, we consider $F(n) = An^2 + Bn + C$. (A polynomial one degree higher than u_n should be used.) Therefore, $F(n+1) = A(n+1)^2 + B(n+1) + C$. According to Theorem 1, we must have

$$\begin{aligned} u_n &= F(n+1) - F(n) \\ n &= \left[A(n+1)^2 + B(n+1) + C\right] - \left[An^2 + Bn + C\right] \\ n &= 2An + (A+B). \end{aligned}$$

Therefore, by equating coefficients of powers of n, we get $2A = 1$, and $A + B = 0$. By solving these simultaneously, we get $A = \frac{1}{2}$ and $B = -\frac{1}{2}$.

Therefore,

$$F(n) = \frac{1}{2}n^2 - \frac{1}{2}n + C,$$

$$\mathrm{F}(n+1) = \frac{1}{2}(n+1)^2,$$
$$-\frac{1}{2}(n+1) + \mathrm{C}, \text{ and } \mathrm{F}(1)=\mathrm{C}.$$

Thus,

$$\begin{aligned}1+2+3+\cdots+n &= \mathrm{F}(n+1) - \mathrm{F}(1)\\ &= \frac{1}{2}(n+1)^2 - \frac{1}{2}(n+1)\\ &= \frac{1}{2}n(n+1).\end{aligned}$$

2. Find the sum of the series $1^2 + 2^2 + 3^2 + \cdots + n^2$. Since $u_n = n^2$, we consider $\mathrm{F}(n) = \mathrm{A}n^3 + \mathrm{B}n^2 + \mathrm{C}n + \mathrm{D}$ [One degree higher than u_n, since the highest power of $\mathrm{F}(n)$ will be annihilated in $\mathrm{F}(n+1) - \mathrm{F}(n)$.] Thus, $\mathrm{F}(n+1) = \mathrm{A}(n+1)^3 + \mathrm{B}(n+1)^2 + \mathrm{C}(n+1) + \mathrm{D}$.
Now,

$$\begin{aligned}u_n = n^2 &= \mathrm{F}(n+1) - \mathrm{F}(n)\\ n^2 &= [\mathrm{A}(n+1)^3 + \mathrm{B}(n+1)^2 + \mathrm{C}(n+1) + \mathrm{D}]\\ &\quad - [\mathrm{A}n^3 + \mathrm{B}n^2 + \mathrm{C}n + \mathrm{D}]\\ n^2 &= 3\mathrm{A}n^2 + (3\mathrm{A} + 2\mathrm{B})n + (\mathrm{A} + \mathrm{B} + \mathrm{C}).\end{aligned}$$

By equating coefficients of powers of n, we get $3\mathrm{A} = 1; 3\mathrm{A} + 2\mathrm{B} = 0$; and $\mathrm{A} + \mathrm{B} + \mathrm{C} = 0$ and solving simultaneously:

$$\mathrm{A} = \frac{1}{3}; \mathrm{B} = -\frac{1}{2}; \mathrm{C} = \frac{1}{6}.$$

Thus, $\mathrm{F}(n) = \frac{1}{3}n^3 - \frac{1}{2}n^2 + \frac{1}{6}\mathrm{n} + \mathrm{D}$

$\mathrm{F}(n+1) = \frac{1}{3}(n+1)^3 - \frac{1}{2}(n+1)^2 + \frac{1}{6}(n+1) + \mathrm{D}$

$\mathrm{F}(1) = \frac{1}{3} - \frac{1}{2} + \frac{1}{6} + \mathrm{D} = \mathrm{D}.$

Hence, $1^2 + 2^2 + 3^2 + \cdots + n^2$

$$\begin{aligned}&= \mathrm{F}(n+1) - \mathrm{F}(1)\\ &= \frac{1}{3}(n+1)^3 - \frac{1}{2}(n+1)^2 + \frac{1}{6}(n+1)\\ &= \frac{n(n+1)(2n+1)}{6}.\end{aligned}$$

3. Find the sum of the series $1^3 + 3^3 + 5^3 + \cdots + (2n-1)^3$. Since u_n is the third degree, $\mathrm{F}(n) = \mathrm{A}n^4 + \mathrm{B}n^3 + \mathrm{C}n^2 + \mathrm{D}n + \mathrm{E}$ and

$$\mathrm{F}(n+1) = \mathrm{A}(n+1)^4 + \mathrm{B}(n+1)^3 + \mathrm{C}(n+1)^2 + \mathrm{D}(n+1) + \mathrm{E}.$$

Thus, $u_n = (2n-1)^3 = \mathrm{F}(n+1) - \mathrm{F}(n)$; or $8n^3 - 12n^2 + 8n^3 - 12n^2 + 6n - 1 = 4\mathrm{A}n^3 + (6\mathrm{A} + 3\mathrm{B})n^2 + 4\mathrm{A}(4\mathrm{A} + 3\mathrm{B} + 2\mathrm{C})n + (\mathrm{A}+\mathrm{B}+\mathrm{C}+\mathrm{D})$. Equating coefficients:

$4\mathrm{A} = 8$, and $\mathrm{A} = 2; 6\mathrm{A} + 3\mathrm{B} = -12$, and $\mathrm{B} = -8$;
$4\mathrm{A} + 3\mathrm{B} + 2\mathrm{C} = 6$; and $\mathrm{C} = 11$;
$\mathrm{A} + \mathrm{B} + \mathrm{C} + \mathrm{D} = -1$, and $\mathrm{D} = -6$.
Therefore, $\mathrm{F}(n) = 2n^4 - 8n^3 + 11n^2 - 6n + \mathrm{E}$;
$\mathrm{F}(n+1) = 2(n+1)^4 - 8(n+1)^3 + 11(n+1)^2 - 6(n+1) + \mathrm{E}$;
and $\mathrm{F}(1) = -1 + \mathrm{E}$.

Thus, $1^3 + 3^3 + 5^3 + \cdots + (2n-1)^3 = \mathrm{F}(n+1) - \mathrm{F}(1) = 2(n+1)^4 - 8(n+1)^3 + 11(n+1)^2 - 6(n+1) + \mathrm{E} - 1 + \mathrm{E}) = 2n^4 - n^2 = n^2\left(2n^2 - 1\right)$.

4. Find the sum of the series $\frac{1}{2} + \frac{1}{4} + \cdots + \frac{1}{2^n}$. Let us consider $\mathrm{F}(n) = \frac{\mathrm{A}}{2^n}$ and therefore, $\mathrm{F}(n+1) = \frac{\mathrm{A}}{2^{n+1}}$. Hence, $u_n = \mathrm{F}(n+1) - \mathrm{F}(n)$. Thus, $\frac{1}{2^n} = \frac{\mathrm{A}}{2^{n+1}} - \frac{\mathrm{A}}{2^n}$, therefore $\mathrm{A} = -2$. Hence, $\mathrm{F}(n+1) = -\frac{1}{2^n}$; and $\mathrm{F}(1) = -1$.

Therefore,

$$\begin{aligned}\frac{1}{2} + \frac{1}{4} + \cdots + \frac{1}{2^n} &= \mathrm{F}(n+1) - \mathrm{F}(1)\\ &= -\frac{1}{2^n} + 1\\ &= 1 - \frac{1}{2^n}.\end{aligned}$$

After sufficient practice, students should be able to fine $\mathrm{F}(n)$ more easily.

Postassessment

1. Find the sum of the series $1 + 8 + 27 + \cdots + n^3$.
2. Find the sum of the series $\frac{1}{5} + \frac{1}{25} + \frac{1}{125} + \cdots + \frac{1}{5^n}$.
3. Find the formula for the sum of a finite arithmetic progression.

Unit 104

A General Formula for the Sum of Series of the Form $\sum_{t=1}^{n} t^r$

The calculation of the sum of convergent series is an important topic. There is not, however, a general formula to calculate the sum of any given convergent series.

This unit will provide you with a general formula to calculate the sum of the series of the specific type $\sum_{t=1}^{n} t^r$.

Performance Objectives

1. *Given some finite series of the form* $\sum_{t=1}^{n} t^r$, *students will find their sum.*
2. *Students will demonstrate an understanding of the technique used to find general formulas for certain specific series.*

Preassessment

Students should know the Binomial Theorem and have a fair knowledge of series and elementary linear algebra.

Teaching Strategies

Review the concept of series and the Binomial Theorem in the following fashion:

A series is a sum of the elements of a given sequence. For example, if we have the sequence of elements $u_1, u_2, \ldots, u_n$, the series is $u_1 + u_2 + \cdots + u_n$. This series can also be represented by the symbol $\sum_{r=1}^{n} u_r$. Thus, the symbol $\sum_{n=1}^{n} k^2$ means $1^2 + 2^2 + \cdots + n^2$.

Another example of series is the one represented by $\sum_{m=0}^{k} \binom{k}{m} a^{k-m} b^m$, where a and b are arbitrary real numbers, k is any positive integer, and $\binom{k}{m} = \frac{k!}{m!(k-m)!}$. This series can be proved to be equal to $(a+b)^k$, and this fact is known as the *Binomial Theorem*.

There is also a less known but important theorem from the theory of series whose proof we are going to give in this unit. We are going to use this theorem and the Binomial Theorem in the development of our discussion.

Lemma 1: Let us consider a finite series $\sum_{r=1}^{n} u_r$. If we can find a function $f(n)$ such that $u_n = f(n+1) - f(n)$, then $\sum_{r=1}^{n} u_r = f(n+1) - f(1)$.

Proof: We have by hypothesis that $u_n = f(n+1) - f(n)$; therefore, if we apply it for $n-1, n-2, \ldots, 3, 2, 1$, we will get the following relations:

$$u_n = f(n+1) - f(n)$$

$$u_{n-1} = f(n) - f(n-1)$$

$$u_{n-2} = f(n-1) - f(n-2)$$

$$\cdots\cdots\cdots\cdots\cdots\cdots\cdots\cdots\cdots$$

$$u_2 = f(3) - f(2)$$
$$u_1 = f(2) - f(1).$$

If we now add these relations, we will get $\sum_{r=1}^{n} u_r$ equal to $f(n+1) - f(1)$, which proves the theorem.

Let $\sum_{t=1}^{n} t^r$ be the series whose sum we want. To apply Lemma 1 to this series, we may consider the arbitrary function $f(n) = \sum_{k=0}^{r+1} b_k n^k$, where b is any real number and n is any positive integer. Thus, $f(n+1) = \sum_{k=0}^{r+1} b_k (n+1)^k$. If we now impose the condition of the hypothesis of Lemma 1 to this function for the series $\sum_{t=1}^{n} t^r$, we have

$$u_n = f(n+1) - f(n)$$

$$n^r = \sum_{k=0}^{r+1} b_k (n+1)^k - \sum_{k=0}^{r+1} b_k n^k$$

$$n^r = \sum_{k=0}^{r+1} b_k [(n+1)^k - n^k].$$

But according to the Binomial Theorem,

$$(n+1)^k = \sum_{m=0}^{k} \binom{k}{m} n^{k-m} \text{ and thus,}$$

$$(n+1)^k - n^k = \sum_{m=1}^{k} \binom{k}{m} n^{k-m}.$$

Therefore in (I), we have

$$n^r = \sum_{k=0}^{r+1} b_k \left[\sum_{m=1}^{k} \binom{k}{m} n^{k-m} \right].$$

This equation leads to the following system of equations:

$$\begin{aligned}
&\binom{r+1}{1} b_{r+1} && = 0\\
&\binom{r+1}{2} b_{r+1} + \binom{r}{1} b_r && = 0\\
&\binom{r+1}{3} b_{r+1} + \binom{r}{2} b_2 + \binom{r-1}{1} b_{r-1} \cdots\cdots\cdots\cdots && = 0\\
&\binom{r+1}{m} b_{r+1} + \binom{r}{m-1} b_r + \binom{r-1}{m-2} b_{r-1} + \cdots\cdots\cdots && = 0\\
&\binom{r+1}{r+1} b_{r+1} + \binom{r}{r} b_r + \cdots + \binom{1}{1} b_1 \cdots\cdots\cdots\cdots && = 0.
\end{aligned}$$

This system of equations can be expressed in matrix form as follows:

$$\begin{bmatrix}
\binom{r+1}{1} & 0 & 0 & 0 & \cdots & 0\\
\binom{r+1}{2} & \binom{r}{1} & 0 & 0 & \cdots & 0\\
\binom{r+1}{3} & \binom{r}{2} & \binom{r-1}{2} & 0 & \cdots & 0\\
. & . & . & . & \cdots & .\\
. & . & . & . & \cdots & .\\
. & . & . & . & \cdots & .\\
\binom{r+1}{m} & \binom{r}{m-1} & \binom{r-1}{m-2} & 0 & \cdots & 0\\
. & . & . & . & \cdots & .\\
. & . & . & . & \cdots & .\\
. & . & . & . & \cdots & .\\
\binom{r+1}{r+1} & \binom{r}{r} & \binom{r-1}{r-1} & \binom{r-2}{r-2} & \cdots & 1
\end{bmatrix}
\begin{bmatrix} b_{r+1}\\ b_r\\ b_{r-1}\\ .\\ .\\ .\\ b_m\\ .\\ .\\ .\\ b_1 \end{bmatrix}
=
\begin{bmatrix} 1\\ 0\\ 0\\ .\\ .\\ .\\ 0\\ .\\ .\\ .\\ 0 \end{bmatrix}.$$

If we call these three matrices A, X, and B, respectively, we have that AX = B. But A is a diagonal matrix, thus, det A = $\prod_{s=1}^{r+1} \binom{s}{1}$ = 0 and therefore A^{-1} exists. (Det A is the determinant of A, which for a simple case is $\begin{vmatrix} a_1 & b_1 \\ a_2 & b_2 \end{vmatrix} = a_1b_2 - a_2b_1$. The II represents "product," in the way the $\sum$ represents "sum.") Thus, we have $X = A^{-1}B$. If a_{ij} represents any element of A^{-1}, then it is implied that $X = A^{-1}B = \begin{pmatrix} a_{11} \\ a_{21} \\ \vdots \\ a_{r+1,1} \end{pmatrix}$ because B is a column vector.

Thus,

$$\begin{pmatrix} b_{r+1} \\ b_r \\ \vdots \\ b_1 \end{pmatrix} = \begin{pmatrix} a_{11} \\ a_{21} \\ \vdots \\ a_{r+1,1} \end{pmatrix}.$$

But this means that $b_{r+2-i} = a_{i,1}$ for all $i \in \{1, 2, 3, \ldots, r+1\}$.

Therefore, for the function $f(n) = \sum_{k=0}^{r+1} b_k n^k$, we have $f(n+1) = \sum_{k=0}^{r+1} b_k(n+1)^k$ and $f(1) = \sum_{k=0}^{r+1} b_k$ where $b_{r+2-i} = a_{i,1}$ for all $i \in \{1, 2, 3, \ldots, r+1\}$.

Thus, by Lemma 1,

$$\sum_{t=1}^{n} t^r = \sum_{k=0}^{r+1} b_k(n+1)^k - \sum_{k=0}^{r+1} b_k, \text{ or}$$

$$\sum_{t=1}^{n} t^r = \sum_{k=0}^{r+1} b_k[(n+1)^k - 1]. \tag{II}$$

But by the Binomial Theorem, we have that

$$(n+1)^k - 1 = \binom{k}{0} n^k + \binom{k}{1} n^{k-1} + \cdots + \binom{k}{k-1} n.$$

Thus, in (II)

$$\sum_{t=1}^{n} t^r = \sum_{k=0}^{r+1} b_k \left[\binom{k}{0} n^k + \binom{k}{1} n^{k-1} + \cdots + \binom{k}{k-1} n\right]$$

or

$$\sum_{t=1}^{n} t^r = n \sum_{k=0}^{r+1} \left[b_k \binom{k}{0} n^{k-1} + b_k \binom{k}{1} n^{k-2} + \cdots + b_k \binom{k}{k-1} \right].$$

And if we call $b_k \binom{k}{j} = c_j$, we have $\sum_{t=1}^{n} t^r = n \sum_{j=0}^{r} c_j n^j$, because when $k = 0, b_0 \binom{0}{j} = c_j = c_0$, which implies that $j = 0$, and when $k = r + 1, b_{r+1} \binom{r+1}{j} = c_j$, which implies that $j = r$.

Thus, we have the following theorem.

Theorem 1: The general formula for series of the form $\sum_{t=1}^{n} t^r$ is $n \sum_{j=0}^{r} c_j n^j$, where $c_j = b_k \binom{k}{j}$ for all $j \in \{0, 1, 3, \ldots, r\}$ and $b_{r+2-i} = a_{i,1}$ for all $i \in \{1, 2, 3, \ldots, r+1\}$.

Example 1

Find $\sum_{t=1}^{n} t^2$.

In this example $r = 2$, therefore we have

$$\begin{pmatrix} \binom{3}{1} & 0 & 0 \\ \binom{3}{2} & \binom{2}{1} & 0 \\ \binom{3}{3} & \binom{2}{2} & \binom{1}{1} \end{pmatrix} \begin{pmatrix} b_3 \\ b_2 \\ b_1 \end{pmatrix} = \begin{pmatrix} 1 \\ 0 \\ 0 \end{pmatrix}.$$

Thus, $\mathrm{A} = \begin{pmatrix} 3 & 0 & 0 \\ 3 & 2 & 0 \\ 1 & 1 & 1 \end{pmatrix}$ and A^{-1} is

$$\mathrm{A}^{-1} = \frac{(\mathrm{Adj.\,A})'}{\mathrm{Det.A}} = \frac{\begin{pmatrix} 2 & -3 & 1 \\ 0 & 3 & -3 \\ 0 & 0 & 6 \end{pmatrix}'}{\begin{vmatrix} 3 & 0 & 0 \\ 3 & 2 & 0 \\ 1 & 1 & 1 \end{vmatrix}}, \quad \text{or}$$

$$A^{-1} = \frac{\begin{pmatrix} 2 & 0 & 0 \\ -3 & 3 & 0 \\ 1 & -3 & 6 \end{pmatrix}}{6}.$$

Therefore,

$$X = \frac{1}{6}\begin{pmatrix} 2 & 0 & 0 \\ -3 & 3 & 0 \\ 1 & -3 & 6 \end{pmatrix}\begin{pmatrix} 1 \\ 0 \\ 0 \end{pmatrix} = \frac{1}{6}\begin{pmatrix} 2 \\ -3 \\ 1 \end{pmatrix}$$

and $\begin{pmatrix} b_3 \\ b_2 \\ b_1 \end{pmatrix} = \frac{1}{6}\begin{pmatrix} 2 \\ -3 \\ 1 \end{pmatrix}$ implies that $\begin{cases} b_3 = \frac{1}{6}(2) \\ b_2 = \frac{1}{6}(-3) \\ b_1 = \frac{1}{6}(1) \end{cases}$.

Thus,

$$f(n) = \frac{1}{6}\left[2n^3 - 3n^2 + n + b_0\right]$$
$$f(n+1) = \frac{1}{6}\left[2(n+1)^3 - 3(n+1)^2 + (n+1) + b_0\right]$$

and,

$$f(1) = \frac{1}{6}(b_0).$$

Hence, the sum of the series will be

$$\begin{aligned} k^2 &= f(n+1) - f(1) \\ &= \frac{1}{6}\left[2(n+1)^3 - 3(n+1)^2 + (n+1)\right] \\ &= \frac{1}{6}n(n+1)(2n+1). \end{aligned}$$

Example 2

Find the sum of $1 + 2 + 3 + \cdots + n$.

In this example, the series is $\sum_{k=1}^{n} k$ and $r = 1$. Thus, the equation will be

$$\begin{pmatrix} \binom{2}{1} & 0 \\ \binom{2}{2} & \binom{1}{1} \end{pmatrix} \cdot \begin{pmatrix} b_0 \\ b_1 \end{pmatrix} = \begin{pmatrix} 1 \\ 0 \end{pmatrix}. \text{ Then, A} = \begin{pmatrix} 2 & 0 \\ 1 & 1 \end{pmatrix}.$$

The inverse of A will then be $\text{A}^{-1} = \begin{pmatrix} \frac{\begin{matrix} 1 & 0 \\ -1 & 2 \end{matrix}}{2} \end{pmatrix}$, and

$$X = \frac{1}{2}\begin{pmatrix} 1 & 0 \\ -1 & 2 \end{pmatrix}\begin{pmatrix} 1 \\ 0 \end{pmatrix} = \frac{1}{2}\begin{pmatrix} 1 \\ -1 \end{pmatrix}.$$

This implies that $b_0 = \frac{1}{2}(1)$ and $b_1 = \frac{1}{2}(-1)$.
Then

$$f(n) = \frac{1}{2}\left(n^2 - n + a_2\right)$$

$$f(n+1) = \frac{1}{2}\left[(n+1) - (n+1) + a_2\right]$$

$$\text{and} \quad f(1) = \frac{1}{2}\left(a_2\right).$$

Then,

$$\sum_{k=1}^{n} k = f(n+1) - f(1)$$

$$= \frac{1}{2}\left[(n+1)^2 - (n+1)\right] = \frac{1}{2}n(n+1).$$

Postassessment

1. Find the sum of $1^3 + 2^3 + \cdots + n^3$.
2. Find the sum of $\sum_{k=1}^{n} k^5$.
3. What are the changes in the general theorem if, in $\sum_{t=1}^{n} t^r, t$ is an even number?

Unit 105

A Parabolic Calculator

After having taught the properties of the parabola to an eleventh grade mathematics class, the teacher might want to discuss some applications of the parabola. The teacher may discuss the reflective properties of a parabolic surface such as a searchlight or the mirror in a telescope. The light source at the focus of a parabolic reflecting surface (Figure 1) reflects its rays off the surface in *parallel paths*. It may be noted that the angle of incidence, ∠FTP, equals the angle of reflection, ∠FTQ.

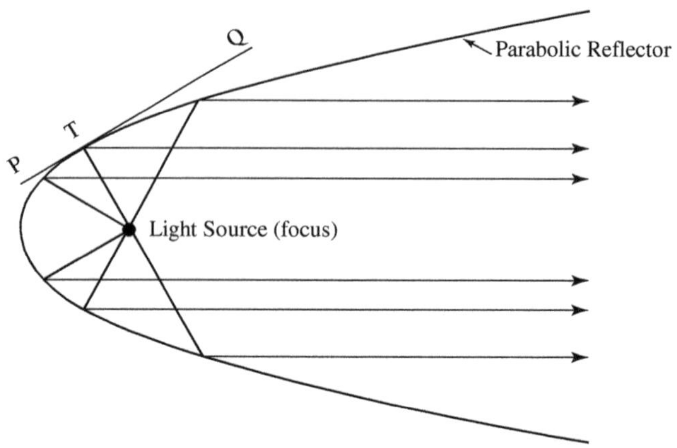

Figure 1.

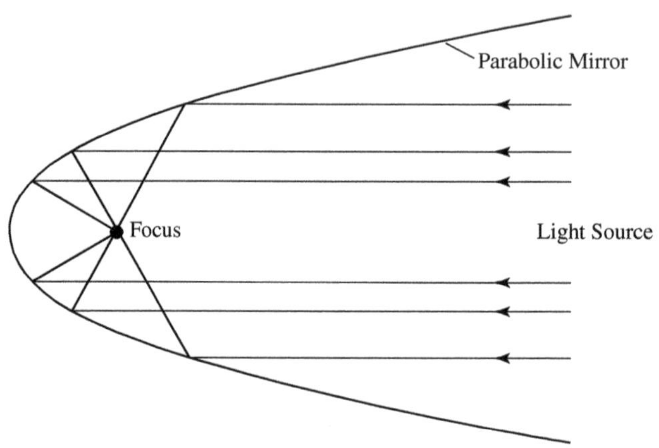

Figure 2.

The same principle is used in telephone (Figure 2) (or radar unit). However, here the rays are generated from external sources and reflected off the mirror (or radar screen) to the focus, which may consists of a camera or other sensing device.

Other applications such as parabolic path of a thrown object may be considered. However, a rather unusual application of the parabola involves its properties on the Cartesian plane. This unit will present a method of using a parabola on the Cartesian plane as a calculating device for performing multiplication and division. The only supplies students will need are graph paper and a straightedge.

Performance Objectives

1. *Students will draw an appropriate parabola and perform a given multiplication with it.*
2. *Students will draw an appropriate parabola and perform a given division with it.*
3. *Students will justify (analytically) why the multiplication method presented in this unit "works."*

Preassessment

Before presenting this unit to the class, the teacher should be sure that students are able to graph a parabola and are able to find the equation of a line, given two points on the line.

Teaching Strategies

On a large sheet of graph paper (preferably one with small squares) have students draw coordinate axes and graph the parabola $y = x^2$. This must be done very accurately. Once this has been completed, students are ready to perform some calculations. For example, suppose they wish to multiply 3×5. They would simply draw the line joining the point on the parabola whose abscissa is 3 with the point whose abscissa is -5. The point product of 3 and 5 is the ordinate of the point where this line intersects the y-axis (Figure 3, AB).

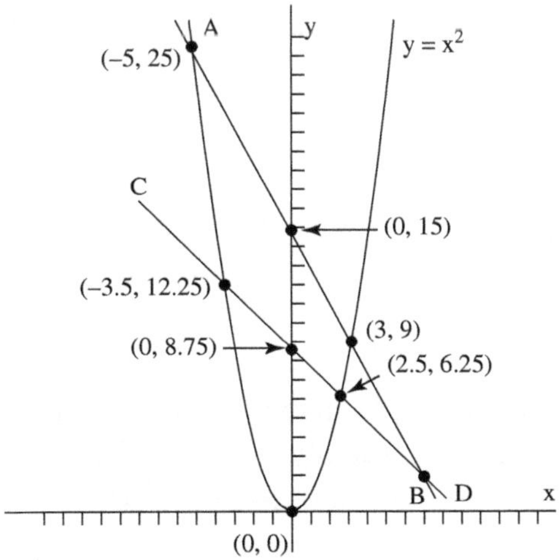

Figure 3.

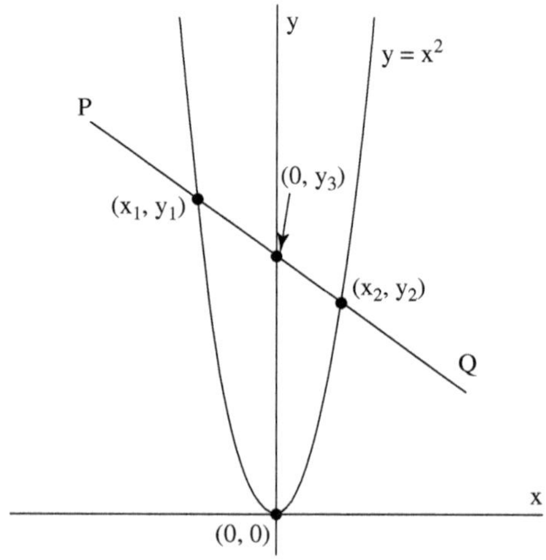

Figure 4.

For further practice, have students multiply 2.5×3.5. Here they must draw the line containing the points $(2.5, 6.25)$ and $(-3.5, 12.25)$. (These are the points on parabola $y = x^2$, whose abscissas are 2.5 and -3.5.) The ordinate of the point where this line (Figure 3, $\overleftrightarrow{CD}$) intersects the Y-axis is the product of 2.5 and 3.5, that is, 8.75. Naturally the size of the graph will determine the degree of accuracy that can be obtained. Students should realize that the points on the parabola whose abscissas were -2.5 and 3.5 could just as well have been used in place of the points whose abscissas were 2.5 and -3.5 in the past example.

At this point, the teacher may ask students how this same can be used to do division. Students, noting that division is the inverse operation of multiplication, should suggest that $\overleftrightarrow{CD}$ could have been used to divide the following: $8.75 \div 3.5$. The other point of intersection that $\overleftrightarrow{CD}$ makes with the parabola, point $(2.5, 6.25)$ yields the answer 2.5.

The teacher would be wise to offer students a variety of drill exercises to better familiarize them with this technique. Students can use a straightedge (without drawing a line) to read the answers from the graph.

After sufficient drill, students may become curious about the reason why this technique of calculation actually "works." To prove that it does work, have the class consider the following general case (Figure 4).

Let $\overleftrightarrow{PQ}$ intersect the parabola $y = x^2$ at points (x_1, y_1) and (x_2, y_2), and intersect the y-axis at $(0, y_3)$. This proof must conclude that $y_3 = |x_1x_2|$.

Proof:

The slope of $\overleftrightarrow{PQ} = \frac{y_2 - y_1}{x_2 - x_1} = \frac{x_2^2 - x_1^2}{x_2 - x_1} = x_1 + x_2$ (since $y_2 = x_2^2$, and $y_1 = x_1^2$).

The slope of $\overleftrightarrow{PQ}$ expressed with any point (x, y) is $\frac{y - y_1}{x - x_1}$.

Therefore, $\frac{y - y_1}{x - x_1} = x_1 + x_2$ is the equation of $\overleftrightarrow{PQ}$.

At the point $(0, y_3)$, $\frac{y_3 - x_1}{0 - x_1} = x_1 + x_2$ and $y_3 = -x_1^2 - x_1 x_2 + y_1$.

But $y_1 = x_1^2$, thus $y_3 = -x_1 x_2$, but this is positive, so $y_3 = |x_1 x_2|$.

With a knowledge of this proof, students may wish to experiment with other parabolas in an attempt to replace $y = x^2$ with a more "convenient" parabola.

This scheme also provides the student with a host of further investigations. For example, it may be used to "construct" a line of length $\sqrt{a}$. The student need only construct a line parallel to the x-axis and intersecting the y-axis at $(0, a)$. The segment of that line that is between the y-axis and the parabola has length $\sqrt{a}$. Further student investigations should be encouraged.

Postassessment

1. Have students draw the parabola $y = x^2$ and then use it to do the following exercises:
   a. 4×5
   b. 4.5×5.5
   c. $4 \div 2.5$
   d. $1.5 \div .5$
2. Have students show how $y = \frac{1}{2}x^2$ may be used to perform multiplication and division operations.

Reference

Posamentier, A. S. and C. Spreitzer, *The Mathematics Everyday Life*, Amherst, New York: Prometheus Books, 2018.

Unit 106

Constructing Ellipses

This unit provides a means of constructing ellipses using a straightedge and a compass.

Performance Objectives

1. *Students will plot points on an ellipse without the use of an equation.*
2. *The relation of the circle to the ellipse will be used by students in the constructions of ellipses.*

Preassessment

Students should have completed tenth year mathematics and be familiar with fundamental trigonometric identities. Knowledge of analytic geometry is helpful, but not necessary. Ask students to construct an ellipse using any method (i.e., analytically or with special tools).

Teaching Strategies

After inspecting student attempts to construct an ellipse, have them consider the accuracy of their work. Some students may have attempted a freehand drawing, while others may have plotted an appropriate curve on a piece of graph paper. This should lead comfortably to Method I.

Method I: Point-by-point construction

One of the definitions of an ellipse is: *the locus of a point, P, which moves such that the sum of its distances from two given fixed points.* F and F', *is a constant.* From Figure 1, the definitions just given implies that

$$\mathrm{PF} + \mathrm{PF}' = \text{a constant.} \tag{1}$$

Customarily, this constant is given the value 2a, and it is not too difficult to derive from this definition the equation of the ellipse:

$$\frac{x^2}{a^2} + \frac{y^2}{b^2} = 1 \tag{2}$$

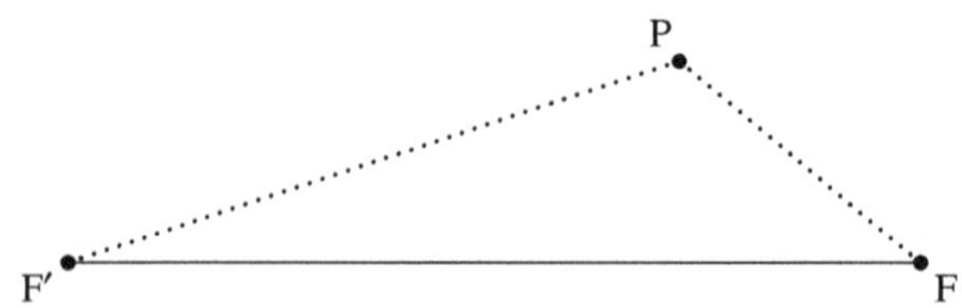

Figure 1.

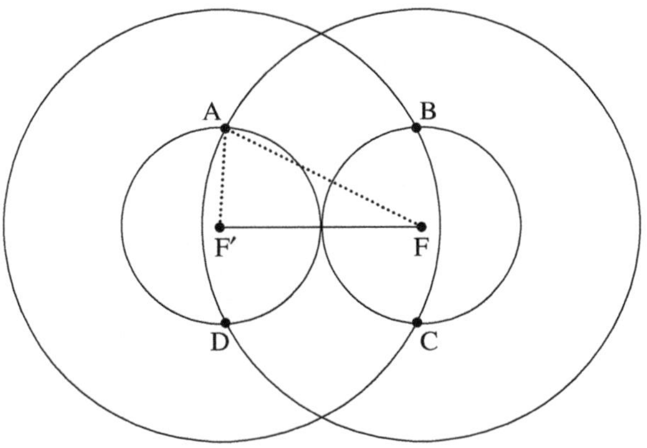

Figure 2.

Indeed, the popular thumbtack and string-loop construction is directly based on this definition (where a string-loop is held taut between two thumbtacks and a pencil that changes position).

Procedure: Place an $8\frac{1}{2}'' \times 11''$ sheet in the horizontal position and center on it a $4''$ horizontal line whose endpoints F and F′ are the ellipse's foci (see Figure 2). Let the constant in (1) equal $6''$. Thus,

$$\text{PF} + \text{PF}' = 6''. \tag{3}$$

With F as center, draw a circle of $4''$ radius. Do likewise with F′ as center. Next, again use F and F′ as centers for circles of $2''$ radius. Note that these four circles intersect in four points A, B, C, and D, which lie on the ellipse. For, if point A is arbitrarily chosen, by construction $\text{FA} = 2''$ and $\text{F}'\text{A} = 4''$; hence $\text{FA} + \text{F}'\text{A} = 6''$, satisfying (3).

A rather accurate construction can be made by using increments of $\frac{1}{2}''$ for the radii of each circle centered on F and F′, starting with $9\frac{1}{2}''$ and $2\frac{1}{2}''$ radii. The next set of four points on the ellipse would be obtained by drawing two circles of $5''$ and $1''$ radius, both on F and F′. These two pairs of circles would be followed by another two pairs of radii $9''$ and $3''$, and so on, as indicated in Figure 3. Indeed, additional ellipses can then be sketched from Figure 3.

Assume that $\text{PF} + \text{PF}' = 5''$. (4)

Then, one need only mark the intersections of those circles centered on F and F′, the sum of whose radii is 5. For example, in constructing (3) by using the suggested $\frac{1}{4}''$ increment, it was necessary at some point to use a radius of $3\frac{1}{2}''$ centered on F and F′ and another radius of $1\frac{1}{2}''$ again centered on F

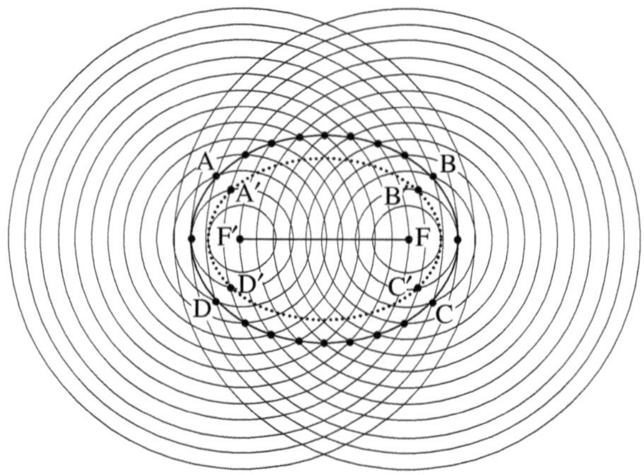

Figure 3.

and F′. The intersection of these four circles, A′, B′, C′, and D′, provide four points on *another* ellipse such that, using A' arbitrarily, $\text{FA}' + \text{F}'\text{A}' = 5''$.

Figure 3 thus shows the two ellipses sketched, both of which have F and F′ as their foci. Other ellipses can be sketched in a similar manner from the same figure.

Method II: Tangent construction

Centered on an $8\frac{1}{2}''$ by $11''$ sheet of paper, have students draw a circle of radius $3''$. Locate a point F in the interior of the circle $2\frac{1}{4}''$ from the center (see Figure 4).

Through F, draw a chord to intersect the circle at P and Q. Using a right triangle template or carpenter's T-square, erect perpendiculars at P and Q to meet the circle at P′ and Q′, respectively. Then $\overleftrightarrow{\text{PP}'}$ and $\overleftrightarrow{\text{QQ}'}$ are each tangents to an ellipse having F as one of its foci. Continue this procedure for many such chords $\overline{\text{PQ}}$, obtaining a diagram similar to that shown in Figure 5.

The proof of this construction is based on the converse of the following theorem: The *locus of the intersection of the tangent to an ellipse with the perpendicular on it from either focus is a circle.* A proof of this theorem can be found in Bowser's *An Elementary Treatise on Analytic Geometry*, pp. 139–140.

Varying the position of F will change the size and shape of the ellipse. Furthermore, the second focus, F′, is located on $\overleftrightarrow{\text{FO}}$ extended through O its own distance.

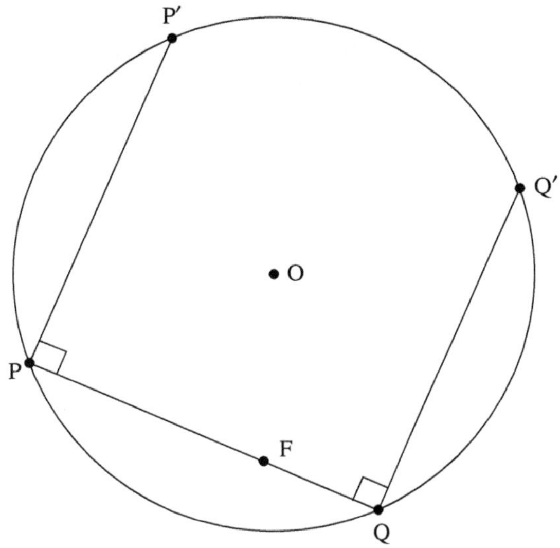

Figure 4.

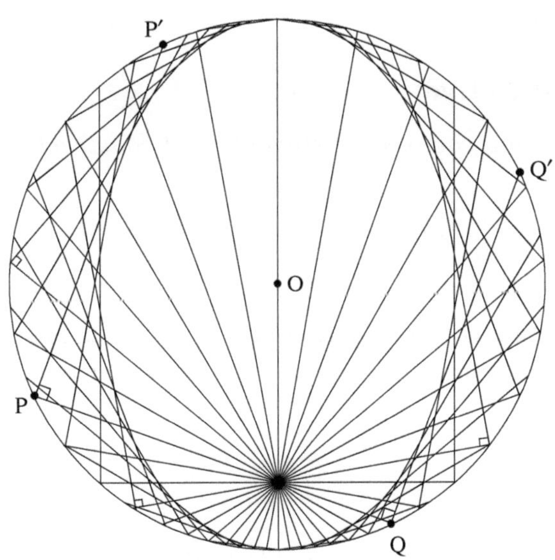

Figure 5.

Postassessment

1. Have students draw Figure 3 on a larger scale, such as FF′ = 6″. Coloring in the numerous regions formed by the intersections of the circles may prove quite satisfying.

2. Have students draw a circle of radius r, with O as center. They then locate a fixed point, F, inside the circle, and draw $\overline{\text{OF}}$. Clearly, $r > \text{OF}$. Then have them draw an arbitrary radius $\overline{\text{OQ}}$. Connect $\overline{\text{FQ}}$, and construct its midpoint, M. At M, they should erect a perpendicular to intersect $\overline{\text{OQ}}$ at P. Then P lies on an ellipse with F as one of its foci. Moreover, $\overleftrightarrow{\text{MP}}$ extended through P is a tangent to the same ellipse. Have students complete this construction and justify it.

Reference

Posamentier, A. S. and H. A. Hauptman. 101+ *Great Ideas for Introducing Key Concepts in Mathematics.* Thousand Oaks, CA: Corwin, 2006.

Unit 107

Constructing the Parabola

Performance Objective

With straightedge and compasses, students will construct a parabola without using an equation.

Preassessment

Ask students to construct the parabola $y = x^2$ on a sheet of graph paper. When this has been done, have them draw any other parabola on a sheet of "blank" paper.

Teaching Strategies

In all likelihood, students will be unable to draw a parabola without use of graph paper. At this point, the teacher may define the parabola in terms of locus. That is, it is the locus of points equidistant from a fixed point and a fixed line. Perhaps students will find this a useful hint in deriving a method of construction of a parabola. After considering students' suggestions, have them consider the following methods.

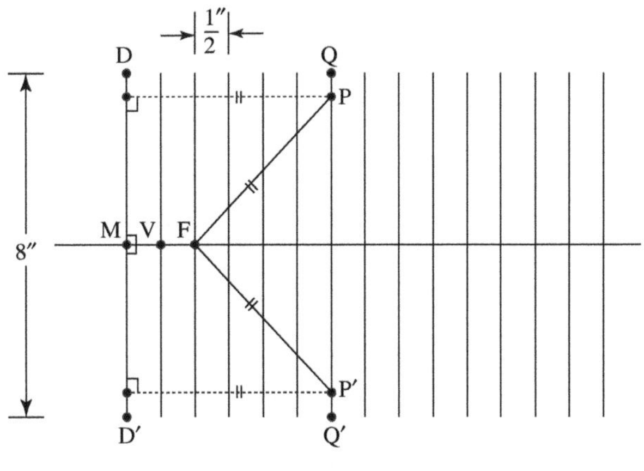

Figure 1.

Method I: Point-by-point construction

Toward the left of an $8\frac{1}{2}$-×-11-inch sheet of paper held horizontally, lightly draw about 15 vertical parallel lines each spaced $\frac{1}{2}$ inch apart (see Figure 1). Each line segment should be 8 inches long. Draw the common perpendicular bisector of these lines. Label your drawing as shown in Figure 1, being sure to place F at the intersection of the third parallel line and the perpendicular bisector. Let $\overleftrightarrow{QQ'}$ be any arbitrary parallel line, say the sixth one from $\overleftrightarrow{DD'}$. By construction, $\overleftrightarrow{QQ'}$ is $(6 \cdot \frac{1}{2}) = 3''$ from $\overleftrightarrow{DD'}$. Maintaining this distance, using F as center, swing an arc with your compasses cutting $\overleftrightarrow{QQ'}$ above and below the perpendicular bisector at P and P′. Then P and P′ are on the parabola with F as the *focus* of the parabola. Repeat this procedure for the other parallel lines, joining all the points so determined (see Figure 2).

Discussion: By construction, the perpendicular distance of either P or P′ from $\overleftrightarrow{DD'}$ is equal to FP or FP′. The definition of a parabola is based on such an equivalence of distance of a varying point from a fixed line and a fixed point. A parabola is the locus of points each of whose distances from a fixed point equals its distance from a fixed line. The line $\overleftrightarrow{DD'}$ is referred to as the *directrix*, and its perpendicular bisector is the *axis* of the parabola. If M is the intersection of the axis and the directrix $\overleftrightarrow{DD'}$, by letting FM $= 2p$, it is not difficult, using the distance formula and the above definition, to show that the parabola's equation is $y^2 = 4px$. (1)

The midpoint, V, of $\overline{\text{MF}}$ is the parabola's *vertex* and is often referred to as the parabola's turning point.

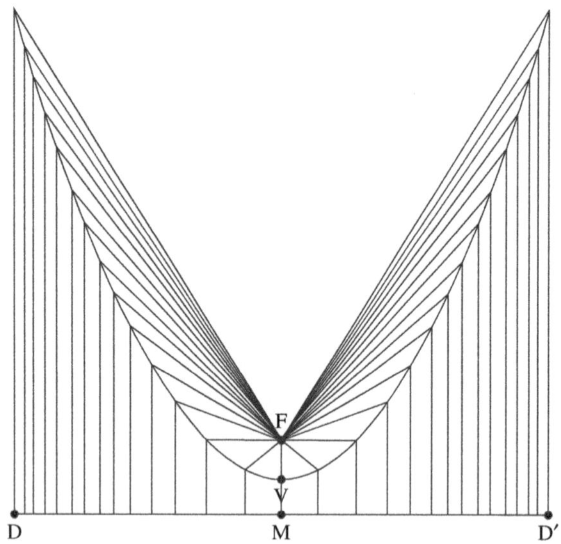

Figure 2.

Method II: Point-by-point construction

Draw a rectangle ABCD (see Figure 3), letting V and G be the midpoints of $\overline{AD}$ and $\overline{BC}$, respectively. Divide $\overline{AB}$ and $\overline{BG}$ into the same *number* of equal parts. Starting from B, let the successive points of equal division be $1, 2, 3, \ldots$ on $\overline{AB}$ and $a, b, c, \ldots$ on $\overline{BG}$. Draw $\overline{aa'}$ perpendicular to $\overline{BG}$. Note that $a', b', c', \ldots$ are points on $\overline{AV}$. Draw $\overline{V1}$, meeting $\overline{aa'}$ at P. Similarly, draw $\overline{bb'}$ perpendicular to $\overline{BG}$ and draw $\overline{V2}$, meeting $\overline{bb'}$ at P′. Continue this for the other points. Then P and P', and other points so obtained are on a parabola with V as its vertex and $\overline{VG}$ as its axis.

Proof: From Figure 3, let $\overleftrightarrow{AD}$ be the y-axis and $\overleftrightarrow{VG}$ the x-axis (see Figure 4). Let b on $\overline{BG}$ now be called Q, b' on $\overline{AV}$ to be labeled Q′. Similarly, let point 2 in Figure 3 be called T. Draw $\overline{VT}$ and $\overline{QQ}'$, meeting at P′. Let $VQ' = GQ = y$, $Q'P' = x$, $VG = h$, and $AV = a$.

By construction (i.e., points T and Q were selected with this property),

$$\frac{AT}{AB} = \frac{GQ}{GB}, \text{ or } \frac{AT}{h} = \frac{y}{a}. \tag{2}$$

From similar triangles VQ′P′ and VAT, $\frac{x}{y} = \frac{AT}{AV} = \frac{AT}{a}$, or $AT = \frac{ax}{y}$. (3)

Substituting this value of AT in (2), $\frac{\frac{ax}{y}}{h} = \frac{y}{a}$, or $\frac{a^2x}{y} = hy$. (4)

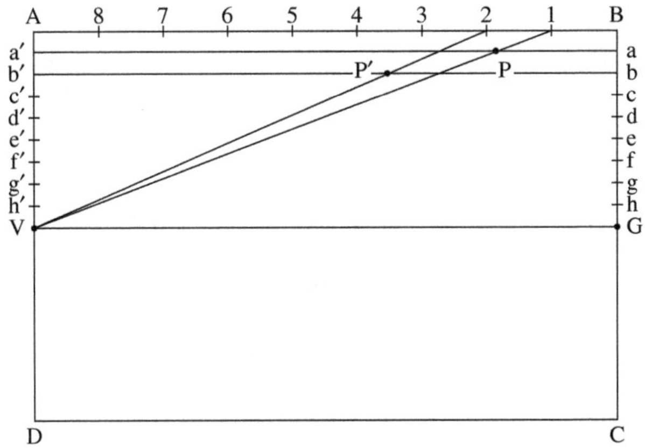

Figure 3.

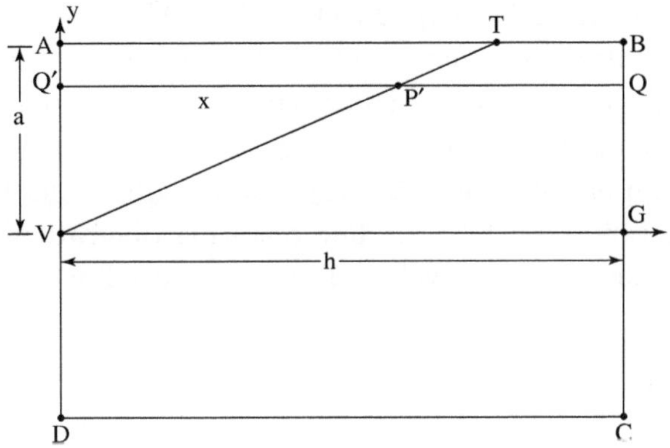

Figure 4.

Solving (4) for y^2 yields $y^2 = \frac{a^2x}{h}$, (5)

which is of the same form as (1) in which $4p = \frac{a^2}{h}$. (6)

Because p is the distance from the vertex to the focus, solving (6) for p will locate the focus in terms of a and h of the original rectangle.

Method III: Envelope construction

Near the bottom edge of a vertically held $8\frac{1}{2}$-×-11-inch sheet of paper, draw a horizontal line the full width of the paper (see Figure 5). Label this line $\overleftrightarrow{AA}$. Draw the perpendicular bisector of $\overline{AA'}$, with midpoint V. Locate F on the

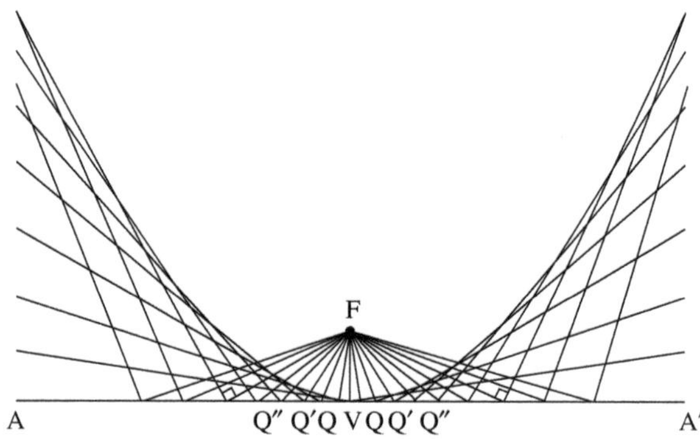

Figure 5.

perpendicular bisector, one inch above V. On $\overline{AA'}$, on either side of V, mark off points $Q, Q', Q^n, \ldots$, gradually increasing, by a very small increment, the successive distances from V to Q, Q to Q', Q' to $Q'', \ldots$ Place a right triangle template or carpenter's T-square such that a leg passes through F and the vertex of the right angle is concurrent with any of the points $Q, Q', Q'', \ldots$ Construct a right angle with one ray $\overrightarrow{FQ}$ and drawing an extended line to the edge of the paper with the other ray of the right angle. Repeat this procedure for other points $Q', Q'', \ldots$. The resulting diagram should be similar to Figure 5, and the shape "enveloped" by all these lines is a parabola.

Postassessment

Have your students draw an angle of any measure, making both sides equal in length. Starting from the vertex, have them measure off equally spaced intervals along each side, labeling the intervals on each side $1, 2, 3, 4, \ldots, 10$, with the vertex labeled 0. Then ask them to connect point 10 on one side with point 1 on the other side. See that they do likewise for points 9 and $2, 8$ and 3, and so forth, always being sure that the sum of the numbers joined is 11. The resulting appearance will be somewhat similar to Figure 5. Students will have drawn an envelope to a parabola. Have students attempt to justify this construction, as well as that for Figure 5.

Reference

Posamentier, A. S. and H. A. Hauptman. 101 + *Great Ideas for Introducing Key Concepts in Mathematics.* Thousand Oaks, CA: Corwin, 2006.

Unit 108

Using Higher Plane Curves to Trisect an Angle

This unit will introduce two higher plane algebraic curves, showing analytically and visually (experimentally) how the trisection is done.

Performance Objectives

1. *Given a certain locus condition, students will learn how to sketch a curve directly from the locus without using an equation.*
2. *Given a polar equation, students will plot the curve on polar coordinate paper.*
3. *Given one of the curves discussed in this unit, students will trisect any angle.*

Preassessment

Students should have done some work with polar coordinates.

Teaching Strategies

The trisection of an angle can be considered a consequence of the following locus problem: Given ΔOAP with fixed base $\overline{\text{OA}}$ and variable vertex P, find the locus of points P such that $m\angle\text{OPA} = 2m\angle\text{POA}$ (see Figure 1). Let O be the pole of a polar coordinate system and $\overleftrightarrow{\text{OA}}$ the initial line, with A having coordinate (2a, 0).

Let $m\angle\text{AOP} = \theta$ and $\text{OP} = r$. Then by hypothesis, $m\angle\text{APO} = 2\theta$, from which it follows $m\angle\text{OAP} = \pi - 3\theta(\pi$ radians $= 180°)$. Extend $\overline{\text{OA}}$ through

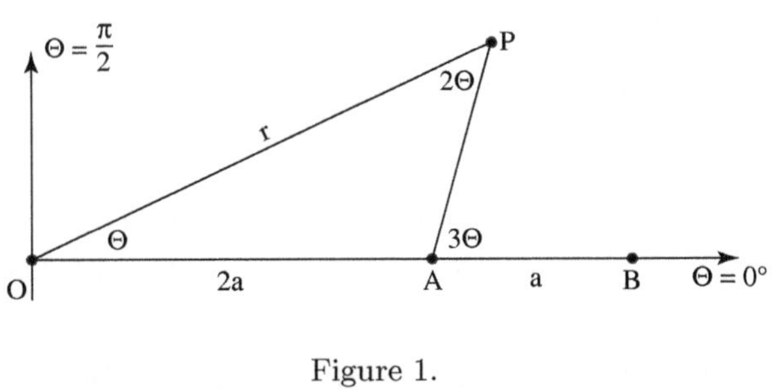

Figure 1.

A a distance of a units, thereby locating point B. Then $m\angle\text{BAP} = 3\theta$. By the law of sines,

$$\frac{r}{\sin(\pi - 3\theta)} = \frac{2a}{\sin 2\theta} \tag{1}$$

from which it follows $r = \dfrac{2a \cdot \sin(\pi - 3\theta)}{\sin 2\theta}$. (2)

Since $\sin(\pi - 3\theta) = \sin 3\theta$ and $\sin 2\theta = 2\sin\theta\cos\theta$, the appropriate substitutions for Equation (2) yield

$$r = \frac{a \cdot \sin 3\theta}{\sin\theta\cos\theta}. \tag{3}$$

With $\sin 3\theta = 3\sin\theta - 4\sin^3\theta = \sin\theta(3 - 4\sin^2\theta)$,

$$\text{(3) yields } r = \frac{3a - 4a \cdot \sin^2\theta}{\cos\theta}. \tag{4}$$

Letting $\sin^2\theta = 1 - \cos^2\theta$ in (4), $r = \dfrac{-a + 4a \cdot \cos^2\theta}{\cos\theta}$, which can easily be simplified, by letting $\dfrac{1}{\cos\theta} = \sec\theta$, to $r = a(4\cos\theta - \sec\theta)$ the required polar equation for the *Trisectrix of Maclaurin.* (5)

By placing P on the other side of $\overleftrightarrow{\text{OA}}$ (Figure 1), a similar derivation will yield

$$r = a(4\cos\theta - \sec\theta). \tag{6}$$

Since $\cos(-\theta) = \cos\theta$ and $\sec(-\theta) = \sec\theta$, it follows that (5) is symmetric with respect to $\overleftrightarrow{\text{OA}}$. Thus, as confirmed by (6), for all points of the locus above $\overleftrightarrow{\text{OA}}$, there are corresponding points below it as well. (These corresponding points are reflections in $\overleftrightarrow{\text{OA}}$.)

To sketch the locus, assigning values to θ in (5) and (6) will, of course, furnish points of the Trisectrix. Have students make an exact copy of ΔAOP on polar coordinate graph paper, so that O is at the origin and $\overleftrightarrow{\text{OA}}$ is on the horizontal axis (at $\theta = 0$ radians). Then students should plot various points of the curve and draw the curve.

Students may now use the original diagram (on regular paper). A novel approach to curve sketching would be to plot points and draw lines directly from the given locus condition. One need only assign various values to θ and a corresponding 3θ value to $\angle$BAP. With $\overleftrightarrow{\text{OB}}$ a fixed base for each of these

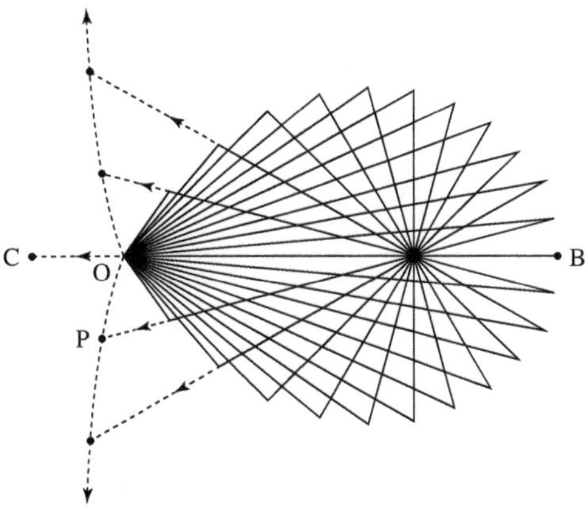

Figure 2.

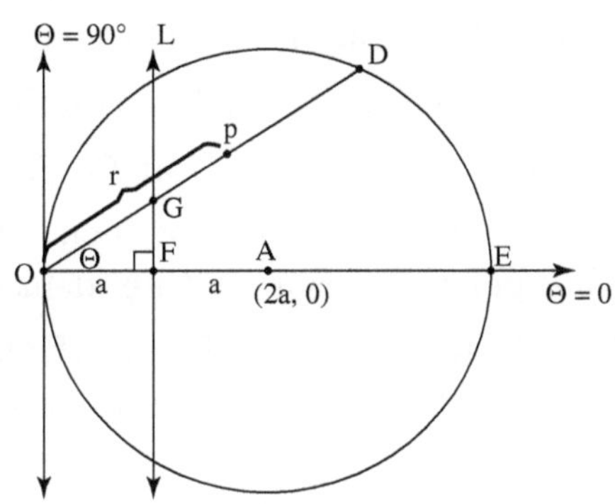

Figure 3.

angles, the intersection of the second side of each angle will yield point P. Figure 2 shows such a construction for values of $0° \leq \theta \leq 55°$ in 5° intervals.

For $\theta = 0°$ and 60°, there is no triangle, merely $\overline{\mathrm{OB}}$. For $\theta > 60°$, consider some arbitrary point C such that $\overleftrightarrow{\mathrm{COA}}$. Let the measure of reflex $m\angle\mathrm{BAP} = 195°$, thus placing P below $\overleftrightarrow{\mathrm{COA}}$. Then $m\angle\mathrm{COP} = 65°$, preserving the trisection property. Note that for $|\theta|\angle 60°$, there will be two asymptotic branches above and below $\overleftrightarrow{\mathrm{OB}}$.

Conversely, given (5), it is a bit more difficult to show that for any point P on the Trisectrix, $m\angle\mathrm{BOP} = \frac{1}{3}m\angle\mathrm{BAP}$. The proof rests on showing that $m\angle\mathrm{OPA} = 2\theta$, from which it follows that $m\angle\mathrm{BAP} = 3\theta$.

Equation (5) for the Trisectrix of Maclaurin can also be obtained from the following locus problem (see Figure 3):

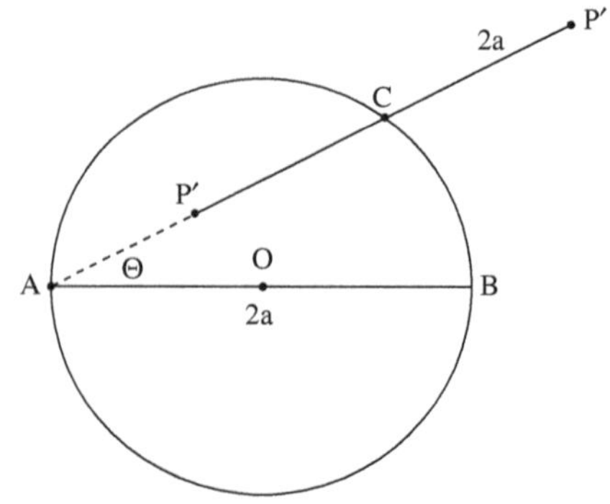

Figure 4.

Given O, the pole of a polar coordinate system and E $(4a, 0)$. At A $(2a, 0)$ as center, draw a circle of radius $2a$. Through F$(a, 0)$, draw a line L perpendicular to $\overleftrightarrow{\text{OE}}$. Locate an arbitrary point D on the "upper" semicircle's circumference and draw $\overleftrightarrow{\text{OD}}$, intersecting L at G. Arbitrarily locate P on $\overleftrightarrow{\text{OD}}$. As D varies along the semicircle, find the locus of P on $\overleftrightarrow{\text{OD}}$ such that OP = GD. The resulting polar equation will be identical to (5). (*Hints for Solution*: Let $m\angle\text{AOD} = \theta, \text{OP} = r = \text{GD}$. Draw $\overleftrightarrow{\text{DE}}$, forming right triangle ODE. Express OD in terms of a and $\cos\theta$, and express OG in terms of a and $\sec\theta$.)

The student now has a method for trisecting an angle (using an additional tool, a curve). Once students have grasped the above, they may consider another curve, the Limacon of Pascal, which also can be used to trisect an angle. However, the following construction for the Limacon should be attempted first.

Draw a circle of diameter AB $= 2a$, center O. Locate an arbitrary point C, distinct from A or B, on the circle's circumference. Place the edge of a ruler at C in such a way that the ruler's midpoint rests on C and that the same edge passes through A. Locate two points P and P′ at a distance of a units on either side of C. Repeat this for different positions of C. P and P′ are then points on the Limacon.

For a visual effect analogous to Figure 2, divide the circle's circumference into 18 evenly spaced arcs. Repeat the procedure in the preceding paragraph for each of these 18 points, thereby yielding 36 points on the Limacon's circumference. Be sure to draw a line to connect all corresponding points P and P′, taking note of the loop that occurs *inside* the Limacon. A suggested diameter for the base circle O is 3″, making P′C = CP = $1\frac{1}{2}''$.

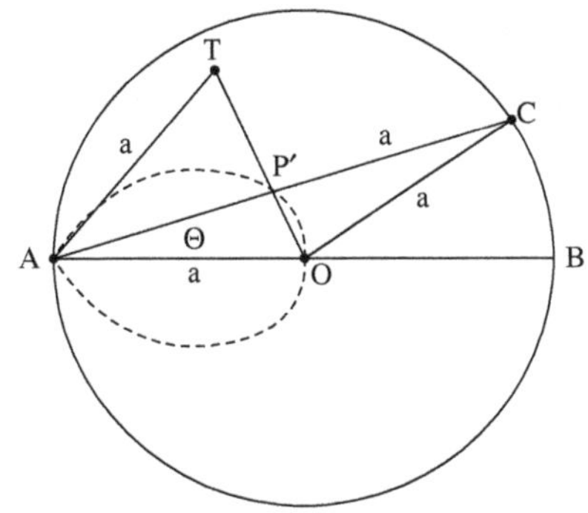

Figure 5.

Figure 5 shows the base circle O, the Limacon's inner loop, as well as a few other points and lines used for the following trisection: Let $\angle$BAT be congruent to the angle to be trisected, making AT $= a$. Draw $\overline{\mathrm{TO}}$, intersecting the loop at P′, and then draw $\overline{\mathrm{AP'}}$. We now show that $m\angle\mathrm{BAT} =$ 3 times $m\angle\mathrm{BAP'}$.

Extend $\overline{\mathrm{AP'}}$ to intersect the circle at C. Because P′ is a point on the Limacon, CP′ $= a$, from the above construction. Draw $\overline{\mathrm{OC}}$ and let $m\angle\mathrm{OAP'} = \theta$. Immediately, OA $=$ OC, $m\angle\mathrm{C} = \theta$, therefore $m\angle\mathrm{AOC} = \pi - 2\theta$. Since ΔOCP′ is isosceles, each of its base angles has measure $\frac{\pi}{2} - \frac{1}{2}\theta$. $m\angle\mathrm{AP'O} = \frac{\pi}{2} + \frac{1}{2}\theta$, and it follows that $m\angle\mathrm{AOP'} = \frac{\pi}{2} - \frac{3}{2}\theta$.

However, ΔATO is also isosceles, hence $m\angle\mathrm{T} = \frac{\pi}{2} - \frac{3}{2}\theta$, from which it follows that $m\angle\mathrm{BAT} = 3\theta$, which was to be proved.

Postassessment

Construct the Limacon of Figure 4 by locating P and P′ a distance of $2a$ units on either side of C.

Unit 109

Constructing Hypocycloid and Epicycloid Circular Envelopes

In this unit, two elementary cycloidal curves will be related to each other. Students will then create a circular envelope that will simultaneously encompass both curves.

Performance Objectives

1. *Students will define hypocycloid and epicycloid.*
2. *Students will construct a hypocycloid and an epicycloid.*
3. *Students will generalize these constructions to other hypocycloids and epicycloids.*

Preassessment

Tenth year mathematics is necessary. A minimal knowledge of polar coordinates is helpful.

Teaching Strategies

Initiate the introduction of hypocycloid and epicycloid curves by rolling varying sized circular discs about the interior and exterior circumference, respectively, of a fixed circular disc of constant radius. If possible, let the radius of the fixed circle be some integral multiple of the radius of the rolling circle. Have your students speculate on the loci obtained by a fixed point on the circumference of the varying sized rolling circles. For the interior rotations, it will be necessary for the fixed circle to be hollowed out. A Spirograph Kit, if available, is an excellent motivational source.

This unit analyzes the case when the interior and exterior rolling circles each have radius $b = \frac{a}{3}$, as shown in Figure 1 (next column). O is the center of the fixed circle, radius a, and is also the pole of a polar coordinate system. C and C′ are the centers, respectively, of the interior and exterior rolling circles. We will assume that both rotating circles are continuously in tangential contact with each other at T. Therefore, $\overleftrightarrow{\text{OT}}$, making an angle of measure θ with the initial line, intersects circles C and C′ at I, T, and I′, and contains the centers, C and C′. It is understood that initially each circle was tangent to the fixed circle at B; furthermore, P and P′ are respectively fixed points on the circumference of circles C and C′. At the instant both circles began their circuits, P and P′ were coincident with B. The partial loci from B to P and P′ are shown in Figure 1.

In Figure 2, we show the complete loci swept out by each fixed point. The locus of P is a *hypocycloid of three cusps*, often called the *deltoid*; whereas the locus of P′ is an *epicycloids of three cusps*. Each circle requires three complete rotations before returning to point B. The fixed circle is then a circumcircle for the deltoid and an incircle for the three-cusped epicycloid.

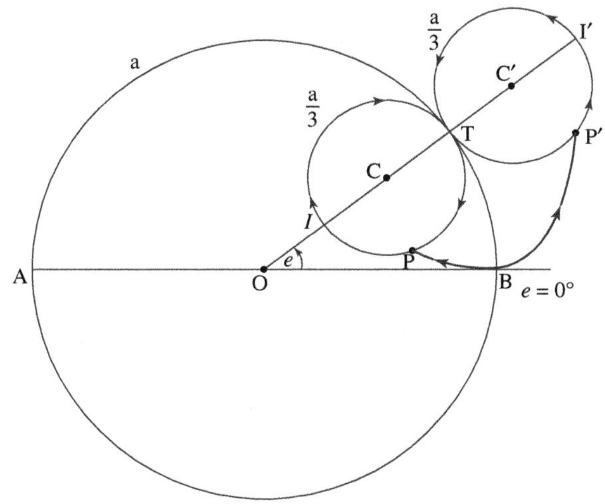

Figure 1.

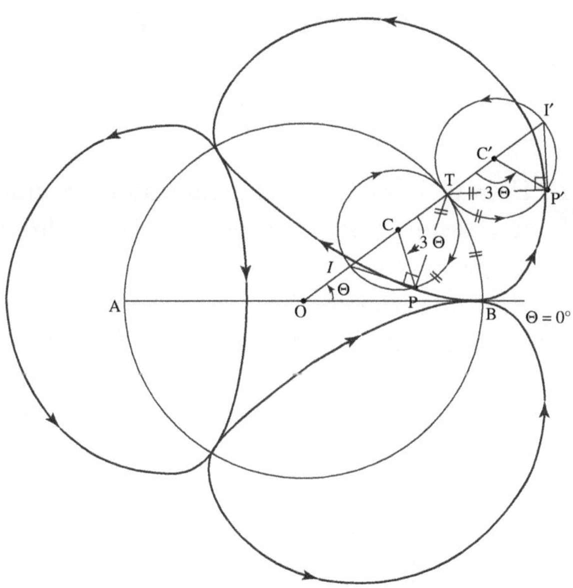

Figure 2.

The following lines are drawn in each circle: $\overline{\text{IP}}, \overline{\text{CP}}$, and $\overline{\text{TP}}$ for circle C; $\overline{\text{TP}'}, \overline{\text{C}'\text{P}'}$, and $\overline{\text{I}'\text{P}'}$ for circle C′. Have your students explain why length $\widehat{\text{TP}}$ = length $\widehat{\text{TB}}$ = length $\widehat{\text{TP}}'$. (1)

Then show that

$$m\angle\text{TCP} = 3\theta = \text{m}\angle\text{TC}'\text{P}'. \tag{2}$$

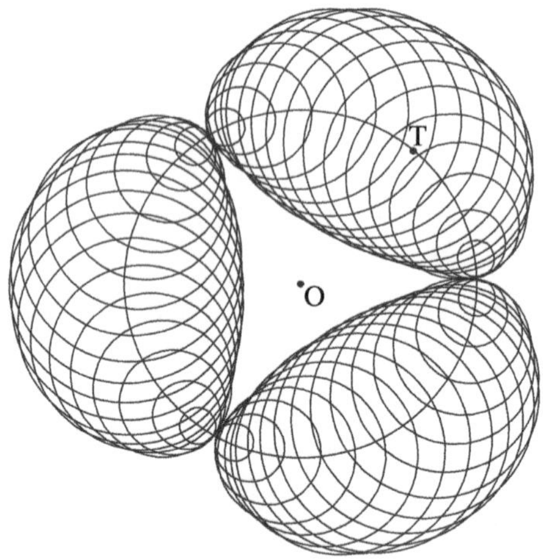

Figure 3.

Since the radii of circles C and C' are equal, $\Delta\text{TPC} \cong \Delta\text{TP}'\text{C}'(\text{SAS})$; and corresponding sides,

$$\text{TP} = \text{TP}'. \tag{3}$$

Observe that $m\angle\text{TPI} = 9° = m\angle\text{TP}'\text{I}'$. Furthermore, we state without proof that $\overline{\text{PI}}$ and $\overline{\text{P'I'}}$ are, respectively, tangents to the deltoid at P and the epicycloids at P'. Further details can be found in Lockwood (1961).

The above equation (3) implies that a circle centered at T, with radius $\text{TP} = \text{TP}'$, will be tangent to each curve at P and P'. Certainly, as T varies, there is a corresponding change in the length of $\overline{\text{TP}}'$ (or $\overline{\text{TP}}$). We show in Figure 3 the final result obtained when circles are drawn for 60 equally spaced positions of T along the fixed circle. We utilize symmetry properties of each curve to minimize the work required to determine the varying lengths of $\overline{\text{TP}}$: since each curve repeats itself every 120°, and $\overline{\text{OI}'}$ divides the curve symmetrically at $m\angle\theta = 60$, it is only necessary to obtain those lengths of $\overline{\text{TP}}$ between $0° \leq \theta \leq 60°$, in intervals of measure 6. These lengths were obtained by making an accurate drawing of the required 10 positions of either rolling circle.

Postassessment

1. Figure 4 shows the results obtained when $b = \frac{a}{4}$. Justify the occurrence of a four-cusped hypocycloid and a four-cusped epicycloid.
2. Generalize for $b = \frac{a}{n}, n = 5, 6, \ldots$

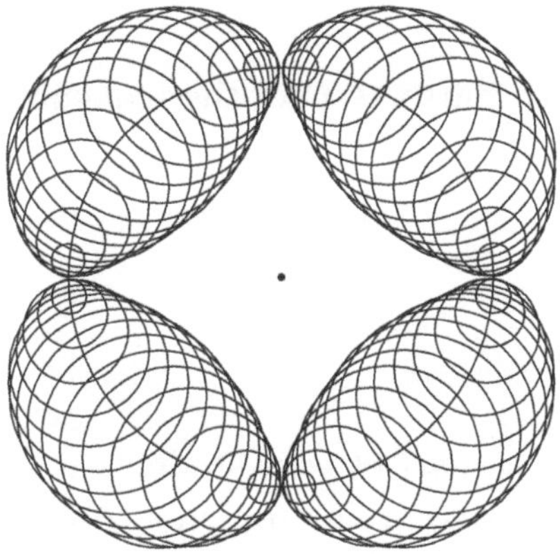

Figure 4.

3. Make a drawing for the case $b = \frac{a}{2}$ (the *nephroid*). Visually show that the locus of a fixed point on the interior rolling circle is a diameter of the fixed circle.
4. Repeat for $b = a$, the *cardioid*. Show that all the circles centered on T pass through a fixed point; i.e., one of the loci degenerates to a point.

Reference

Beard, R. S. *Patterns in Space.* Creative Publications, 1973.
Lockwood, E. H. *A Book of Curves.* Cambridge University Press, 1961.
Posamentier, A. S. and C. Spreitzer, *The Mathematics Everyday Life*, Amherst, New York: Prometheus Books, 2018.

Unit 110

The Harmonic Sequence

This unit is best presented to a class after arithmetic and geometric sequences have been mastered.

Performance Objectives

1. *Students will define harmonic sequence.*
2. *Students will illustrate a harmonic sequence geometrically.*
3. *Students will solve simple problems with harmonic sequences.*

Preassessment

Ask students to find the fourth term of the sequence:

$$1\frac{1}{3}, 1\frac{11}{17}, 2\frac{2}{13}.$$

Teaching Strategies

A natural response on the part of your students is to try to find the fourth term of the above sequence by trying to find a common difference, and when that fails, a common ratio. In a short time, your students will feel frustrated. This will offer you a good opportunity to motivate your students toward a "new" type of sequence. Ask students to write each term in improper fraction form and then write its reciprocal. This will yield: $\frac{3}{4}, \frac{17}{28}, \frac{13}{28}$ or $\frac{21}{28}, \frac{17}{28}, \frac{13}{28}$. Further inspection of this new sequence will indicate it to be an arithmetic sequence with a common difference of $\frac{-4}{28}$. Students will now easily obtain the required fourth term, $\frac{1}{\frac{9}{28}} = \frac{28}{9} = 3\frac{1}{9}$.

Students should now be motivated to learn more about the harmonic sequence.

Consider three or more terms in an arithmetic sequence; for example, $a_1, a_2, a_3, \ldots, a_n$. The sequence of reciprocals of these terms, $\frac{1}{a_1}, \frac{1}{a_2}, \frac{1}{a_3}, \ldots, \frac{1}{a_n}$ is called a harmonic sequence. The term *harmonic* comes from a property of musical sounds. If a set of strings of uniform tension whose lengths are proportional to $1, \frac{1}{2}, \frac{1}{3}, \frac{1}{4}, \frac{1}{5}, \frac{1}{6}$ are sounded together, the effect is said to be "harmonious" to the ear. This sequence is harmonic, as the reciprocals of the terms from an arithmetic sequence, $1, 2, 3, 4, 5, 6$.

There is no general formula for the sum of the terms in a harmonic series. Problems dealing with a harmonic sequence are generally considered in terms of the related arithmetic sequence.

Two theorems would be useful to consider:

Theorem 1: If a constant is added to (or subtracted from) each term in an arithmetic sequence, then the new sequence is also arithmetic (with the same common difference).

Theorem 2: If each term in an arithmetic sequence is multiplied (or divided) by a constant, the resulting sequence is also arithmetic (but with a different common difference).

The proofs of these theorems are left as exercises.

The proofs of these theorems are simple and straightforward and do not merit special consideration here. However, the following example will help students gain facility in work with harmonic sequences.

Example

If a, b, c forms a harmonic sequence, prove that $\frac{a}{b+c}, \frac{b}{c+a}, \frac{c}{a+b}$ also forms a harmonic sequence.

Solution: Since $\frac{1}{a}, \frac{1}{b}, \frac{1}{c}$ forms an arithmetic sequence, $\frac{a+b+c}{a}, \frac{a+b+c}{b}, \frac{a+b+c}{c}$ also forms an arithmetic sequence. This may be written as $1+\frac{b+c}{a}, 1+\frac{a+c}{b}, 1+\frac{a+b}{c}$. Therefore, $\frac{b+c}{a}, \frac{a+c}{b}, \frac{a+b}{c}$, forms an arithmetic sequence. Thus, $\frac{a}{b+c}, \frac{b}{a+c}, \frac{c}{a+b}$ forms a harmonic sequence.

Perhaps one of the more interesting aspects of any sequence is to establish a geometric model of the sequence.

One geometric interpretation of a harmonic sequence can be taken from the intersection points of the interior and exterior angle bisectors of a triangle with the side of the triangle.

Consider ΔABC, where $\overline{\text{AD}}$ bisects $\angle$BAC and $\overline{\text{AE}}$ bisects $\angle$CAF and B, D, C, and E are collinear (see Figure 1). It can easily be proved that for exterior angle bisector $\overline{\text{AE}}$, $\frac{\text{BE}}{\text{CE}} = \frac{\text{AB}}{\text{AC}}$ (i.e., draw $\overline{\text{GC}}//\overline{\text{AE}}$; AG = AC; also $\frac{\text{BE}}{\text{CE}} = \frac{\text{AB}}{\text{AG}} = \frac{\text{AB}}{\text{AC}}$).

Similarly for interior angle bisector $\overline{\text{AD}}$,

$$\frac{\text{BD}}{\text{CD}} = \frac{\text{AB}}{\text{AC}}$$

(The proof is done by drawing $\overline{\text{CF}}//\overline{\text{AD}}$; AF = AC; $\frac{\text{BD}}{\text{CD}} = \frac{\text{AB}}{\text{AF}} = \frac{\text{AB}}{\text{AC}}$.)

Therefore, $\frac{\text{BE}}{\text{CE}} = \frac{\text{BD}}{\text{CD}}$ or $\frac{\text{CD}}{\text{CE}} = \frac{\text{BD}}{\text{BE}}$. It is then said that the points B and C separate the points D and E harmonically.

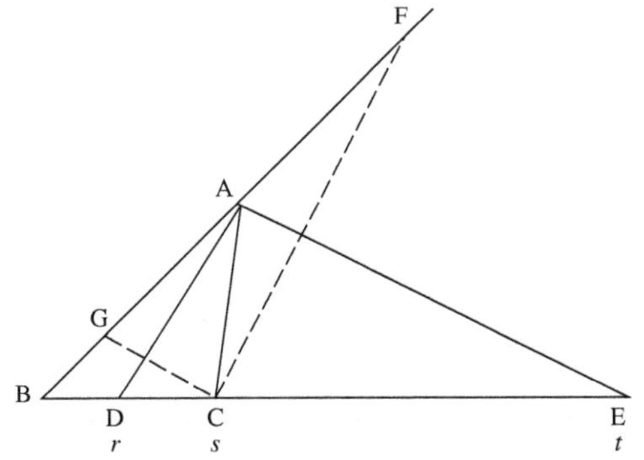

Figure 1.

Now suppose $\overleftrightarrow{\text{BDCE}}$ is a number line with B as the zero point, point D at coordinate r, point C at coordinate s, and point E at coordinate t. Therefore, $\text{BD} = r, \text{BC} = s$, and $\text{BE} = t$. We shall show that r, s, t forms a harmonic sequence. Since $\frac{\text{CD}}{\text{CE}} = \frac{\text{BD}}{\text{BE}}, \frac{\text{BC}-\text{BD}}{\text{BE}-\text{BC}} = \frac{\text{BD}}{\text{BE}}$, or $\frac{s-r}{t-s} = \frac{r}{t}$. Therefore, $t(s-r) = r(t-s)$ and $ts - tr = rt - rs$. Dividing each term by rst, we get $\frac{1}{r} - \frac{1}{s} = \frac{1}{s} - \frac{1}{t}$, which indicates that $\frac{1}{t}, \frac{1}{s}, \frac{1}{r}$ forms an arithmetic sequence. Thus, r, s, t forms a harmonic sequence.

Students should now have reasonably good insight into a harmonic sequence.

Postassessment

1. Set up and equation with the terms of the harmonic sequence a, b, c. (Use the definition.)
2. Find the 26th term of the sequence:

$$2\frac{1}{2}, 1\frac{12}{13}, 1\frac{9}{16}, 1\frac{6}{19} \cdots$$

3. Prove that if a^2, b^2, c^2 forms an arithmetic sequence, then $(b+c), (c+a)$, $(a+b)$ forms a harmonic sequence.

References

Posamentier, A. S. and C. Spreitzer, *The Mathematics Everyday Life*, Amherst, New York: Prometheus Books, 2018.

Unit 111

Transformations and Matrices

This unit will algebraically formalize a discussion of geometric transformations by the use of matrices.

Performance Objectives

1. *Given a particular geometric transformation, students will name the* 2×2 *matrix that effects that transformation.*
2. *Given certain* 2×2 *matrices, students will, at a glance, name the transformation each matrix effects.*

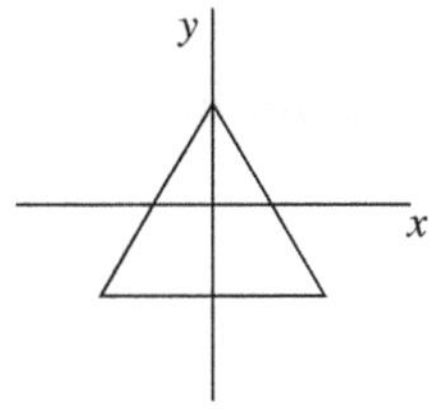

Figure 1.

Preassessment

1. On a Cartesian plane, graph ΔABC with vertices at A$(2,2)$, B$(4,2)$, C$(2,6)$. Write the coordinates of A$'$, B$'$, C$'$ that result when ΔABC undergoes each of the following transformations:

   a. Translation by -5 units in the x direction and 2 units in the y direction.
   b. Reflection across the x-axis.
   c. Rotation of $90°$ about the origin.
   d. Enlargement scale factor 2 with center at $(2,2)$.

2. Given an equilateral triangle as shown in Figure 1 with each vertex the same distance from the origin, list geometric transformations that leave the position of the triangle unchanged, assuming the vertices are indistinguishable.

Teaching Strategies

The teacher should first make students aware of a matrix array of numbers. Tell the students that a matrix of size $a \times b$ has a rows and b columns enclosed in brackets. When $a = b$, the matrix is said to be *square*. The class should see that adding matrices involves adding the numbers in corresponding positions in each matrix, for example $\begin{bmatrix} a \\ b \end{bmatrix} + \begin{bmatrix} c \\ d \end{bmatrix} = \begin{bmatrix} a+c \\ b+d \end{bmatrix}$, and the students must see that matrices must be of the same size (dimension) to be added.

When showing students how to multiply matrices, you should use the following general form. Note carefully the column–row relationship between the two matrix factors in the product.

$$\begin{bmatrix} a & b \\ c & d \end{bmatrix} \cdot \begin{bmatrix} x \\ y \end{bmatrix} = \begin{bmatrix} ax+by \\ cx+dy \end{bmatrix}.$$

Students may now describe the position of a point either by its coordinate (x,y) or by a 2×1 matrix, called a *position vector*, $\begin{bmatrix} x \\ y \end{bmatrix}$, which represents the vector from the origin to the point.

You may find it effective to use the phrase "is mapped onto" when describing the effect of a transformation. The symbol for this phrase using matrices is " →."

Translations provide a simple introduction into the use of matrices.

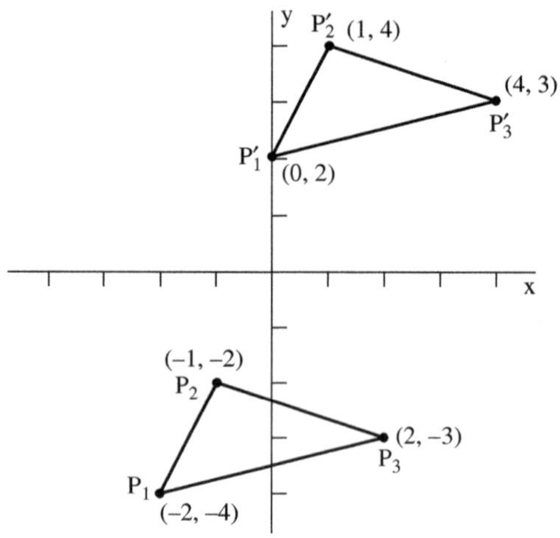

A translation of the triangle $P_1P_2P_3$ to $P_1'P_2'P_3'$ can be generalized in matrix form

$$\begin{bmatrix} x \\ y \end{bmatrix} \to \begin{bmatrix} x \\ y \end{bmatrix} + \begin{bmatrix} 2 \\ 6 \end{bmatrix}.$$

$\begin{bmatrix} 2 \\ 6 \end{bmatrix}$ represents here a "translation vector," meaning it translates (x, y) 2 units in the x direction and 6 units in the y directions. Students should readily see that by matrix addition each point P_1, P_2, P_3 is mapped onto P_1', P_2', P_3' respectively. That is, $P_1\begin{bmatrix} -2 \\ -4 \end{bmatrix} + \begin{bmatrix} 2 \\ 6 \end{bmatrix} = P_1'\begin{bmatrix} 0 \\ 2 \end{bmatrix}$, $P_2\begin{bmatrix} -1 \\ -2 \end{bmatrix} + \begin{bmatrix} 2 \\ 6 \end{bmatrix} = P_2'\begin{bmatrix} 1 \\ 4 \end{bmatrix}$, $P_3\begin{bmatrix} 2 \\ -3 \end{bmatrix} + \begin{bmatrix} 2 \\ 6 \end{bmatrix} = P_3'\begin{bmatrix} 4 \\ 3 \end{bmatrix}$. Matrix addition of 2×1 position vectors can therefore describe any translation in the two-dimensional plane. Students should be given several examples and exercises where a particular point (x_1, y_1) is translated into any other point (x_2, y_2) by a suitable choice of 2×1 matrix $\begin{bmatrix} x \\ y \end{bmatrix}$ such that

$$\begin{bmatrix} x_1 \\ y_1 \end{bmatrix} + \begin{bmatrix} x \\ y \end{bmatrix} = \begin{bmatrix} x_2 \\ y_2 \end{bmatrix}.$$

Rotations, reflections, and *enlargements* are more interesting, and to describe them algebraically requires 2×2 matrices. Students should first be given two or three example of the type $\begin{bmatrix} 2 & 3 \\ -1 & 2 \end{bmatrix} \cdot \begin{bmatrix} 3 \\ 2 \end{bmatrix} = \begin{bmatrix} 12 \\ 1 \end{bmatrix}$ for two reasons.

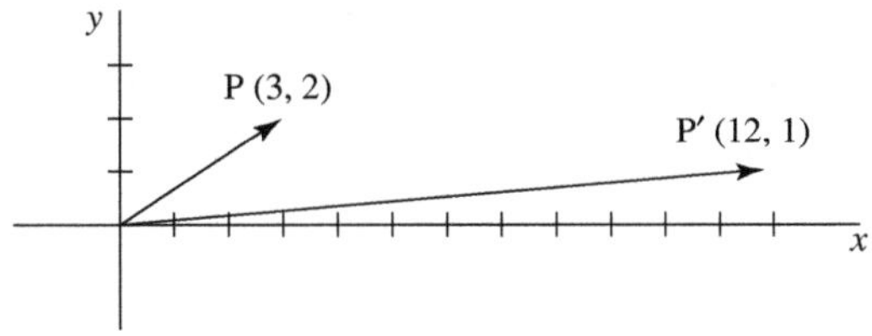

Figure 2.

For one, they may need practice in matrix multiplication, a skill they must master before embarking on the rest of this topic, and second, and most important in this strategy, is that students begin to think of the matrix $\begin{bmatrix} 2 & 3 \\ -1 & 2 \end{bmatrix}$ (or any 2×2 matrix) as a transformation of the point $\mathrm{P}(3, 2)$ onto the point $\mathrm{P}'(12, 1)$. Each example of this type should be accompanied by an illustration, as in Figure 2.

When the students are familiar with the notion that any 2×2 matrix represents a transformation, the teacher should be ready to show them that some 2×2 matrices represent special transformations with which they are familiar. For example, the teacher may give them the matrix and point, respectively (which follow):

1. $\begin{bmatrix} -1 & 0 \\ 0 & 1 \end{bmatrix} \cdot \overset{\mathrm{P}_1}{\begin{bmatrix} 2 \\ 3 \end{bmatrix}} = \overset{\mathrm{P}_1'}{\begin{bmatrix} -2 \\ 3 \end{bmatrix}}$

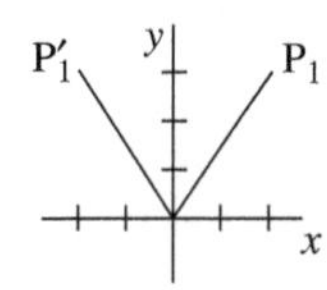

and

2. $\begin{bmatrix} 0 & -1 \\ 1 & 0 \end{bmatrix} \cdot \overset{\mathrm{P}_2}{\begin{bmatrix} 3 \\ 1 \end{bmatrix}} = \overset{\mathrm{P}_3'}{\begin{bmatrix} -1 \\ 3 \end{bmatrix}}$

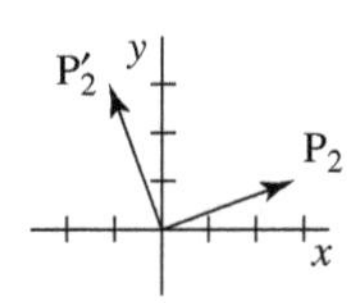

and

3. $\begin{bmatrix} 3 & 0 \\ 0 & 3 \end{bmatrix} \cdot \overset{\mathrm{P}_3}{\begin{bmatrix} -2 \\ -1 \end{bmatrix}} = \overset{\mathrm{P}_3'}{\begin{bmatrix} -6 \\ -3 \end{bmatrix}}$

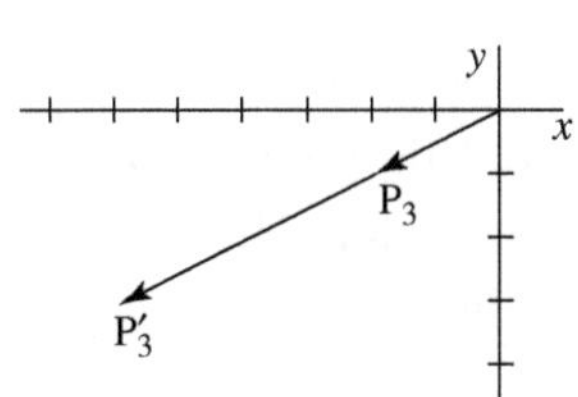

Students may have recognized the transformations in these examples as (1) a reflection (across the y axis), (2) a positive rotation of 90°, and (3) an enlargement, scale factor 3. To emphasize the point that it was not merely by chance that the matrices accomplished these transformations, ask the students to take the general point $\begin{bmatrix} x \\ y \end{bmatrix}$ and multiply it by each of the matrices. The students' answers will be, respectively, $\begin{bmatrix} -x \\ y \end{bmatrix}$, $\begin{bmatrix} -y \\ x \end{bmatrix}$, $\begin{bmatrix} 3x \\ 3y \end{bmatrix}$. Students will therefore see (the second case may require some insight) that the matrices $\begin{bmatrix} -1 & 0 \\ 0 & 1 \end{bmatrix}$, $\begin{bmatrix} 0 & -1 \\ 1 & 0 \end{bmatrix}$, and $\begin{bmatrix} 3 & 0 \\ 0 & 3 \end{bmatrix}$ do indeed accomplish, in the general case, the transformations they recognized in the particular examples.

To achieve performance objectives, whereby matrices provide a handy tool in transformational work, show what 2×2 matrices do to the unit vectors $i \begin{bmatrix} 1 \\ 0 \end{bmatrix}$ and $j \begin{bmatrix} 0 \\ 1 \end{bmatrix}$.

Choose any 2×2 matrix, as in the previous example, $\begin{bmatrix} 2 & 3 \\ -1 & 2 \end{bmatrix}$, and ask your students to multiply the unit vectors i and j by this matrix.

$$\begin{bmatrix} 2 & 3 \\ -1 & 2 \end{bmatrix} \cdot \begin{bmatrix} 1 \\ 0 \end{bmatrix} = \begin{bmatrix} 2 \\ -1 \end{bmatrix},$$

$$\begin{bmatrix} 2 & 3 \\ -1 & 2 \end{bmatrix} \cdot \begin{bmatrix} 0 \\ 1 \end{bmatrix} = \begin{bmatrix} 3 \\ 2 \end{bmatrix}.$$

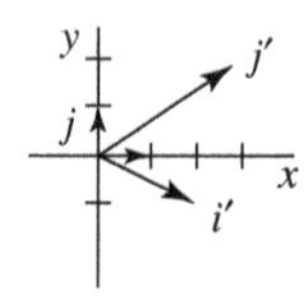

Use another example:

$$\begin{bmatrix} -1 & 0 \\ 0 & -1 \end{bmatrix} \cdot \begin{bmatrix} 1 \\ 0 \end{bmatrix} = \begin{bmatrix} -1 \\ 0 \end{bmatrix},$$

$$\begin{bmatrix} -1 & 0 \\ 0 & -1 \end{bmatrix} \cdot \begin{bmatrix} 0 \\ 1 \end{bmatrix} = \begin{bmatrix} 0 \\ -1 \end{bmatrix}.$$

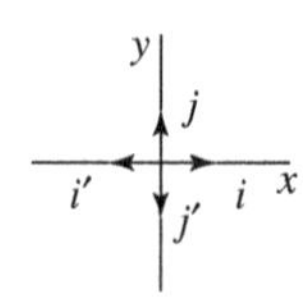

Continue with more examples until it is obvious to the students that in any 2×2 matrix $\begin{bmatrix} a & b \\ c & d \end{bmatrix}$, multiplying by $\begin{bmatrix} 1 \\ 0 \end{bmatrix}$ gives the first column $\begin{bmatrix} a \\ c \end{bmatrix}$ and multiplying by $\begin{bmatrix} 0 \\ 1 \end{bmatrix}$ gives the second column $\begin{bmatrix} b \\ d \end{bmatrix}$. In other words, the matrix maps the base vectors $\begin{bmatrix} 1 \\ 0 \end{bmatrix}$ and $\begin{bmatrix} 0 \\ 1 \end{bmatrix}$ onto $\begin{bmatrix} a \\ c \end{bmatrix}$ and $\begin{bmatrix} b \\ d \end{bmatrix}$, respectively.

When this conclusion is reached and appreciated by each student, the class then holds the key to reaching the performance objectives. Check whether the class can answer each of the following two types of problems:

1. Which transformation is performed by applying the matrix $\begin{bmatrix} 2 & 0 \\ 0 & 2 \end{bmatrix}$ to the unit vectors? Show them that $\begin{bmatrix} 2 & 0 \\ 0 & 2 \end{bmatrix} \cdot \begin{bmatrix} 1 \\ 0 \end{bmatrix} = \begin{bmatrix} 2 \\ 0 \end{bmatrix}$ and so $\begin{bmatrix} 1 \\ 0 \end{bmatrix}$ has been

mapped onto $\begin{bmatrix} 2 \\ 0 \end{bmatrix}$, and $\begin{bmatrix} 2 & 0 \\ 0 & 2 \end{bmatrix} \cdot \begin{bmatrix} 0 \\ 1 \end{bmatrix} = \begin{bmatrix} 0 \\ 2 \end{bmatrix}$ and so $\begin{bmatrix} 0 \\ 1 \end{bmatrix}$ has been mapped onto $\begin{bmatrix} 0 \\ 2 \end{bmatrix}$. Conclude that $\begin{bmatrix} 2 & 0 \\ 0 & 2 \end{bmatrix}$ is therefore an enlargement, scale factor 2.

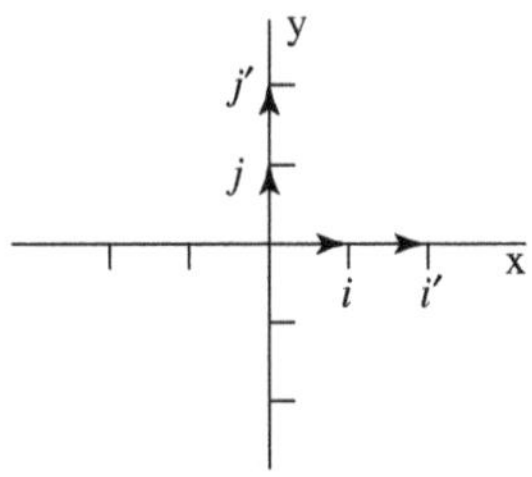

2. Which matrix effects the transformation: rotation by 180°, centered at the origin? Ask the students to draw the effect of 180 rotation the base vectors.

   Check with them that $\begin{bmatrix} 1 \\ 0 \end{bmatrix} \rightarrow \begin{bmatrix} -1 \\ 0 \end{bmatrix}$ and $\begin{bmatrix} 0 \\ 1 \end{bmatrix} \rightarrow \begin{bmatrix} 0 \\ -1 \end{bmatrix}$. Ask them what 2×2 matrix will give the results $\begin{bmatrix} a & d \\ c & d \end{bmatrix} \cdot \begin{bmatrix} 1 \\ 0 \end{bmatrix} = \begin{bmatrix} -1 \\ 0 \end{bmatrix}$ and $\begin{bmatrix} a & b \\ c & d \end{bmatrix} \cdot \begin{bmatrix} 0 \\ 1 \end{bmatrix} = \begin{bmatrix} 0 \\ -1 \end{bmatrix}$. They should by now know that $\begin{bmatrix} -1 \\ 0 \end{bmatrix}$ gives the first column of the 2×2 matrix and $\begin{bmatrix} 0 \\ -1 \end{bmatrix}$ gives the second column.

Therefore, the desired matrix is $\begin{bmatrix} -1 & 0 \\ 0 & -1 \end{bmatrix}$. Numerous examples and exercises should follow, such as those found in the postassessment.

Depending on his or her aims, the teacher may want to go on at this point to discuss inverse transformations (i.e., ones that reverse the effect of the original transformation) and transformations followed in the same problem by other transformations. The teacher should be aware here too of the value of using matrices, for the inverse of a matrix represents the inverse of a transformation, and multiplying two 2×2 matrices represents the effect of one transformation followed by another.

Postassessment

1. What transformation does the matrix

   a. $\begin{bmatrix} 0 & 1 \\ -1 & 0 \end{bmatrix}$

   b. $\begin{bmatrix} 2 & 0 \\ 0 & -2 \end{bmatrix}$

   c. $\begin{bmatrix} 0 & -1 \\ -1 & 0 \end{bmatrix}$ represent?

2. Find the matrix of each of the following transformations:

   a. Reflection across the line $y = x$.
   b. Enlargement with center at the origin, scale factor $\frac{1}{2}$.
   c. Rotation by $-90°$.

Unit 112

The Method of Differences

Many students familiar with arithmetic and geometric progressions will welcome the opportunity to extend their knowledge of sequences and series to a much broader class of simple function.

Performance Objectives

1. *Given sufficient terms of a sequences whose* nth *term is a rational, integral function of* n, *students will form an array consisting of the successive order of differences.*
2. *Given such an array, students will then use the method to differences to find expressions for the* nth *term and the sum of the first* n *terms.*

Preassessment

Students should be acquainted with the Binomial Theorem for positive, integral exponents as it is ordinarily taught in high schools.

Teaching Strategies

Begin the lesson by challenging the class to find the general term of the sequence $2, 12, 36, 80, 150,\ 252\ldots.$ After initial efforts of most students to find the familiar arithmetic and geometric progressions have proved unsuccessful, hint that sequences of this sort might be generated by single polynomials, e.g., $n^2(1, 4, 9, \ldots)$, or $n^3(1, 8, 27, \ldots)$. One student or another may shortly recognize that the nth term is given by $n^3 + n^2$.

Elicit that an infinite number of such sequences could be produced by using familiar polynomial functions. Then explain that a simple method exists for finding both the general term and the sum of these sequences. It is called the Method of Differences, and although it is not generally taught to high school students, it is nonetheless well within their grasp.

Have the class form the "difference between successive terms" of the above sequence, and then continue the process as shown in the following:

$$\begin{array}{ccccccccccc} 2 & & 12 & & 36 & & 80 & & 150 & & 252\cdots \\ & 10 & & 24 & & 44 & & 70 & & 102\cdots & \\ & & 14 & & 20 & & 26 & & 32\cdots & & \\ & & & 6 & & 6 & & 6\cdots & & & \end{array} \tag{1}$$

Observe that we reach a line of differences in which all terms are equal. To test whether this occurrence is merely accidental, have the students form sequences from polynomials such as $n^3 + 5n, 2n^3 + 3$, and so forth, and then repeat the process of taking successive differences. A consensus will soon emerge that an eventual line of equal terms is indeed characteristic of such sequences. The formal proof of this proposition (although simple) is not necessary at this time. Sufficient motivation will have been produced to examine the general case shown as follows:

$$\begin{array}{ll} \text{Given sequence:} & U_1, U_2, U_3, U_4, U_5, U_6, \cdots \\ \text{1st order of difference:} & \Delta U_1, \Delta U_2, \Delta U_3, \Delta U_4, \Delta U_5, \cdots \end{array} \tag{2}$$

$$\begin{array}{ll} \text{2nd order of difference:} & \Delta_2 U_1, \Delta_2 U_2, \Delta_2 U_3, \Delta_2 U_4, \ldots \\ \text{3rd order of difference:} & \Delta_3 U_1, \Delta_3 U_2, \Delta_3 U_3, \ldots \\ & \quad\quad\quad \ldots . \end{array}$$

The notation will be self-evidence to all who have written out several previous arrays. Thus, $\Delta U_3 = U_4 - U_3, \Delta_2 U_3 = \Delta U_4 - \Delta U_3$, etc. If the delta symbol, Δ, appears too forbidding for some, it can simply be replaced by the letter D.

From the method of forming each entry in (2), it can be seen that any term is equal to the sum of the term immediately preceding it added to the term below it on the left.

Using nothing more than this simple observation, we will now express each term of the given sequence as a function of the descending terms making up the left-hand boundary.

Thus, $\boxed{U_2 = U_1 + \Delta U_1}$. (3)

Also, $U_3 = U_2 + \Delta U_2$ with $\Delta U_2 = \Delta U_1 + \Delta_2 U_1$.

$U_3 = (U_1 + \Delta U_1) + (\Delta U_1 + \Delta_2 U_1)$

$$\boxed{U_3 = U_1 + 2\Delta U_1 + \Delta_2 U_1}. \tag{4}$$

By referring to (2) students should be able to follow the reasoning that leads to an expression for U_4 in terms of U_1.

$U_4 = U_3 + \Delta U_3$; however, $\Delta U_3 = \Delta U_2 + \Delta_2 U_2$.

But, $\Delta U_2 = \Delta U_1 + \Delta_2 U_1$, and $\Delta_2 U_2 + \Delta_2 U_1 + \Delta_3 U_1$.

Therefore, $\Delta U_3 = \Delta U_1 + 2\Delta_2 U_1 + \Delta_3 U_1$.

Now, using (3) and (4):

$U_4 = (U_1 + 2\Delta U_1 + \Delta_2 U_1) + (\Delta U_1 + 2\Delta_2 U_1 + \Delta_3 U_1)$

$$\boxed{U_4 + U_1 + 3\Delta U_1 + 3\Delta_2 U_1 + \Delta_3 U_1}. \tag{5}$$

Calling attention to the boxed expressions for U_2, U_3, U_4 the teacher can elicit the fact that the numerical coefficients involved are those of the Binomial Theorem. Note, however, that the coefficients used for the *fourth* term $(1, 3, 3, 1)$ are those found in a binomial expansion for the exponent *three*. If this remains true generally, we shall be able to write

$$\boxed{\begin{aligned} \text{Un} = U_1 + (n-1)\Delta U_1 + \frac{(n-1)(n-2)}{1 \cdot 2}\Delta_2 U_1 \\ + \cdots +_{n-1} C_r \Delta_r U_1 + \cdots + \Delta_{n-1} U_1. \end{aligned}} \tag{6}$$

If desired, the formal proof of (6) can be easily obtained by mathematical induction once the identity ${}_nC_r + {}_nC_{r-1} = {}_{n+1}C_r$ is established.

Some teachers may wish to rewrite (6) so as to resemble the notation typically used in treating arithmetic progressions. To do this, let the first term of the sequence be "a" and the first terms of each successive order of difference are $d_1, d_2, d_3 \ldots$. Then the nth term will be as follows:

$$\ell = a + (n-1)d_1 + \frac{(n-1)(n-2)}{1 \cdot 2} d_2 + \cdots \tag{7}$$

Finding the sum of the first n *terms*

Examine the following array in which $U_1, U_2 \ldots$ are again the terms of the given sequence we wish to sum.

$$\begin{array}{c} S_1 \quad S_2 \quad S_3 \quad S_4 \quad S_5 \cdots \\ U_1 \quad U_2 \quad U_3 \quad U_4 \cdots \\ \Delta U_1 \quad \Delta U_2 \quad \Delta U_3 \cdots \end{array} \tag{8}$$

Observe that the S-terms are formed by relations such as

$$S_2 = 0 + U_1 = U_1$$

$$S_3 = S_2 + U_2 = U_1 + U_2$$

$$S_4 = S_3 + U_3 = U_1 + U_2 + U_3$$

$$S_5 = S_4 + U_4 = U_1 + U_2 + U_3 + U_4.$$

Thus, if we can find an expression for S_{n+1}, we will have also found the sum of the first n terms. To find $S_n + 1$, one simply applies the previously determined Equation (6) to the above array (8). Before doing so, students should carefully compare (8) with (2). Then it should become evident that the proper application of Equation (6) yields

$$S_{n+1} = 0 + nU_1 + \frac{n(n-1)}{1 \cdot 2}\Delta U_1 + \cdots + \Delta_n U_n \quad \text{or}$$

$$\boxed{U_1 + U_2 + \cdots U_n = nU_1 + \tfrac{n(n-1)}{1\cdot 2}\Delta U_1 + \cdots + \Delta_n U_n}. \tag{9}$$

As an illustration, let us sum the first n squares of the integers.

$$\begin{array}{ccccccccc} 1, & & 4, & & 9, & & 16, & & 25\cdots \\ & 3, & & 5, & & 7, & & 9\cdots & \\ & & 2, & & 2, & & 2\cdots & & \end{array}$$

$$\begin{aligned} \text{Sum of } n^2 &= n \cdot 1 + \frac{n(n-1)}{1 \cdot 2} \cdot 3 + \frac{n(n-1)(n-2)}{1 \cdot 2 \cdot 3} \cdot 2 \\ &= \frac{6n + 9n(n-1) + 2n(n-1)(n-2)}{(6)} \\ &= \frac{n}{6}(n+1)(2n+1). \end{aligned}$$

Students may readily confirm the validity of this expression. Again, as before, the teacher may elect to rewrite (9) in terms of $a, d_1, d_2, \ldots$ and so forth.

Postassessment

Have students find the nth term and the sum of n terms for

1. $2, 5, 10, 17, 26 \ldots$
2. $1, 8, 27, 64, 125 \ldots$
3. $12, 40, 90, 168, 280, 432 \ldots$

Have students create sequences of their own from simple polynomials and then challenge their classmates to discover the general term.

Unit 113

Probability Applied to Baseball

Each year the first months of school coincide with the last months of major league baseball. The playing of the World Series in October normally provides a source of distraction. For mathematics teachers, however, this event can be harnessed to provide useful motivation for a host of probability applications that have high intrinsic academic value.

Performance Objectives

1. *Given odds for opposing World Series teams, students will calculate the expected number of games to be played.*
2. *Given the batting average of any hitter, students will estimate the probability of his attaining any given number of hits during a game.*

Preassessment

Previous study of permutations and probability is not necessary if some introductory discussion is provided. The topic is thus suitable for younger senior high school students (who have not yet learned the Binomial Theorem). Older students need even less preparation.

Teaching Strategies

The lesson should begin with an informal, spirited discussion of which team is likely to win the World Series. Newspaper clippings of the "odds" should be brought in. This leads directly to the question of how many games will be required for a decision.

Length of series

If the outcome of a World Series is designated by a sequence of letters representing the winning team (NAANAA means National League won the first and fourth games while losing the rest), challenge the class to find the total number of possible outcomes.

Discuss the solution in terms of "permutations of objects that are not all different." Observe that separate cases of four, five, six, or seven objects must be considered. Elicit that a constraint in the problem is that the winning team must always win the last game. The results can be tabulated as shown in Table 1.

As to the probabilities that the series will actually last four, five, six, or seven games, these clearly depend on the relative strengths of the teams. Most students will recognize intuitively that the prospects for a long series increase when the teams are closely matched, and vice versa.

Table 1.

No. of games played	4	5	6	7
No. of sequences	2	8	20	40

If newspapers "odds" are available, these should be translated into the probability of As winning (p) and Ns winning ($q = 1 - p$). If the odds are $m : n$, then $p = \frac{m}{(m+n)}$.

After a brief discussion reviewing the principle that governs the probability of independent events (perhaps illustrated by coin throws), it should become clear that the probability of an American League sweep is P(A in 4) $= p \cdot p \cdot p \cdot p = p^4$. Similarly P(N in 4) $= q^4$ and the overall probability of a four-game series is simply $p^4 + q^4$.

It is rewarding for students to find their intuitions confirmed, so it will be illuminating to substitute the various values for p and q that result from different odds, as shown by Table 2.

Table 2.

If odds favoring A are	1:1	2:1	3:1
P (4-game series)	.13	.21	.32

Students should be encouraged to extend this table of values.

Before calculating the likelihood of a five-game series, recall work done at the outset to determine the number of possible five-game sequences. There were eight such sequences (NAAAA, ANAAA, AANAA, AAANA, ANNNN, and so forth) and the probability associated with *each* of the first four is given by $q \cdot p \cdot p \cdot p = p \cdot q \cdot p \cdot p \cdot p = p \cdot p \cdot q \cdot p \cdot p = p \cdot p \cdot p \cdot q \cdot p = p^4 q$.

Since these outcomes are mutually exclusive, the probability of $P(A \text{ in } 5) = 4p^4 q$. In identical fashion $P(N \text{ in } 5) = 4pq^4$ and the overall chance for a five-game series is $4p^4q + 4pq^4 = 4pq^4 = 4pq\left(p^3 + q^3\right)$. Similarly, by falling back on the work done on permutations at the start of the lesson, it is simple to show that $P(\text{ six-game series }) = 10p^2q^2\left(p^2 + q^2\right)$ and $P(\text{ seven-game series }) = 20p^3q^3(p + q) = 20p^3q^3[p + q = 1]$.

As more complete information is derived, Table 2 can be extended to five, six, and seven games for various initial odds.

At this point, students may be introduced to (or reminded of) the important concept of Mathematical Expectation, E(X). With probabilities available for each outcome, E(X) for the length of the series can be calculated.

Table 3.
Odds 1:1

X- No. of games	4	5	6	7
P(X)	.13	.24	.31	.31

$$\mathrm{E(X)} = \sum \mathrm{X_i P\,(X_i)} = 5.75$$

$$\left[\sum \text{ notation can be avoided}\right]$$

Batting probabilities

Most students who follow baseball believe they have a clear understanding of the meaning of "batting average," and its implications for a hitter going into a game. Challenge the class to estimate the chances of a player getting at least one hit in four times at bat if his season-long batting average has been .250. Some may feel there is a virtual certainty of a hit since $.250 = \frac{1}{4}$.

Begin the analysis, as before, by using a sequence of letters to denote the hitter's performance (NHNN means a hit the second time up). Again, calculate the total number of possible sequences, which will be seen to be 16. Weaker students would be advised to write out each of these permutations.

Select the simplest case of NNNN. From previous work, the probability of this outcomes should be evident to the class as P (hitless) $= \frac{3}{4}\cdot\frac{3}{4}\cdot\frac{3}{4}\cdot\frac{3}{4} = \frac{(3)^4}{(4)} = \frac{81}{256} = 0.32$. (Recall: P(H) $= \frac{1}{4}$ and P(N) $= \frac{3}{4}$.) since all other sequences involve at *least* one hit, their combined probability is $1 - .32 = 0.68$. Thus, there is only a 68 percent chance that the batter will achieve at least one hit in four times. Although this result is not startling, it may definitely entail some readjustment in the thinking of some students.

In a similar manner, probabilities can be calculated for the cases of one hit (four possible sequences), two hits (six possible sequences), and so on.

The above topics represent examples of Binomial Experiments and Bernoulli Trials. Discuss with the class the definition of a Bernoulli Trial and the criteria for a Binomial Experiment, using illustrations from dice tossing, coin throwing, and so forth. Elicit their estimation of whether these concepts might be usefully applied to other real-life events such as gamete unions in genetics, the success of medical procedures such as surgery, and finally, the success of a batsman in baseball.

There are inherent limitations in attempting to treat baseball performances as Bernoulli Trials, especially in a unique situation like a World Series. Nonetheless, there is value in having students acquire a sense of "first-order approximations." At the same time, their insight into the applications of mathematics can be deepened by confronting a topic with which they feel familiar and are competent to evaluate.

Postassessment

Have students do the following:

1. Extend Table 2 for P(five-game series), P(6), P(7) for the odds shown, as well as other realistic odds.
2. Reconstruct Table 3 for odds of 2:1, 3:1 and then calculate E(X) for these cases.

References

Posamentier, A. S. and C. Spreitzer, *The Mathematics Everyday Life*, Amherst, New York: Prometheus Books, 2018.

Posamentier, A. S. and Peter Poole, *Understanding Mathematics Through Problem Solving*, Hackensack, New Jersey: World Scientific Publishing, 2020.

Unit 114

Introduction to Geometric Transformations

Beginning with an introduction to the three basic rigid motion transformations, this unit will show how a group can be developed, where the elements are transformations.

Performance Objectives

1. *Students will define translation, rotation, and reflection.*
2. *Students will identify the appropriate transformation from a diagram showing a change of position.*
3. *Students will test the group postulates for a given set of transformations under composition.*

Preassessment

This unit should be presented when students have mastered the basics of geometry. They should be familiar with the concept of group, but do not need to have had any exposure to transformations prior to this unit. A knowledge of functions is also helpful for this unit.

Teaching Strategies

The first part of this unit will concern itself with a brief introduction to the three basic rigid motion transformations: translations, rotations, and reflections.

Students should recall that a one-to-one and onto function is a congruence. That is, $\overline{\mathrm{AB}} \xrightarrow[\text{onto}]{\text{1-1}} \overline{\mathrm{CD}}$ implies that $\overline{\mathrm{AB}} \cong \overline{\mathrm{CD}}$.

Translations

Consider $T{:}\alpha \xrightarrow[\text{onto}]{\text{1-1}} \alpha$, that is, a mapping of the entire plane onto itself in the direction of a given vector v.

In Figure 1, above each point in the plane is taken to a new point in the plane in the direction and distance of the translation vector v. Here $\mathrm{T(B)} = \mathrm{B'B'}$ is the image of B under the translation.

Points along line ℓ, which is parallel to v, are mapped onto other points of ℓ. To insure a good understanding of this type of transformation, ask your students the following questions:

1. Which lines are mapped onto themselves? (those parallel to the translation vector)
2. Which points are mapped onto themselves? (none)
3. Which vector determines the inverse of T? (the negative of v)

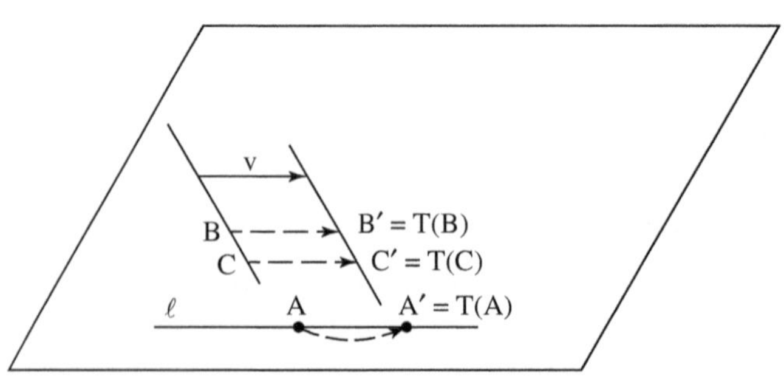

Figure 1.

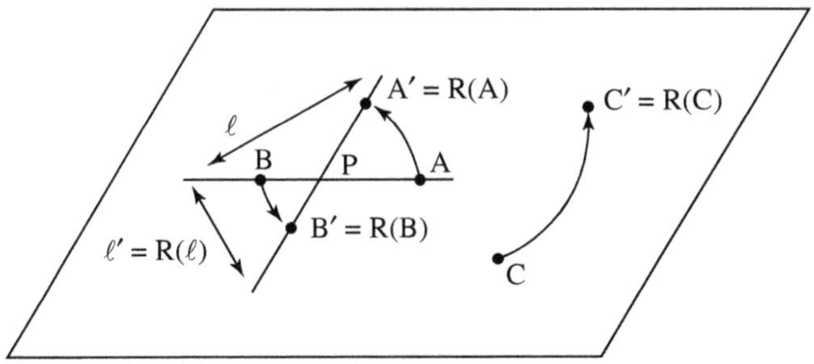

Figure 2.

Rotations

Consider R:$\alpha \xrightarrow[\text{onto}]{\text{1-1}} \alpha$, that is a mapping of the entire plane onto itself as determined by a rotation of any angle about a point. We shall agree to consider only counterclockwise rotations unless specified otherwise.

In Figure 2, R represents a rotation of 90° about P. The following questions should help your students understand this transformation:

1. Are any points mapped upon themselves by a rotation R? (yes, P)
2. Are any lines mapped upon themselves by a rotation R? (No, unless the rotation is 180°, written R_{180}, then any line through the center of rotation (P) is mapped upon itself.)
3. If ℓ is in the plane, how are ℓ and $\ell' = R_{90}(\ell)$ related? (perpendicular)
4. What is the inverse of R_{90}?(either R_{270}, or R_{630}, etc., or R_{-90})
5. What is the inverse of R_{180}?(R_{180})
6. If R_aR_b means a rotation of $b°$ followed by a rotation of $a°$, describe $R_a^2, R_b^3, R_a^4 (R_{2a}, R_{3b}, R_{4a})$.
7. Simplify $R_{200} \cdot R_{180}$. $(R_{380} = R_{380-360} = R_{20})$
8. Simplify $R_{90} \cdot R_{270}$. $(R_{360} = R_0)$
9. Simplify R_{120}^4. $(R_{480} = R_{480-360} = R_{120})$

Reflections

Consider M_ℓ:$\alpha \xrightarrow[\text{onto}]{\text{1-1}} \alpha$, a mapping of the entire plane onto itself as determined by a reflection in a point or a line.

To find the reflection of a point A in a given point P, simply locate the point A′ on ray $\overrightarrow{AP}$ (on the opposite side of P as is A) so that A′P = AP. In Figure 3(a), A′ is the image (or reflection) of A.

To find the reflection of a point A in a given line ℓ, locate the point A′ on the line perpendicular to ℓ and containing A, at the same distance from ℓ

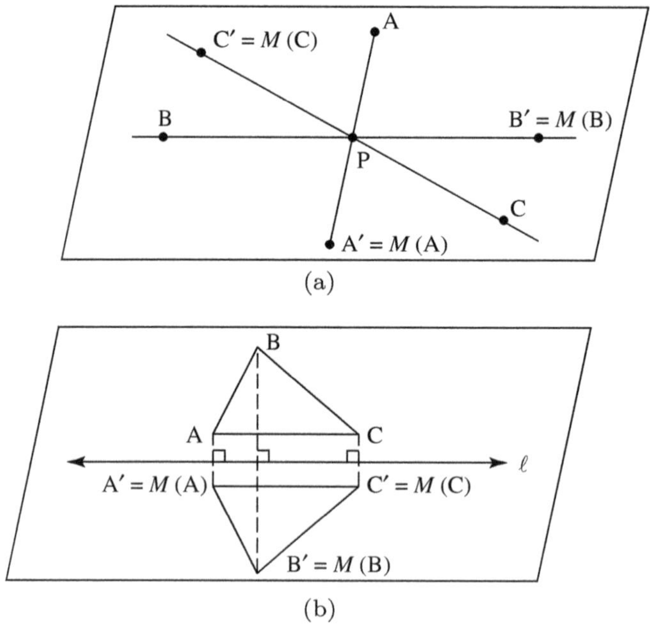

Figure 3.

as A, but on the opposite side. In Figure 3(b), the points of a triangle (and hence the triangle itself) are reflected in ℓ.

Once again some questions for your students:

1. What is the inverse of M_ℓ? (M_ℓ)
2. How does an image differ from its pre-image? (different orientation or "mirror image")
3. How does the reflection of a line in a given point change the line's orientation? (changes the order of points on the line from given to the reverse of that)
4. Describe each of the following:
   a. $M_\ell(m)$, where $\ell // m$. ($m' // \ell$ on the opposite side of ℓ as is m)
   b. $M_\ell(n)$, where $\ell \perp n$. (n' is the same line as n)
   c. $M_\ell(k)$, where ℓ is oblique to k. (k' forms the same angle with ℓ as does k at the same point as k but on the opposite side of ℓ)

Groups

To discuss a group of transformations, it would be helpful to review the definition of a group.

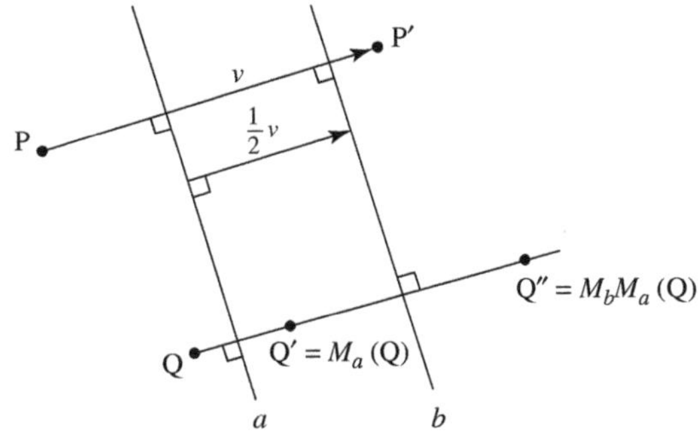

Figure 4.

1. A set with one operation.
2. Associative property must hold true.
3. An identity element must exist.
4. Every element must have an inverse.

To consider the elements as the three types of transformations would be confusing. We shall therefore show that (I) any translation is the product of two reflections, and (II) any rotation is the product of two reflections. This will enable us to work exclusively with reflections. The word *product* as it was used above refers to the "composition" of transformations; that is, one transformation performed after the other.

1. To show that any translation T_v is equivalent to the composition of two reflections consider the following Figure 4.

At either end of any vector $\frac{1}{2}$ v, consider the lines perpendicular to v. By reflecting any point Q in line a and then in line b, Q^n is obtained, which is $T_v(Q)$. That is, $M_a(Q) = Q'$ and $M_b(M_a(Q)) = Q'' = M_bM_a(Q)$.

2. To show that any rotation R_θ is equivalent to the composition of two reflections consider the following Figure 5.

Through the center of rotation, C, draw two lines forming an angle of measure $\frac{1}{2}\Theta$. Select any point P and reflect it through line a and then reflect that image through line b.

For convenience, we shall use line a before b. Using the two pairs of congruent triangles in Figure 5, we can easily prove that $M_bM_a(P) = P''$ is in fact equal to $R_\Theta(P)$.

Now that we can replace translations and rotations with combinations of reflections, ask your students to verify that a group of transformations is at hand. They must demonstrate *all* four properties listed above.

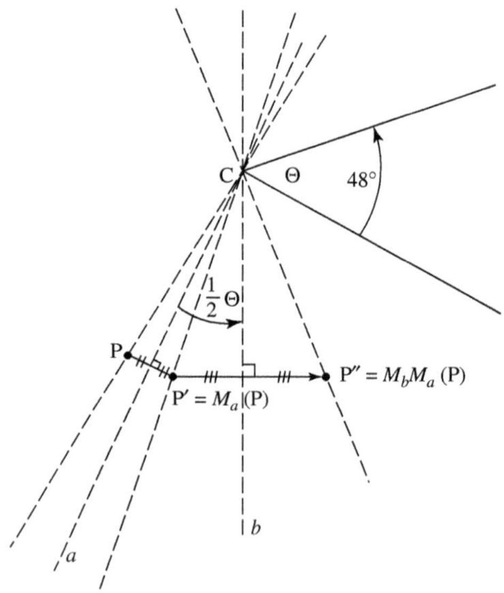

Figure 5.

Postassessment

1. Define translation, rotation, and reflection.
2. Describe each of the following as a single transformation.
3. Show that reflections form a group under the operation of composition.

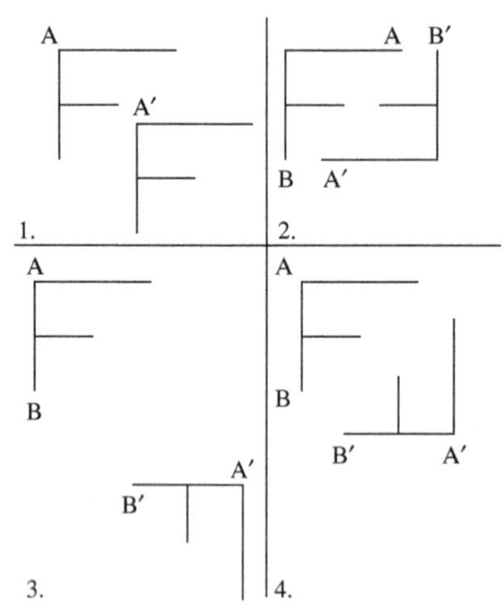

Reference

Maresch, G. and A. S. Posamentier, *Solving Problems in Our Spatial World*, Hackensack, New Jersey: World Scientific Publishing, 2019.

Unit 115

The Circle and the Cardioid

Performance Objectives

1. *Given a circle, students will be able to draw a cardioid without using an equation.*
2. *Students will be able, by experimentation, to generate curves other than the cardioid.*

Preassessment

Introduce students to the cardioid by having them set up a table of values for $r = 2a(1 + \cos\theta)$. Then have students locate the corresponding points on a polar coordinate graph. After they have constructed this curve, which was first referred to as a *cardioid* (heart-shaped) by de Castillon in 1741, students should be enticed to consider some rather unusual methods for constructing this curve.

Teaching Strategies

Method I. Ask the students to draw a base circle, O, of diameter 3 inches, centered evenly on an $8\frac{1}{2} \times 11$-inch sheet. With a protractor, divide the circumference into 36 equally spaced points (see Figure 1). Through any of these 36 points, T, construct a tangent, t, to the circle. This need not necessarily be constructed in the classical way using straightedge and compasses, but rather with a right triangle template, or a carpenter's T-square, in which one leg of the triangle passes through the center of the circle. Drawing a tangent to a circle in this manner is based on the fact that the tangent to a circle is perpendicular to the radius at the point of tangency. From a *fixed* point, A, on the circle's circumference (where A is one of the 36 points), drop a perpendicular to meet t at P. Now construct tangents to all the remaining points (except through A), to each tangent repeating the preceding step of dropping a perpendicular from A to t. The resulting figure will appear as shown in Figure 2; the locus of all such points P is a cardioid. Point A is then the cardioid's *cusp.* In Figure 2, certain construction lines from Figure 1 were eliminated so as to enhance the final drawing. Also,

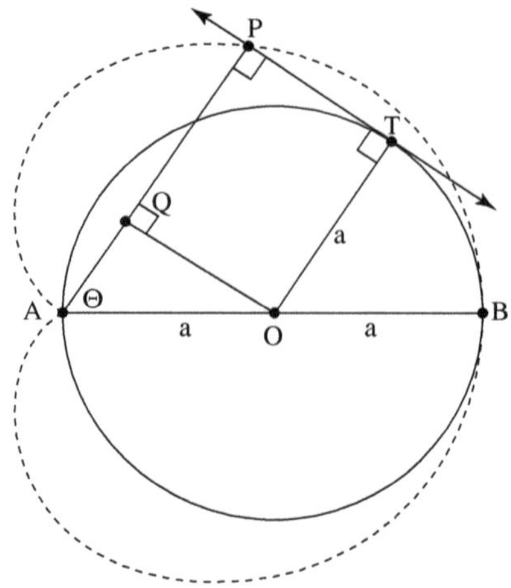

Figure 1.

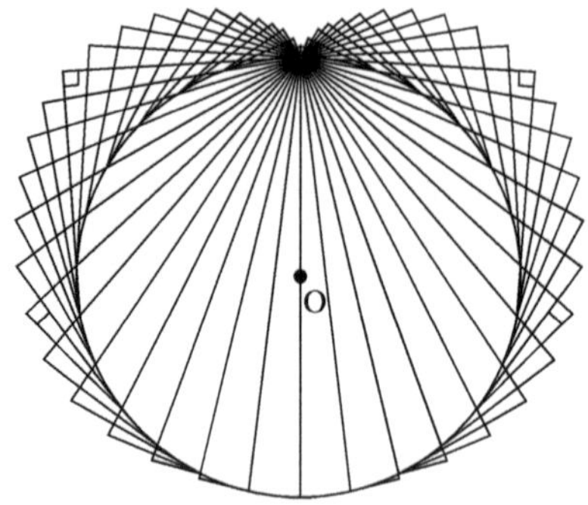

Figure 2.

48 points were used along the base circle in Figure 2 for a more compact appearing effect. Finally, note that the orientation of the cardioid in Figure 2 differs from that in Figure 1 by a 90° clockwise rotation of the cardioid about A.

Proof: In Figure 1, let A be the fixed point on the circle O. Let A also be the pole, and diameter $\overline{AB}$ the initial line, of a polar coordinate system. Draw $\overline{OT}$. Thus, $\overline{AP}//\overline{OT}$. From O, drop a perpendicular to meet $\overline{AP}$ at Q.

Then $\overline{OQ}// \overleftrightarrow{TP}$ and OTPQ is a rectangle. Let OT = OP = a, r = AP and $m\angle BAP = \theta$. Then, from Figure 1,

$$r = AQ + QP. \tag{1}$$

In right $\triangle AQO, \cos\theta = \frac{AQ}{a}$, so that

$$AQ = a \cdot \cos\theta. \tag{2}$$

With QP $= a$, and using (2), (1) becomes

$$r = a \cdot \cos\theta + a, \text{ or}$$
$$r = a(1 + \cos\theta), \tag{3}$$

which is the polar equation of a cardioid.

The procedure given above is an example of how *a pedal* curve to a given curve is generated (i.e., the cardioid is referred to as the *pedal* to circle O with respect to A). All pedal curves are obtained in this manner: some arbitrary fixed point is chosen, often on the curve itself, and from that point perpendiculars are dropped to various tangents to the particular curve. The locus of the intersection of the perpendiculars to each tangent from a fixed point defines the pedal curve. Though the tangent to a circle can easily be constructed, and hence a pedal curve drawn, further pedal constructions of a visual nature can prove challenging.

Method II. As in Method I, draw a base circle of diameter 3 inches, except that the circle should be placed about one inch left of center on a vertically held $8\frac{1}{2} \times 11$-inch sheet (see Figure 3). Divide the circle into 18 equally spaced points. Label the diameter AB = 2a, with A once again a fixed point. Let M be one of these 18 points, distinct from A. Place the edge of a marked straightedge on M, being sure that the straightedge also passes through A. Locate two points of $\overrightarrow{AM}$, P and P′, on either side of M, at a distance of 2a units from M. Thus, M is the midpoint of $\overrightarrow{PP'}$. Continue for all such points M, as M is allowed to move to each of the remaining points (of the originally selected 18 points) around the circumference. The locus of all such points P and P′ is a cardioid (see Figure 4).

Proof: In Figure 3, draw $\overline{MB}$, let AP = r, and let $m\angle BAM = \theta$. Since $\triangle AMB$ is a right triangle,

$$AM = 2a \cdot \cos\theta. \tag{4}$$

But

$$r = AM + MP, \tag{5}$$

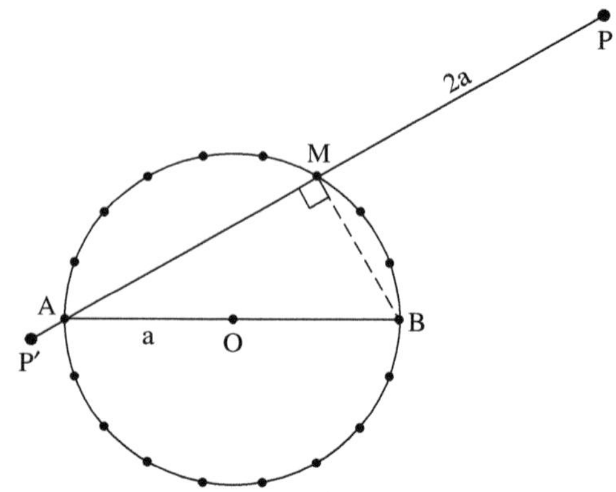

Figure 3.

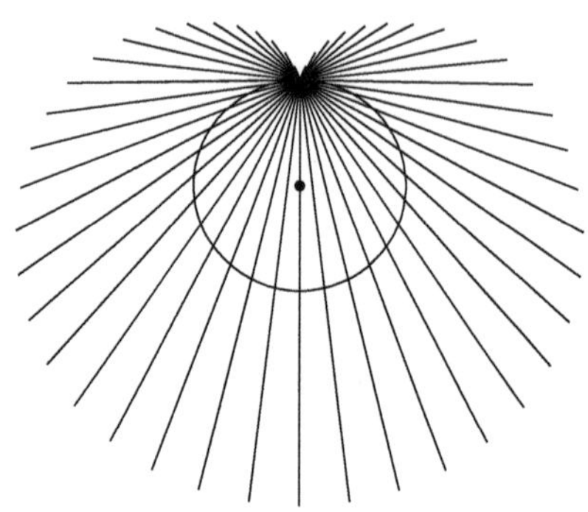

Figure 4.

and by construction, MP $= 2a$. Substituting this, as well as (4), into (5), we get

$$r = 2a \cdot \cos\theta + 2a,$$

$$\text{or} \quad r = 2a(1 + \cos\theta), \tag{6}$$

a form identical with (3) except for the constant $2a$. For $\theta + 180°$, we obtain P′, so that r = AP′. Since $\cos(\theta + 180°) = -\cos\theta$, we would have obtained

$$r = 2a(1 - \cos\theta) \tag{7}$$

had we repeated the above steps.

The cardioid construction given here is an example of a *conchoid* curve. Using this technique with an ellipse, the choice of a fixed point could be an extremity of the major or minor axis. Varying the length of the line on

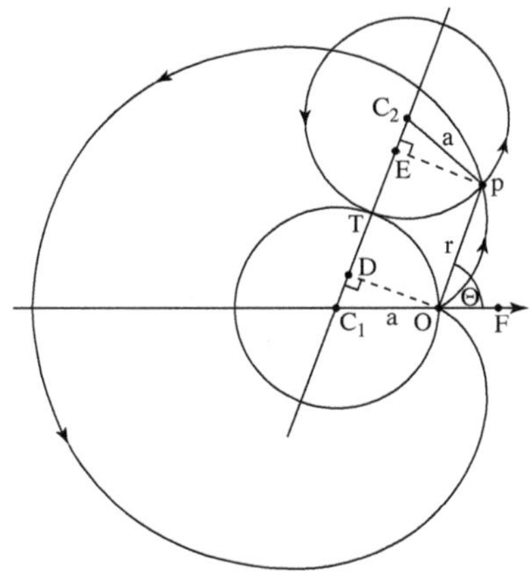

Figure 5.

either side of M to equal the major or minor axis creates several interesting combinations of conchoid curves for the ellipse.

Method III. The cardioid can also be generated as the locus of a point, P, on the circumference of a rolling circle which rolls, without slipping on a fixed circle of equal diameter.

Proof: Let the fixed circle, C_1, of a radius a, be centered at the pole of a polar coordinate system. Let O be the intersection of circle C_1 with the initial line. Circle C_2, whose initial position was externally tangent to circle C_1 at O, has now rolled to the position shown in Figure 5, carrying with it fixed point P. Thc locus of P is desired.

Circle C_2 is now tangent to circle C_1 at T. A line joining C_1 to C_2 will pass through T. Because $\widehat{\text{PT}} \cong \widehat{\text{OT}}, m\angle \text{OC}_1\text{T} = m\angle \text{PC}_2\text{T}$. If $\text{C}_2\text{P} = a$ is drawn, as well as $\overline{\text{OP}}$, we obtain isosceles trapezoid $\text{OC}_1\text{C}_2\text{P}$, and hence $m\angle\text{POF} = \theta$.

Drop perpendiculars from O and P to $\overleftrightarrow{\text{C}_1\text{C}_2}$, meeting $\overleftrightarrow{\text{C}_1\text{C}_2}$, at D and E, respectively, Immediately, $\Delta\text{ODC}_1 \cong \Delta\text{PEC}_2$, from which it follows that $\text{C}_1\text{D} = \text{EC}_2$ and $\text{DE} = \text{OP}$. In triangles ODC_1 and PEC_2,

$$\text{C}_1\text{D} = a \cdot \cos\theta, \tag{8}$$

and

$$\text{C}_2\text{E} = a \cdot \cos\theta. \tag{9}$$

Now

$$\mathrm{C_1C_2} = 2a = \mathrm{C_1D} + \mathrm{DE} + \mathrm{C_2E}. \tag{10}$$

Substituting (8) and (9) into (10), and remembering that DE = OP = r,

$$2a = a \cdot \cos\theta + \mathrm{r} + a \cdot \cos\theta, \tag{11}$$

which yields, upon simplification,

$$r = 2a(1 - \cos\theta). \tag{12}$$

This is identical with (7).

The concept of a circle rolling smoothly on the circumference of another, fixed circle has been well studied. The locus of a point on the circumference of the rolling circle gives rise to a curve called an *epicycloid*, of which the cardioid is a special case. Changing the ratio of the fixed circle results in a variety of well-known higher plane curves.

Postassessment

Refer to Method I for drawing the following two variations, and in so doing, possibly obtain equations similar to (3):

1. Choose the fixed point A outside the circle at a distance of $2a$ units from O.
2. Choose A *inside* the circle, at a distance of $\frac{a}{2}$ units from O.

Unit 116

Complex-Number Applications

The number system we presently use has taken a long time to develop. It proceeded according to the necessity of the time. To early humans, counting numbers was sufficient to meet their needs. Simple fractions, such as the unit fractions employed by the Egyptians followed. Although the early Greeks did not recognize irrational numbers, their necessity in geometric problems brought about their acceptance. Negative numbers, too, were used when their physical application became apparent, such as their use in temperature. Complex numbers, however, are studied because the real number system is

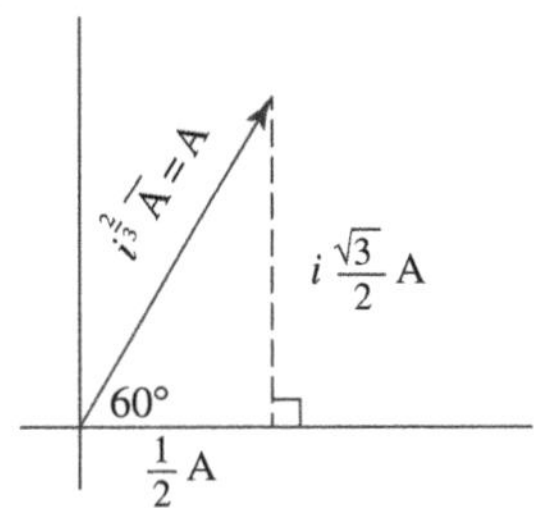

Figure 1.

not algebraically complete without them. Their application to the physical world is not explored by most mathematics students. This unit introduces students to some physical applications of complex numbers.

Performance Objective

Students will be able to solve some physics problems involving complex numbers and vector quantities.

Preassessment

Students should be familiar with operations with complex numbers and vector analysis. A knowledge of basic physics is also recommended.

Teaching Strategies

In algebra, a complex plane is defined by two rectangular coordinate axes in which the real parts of the complex numbers are plotted along the horizontal axis and the imaginary parts are plotted along the vertical axis. This complex plane can be developed if we take an approach in which i is treated as the "sign of perpendicularity" as an operator functioning to rotate a vector through an angle of 90°.

To develop this idea, we begin with any vector quantity, A, which is represented by a vector $\longrightarrow$ whose length indicates the magnitude and whose arrow tip indicates sense. (To distinguish between vector and scalar quantities, a bar is placed over the symbol to indicate a vector quantity, $\overline{A}$, while a symbol without a bar indicates a scalar quantity, A). Now, if $\overline{A}$ is operated upon by -1 (multiplied by -1), we have $-\overline{A}$ whose graphical representation is $\longleftarrow$. Thus, operating upon the vector $\overline{A}$ by -1 rotates it through 180° in the positive sense. Now, since $i^2 = -1$, i must represent rotating the vector through an angle of 90°, since two applications of 90° will result

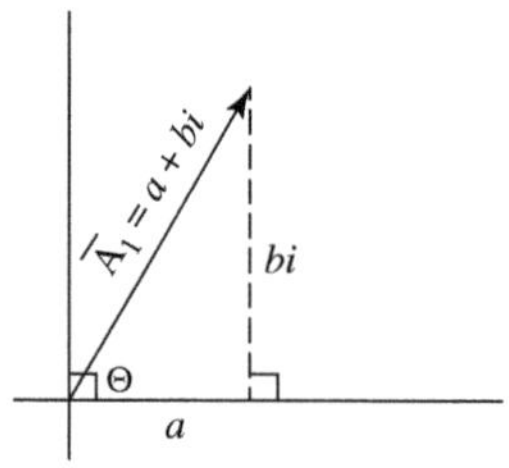

Figure 2.

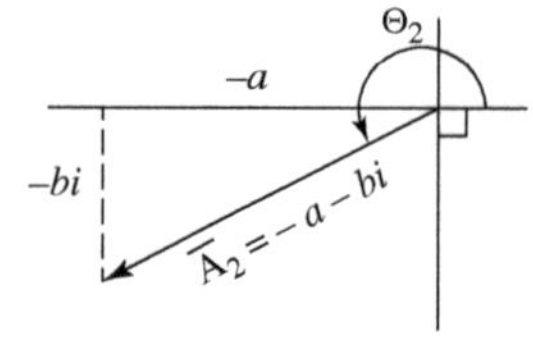

Figure 3.

in a rotation of 180°. Therefore, operating upon a vector by i^3 rotates the vector through 270°, and so on. Similarly we can consider using as operators higher roots of −1, which will rotate the vector through a smaller angle. So $\sqrt[3]{-1} = \left(i^2\right)^{\frac{1}{3}} = i^{\frac{2}{3}}$ will rotate a vector through an angle of 60° since three applications of this operator is equivalent to operating by −1. We can show this in a vector diagram. Given a vector $\overline{\mathrm{A}}$, with magnitude A, we operate upon it by $i^{\frac{2}{3}}$, i.e., we rotate the vector through 60°. This, of course, does not change its magnitude, A. Therefore, the real component is $\mathrm{A}\cos 60° = \frac{1}{2}\mathrm{A}$ and the imaginary component is $\mathrm{A}\sin 60° = \sqrt{\frac{3}{2}}\mathrm{A}$. So $\sqrt[3]{-1\overline{\mathrm{A}}} = i^{\frac{2}{3}}\overline{\mathrm{A}} = \mathrm{A}\cos 60° + i\mathrm{A}\sin 60°$, which indicates the position of the vector. In this way, $\sqrt[n]{-1\overline{\mathrm{A}}} = i^{\frac{2}{n}}\overline{\mathrm{A}} = \mathrm{A}\ \mathrm{Cos}\frac{\pi}{n} + i\mathrm{A}\ \mathrm{Sin}\frac{\pi}{n}$. To generalize, to rotate a vector A through the angle Θ, we use the operator $\cos\Theta + i\sin\Theta$.

Now, if we are given a vector $\overline{\mathrm{A}}_1 = a + bi$, we can graph it on the complex plane. The position of the vector is given by $\Theta_1 = \arctan\frac{b}{a}$. Its magnitude is $\mathrm{A}_1 = \sqrt{a^2 + b^2}$. Vector $\overline{\mathrm{A}}_1$ is pictured in Figure 2. Similarly vector $\overline{\mathrm{A}}_2 = -a - bi$ has magnitude $\mathrm{A}_2 = \sqrt{(-a)^2 + (-b)^2}$, the same as that of vector A_1. Its position is given by $\Theta_2 = \arctan\frac{-b}{-a}$ and is located in the third quadrant as pictured in Figure 3.

Now, we can explore a physical interpretation of these operators. In physics books, $\sqrt{-1}$ is represented by the letter j and electrical current by the letter I. Since this unit is written for mathematics students, we used i to represent $\sqrt{-1}$ and, for the sake of clarity, we will use the letter J to represent the electrical current.

In the study of alternating currents, we have for the current the vector $\overline{\mathrm{J}} = j_1 + ij_2$. The voltage $\overline{\mathrm{E}}$ of this frequency can be represented by $\mathrm{E} = \varepsilon_1 + \varepsilon_2 i$. The impedance (effective resistance of the current) of the circuit, *not* a vector, can be represented by $\mathrm{Z} = r \pm ix$, where r is the ohmic resistance and x is the reactance. The angle between $\overline{\mathrm{E}}$ and $\overline{\mathrm{J}}$ is the phase angle (the angle by which the current lags behind the electromotive force, emf) and is represented by $\phi = \arctan\frac{r}{x}$.

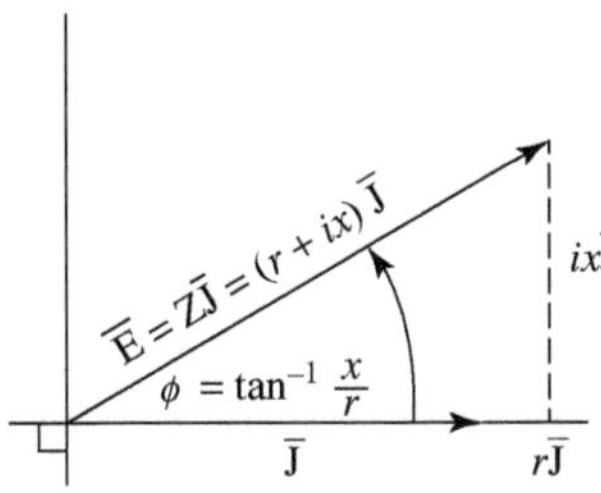

Figure 4.

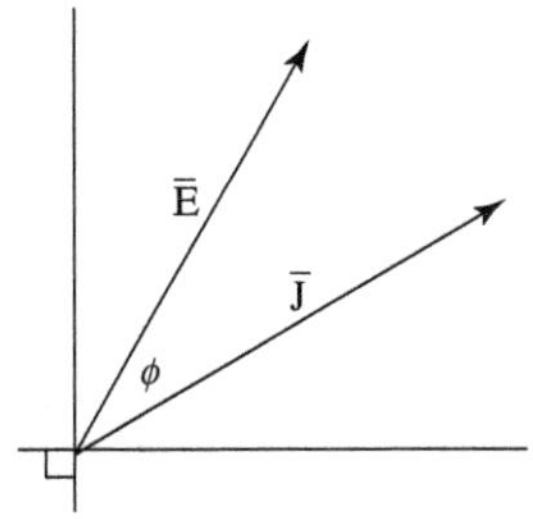

Figure 5.

We can obtain mathematically the product of two impedances, but it has no physical meaning. The product of the voltage and the current, although it has no physical meaning, is referred to as apparent power. If, however, we take the product of the current and the impedance $Z\bar{J} = (r + ix)(j_1 + ij_2) = (rj_1 + xj_2) + i(rj_2 + xj_1)$, we have a voltage of the same frequency as the current, the *actual voltage*, i.e., $Z\bar{J} = \bar{E}$. This is Ohm's law in the complex form (Ohm's law states that for direct currents the voltage is equal to the product of the resistance and the current). So, in direct currents one deals with scalar quantities, whereas in alternating current circuits the quantities are vectors expressable as complex numbers obeying the laws of vector algebra. In the following diagrams for Ohm's law, the first diagram (Figure 4) has $\bar{J}$ on the real axis whereas the second diagram (Figure 5) does not.

Let us now try some problems.

Example 1

Let $r = 5$ ohms, $x = 4$ ohms, and the current J is 20 amperes. Take J on the real axis.

Given this information we have the impedance, $Z = 5+4i$. The inductive circuit is $\bar{E} = \overline{JZ} = 20(5 + 4i) = 100 + 80i$. Therefore, $E = \sqrt{100^2 + 80^2} = 128$ volts. The angle E makes with the real axis is $\Theta = \arctan \frac{80}{100} = 38°40'$, which also happens to be the phase angle for this problem. The vector diagram is shown in Figure 6.

Example 2

Let us change the above example slightly by letting the angle that the current vector $\bar{J}$ makes with the real axis be 30°. We will let the remaining data be as before.

We still have $Z = 5 + 4i$.

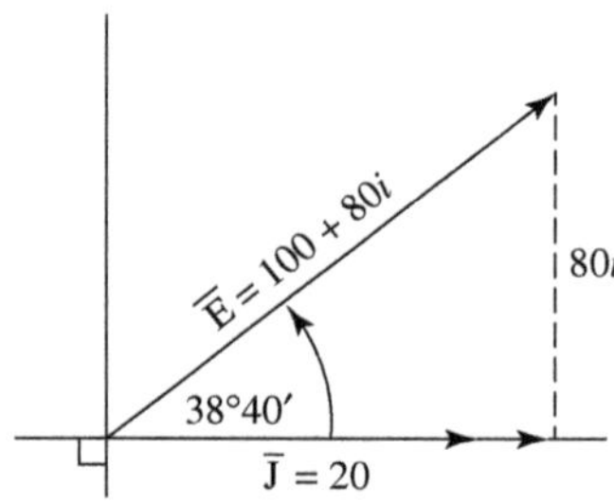

Figure 6.

We now have $\overline{\text{J}} = 20\,(\cos 30° + i \sin 30°) = 20\ (.866 + .5i) = 17.32 + 10i$
$\overline{\text{E}} = \overline{\text{J}}Z = (17.32 + 10i) \cdot (5 + 4i) = 46.6 + 119.28i$.

$\text{E} = \sqrt{(46.6)^2 + (119.28)^2} = 128$ volts, just as before. The angle E makes with the real axis is $\theta = \text{arc tan}\frac{119.28}{46.6} = 68°40'$.

The phase angle ϕ remains the same, $\phi = \text{arc tan}\frac{4}{5} = 38°40'$.

The vector diagram for this problem is:

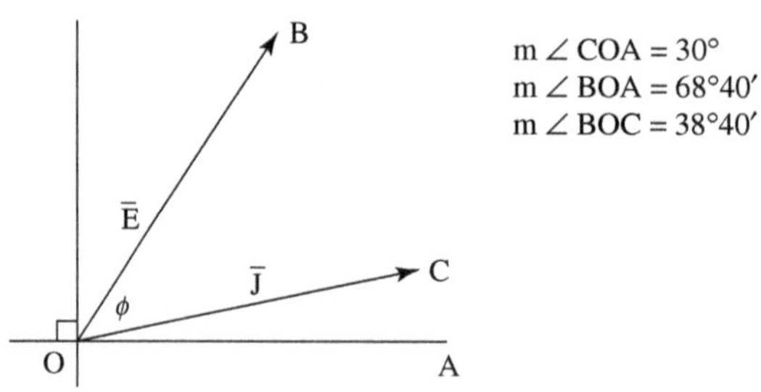

Figure 7.

Students should now be able to solve similar physics problems involving complex numbers.

Postassessment

1. Let $r = 3$ ohms and $x = 4$ ohms. Take $\overline{\text{J}}$ on the real axis, J = 3. Find the impedance Z, the complex expression for $\overline{\text{E}}$ and the magnitude of E. What is Θ, the angle $\overline{\text{E}}$ makes with the real axis? What is ϕ, the phase angle? Draw a vector diagram for this problem.
2. Use the data for the above problem, letting $\overline{\text{J}}$ make an angle of 20° with the real axis. Recalculate all quantities, both the complex expressions and the magnitudes. Draw the vector diagram.

3. Let $\overline{\mathrm{E}} = 4 + 14i$ and $\overline{\mathrm{J}} = 2 + 3i$. Find the complex expression for Z and its magnitude. What is the phase angle? Draw a vector diagram.

Reference

Suydam, V. A. *Electricity and Electromagnetism.* New York: D. Van Nostrand Company, 1940.

Unit 117

Hindu Arithmetic

The mathematics curriculum may be enriched for students on many levels by the study of a number system and its arithmetic. An investigation into the mechanics of the system, its contribution to our own system, and other suitable tangents may be entered upon by a student on a high level. For other students, it can serve as practice in basic skills with integers, as students check the answers they obtain in working out problems. This unit introduces students to the ancient Hindu numerical notation and system of adding, subtracting, multiplying, and dividing (ca. 900, India).

Performance Objectives

1. *Given an addition problem, students will find the answer using a Hindu method.*
2. *Given a subtraction problem, students will find the answer using a Hindu method.*
3. *Given a multiplication problem, students will find the answer using a Hindu method.*
4. *Given a division problem, students will find the answer using a Hindu method.*

Preassessment

Students need only be familiar with the basic operations of integers, i.e., addition, subtraction, multiplication, and division.

Teaching Strategies

The symbols of the nine numerals used in Hindu reckoning in ascending order are १, २, ३, ४, ५, ६, ७, ८, ९, ०. This system included a symbol for zero, ०. This, then, was a positional system, as in our own, as oppposed to a grouping system such as that used by the ancient Egyptians. Therefore, similar to our modern system, 5639 would be represented as ५६३९.

At this point, we can discuss the importance of the Hindu symbol for zero. A comparison with the Egyptian numeral system might be instructive (see Unit 14 on Ancient Egyptian Arithmetic). The historical importance of zero could be investigated. Without a zero, numbers were cumbersome, and intricate computations were difficult. In the Hindu system, record keeping and other computations necessary for commerce, astronomical computations, and mathematical tables were advanced, for with the existence of a placeholder numbers are easier to write and read, and can be manipulated with greater ease.

Hindu computation was generally written upon surfaces where corrections and erasures were easily done. For our purposes, instead of erasing numbers we will cross-out numbers so that the methods discussed will be more easily followed.

Hindu addition, though set up vertically as in our method, was done from left to right. Consider the problem: $6537+886$. The addition begins on the left with 8 being added to the 5. The 1 of this 13 is added to the numeral on the left, 6, changing it to 7, and the 5 is now changed to a 3. This process continues, from left to right. Thus, the solution would look like this:

7	4	2	
7	~~3~~	~~1~~	3
~~6~~	~~5~~	~~3~~	~~7~~
	8	8	6

The result is 7423.

Subtraction was also done from left to right, the larger number being placed above the smaller one. To subtract 886 from 6537, we begin by subtracting 8 from 5. Since this is not possible, we subtract the 8 from 65, leaving 57. We put the 5 in place of the 6, and the 7 in place of the 5. We continue this process by subtracting 8 from 3 using the described method. The solution of the entire problem would look like this:

	6		
5	~~7~~	5	1
~~6~~	~~5~~	~~3~~	~~7~~
	8	8	6

The result is 5651.

To multiply as the Hindus did, we begin by placing the units' digit of the multiplier under the highest place of position of the multiplicand. To multiply 537 by 24, we begin this way:

$$\begin{array}{cccc} & 5 & 3 & 7 \\ 2 & 4 & & \end{array}$$

We multiply 2 by 5 and place the resulting 0 above the 2, and the 1 to the left. We now multiply 4 by the 5, placing the resulting 0 in place of the 5 above the 4, and adding the 2 to the 0, therefore, we now have 2 in the next place:

$$\begin{array}{ccccc} & 2 & 0 & & \\ 1 & \not{0} & \not{5} & 3 & 7 \\ & 2 & 4 & & \end{array}$$

Now that we have finished with the 5, we shift the multiplier one place to the right, the 4 now being below the 3, indicating 3 is the number we are now concerned with.

$$\begin{array}{ccccc} & 2 & 0 & & \\ 1 & \not{0} & \not{5} & 3 & 7 \\ & & 2 & 4 & \end{array}$$

We multiply as before, first the 2 by the 3, then the 4, and when finished shift again to the right. When we have done the entire problem, it will look like this:

$$\begin{array}{ccccc} & & 8 & & \\ & & \not{7} & 8 & \\ & & 6 & 6 & \\ & 2 & \not{0} & \not{2} & 8 \\ 1 & \not{0} & \not{5} & \not{3} & 7 \\ & \not{2} & \not{4} & & \\ & & \not{2} & \not{4} & \\ & & & 2 & 4 \end{array}$$

The answer is 12,888.

It can be seen that crossing out, instead of erasing, requires space. As division is the most complex of the basic operations, the problem will be done step by step, instead of crossing out, substituting new results for the old numbers.

To divide, we place the divisor below the dividend, aligning them on the left. Thus, we begin the problem $5832 \div 253$ as $\begin{array}{l} 5\,8\,3\,2 \\ 2\,5\,3 \end{array}$. As 253 is below 583, we seek a number to multiply 253 by, so that the product will be as close as possible to 583 without exceeding it. The number we seek here is 2, and it is placed thus:

$$\begin{array}{cccc} & & 2 & \\ 5 & 8 & 3 & 2 \\ 2 & 5 & 3 & \end{array}$$

We now multiply 253 by 2 (as the Hindus did) and subtract the result from 583 (as the Hindus did). This gives us 77, which we put in the place of 583. So we now have

$$\begin{array}{cccc} & & 2 & \\ & 7 & 7 & 2 \\ 2 & 5 & 3 & \end{array}$$

The divisor is shifted to the right to get

$$\begin{array}{ccc} & 2 & \\ 7 & 7 & 2 \\ 2 & 5 & 3 \end{array}$$

The process continues as above until we reach the results

$$\begin{array}{c} 23 \\ 13 \\ 253 \end{array}$$

which shows the quotient to be 23 and the remainder 13.

Investigate these processes with your students to the extent you feel necessary. You will find instructive the similarity of these algorithms to ours.

Postassessment

1. Have students write the following numbers using Hindu numerals:
   a. 5342 b. 230796
2. Have students solve the following problems using Hindu methods:
   a. $3567 + 984$ b. $8734 - 6849$ c. 596×37 d. $65478 \div 283$

References

Eves, H. *An Introduction to the History of Mathematics.* 4th edn. New York: Holt, Rinehart and Winston, 1976.

Van der Waerden, B. L. *Science Awakening.* New York: John Wiley & Sons, 1963.

Unit 118

Proving Numbers Irrational

When high school students are introduced to irrational numbers, they are usually asked to accept the fact that certain numbers like $\sqrt{2}$, $\sin 10^\circ$, and so on are irrational. Many students wonder, however, how it could be proved that a given number is irrational. This unit presents a method to prove the irrationality of certain algebraic numbers.

Performance Objectives

1. *Given certain algebraic numbers students will be able to prove their irrationality.*
2. *Students will find some specific patterns that will determine in advance whether a given algebraic number is irrational.*

Preassessment

Students should be familiar with the concepts of irrational numbers and algebraic numbers. They should also have a general background in algebraic equations, radicals, trigonometry, and logarithms.

Teaching Strategies

Begin the lesson by asking students to give examples of irrational numbers. Ask them how they are sure these numbers are irrational. Then have them define irrational numbers. Students will be curious enough at this point to want to investigate the following theorem.

Theorem: Consider any polynomial equation with the integer coefficients $a_n x^n + a_{n-1}x^{n-1} + \cdots + a_1 x + a_0 = 0$. If this equation has a rational root $\frac{p}{q}$, where $\frac{p}{q}$ is in its lowest terms, then p is a divisor of a_0 and q is a divisor of a_n.

Proof: Let $\frac{p}{q}$ be a root of the given equation. Then it satisfies the equation and we have

$$a_n \left(\frac{p}{q}\right)^n + a_{n-1}\left(\frac{p}{q}\right)^{n-1} + \cdots + a_1\left(\frac{p}{q}\right) + a_0 = 0. \qquad \text{(I)}$$

We now multiply (I) by q^n to obtain

$$a_n p^n + a_{n-1}p^{n-1}q + \cdots + a_1 pq^{n-1} + a_0 q^n = 0.$$

This equation can be rewritten as

$$a_n p^n = -a_{n-1}p^{n-1}q - \cdots - a_1 pq^{n-1} - a_0 q^n$$

or

$$a_n p^n = q(-a_{n-1}p^{n-1} - \cdots - a_1 pq^{n-2} - a_0 q^{n-1}).$$

This shows that q is a divisor of $a_n p^n$. But if $\frac{p}{q}$ is in its lowest terms, then p and q are relatively prime and therefore q is a divisor of a_n. Likewise, if we rewrite equation (I) as

$$a_0 q^n = p(-a_1 q^{n-1} - \cdots - a_{n-1}p^{n-2}q - a_n p^{n-1})$$

we see that p is a divisor of $a_0 q^n$. Again, because p and q are relatively prime, we have that p is a divisor of a_0.

This completes the proof of the theorem.

Example 1

Prove that $\sqrt{5}$ is irrational.

$\sqrt{5}$ is a root of $x^2 - 5 = 0$. Then according to the notation used for the theorem, $a_2 = 1$ and $a_0 = -5$. Now any rational root, $\frac{p}{q}$, of this equation will have to be of such a nature that p will have to divide -5, and q will have to divide 1. This is so because of the previous theorem. But the only divisors of 1 are $+1$ and -1. Thus, q must be either $+1$ or -1, and the rational root of the equation must be an integer. This integer p, according to the theorem must divide -5, and the only divisors of -5 are $-1, 1, 5$, and -5. However, none of these is a root of the equation $x^2 - 5 = 0$, that is $(1)^2 - 5 = 0; (-1)^2 - 5 = 0(5)^2 - 5 = 0$; and $(-5)^2 - 5 = 0$, are *all false.* Hence, $x^2 - 5 = 0$ has no rational root, and $\sqrt{5}$ is therefore an irrational number.

Example 2

Prove that $\sqrt[3]{2}$ is irrational.

$\sqrt[3]{2}$ is a root of $x^3 - 2 = 0$. Then p must divide -2, and q must divide 1. Thus, if this equation has a rational root, this root must be an integer and a divisor of -2. Now the only divisors of -2 are: $2, -2, 1$, and -1. But none of these is a root of the equation $x^3 - 2 = 0$, because $(2)^2 - 2 = 0$, $(-2)^2 - 2 = 0, (1)^2 - 2 = 0$, and $(-1)^2 - 2 = 0$ are all false. Hence, $\sqrt[3]{2}$ is irrational.

Example 3

Prove that $\sqrt{2}+\sqrt{3}$ is irrational.

If we write $x=\sqrt{2}+\sqrt{3}$, we have that $x-\sqrt{2}=\sqrt{3}$. Now, square both sides and obtain $x^2-1=2x\sqrt{2}$. Squaring again gives us $x^4-2x^2+1=8x^2$ or $x^4-10x^2+1=0$.

This equation has been so constructed that $\sqrt{2}+\sqrt{3}$ is a root. But the only possible rational roots of this equation are those integers that are divisors of 1, that is, -1 and 1. But none of these is root of the equation because $(1)^4-10(1)^2+1=0$ and $(-1)^4-10(-1)^2+1=0$ are both false. Hence, this equation has no rational roots and consequently $\sqrt{2}+\sqrt{3}$ is irrational.

Example 4

Prove the sin 10° is irrational.

We have the identity $\sin 3\theta = 3\sin\theta - 4\sin^3\theta$. Now if we replace θ by 10°, and notice that

$$\sin 30^\circ = \frac{1}{2}, \text{ we get}$$

$$\frac{1}{2} = 3\sin 10^\circ - 4\sin^3 10^\circ.$$

If we now make $\sin 10^\circ = x$, we obtain

$$\frac{1}{2} = 3x - 4x^3 \text{ or}$$

$$8x^3 - 6x + 1 = 0.$$

According to the theorem, p must be a divisor of 1, and q must be a divisor of 8, thus the only possible rational roots are $\pm\frac{1}{8}, \pm\frac{1}{4}, \pm\frac{1}{2}$, and ± 1. But none of these eight possibilities is a root of the equation, as can be seen by substitution into the equation obtained. Therefore, this equation has no rational roots, and since $\sin 10^\circ$ is a root of the equation, it must be irrational.

Students should now be able to prove irrational those numbers that occur most frequently in high school textbooks and that students are *told* are irrational. It is important to have students understand why a mathematical concept is true after they have comfortably worked with the concept. All too often students accept the irrationality of a number without question. This unit provides a method that should bring some true understanding to the average high school mathematics student. In addition to the problems posed in the *Postassesment*, students should be encouraged to use the technique presented here when the need arises.

Postassessment

Those students who have mastered the technique learned through the previous examples, should be able to complete the following exercises:

1. Prove that $\sqrt{2}$ is irrational.
2. Prove that $\sqrt[3]{6}$ is irrational.
3. Prove that $\sqrt[3]{3} + \sqrt{11}$ is irrational.
4. Prove that $\text{Cos}\,20°$ is irrational.
5. Prove that a number of the form $\sqrt[n]{m}$, where n and m are natural numbers, is either irrational or an integer.

Unit 119

How to Use a Computer Spreadsheet to Generate Solutions to Certain Mathematics Problems

This unit presents some simple examples of how spreadsheets, such as Microsoft Excel, Google Sheets, can be used to generate solutions to certain mathematics problems. A computer with an appropriate spreadsheet must be available and pupils should be familiar with its operation. Secondary school students of any grade should find this challenging as well as fascinating.

Performance Objectives

1. *Students will generate a Fibonacci sequence on a spreadsheet.*
2. *Students will create a Pascal triangle on a spreadsheet.*
3. *Students will list other mathematical problems appropriate for spreadsheet solution.*

Preassessment

Students need to review Enrichment Units 85 (Fibonacci sequence) and 99 (Pascal's pyramid — especially the first part, which discusses the Pascal triangle). Students should also be familiar with basic operations on a microcomputer and electronic spreadsheets.

Teaching Strategies

An electronic spreadsheet is an array that appears on the screen of a microcomputer. Most spreadsheets have built-in mathematical functions so that elements in the ith row, j^{th} column can be easily accessed for any given i or j. For example, show students how to use the functions that determine how maximum value, minimum value, average value, median, mode, standard deviation, and so on, for a set of numbers listed on a spreadsheet. Point out that many other mathematical applications may be found for spreadsheets in addition to those already built into the program.

One interesting application is to generate a Fibonacci sequence as well as a sequence of ratios of successive pairs of numbers. Pay special attention to the formula used to generate a Fibonacci sequence as it is listed in Enrichment Unit 85: $f(n) = f(n-1) + f(n-2)$.

One way this formula can be translated to "spreadsheet language" is, "For a given row, the number in the nth column is equal to the sum of the numbers in the two previous columns."

Together with the students, use "relative referencing" to develop the formula that a given cell's contents should equal the sum of the entries of the two columns to its left in the same row. Thus, if the initial numbers 1, 1 are entered in cells A1 and B1, then cell C1 will contain the formula

$$= \text{SUM(A1, B1)}$$

or

$$= \text{Al} + \text{Bl}.$$

This technique, in conjunction with the spreadsheet's feature for copying and updating a formula (fill handle in Excel) may be used to write as many terms as will fit onto one row. Then point and click to continue the formula onto the next row.

A second sequence of ratios of successive pairs of terms, as indicated in the enrichment unit, may be generated as follows. If the initial numbers 1, 1 are entered in cells Al and B1, then cell B2 will contain the following formula:

$$\frac{\text{C1}}{\text{B1}}.$$

The following sample was produced on the IBM-compatible computer using Microsoft Excel.

	A	B	C	D	E	F	G	H	I
1	1	1	2	3	5	8	13	21	34
2		2.000	1.500	1.667	1.600	1.625	1.615	1.619	
3									

Now suggest a second application that might be of interest: the Pascal triangle. After some discussion of Enrichment Unit 99, especially the rule for generating the triangle, suggest that the triangle be written as follows:

1	1				
1	2	1			
1	3	3	1		
1	4	6	4	1	
1	5	10	10	5	1

....

Ask pupils to suggest an appropriate spreadsheet formula that would generate this triangle. Point out that the first and last entry of each row is 1 and that each of the other entries is the sum of the number in the row above and the number in that same row but one column to its left.

Postassessment

Ask the class to prepare a list of mathematical topics that might be appropriate for development via a spreadsheet, and solve some of them. You might suggest topics taken from the enrichment units in this book.

The following could prove to be challenging:

Magic squares
Palindromic numbers
The Sieve of Eratosthenes
Solving a quadratic equation
Continued fractions

Unit 120

The Three Worlds of Geometry

The inquisitive nature of humans causes them to probe deeply into that which troubles them. This unit presents the startling results achieved after 20 centuries of probing into a seemingly minor problem.

Many people throughout history were responsible for the geometry we know today. One man, however, stands above all others. That man is *Euclid*, the brilliant Greek mathematician of antiquity who developed and wrote the first geometry text, the *Elements*, ca. 300 B.C. The significance of this treatise was that it showed the capability of the human mind to arrive at nontrivial conclusions by reasoning power alone — a power that no other creature possesses.

In the *Elements*, Euclid developed geometry as a postulational system based on five postulates:

1. A straight line may be drawn from any point to any other point.
2. A line segment may be extended any length along a straight line.
3. A circle may be drawn from any center at any distance from that center.
4. All right angles are congruent to one another.
5. If a straight line intersects two other straight lines, and makes the sum of the interior angles on the same side less than two right angles, the straight lines, if extended indefinitely, will meet on that side which has the angles whose sum is less than two right angles.

It was the length and relative complexity of this fifth postulate that led to its intensive investigation and analysis by scholars throughout the ages. Some of the fruits of these investigations are presented in this unit.

Performance Objectives

1. *Students will define the* Saccheri quadrilateral *and use it in formal proofs.*
2. *Students will compare and contrast the existence of parallel lines in the models of Euclid, Riemann, and Bolyai–Lobachevsky.*
3. *Students will learn how to prove that the sum of the measures of the angles of a triangle may be more than, less than, or equal to* 180°.

Preassessment

Students should be familiar with the traditional high school geometry course, especially the theorems related to parallel and perpendicular lines, the exterior angle of a triangle, geometric inequalities, and direct and indirect proofs.

Teaching Strategies

In the early part of the 19th century, Playfair's postulate was shown to be a simpler, logical equivalent to Euclid's fifth postulate: *Through a point not on a given line, one and only one line can be drawn parallel to the given line.* (*When speaking of parallels in this unit, we use Euclid's definition: Parallel lines are straight lines that, being in the same plane and being extended indefinitely in both directions, do not meet one another in either direction.*)

An analysis of Euclid's fifth postulate yields three possible variations. We call them "worlds" and now compare them:

Euclid's postulate: Through a point not on a given line, one and only one line can be drawn parallel to the given line (Figure 1).

Riemann's postulate [in honor of Bernhard Riemann, German mathematician (1826–1866)]: Two straight lines always intersect one another (Figure 2).

Bolyai and *Lobachevsky's postulate*: In honor of mathematicians, Janos Bolyai (1802–1860), Hungarian, and Nikolai Lobachevsky (1793–1856), Russian: Through a point not on a given line, more than one line can be drawn not intersecting the given line (Figure 3).

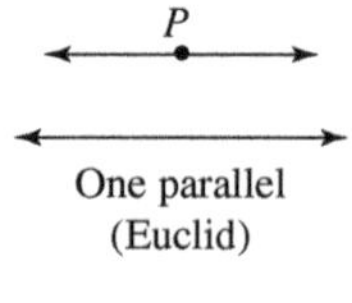

Figure 1.

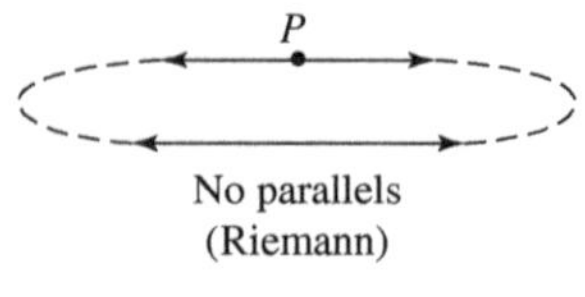

Figure 2.

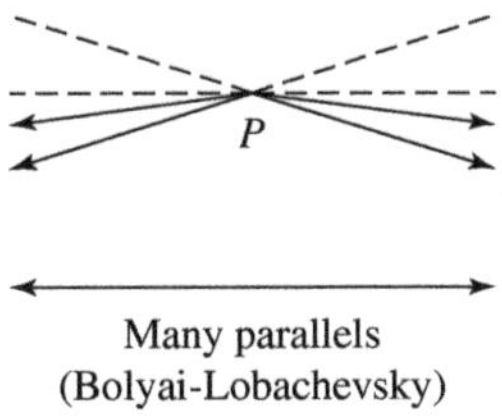

Figure 3.

The three worlds

The teaching unit that you will present to your students will include some historical background about Euclid's fifth postulate. Girolamo Saccheri (1667–1733), an Italian monk-mathematician, devised this quadrilateral to help him in his attempt to prove that Euclid's fifth postulate was in reality a theorem based on the other four postulates and thus not independent of them. He failed, but during the course of his efforts, he developed *other* perfectly consistent postulational systems and thus, without realizing it, other types of geometries — forerunners of what we now call non-Euclidean geometry.

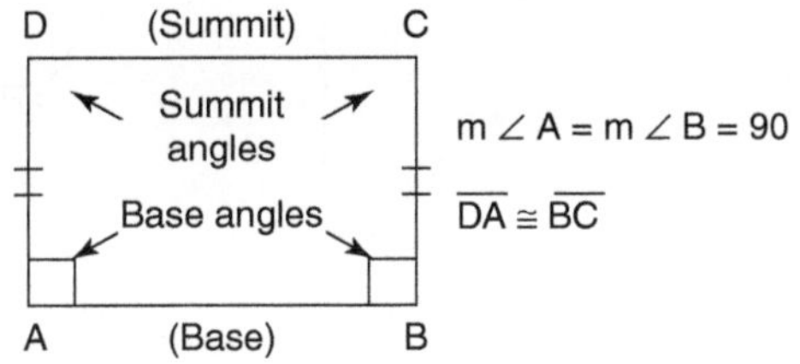

Now have the students use the following outline to complete the proof that the summit angles of a Saccheri quadrilateral are congruent.

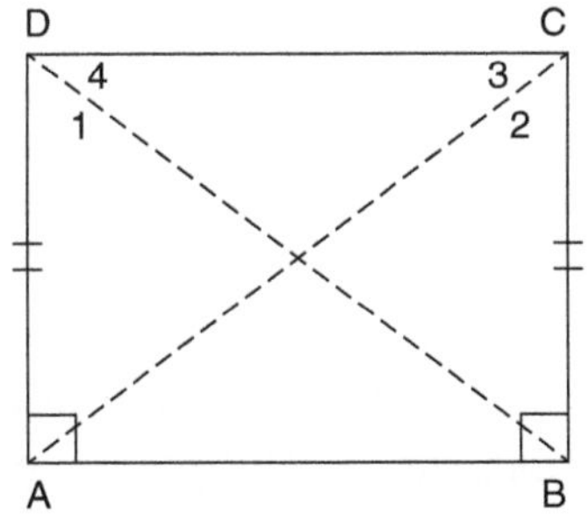

Given: Saccheri quadrilateral ABCD
Prove: $\angle D \cong \angle C$

1. Draw $\overline{BD}$ and $\overline{AC}$
2. Prove $\triangle ABD \cong \triangle ABC$
3. $\therefore \angle 1 \cong \angle 2$, and $\overline{BD} \cong AC$
4. Now prove $\triangle DCA \cong \triangle DCB$
5. $\therefore \angle 4 \cong \angle 3$
6. $\therefore \angle D \cong \angle C$

Next have the students show, using only the reasons with which they are familiar from high school geometry, that in the world of Euclid, the summit angles of a Saccheri quadrilateral are *right*. However, they can also now show that the summit angles of a Saccheri quadrilateral are *obtuse* in the world of Riemann (where all lines meet) — *still using the same geometry they have been using all along.*

Given: Saccheri quadrilateral ABCD
Prove: $\angle 1$ and $\angle D$ are obtuse

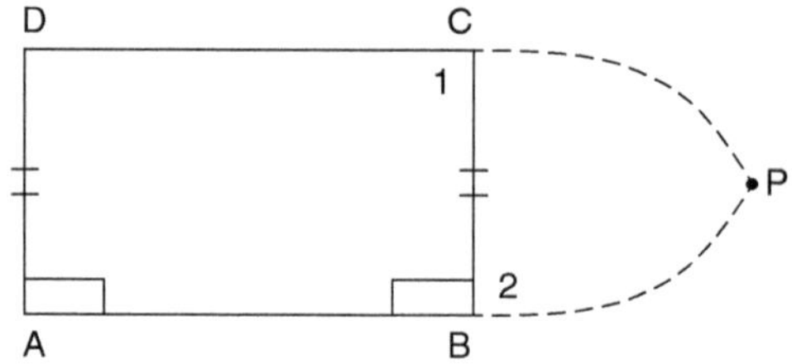

1. Extend $\overline{AB}$ and $\overline{DC}$ until they meet at point P. (Why can this be done? Remember that this is the world where all lines meet.)
2. $m\angle 2 = 90$.
3. $m\angle 1 > m\angle 2$. (Recall a theorem about the exterior angle of a triangle.)
4. $\therefore \angle 1$ is obtuse.
5. But $\angle 1 \cong \angle D$.
6. $\therefore$ Both $\angle 1$ and $\angle D$ are obtuse.

At this point there can hardly be any doubt about the size of the summit angles of a Saccheri quadrilateral in the Bolyai–Lobachevsky world. Clearly they must be acute.

Next, show pupils how to develop a proof about the sum of the measures of the angles of a triangle as summarized in this chart:

Kind of world	Sum of the measure of the angles of a triangle
Riemann (no parallels)	Is more than 180°
Euclid (one parallel)	Is equal to 180°
Bolyai–Lobachevsky (many parallels)	Is less than 180°

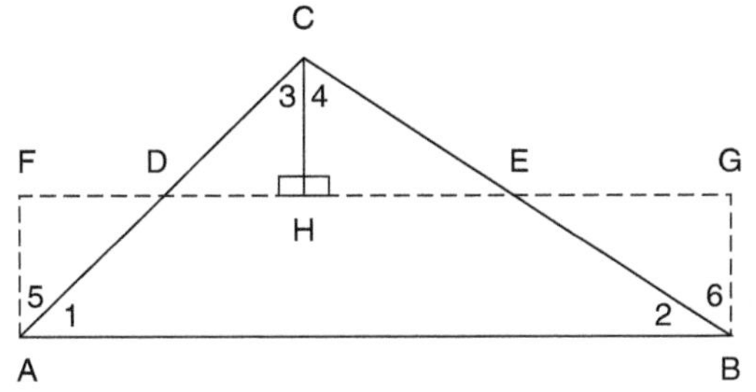

Given: $\triangle ABC$

1. Let D be the midpoint of $\overline{AC}$, and let E be the midpoint of $\overline{BC}$.
2. Draw $\overline{DE}$.
3. Draw $\overline{CH} \perp \overline{DE}$.
4. Mark off $\overline{DF} \cong \overline{DH}$ and $\overline{EG} \cong \overline{HE}$.
5. Draw $\overline{FA}$ and $\overline{BG}$.
6. $\triangle FDA \cong \triangle CDH$ and $\Delta CHE \cong \triangle BGE$.
7. Show that FGBA is a Saccheri quadrilateral with base $\overline{FG}$.
8. $\angle 5 \cong \angle 3$ and $\angle 6 \cong \angle 4$.

   The sum of the measure of the angles of $\triangle ABC$.
9. $= m\angle 1 + m\angle 2 + (m\angle 3 + m\angle 4)$.
10. $= m\angle 1 + m\angle 2 + (m\angle 5 + m\angle 6)$.
11. $= m\angle 1 + m\angle 5 + (m\angle 2 + m\angle 6)$
12. $= m\angle FAB + m\angle GBA$.
13. $=$ the sum of the measures of the summit angles of Saccheri quadrilateral FGBA.

Your students will not appreciate why Saccheri felt he failed in what he had set out to do. He did, after all, prove theorems that appeared to arrive at contradictory conclusions. Instead, he turned out to be one of the unsung heroes of mathematics.

Postassessment

Show how residents of all three worlds can complete the following proofs.

1.

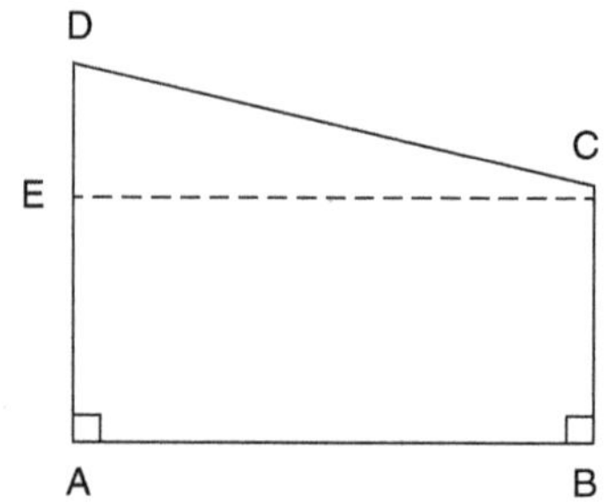

Given: $\overline{DA} \perp \overline{AB}; \overline{CB} \perp \overline{AB}; DA > CB$
Prove: $m\angle BCD > m\angle D$

2. The converse of (1) can also be proved by residents of each of the worlds. Show how this can be done by continuing the outlined proof.

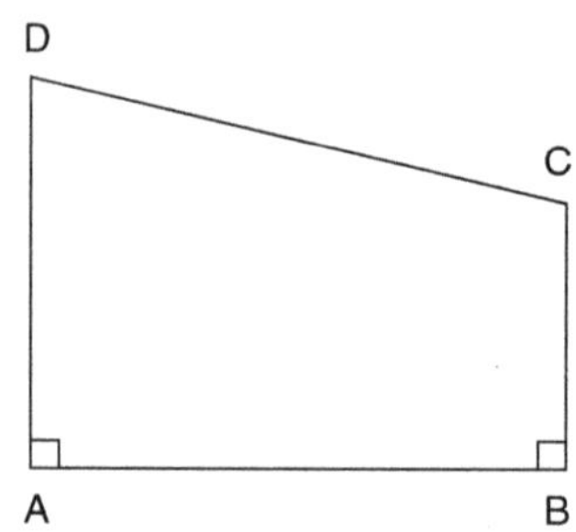

Given: $\overline{DA} \perp \overline{AB};\ \overline{CB} \perp \overline{AB};\ m\angle C > m\angle D$
Prove: BDA > CD
Hint: Use *reductio ad absurdum.*

3. Complete the proof that the line joining the midpoints of the base and summit of a Saccheri quadrilateral (the *midline*) is perpendicular to both of them.

Given: Saccheri quadrilateral ABCD; M and N are midpoints ($\overline{MN}$ is midline)
Prove: $\overline{MN} \perp \overline{AB}$ and $\overline{DC}$

1. Draw $\overline{DM}$ and $\overline{MC}$.
2. Prove $\triangle AMD \cong \triangle BMC$ and $\triangle DNM \cong \triangle CNM$.
3. To be completed by student.

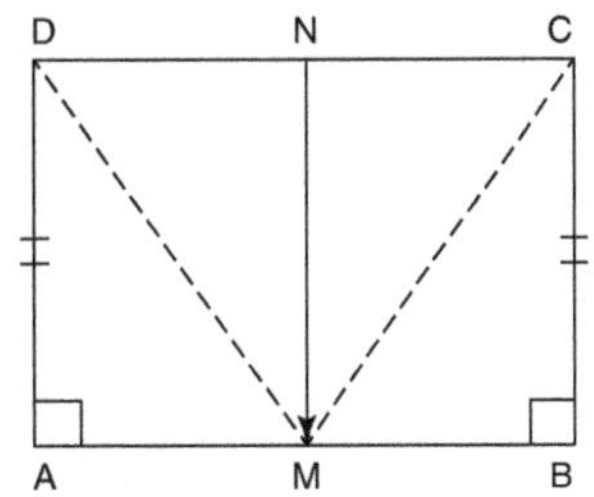

4. Only residents of Riemann's world (where the summit angles of a Saccheri quadrilateral are obtuse) can now prove this statement: The measure of the summit of a Saccheri quadrilateral is less than the measure of its base. An outline of the proof follows:

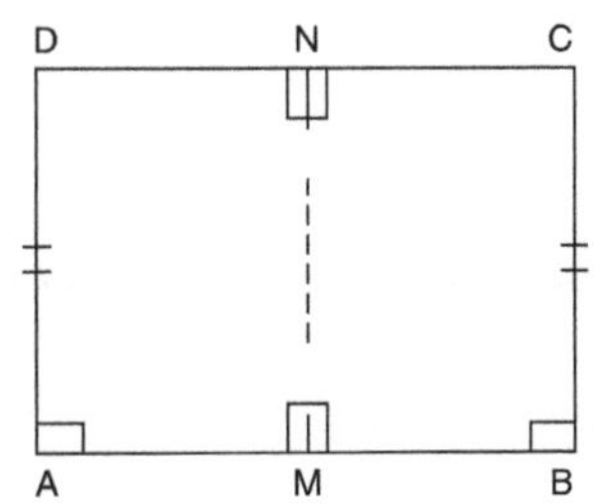

World of Riemann

Given: Saccheri quadrilateral ABCD
Prove: DC < AB

1. Draw the midline $\overline{MN}$.
2. m∠BMN = m∠MNC = 90.
3. ∠C is obtuse. (Why?)
4. In quadrilateral MNCB (with base $\overline{MN}$), m ∠C > m ∠B. (Why?)
5. ∴ NC < MB. (Why?)
6. ∴ DC < AB. (Why?)

5. Show how residents of the world of Bolyai–Lobachevsky (where the summit angles of a Saccheri quadrilateral are acute) can prove that the measure of the summit of a Saccheri quadrilateral is greater than the measure of the base.
6. Show how resident of the world of Euclid (where the summit angles of a Saccheri quadrilateral are right) can prove that the measure of the summit of a Saccheri quadrilateral is equal to the measure of the base.

Reference

Wolfe, H. E. *Introduction of Non-Euclidean Geometry.* New York: Dryden Press, 1945.

Unit 121

πie Mix

Leonhard Euler (1707–1783), a Swiss mathematician, startled the mathematical world when he discovered an expression that combined into a single formula theretofore seemingly unrelated numbers such as π, i, e, and 1. This unit demonstrates that formula and indicates how he might have developed it.

Performance Objectives

1. *Students will learn that* e^x, *sin* x, *and cos* x *may be represented by means of a power series.*
2. *Students will see the consequences of probing into a previously uncharted course in mathematics.*
3. *Students will use Euler's formula to derive two trigonometric identities.*

Preassessment

Students should be able to evaluate powers of the imaginary number i. They should be familiar with factorial notation. Students must also be familiar with the natural logarithm base, e, and trigonometric identities for sine and cosine of the sum of two angles.

Teaching Strategies

Tell students that for any real x, it can be proved in calculus that certain functions, under given conditions, may be represented as infinite series of powers. For example,

$$\sin x = x - \frac{x^3}{3!} + \frac{x^5}{5!} - \frac{x^7}{7!} + \frac{x^9}{9!} - \frac{x^{11}}{11!} + \cdots$$

$$\cos x = 1 - \frac{x^2}{2!} + \frac{x^4}{4!} - \frac{x^6}{6!} + \frac{x^8}{8!} - \frac{x^{10}}{10!} + \cdots$$

$$\text{and } e^x = 1 + x + \frac{x^2}{2!} + \frac{x^3}{3!} + \frac{x^4}{4!} + \frac{x^5}{5!} + \frac{x^6}{6!} + \cdots$$

Euler took a bold step when he questioned the hypothesis that x must be real, because if we substitute for x the imaginary number $i\theta$, where θ is real and $i = \sqrt{-1}$, an interesting thing happens:

$$e^{i\theta} = 1 + i\theta + \frac{(i\theta)^2}{2!} + \frac{(i\theta)^3}{3!} + \frac{(i\theta)^4}{4!} + \frac{(i\theta)^5}{5!} + \frac{(i\theta)^6}{6!} + \cdots$$

Recalling that $i^2 = -1$, $i^3 = -i$, and $i^4 = 1$, we can simplify the terms of the series until we get

$$\begin{aligned} e^{i\theta} &= 1 + \mathrm{i}\theta - \frac{(\theta)^2}{2!} - i\frac{(\theta)^3}{3!} + \frac{(\theta)^4}{4!} + \frac{(\theta)^5}{5!} - \frac{(\theta)^6}{6!} - \cdots \\ &= \left[1 - \frac{(\theta)^2}{2!} + \frac{(\theta)^4}{4!} - \frac{(\theta)^6}{6!} + \cdots\right] \\ &\quad + i\left[\theta - \frac{(\theta)^3}{3!} + \frac{(\theta)^5}{5!} - \ldots\right] \\ &= \cos\theta + i\sin\theta. \end{aligned}$$

Recalling again that $\cos 2\pi = 1$, and $\sin 2\pi = 0$, we may conclude that $e^{2\pi i} = 1$. It is *this* formula that caused a stir. We now turn to an unanticipated result.

Letting $\theta = x + y$ gives us

$$e^{i(x+y)} = \cos(x+y) + i\sin(x+y). \tag{1}$$

But also,

$$\begin{aligned} e^{i(x+y)} &= e^{ix}e^{iy} \\ &= (\cos x + i\sin x)(\cos y + i\sin y) \\ &= (\cos x\cos y - \sin x\sin y) + i(\sin x\cos y + \cos x\sin y). \end{aligned} \tag{2}$$

Equating the *real* and *imaginary* parts of (1) and (2), we get

$$\cos(x+y) = \cos x\cos y - \sin x\sin y$$

and

$$\sin(x+y) = \sin x\cos y - \cos x\sin y.$$

These are easily recognized as familiar trigonometric formulas.

Postassessment

1. Use the Maclaurin series approach to derive formulas for $\cos(x - y)$ and $\sin(x - y)$.
2. Show that $e^{\pi i} + 1 = 0$.
3. Show how $e^{i\theta}$ may represent an operator that rotates a complex number counterclockwise through an angle θ along a unit circle.
4. Show the connection between Euler's formula and DeMoivre's theorem for finding powers and roots of a complex number.

Unit 122

Graphical Iteration

This unit focuses on chaos theory and its connection to the secondary school curriculum. It offers an opportunity for students to explore a current area of interest in mathematics through the powers of the graphing calculator and the computer.

Performance Objectives

1. *Given a graphing calculator or computer, students will exhibit graphical iteration under a quadratic.*
2. *Students will investigate the iterative behavior under the parabola* $f(x) = ax(1 - x)$ *for different values of a from 1 to 4 with various initial iterates in the interval from 0 through 1.*

Preassessment

Students should be familiar with the role the coefficient a plays in the shape of the quadratic $f(x) = ax(1 - x)$ and recognize the x-intercepts of the function. They should also have an initial understanding of the nature of iteration, where the output, $f(x_0)$, for the initial iterate, x_0, becomes the new next iterate, x_1.

Teaching Strategies

Begin with a discussion of the characteristics of the familiar general parabola, $f(x) = ax^2 + bx + c$, and then the specific quadratics, $f(x) = ax(1 - x)$. Illustrate how the reflecting line can geometrically transform an input x_i and its output $f(x_i)$ into a new input x_{i+1}. As various values of the coefficient a are used, note how the quadratic intersects the diagonal, reflecting line, $f(x) = x$, in different places as the coefficient a increases from 1 through 4.

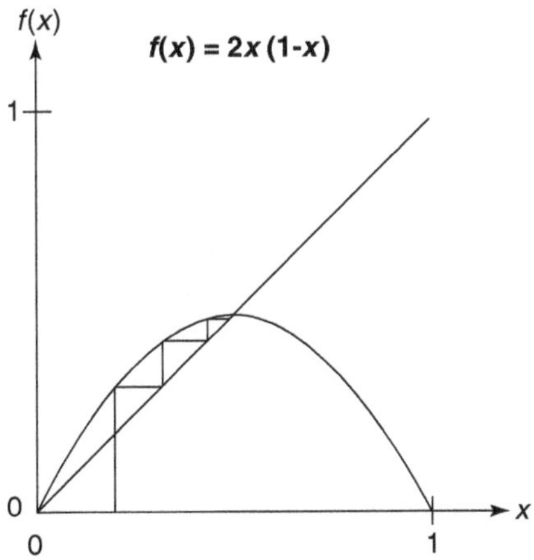

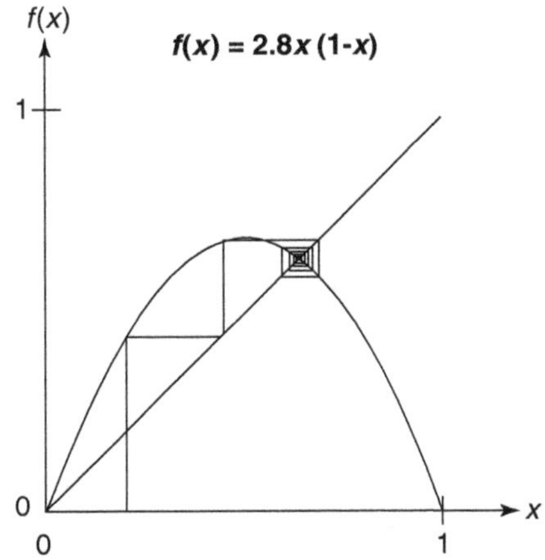

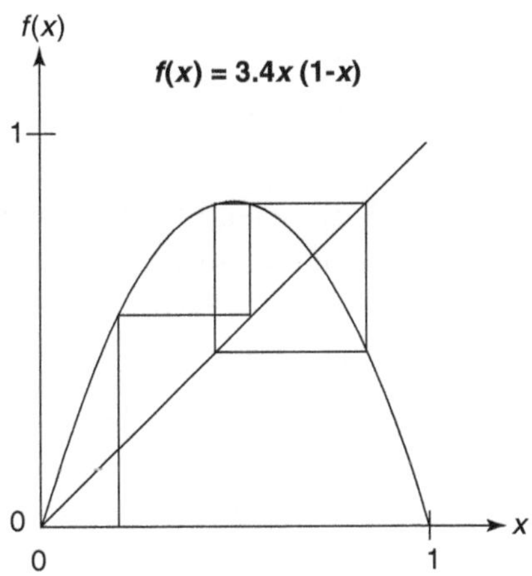

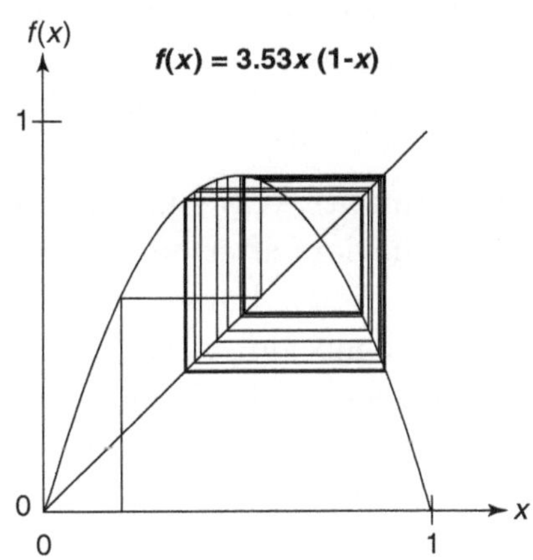

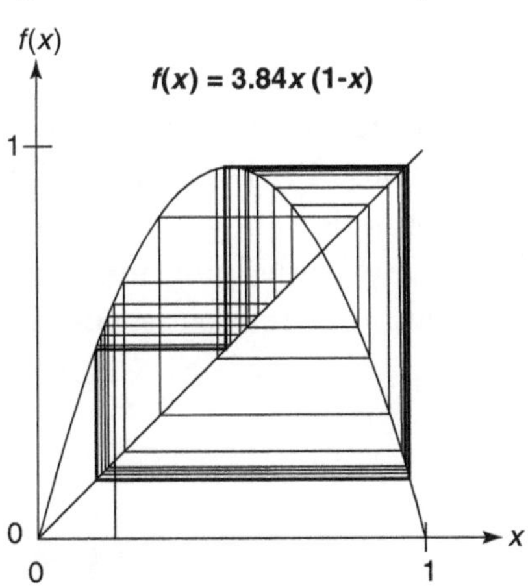

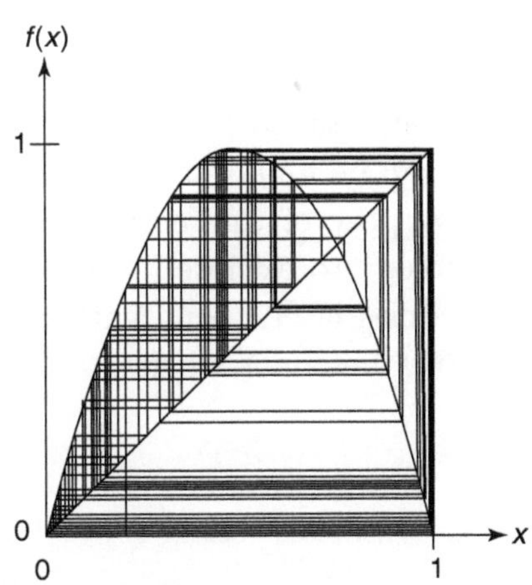

As one explores the different iteration characteristics, various behaviors become apparent. Have students explore and discover that the starting value of the initial iterate does not affect the long-term behavior of the iteration, even though the early values will differ. For simplicity, the graphs shown here all start with the initial iterate 0.2.

A brief summary of the illustrated iteration behaviors is given here:

For $a = 2$, the iterates staircase into the intersection point, $x = 0.5$.

For $a = 2.8$, the iterates ultimately spiral into the fixed point, $x = 0.643$.

For $a = 3.4$, a period-2 behavior appears between $x = 0.452$ and 0.842.

For $a = 3.53$, a period-4 behavior begins to develop, eventually moving among $x = 0.369, 0.822$, 0.517, and 0.881.

For $a = 3.84$, surprising period-3 behavior appears to emerge around $x = 0.149, 0.488$, and 0.959.

For $a = 4$, pure chaotic behavior appears.

There are many levels of explanation and interpretation associated with what we call chaotic behavior. It is often characterized by three separate but connected ideas: mixing, sensitivity, and periodicity.

For *mixing*, in every interval, however small, there exists a point that, through iteration, will reach and mix through all such intervals between 0 and 1.

For *sensitivity*, very small differences in iterates may lead to dramatically different behaviors after only a small number of successive iterations.

For *periodicity*, buried within the apparent chaotic behavior, certain points do not mix but visit repeatedly only a small number of locations.

Students should be encouraged to explore these properties as their own level of interest motivates them.

What governs this rather strange, changing iteration behavior under this rather simple quadratic? And where do the transitions from fixed point, to period-2, to period-4, to chaotic behavior occur? Are there any surprises in between, such as a period-3 behavior momentarily appearing after signs of chaos have already emerged? Some directions toward answering these questions can be found in the Feigenbaum plot described briefly in Unit 123.

The purpose of this activity is one of motivation, having students utilize technology to explore iteration behavior. The underlying mathematical analysis comes later, along with more exploration of related issues, such as the iteration behavior when $a > 4$.

Postassessment

With their graphing calculators or computers, students should be able to meaningfully perform these skills:

1. Recognize the different iteration behaviors that emerge for various values of a in the interval from 1 through 4 for $f(x) = ax(1 - x)$.
2. Find the specific values for fixed point and periodic iteration behavior.

References

Peitgen, H., H. Jurgens, D. Saupe, E. Maletsky, T. Perciante, and L. Yunker. *Fractals for the Classroom: Strategic Activities, Volume Two.* New York: Springer-Verlag, 1992.

Posamentier, A. S. and I. Lehmann, *The Secrets of Triangles: A Mathematical Journey,* Amherst, New York: Prometheus Books, 2012.

Unit 123

The Feigenbaum Plot

This unit shows how the Feigenbaum plot details the bifurcation points in the changing iteration behavior under the quadratic $f(x) = ax(1 - x)$. Specific values of the coefficient a are plotted against the corresponding attractors of the fixed and periodic points. Regions of chaotic behavior are readily visible.

Performance Objectives

1. *Students will read specific attractor values in the interval from 0 to 1 for x for various values of the coefficient a from 1 to 4.*
2. *Students can match the information in this unit against the data collected in Unit 122 and explore again those iterations, this time with greater insight into the structure of the differing behaviors.*

Preassessment

Students should be familiar with the general iteration behavior of the quadratic $f(x) = ax(1-x)$ as it moves from predictable, fixed-point behavior at $a = 1$ to totally unpredictable, chaotic behavior at $a = 4$ for x iterates between 0 and 1.

Teaching Strategies

Students expect to find the variable x on the horizontal axis as it was in the last unit on graphical iteration. Thus, the orientation of the Feigenbaum plot may cause some initial concern. The horizontal axis shows the parameter a ranging from 1 to 4, with the corresponding x-values of the attractions on the vertical axis and ranging from 0 to 1.

The spiral moves in to the fixed-point attractor of $x = 0.643$ when $a = 2.8$. This value of x can now be read directly from the Feigenbaum plot, as shown. Likewise, we can read the period-2 attractors of $x = 0.452$ and 0.842 when $a = 3.4$.

The Feigenbaum plot clearly shows the bifurcation points where the period is doubled from 1 to 2, from 2 to 4, from 4 to 8, and so on. Students should explore these regions again using Unit 122, but they should not be discouraged if precise separation points cannot be found. Remember, the parameter a is a real number, not restricted by the finite arithmetic of the graphing calculator or computer.

Students will quickly discover where the period-3 window briefly appears in one of the gaps buried in surrounding chaotic behavior. The values of the period-3 attractors, $x = 0.149, 0.488$, and 0.959, are supported here for $a = 3.84$.

This plot is named after American physicist Mitchell Feigenbaum, who developed it while working at the Los Alamos Laboratory in the 1970s. Underlying this work is his discovery that the ratios of distances between successive bifurcation points converges, surprisingly, to a constant that now bears his name. This universal Feigenbaum constant is

$$\delta = 4.669202\ldots$$

and it appears in many different iteration situations in mathematics and science.

Feigenbaum Plot

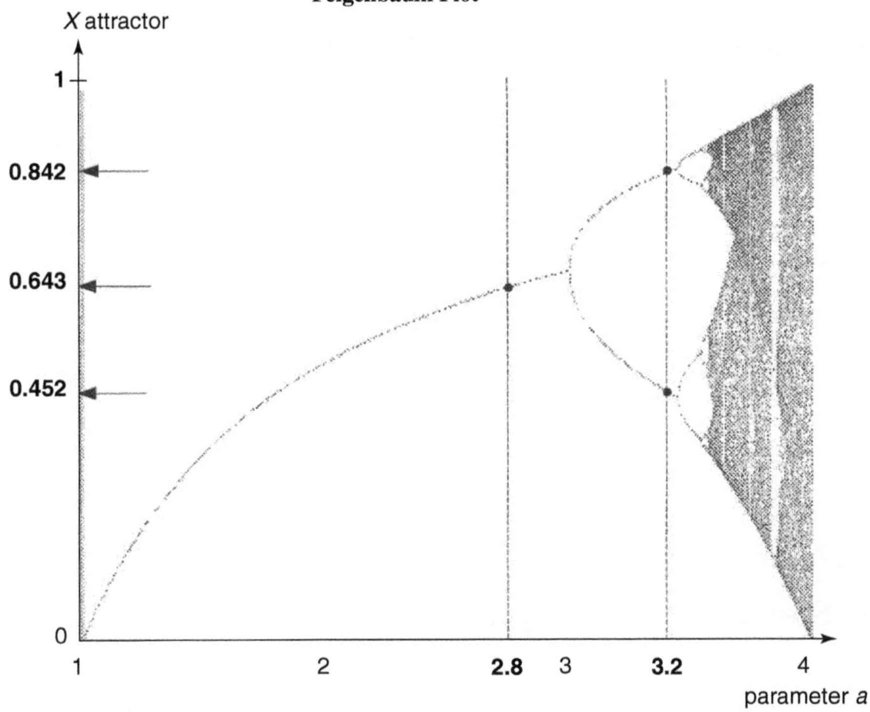

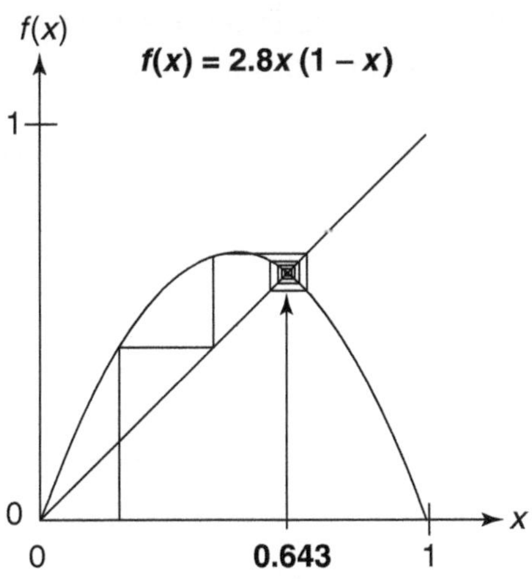

Fixed-point attractor at $x = 0.643$ when $a = 2.8$.

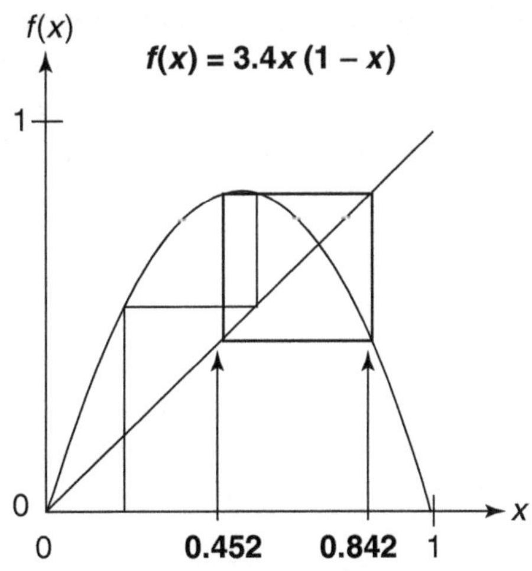

Period-2 attractors at $x = 0.452$ and 0.842 when $a = 3.4$.

Postassessment

Students should be able to explain the following:

1. Period-doubling bifurcation as seen in the Feigenbaum plot.
2. The connection between the Feigenbaum plot and the iteration behavior for the quadratic $f(x) = ax(1 - x)$.

References

Gleick, J. *Chaos: Making a New Science.* New York: Penguin Books, 1987.
Peitgen, H., H. Jurgens, D. Saupe, E. Maletsky, T. Perciante, and L. Yunker. *Fractals for the Classroom: Strategic Activities, Volume Two.* New York: Springer-Verlag, 1992.

Unit 124

The Sierpinski Triangle

Now that the age of technology is upon us, iteration has taken on a new level of importance in mathematical thinking. This unit reflects how that attention can be directed into the school curriculum through geometry. A simple geometric process, repeated over and over, can transform a plain triangular region into an elegant, abstract fractal structure, the Sierpinski triangle.

Performance Objectives

1. *Students will exercise geometric iteration, first seeing and then visualizing the geometric changes in successive stages of the structure, and then be able to express these changes in both numeric and algebraic form.*
2. *Students will be able to define and illustrate self-similarity.*

Preassessment

Students should have experience with scaling and similarity and with pattern recognition in various forms, and they should be familiar with the general notions of iteration and recursive thinking.

Teaching Strategies

In the late 1800s and early 1900s, mathematicians sought to create new kinds of geometric structures that possessed unique properties. Many of their results are recognized and classified today as fractals. One such creation from that period was the Sierpinski triangle, named after Polish mathematician Waclaw Sierpinski.

Have each student start with a triangular piece of paper. Connect the midpoints of the sides of the triangle to form four similar triangles at half linear size. Cut them apart. Keep the three corner triangles, and remove the middle one. Think of this as stage 1. Apply the same algorithm again on each of the three new, smaller triangles to get stage 2. Then apply the algorithm again on the nine still smaller triangles to get stage 3. Imagine the iteration process continued through stage 4.

Each stage contains three times as many triangular parts as the preceding stage. It is thus apparent, from this approach, that each successive stage requires three times as many applications of the algorithm. There are always more and more applications on smaller and smaller parts. As far as the cutting goes, the triangular pieces soon get too small in size and too large in number. Is there another view of the process where the iteration rule remains exactly the same throughout and is always applied exactly once in going from one stage to the next? The answer is yes.

Have your students think globally of the whole structure at each stage, not of the ever-increasing number of smaller and smaller parts. Here is one possible scenario:

Take any stage of the figure to the copy machine.
Set the machine at 50%, reducing linear dimensions to half.
Make three copies at this half size.
Use them to build the next stage of the structure.

Let this be the iteration algorithm. Have your students actually build the first several stages this way, repeating the exact same process over and over again. Then have them imagine the iteration continuing, and let them visualize how the figure changes, becoming more and more delicate with increasing complexity at each successive stage.

Viewing the building process through these rules, the notion of self-similarity becomes very apparent. Successive structures contain more and more copies of the original stage-0 structure at more and more different scales. But is only the limit figure that truly exhibits self-similarity. Finite stages contain copies *essentially* like the whole, but only the limit figure contains *exact* copies of the whole at all scales!

The Sierpinski triangle is that limit figure. It is an abstract, infinitely complex fractal structure, where all the small triangular regions have reduced themselves to points. Students need to know that it can only be seen in the mind. What the eye sees, at its best, is only some limited finite stage in the development of the Sierpiński triangle.

Within this geometric iteration process, however, lie many mathematical connections. The following table shows how this building process can be related to number patterns, perimeter and area, exponents, geometric series, and limits, to name a few. Indeed, fractals, and the Sierpinski triangle in particular, offer one of the most powerful examples of the kind of mathematical connections referred to in the *Curriculum and Evaluation Standard of School Mathematics* of the National Council of Teachers of Mathematics.

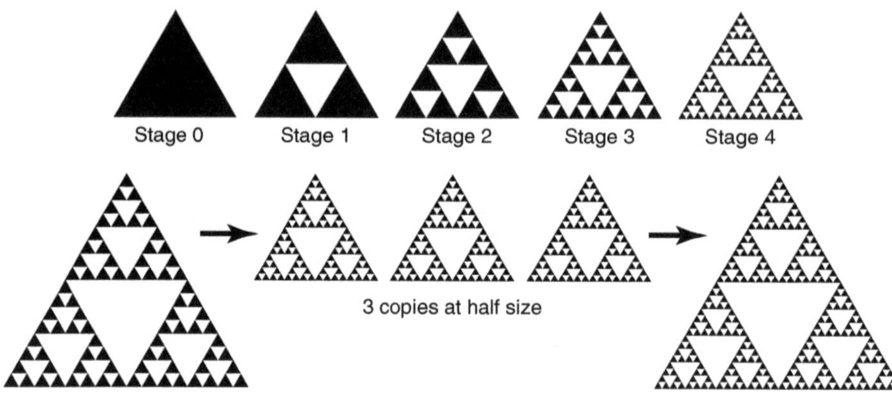

Stage	0	1	2	3	4	n
Number of triangles	1	3	9	27	81	3^n
Area	1	$\frac{3}{4}$	$\frac{9}{16}$	$\frac{27}{64}$	$\frac{81}{256}$	$\left(\frac{3}{4}\right)^n$
Perimeter	1	$\frac{3}{2}$	$\frac{9}{4}$	$\frac{27}{8}$	$\frac{81}{16}$	$\left(\frac{3}{2}\right)^n$

The perimeter and area at stage 0 are defined to be 1 unit and 1 square unit, respectively. This allows the student to focus on the constant multiplier in both sequences and to make the direct connection to geometric sequences. At a different level, the student might be asked to compute the changing perimeter and area starting with an equilateral triangle measuring 4 inches on each side.

Note that the areas in the table refer to those of the shaded triangular regions remaining at each and every stage. The students should see these as converging to 0. In that sense, the limiting area of the Sierpinski triangle is 0! On the other hand, the perimeters in the table refer to the distances around each and every triangular piece at each stage. Here, the students should see these as diverging. In that sense, the limiting perimeter is infinite! Put together, these two different behaviors for the area and perimeter of the same figure give yet another glimpse of the uniqueness of this structure.

Postassessment

Students should be able to explain the following:

1. A building algorithm that, when iterated, generates the Sierpinski triangle.
2. The nature of self-similarity as found in the Sierpinski triangle.
3. How the perimeter and area change as successive stages are generated.

Reference

Peitgen, H., H. Jurgens, D. Saupe, E. Maletsky, T. Perciante, and L. Yunker. *Fractals for the Classroom: Strategic Activities, Volume One.* New York: Springer-Verlag, 1991.

Unit 125

Fractals

It was not quite 25 years ago that Benoit Mandelbrot coined the word *fractal.* At that time, it would have been hard to believe that this topic would move so fast and reach so many in such a short time. Nor would one likely have foreseen their being attracted, connected, and embedded in the curriculum of school mathematics so quickly. But technology, plus an urge to infuse our teaching with new ideas, has made that possible. Today, there are many software packages on the market that can put the dynamics and aesthetics of fractals directly before your students. It is quite another matter for students to see what mathematics underlies these fascinating structures. Much of that can be done in your classroom with assorted hands-on activities and experiences.

This activity expands and extends the iterative geometric generation of the Sierpinski triangle into a whole family of fractal structures.

Performance Objectives

1. *Students will generate successive stages of various fractals based on an adaptation of the building code for the Sierpinski triangle.*
2. *Students will recognize self-similarity in fractals of this type, and find from that what their building codes are.*

Preassessment

Students should have experience with scaling, similarity, self-similarity, and the geometric transformations of rotations and reflections.

Teaching Strategies

Consider modifying the building blocks of the Sierpinski triangle to be centered around squares instead of triangles. For many students, this first step may be the hardest of all to take. How can the Sierpinski triangle emerge from a process that involves only squares?

Every finite stage of this developing fractal consists of small square regions: the higher the stage, the smaller the squares. But in the limit, each of these small squares approaches a point. Whether squares or triangles, in the limit, both approach points. The fact is, the limit figure, generated from squares or from triangles, is the same fractal structure, the Sierpinski triangle. Both sequences of figures, although always different, are approaching the same attractor.

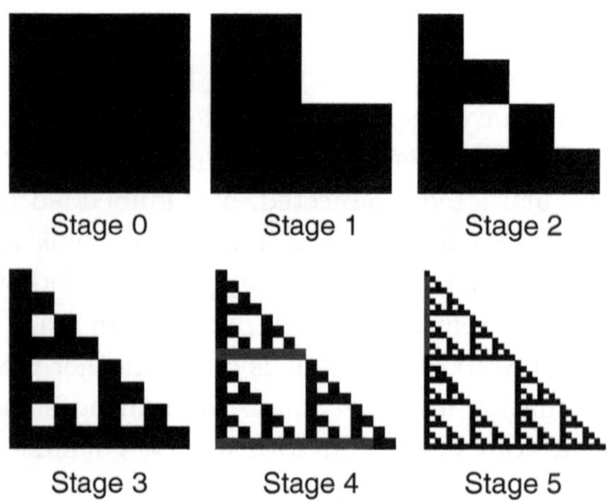

Here is a model of the building code. Iterated over and over, the Sierpinski triangle emerges.

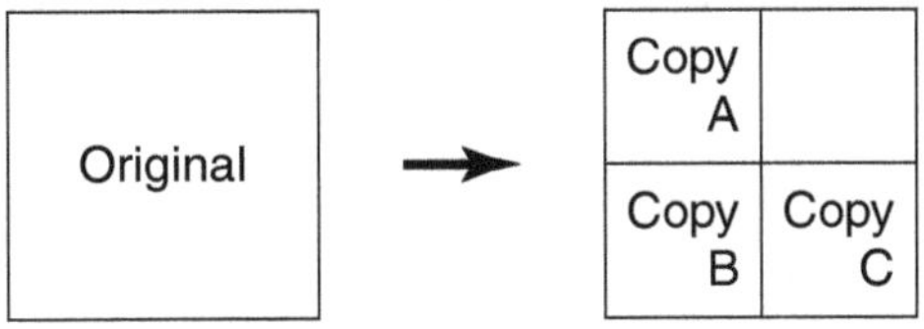

BUILDING CODE

Reduce to $\frac{1}{2}$ size.
Make 3 copies.
Rebuild.

Once this idea is established, incorporate the transformations of a square into the process, and a whole family of Sierpinski-like fractals can be constructed. The result will be that each of your students can explore her or his own personal fractal.

With this code, cell A is rotated 270° and cell B is rotated 180°, each clockwise. Students will quickly see that a very different structure begins to emerge.

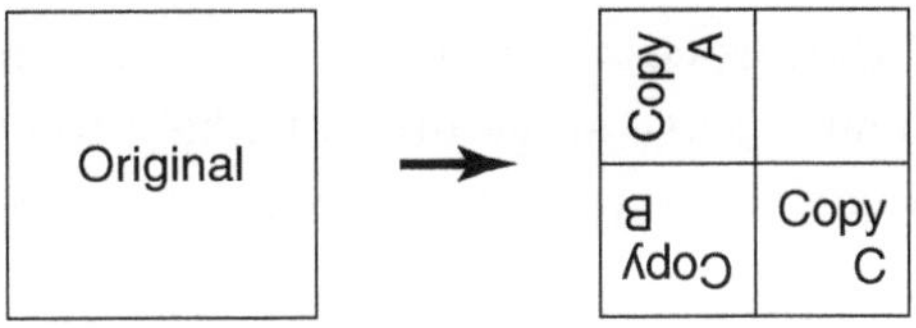

BUILDING CODE

Reduce to $\frac{1}{2}$ size.
Make 3 copies.
Rebuild, rotating copies
A and B as shown.

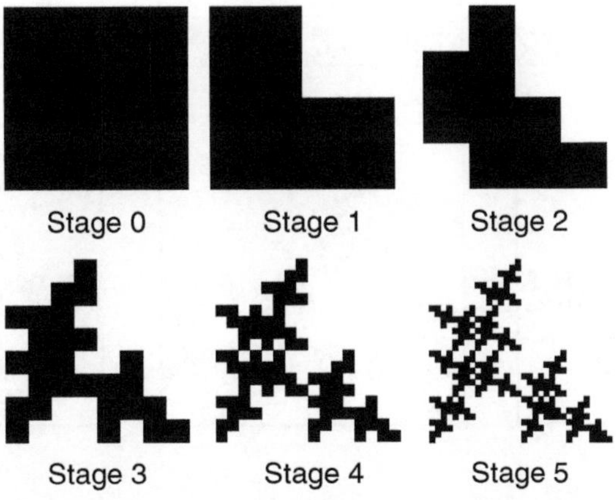

Encourage students to make up their own codes and create the first few stages of the corresponding fractal. Finely ruled graph paper can be used, or you can supply grids of this type for students to use.

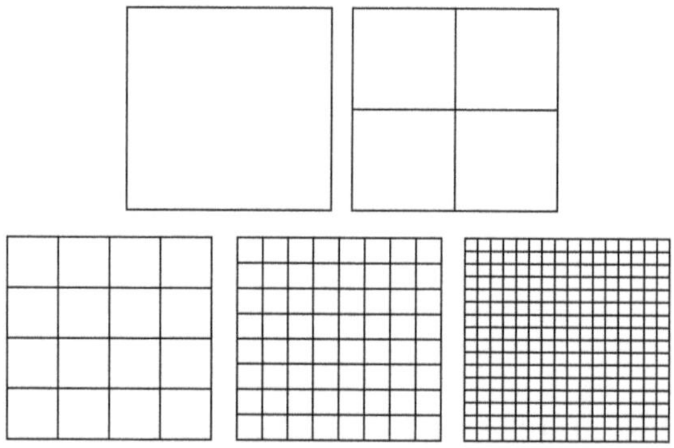

Software packages which contain draw programs that can be used very effectively in generating these structures by this process. Their snap-on grids enable accurate constructions.

One word of caution. Whether drawing by hand or using a computer to create the graphics, remember that it is the whole figure at each stage that is reduced, replicated, and rebuilt through the geometric transformations. Students who apply the process incorrectly to more and more smaller and smaller parts at each successive stage are not likely to create correct figures. At each and every stage, make just three reduced copies of the whole before you rebuild.

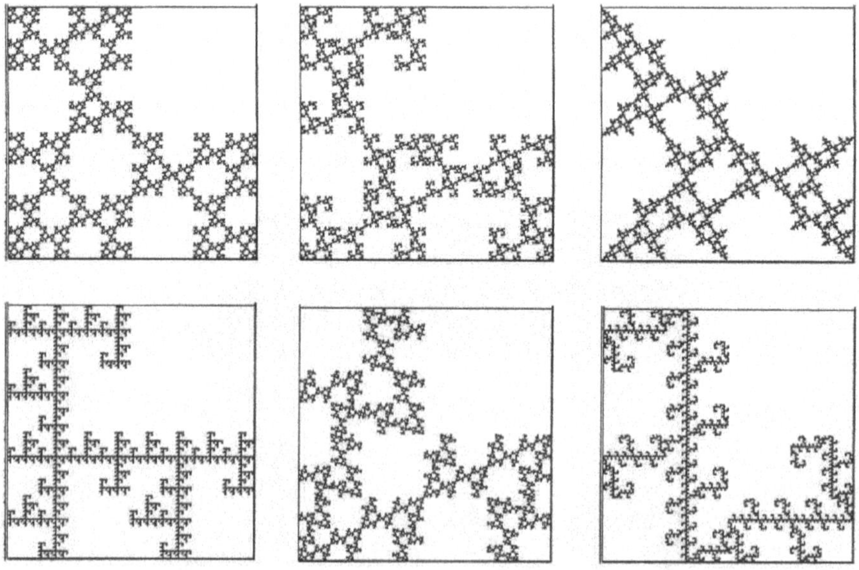

A second part of this activity is to have students look at the detailed stages of others and see if they can spot the building codes that were used. This can be a very powerful and challenging visual experience for some. Recall, there are eight possible transformations of the square.

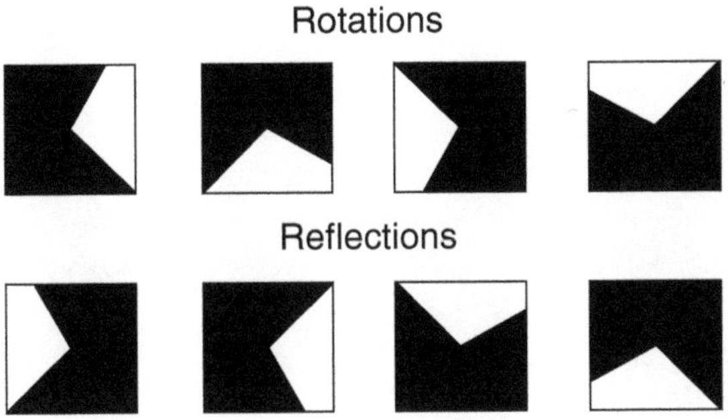

Start with structures that involve only rotations first. For many students, they are the easier ones to see. Save the reflections for later, when your students have had more practice. See if your students can write out the building codes used to create these fractals. Only rotations were used.

Postassessment

Students should be able to do the following:

1. Follow a given building algorithm through several successive stages of development.
2. Use self-similarity to identify the building code from a constructed fractal of this type.

References

Peitgen, H., H. Jurgens, D. Saupe, E. Maletsky, T. Perciante, and L. Yunker. *Fractals for the Classroom: Strategic Activities, Volume Three.* New York: Springer-Verlag, 1997.

Posamentier, A. S. and I. Lehmann, *The Secrets of Triangles: A Mathematical Journey*, Amherst, New York: Prometheus Books, 2012.

Lightning Source UK Ltd.
Milton Keynes UK
UKHW032154070621
385101UK00002B/18

9 789811 240423